Higher Segal Spaces and Applications

CONTEMPORARY MATHEMATICS

838

Higher Segal Spaces and Applications

Workshop
Higher Segal Spaces and their Applications to Algebraic K-Theory, Hall Algebras, and Combinatorics
January 21–26, 2024
Banff International Research Station, Banff, Alberta, Canada

Julia E. Bergner
Joachim Kock
Maru Sarazola
Editors

AMS AMERICAN MATHEMATICAL SOCIETY

2020 *Mathematics Subject Classification.* Primary 18N50, 05Exx, 16Gxx, 16Txx, 19Dxx, 53Dxx.

Library of Congress Cataloging-in-Publication Data

Cataloging-in-Publication Data has been applied for by the AMS.
See http://www.loc.gov/publish/cip/.

∞ The paper used in this book is acid-free and falls within the guidelines
established to ensure permanence and durability.
Visit the AMS home page at `https://www.ams.org/`

10 9 8 7 6 5 4 3 2 1 31 30 29 28 27 26

Contents

Preface

We are pleased to present these ten expository papers on higher Segal spaces, originating from the "service talks" at the *Workshop on Higher Segal Spaces and their Applications to Algebraic K-Theory, Hall Algebras, and Combinatorics*, which took place at the Banff International Research Station, January 21–26, 2024.

Background. Segal spaces are used in homotopy theory, algebraic geometry, logic, and several other fields of modern mathematics, and in particular play an important role in many approaches to higher category theory. While ordinary categories can be described by a binary associative composition law, higher categories involve compositions and associativity that might be defined or hold up to (higher) homotopy. Segal spaces model such weak categories that have spaces of both objects and morphisms, and encode the homotopical ambiguity in the composition as higher simplices in a simplicial object. For this reason, the approach is well-suited to homotopy theory, not only for describing topological examples, such as cobordism categories, but also for investigating them using homotopy-theoretic tools such as model categories.

Generalizing this approach, 2-Segal spaces encode a still weaker structure, where composition can be multi-valued, or sometimes not defined at all, but is still associative in a suitable sense. Alternatively, such a structure can be regarded as encoding decomposition rather than composition. The definition of 2-Segal spaces, as well as k-Segal spaces for $k > 2$, was first given by Dyckerhoff and Kapranov in 2012, motivated by an ample range of applications in homological algebra, representation theory, and geometry. Shortly after, the equivalent notion of decomposition space was developed in combinatorics by Gálvez, Kock, and Tonks in 2014. It is characteristic of the richness of the theory that although the two notions are equivalent, the formulations, settings, and directions of development were rather disjoint. Over the past years, as the theory has developed further, the two viewpoints are converging, while at the same time the scope of the theory has widened, expanding into new applications. It is now quite a lively area of research, especially in view of the diverse background of people using and developing the theory.

The workshop in Banff. In 2017, Julie Bergner and Mark Penney took the initiative of organizing the first international workshop on the topic of higher Segal spaces, to be held at CMO/BIRS in Oaxaca in 2020. Unfortunately the workshop had to be cancelled due to the pandemic; given that one of the main purposes of the workshop was to bring everyone in the field together in person, it did not seem well-suited to an online format. The workshop finally took place in 2024, now at BIRS in Banff, with a modified team of organizers, with Joachim Kock and Maru Sarazola replacing Mark Penney. While the initial goal was to bring together researchers working with higher Segal spaces in disparate areas such as algebraic K-theory,

Hall algebras, and combinatorics, the list of application directions broadened still further to include group theory, mathematical physics, and symplectic geometry. Significantly, many of the participants were newcomers to the theory of higher Segal spaces, drawn by curiosity and the desire to get more involved in this growing research area.

In view of the mathematical diversity of the participants, it was decided that most of the talks should be introductory talks and survey talks, so as to give all participants a solid background and an overview of the different aspects and directions of applications. The speakers were enthusiastic about giving such "service" talks. The remaining talks were more specialized, mostly given by young researchers. There was also room in the schedule for open problem sessions and discussion sessions, as well as further talks scheduled along the way, on topics deemed to deserve further treatment.

Overall the workshop went very well in an enthusiastic and friendly atmosphere, and in particular the background talks were very well received. It was readily agreed that they should form the basis for a proceedings volume of expository nature with the aim of providing a reference and resource for people in different areas wanting to learn more about the theory and its range of applications. The outcome is the present volume, which gathers ten expository papers on higher Segal spaces, written by experts in the field.

Short outline of the ten papers.

The order of the 10 contributions reflects the logical order, which was essentially also the order of talks at the workshop, except for the Gálvez–Kock–Tonks contribution, placed last for its different origin.

The first two papers by Walker Stern and Philip Hackney present the basic definitions and properties, the former using the 2-Segal space approach, and the latter from the perspective of decomposition spaces, including the equivalence of the two definitions. Both of these papers give a gentle introduction, which should be accessible to a newcomer to the field, requiring only some elementary category theory and familiarity with the definition of simplicial sets.

With that foundation in place, we move on to discuss some of the main areas of application, beginning with papers by Viktoriya Ozornova and Martina Rovelli addressing the deep connection between 2-Segal spaces and the $S_\bullet$-construction, of fundamental interest in algebraic K-theory. These papers are followed by the one introducing the theory of Hall algebras arising from 2-Segal spaces by Benjamin Cooper and Matthew Young. These two applications are central to the theory, and the three papers explain the role of 2-Segal spaces in these areas.

We then turn to the theory of k-Segal spaces for $k > 2$ in the paper of Tobias Dyckerhoff. While this theory is still considerably less developed than that of 2-Segal spaces, there is much interest in understanding these generalizations, and this paper should facilitate more activity in that direction.

The next three papers introduce further applications and refinements of the theory. The first is an introduction to the theory of CGW categories by Maru Sarazola, Brandon Shapiro, and Inna Zakharevich. These structures are expected to be closely related to 2-Segal spaces, although the precise relationship is still under investigation. The second paper in this part is by Hiro Tanaka and introduces applications to sympletic geometry via a connection between 2-Segal spaces and Fukaya categories. Finally, in the third paper, Julie Bergner and Walker Stern give an expository treatment of cyclic spaces, culminating in a model structure for cyclic 2-Segal spaces.

The last paper in this volume, written by Imma Gálvez-Carrillo, Joachim Kock, and Andrew Tonks, has been available online for some time but has now been completed for publication here. It gives a wealth of applications of decomposition spaces to classical combinatorics, and is a valuable resource for understanding that motivation for the subject.

Thanks. We thank the K-theory Foundation for sponsoring the travel of a number of young participants to the workshop. We are very grateful to BIRS for hosting the workshop, providing a superb setting and excellent working conditions (with such an inspirational backdrop). We thank the BIRS staff for the smooth running of the workshop. We are grateful to Christine Thivierge and Bridget Manzella at the AMS for their support in putting together this volume. Last but not least, we are very thankful to all the participants of the workshop for making it so enjoyable, to the contributors to the volume for the effort they put into writing up the papers, and to the anonymous referees for their expert advice and attention to detail. The enthusiasm of everybody made the task of the organizers/editors very easy.

Julie Bergner, Joachim Kock, and Maru Sarazola
May 2025

Group photo from the Workshop on Higher Segal Spaces and their Applications to Algebraic K-Theory, Hall Algebras, and Combinatorics, which took place at the Banff International Research Station, January 21–26, 2024 (24w5266). From the top left: Jonathan Campbell, Philip Hackney, Martina Rovelli, Chris Grossack, Julie Bergner, Walker Stern, Teresa Hoekstra Mendoza, Maxine Calle, Justin Lynd, Matthew Young, Tobias Dyckerhoff, Joachim Kock, Rajan Mehta, Charles Weibel, Brandon Shapiro, Maru Sarazola, Inna Zakharevich, Andrew Tonks, Imma Gálvez-Carrillo, Ayelet Lindenstrauss, Carmen Rovi, Viktoriya Ozornova, Jonte Gödicke, Felix Nass, Ivan Contreras, Tanner Carawan, Mikhail Gorsky, Benjamin Cooper. (Absent from the photo: Hiro Lee Tanaka)

Contemporary Mathematics
Volume **838**, 2026
https://doi.org/10.1090/conm/838/16735

Perspectives on the 2-Segal conditions

Walker H. Stern

ABSTRACT. In this article we explain and motivate the 2-Segal conditions, focusing on the underlying intuition. We first discuss higher associativity, and show how contemplating the associativity of partially- and multiply-defined multiplications or compositions leads naturally to the 2-Segal conditions. We then explain how these conditions may be reinterpreted geometrically in terms of a kind of state sum. We make both of these perspectives precise, and indicate some of the many directions in which they may be generalized. Throughout, we attempt to emphasize the elementary nature of these intuitions, focusing on simplicial sets, 1-categories, and pictorial and diagrammatic arguments wherever possible.

1. Background, or what to expect from this paper

What we here term the *2-Segal conditions* were, in fact, independently arrived at by two different groups: by Dyckerhoff and Kapranov in [**7**] using the name "2-Segal spaces"; and by Galvez-Carrillo, Kock, and Tonks in [**10**] under the name "decomposition spaces". As indicated by the choice of terminology, this article is primarily focused on the perspectives which arise in the first of these sources, leaving discussions of the second to papers later in this volume (in particular, [**15**] will focus on exploring the perspective of [**10**]).

In their greatest generality, the 2-Segal conditions pertain to simplicial objects

$$X\colon \Delta^{\mathrm{op}} \longrightarrow \mathcal{C}$$

with values in an ∞-category $\mathcal{C}$. However, the technicalities necessitated by this (and by the closely related model-categorical perspective) tend to greatly obscure the intuitions involved. As such, we will focus exclusively on the case in which $\mathcal{C}$ is a 1-category. Indeed, throughout this article, we will limit ourselves to *simplicial sets.*

In very broad strokes, there are two directions one can take to understand the 2-Segal conditions. On the one hand, the 2-Segal conditions can be arrived at by taking the *spines* — 1-dimensional simplicial subsets of Δ^n — which give rise to the Segal conditions, and replacing them with 2-dimensional simplicial subsets corresponding to triangulations of polygons. On the other hand, the 2-Segal conditions can be arrived at as the minimal set of conditions necessary to require the associativity of a not-necessarily-well-defined operation. We will refer to the former intuition as *the geometric perspective* and the latter as *the algebraic perspective.*

These two perspectives are closely linked. The algebraic perspective focuses on associativity, which, in turn, can be viewed as the requirement that all n-ary composites of a binary multiplication corresponding to rooted binary trees are equal. However, rooted binary trees are precisely dual to triangulations of polygons with ordered vertex sets, allowing for an easy graphical connection between these two conditions.

In this article, we begin with an intuitive exploration of the algebraic perspective, and situate the associativity guaranteed by the 2-Segal conditions in the context of nerves of categories. This leads to the interpretation of the 2-Segal conditions as a relaxation of the 1-Segal conditions which forgets that the composition operation in the nerve of a category must be a map of sets, but remembers that it must be associative. From there we explore the geometric perspective, explaining the connection between associativity, trees, and polygonal subdivisions.

The last two sections are devoted to making these perspectives rigorous. In section 4, we define membrane sets and rigorously formulate the geometric version of the 2-Segal conditions. We then show that it is equivalent to the original definition given in section 2. After a brief digression on the coskeletalness of 2-Segal simplicial sets, we then turn to the algebraic perspective in section 5. There, we prove that 2-Segal objects are equivalent to a particular coherently associative structure — algebras in the bicategory of spans of sets.

While this article is intended to act as a gentle introduction to the 2-Segal conditions, assuming some technical background is inevitable. In particular, the entire article will assume some familiarity with basic 1-category theory, the simplex category Δ, and the basics of simplicial sets. This is the only technical background truly required to read §2, §3, and §4. Occasional remarks and footnotes may point to higher categorical intuitions or the associated literature, but these can be safely ignored should the reader so wish. The one exception to this principle is section 4.1, which requires some knowledge of truncation and coskeleton functors. Textbook treatments of the coskeleton can be found in [**13**, §VII.1], [**24**, Section 0AMA], and [**20**, Tag 051Z].

The final section, §5, amounts to a recapitulation, in a drastically simplified setting, of one of the main results of [**26**]. While it is, in principle, possible to read §5 without knowledge beyond what is listed above, the meanings of all of the constructions and definitions in the section will be somewhat opaque to a reader without some understanding of 2-categories and bicategories.

2. The algebraic perspective: (higher) associativity

The first perspective which will lead us to the 2-Segal conditions is the study of associativity. In particular, the consideration of operations which are not well-defined functions (i.e., which may be partially defined, multiply defined, or both) will lead inexorably to the 2-Segal conditions as an avatar of associativity. Before reaching for this high level of generality, however, let us take a step back, and look at associativity in a more familiar context: that of the composition in a category.

Given a category $\mathcal{C}$, we may define its *nerve* to be the simplicial set

$$N(\mathcal{C})\colon \Delta^{\mathrm{op}} \longrightarrow \mathsf{Set}$$

which sends the ordinal $[n]$ to the set $\mathrm{Hom}_{\mathsf{Cat}}([n], \mathcal{C})$. Conventionally, we picture the first few sets in the nerve of a category as follows

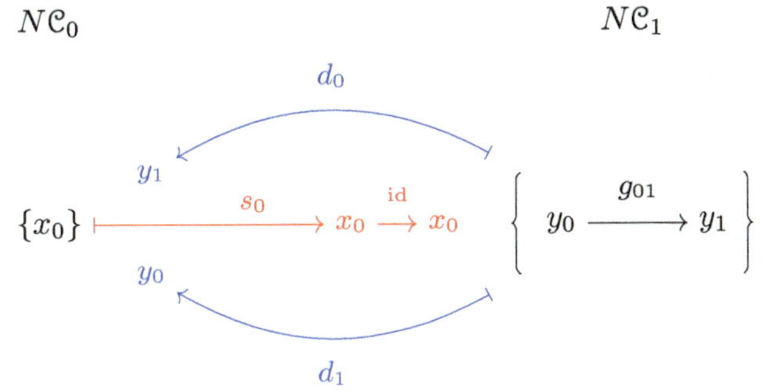

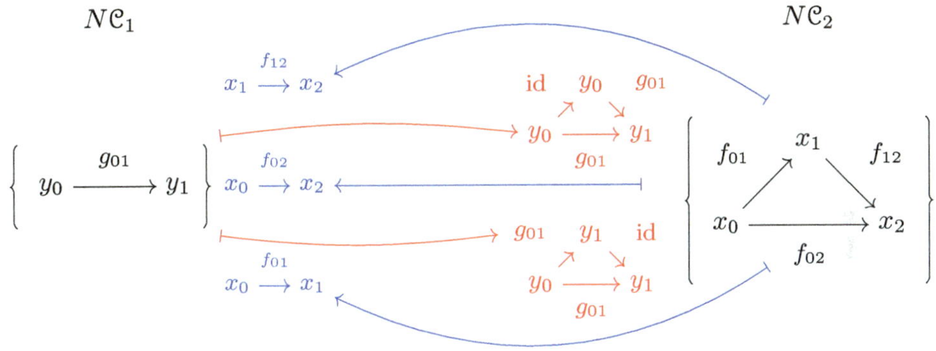

In this section, we will attempt to untangle the reasons for which the nerve of a category comes equipped with an *associative* composition.

However, any analysis of associativity must first address the question of why the nerve of a category comes equipped with any kind of composition at all. The starting point in our answer to this question is the well-known NERVE THEOREM.

THEOREM 2.1. *The nerve functor*

$$N\colon \mathsf{Cat} \longrightarrow \mathsf{Set}_\Delta$$

is fully faithful. A simplicial set X is isomorphic to the nerve of a category if and only if, for every $n \geq 2$, the map

$$X_n \longrightarrow \underbrace{X_1 \times_{X_0} \cdots \times_{X_0} X_1}_{\times n}$$

which sends a simplex to its spine[1] *is a bijection.*

A variant of this theorem first appeared in [**14**, Prop. 4.1], as part of Grothendieck's studies in descent theory. We will henceforth follow the convention that we call the maps

$$X_n \longrightarrow \underbrace{X_1 \times_{X_0} \cdots \times_{X_0} X_1}_{\times n}$$

[1] For a definition of the spine of a simplex, see the next page.

the *1-Segal*[2] *maps* and the condition that the 1-Segal maps be bijections as the *1-Segal conditions.* We will call a simplicial set which satisfies the 1-Segal conditions a *1-Segal simplicial set.* The reason for this naming convention is the use, by Graeme Segal, of the conditions which now bear his name in [**23**].

To understand the 1-Segal conditions, we must first carefully analyze the 1-Segal maps and unpack their relation to the defining features of categories. Let us start by describing the *spine* of a standard n-simplex. We define a 1-dimensional simplicial subset $\mathrm{Sp}(\Delta^n) \subset \Delta^n$ to consist of those simplices $[k] \to [n]$ which factor through one of the subsets $\{i, i+1\} \subset [n]$ for $0 \leq i \leq n-1$. More intuitively, this simplicial set consists of precisely n 1-simplices, glued end to end, i.e.,

$$\mathrm{Sp}(\Delta^n) \cong \Delta^1 \coprod_{\Delta^0} \cdots \coprod_{\Delta^0} \Delta^1.$$

To make more explicit the inclusion of the spine into Δ^n, we can write

$$\mathrm{Sp}(\Delta^n) \cong \Delta^{\{0,1\}} \coprod_{\Delta^{\{1\}}} \cdots \coprod_{\Delta^{\{n-1\}}} \Delta^{\{n-1,n\}},$$

implicitly identifying the symbol Δ^I, where I is an ordinal isomorphic to $[m]$, with Δ^m. This notational convention will be of great use to us later.

The spines of the 2- and 3-simplices can be easily pictured.

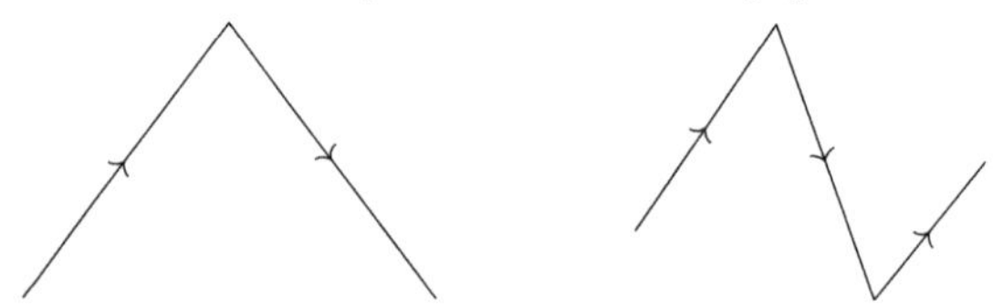

The 1-Segal maps are then given by restricting a simplex $\sigma \in X_n$ to its spine. More formally, they are the maps

$$X_n \cong \mathrm{Hom}_{\mathsf{Set}_\Delta}(\Delta^n, X) \longrightarrow \mathrm{Hom}_{\mathsf{Set}_\Delta}(\mathrm{Sp}(\Delta^n), X) \cong \underbrace{X_1 \times_{X_0} \cdots \times_{X_0} X_1}_{\times n}.$$

For instance, on a 3-simplex in $N\mathcal{C}$, the 1-Segal map acts as

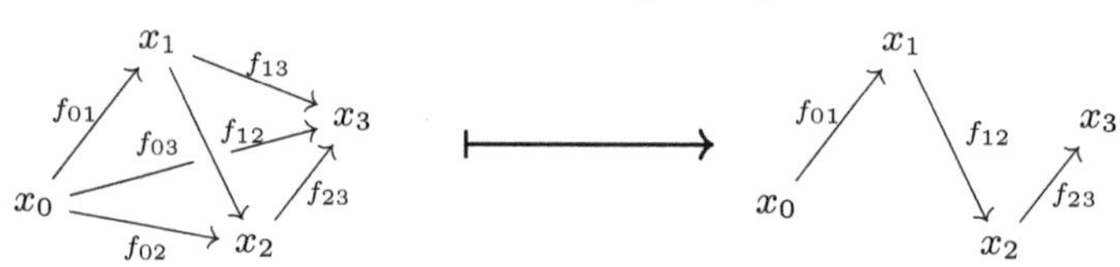

To understand the nerve theorem, we must unpack how a category arises from a 1-Segal simplicial set X. We can easily see that to identify X with the nerve of a category $\mathcal{C}$, we must define the sets of objects and morphisms in $\mathcal{C}$ to be $\mathrm{Ob}(\mathcal{C}) = X_0$ and $\mathrm{Mor}(\mathcal{C}) = X_1$, respectively. Similarly, the source and target maps from $\mathrm{Mor}(\mathcal{C})$ to $\mathrm{Ob}(\mathcal{C})$ must be identified with d_1 and d_0, respectively, and the identity morphism on an object $x \in X_0$ must be given by $s_0(x)$. However, two important questions remain.

Firstly: how does composition arise from the 1-Segal conditions? Given two "morphisms" $f, g \in X_1$, they are composable precisely when the source of f matches

[2]We amend the convention that these be called simply the *Segal maps* because we will, by necessity, need to discuss *2-Segal* maps later in this work.

the target of g, i.e., when $d_1(f) = d_0(g)$. That is, precisely when they define an element $(g, f) \in X_1 \times_{X_0} X_1$. The composition is thus defined by a *span* of sets

$$X_1 \times_{X_0} X_1 \xleftarrow{(d_2,d_0)} X_2 \xrightarrow{d_1} X_1$$

so that the composite $f \circ g$ is the d_1 face of the 2-simplex with spine (g, f). The 1-Segal conditions require that the leftward arrow in the span is a bijection, and so, inverting it, we get a well-defined composition map

$$X_1 \times_{X_0} X_1 \underset{\cong}{\xleftarrow{(d_2,d_0)}} X_2 \xrightarrow{d_1} X_1.$$

We can also interpret the span which defines the composition pictorially:

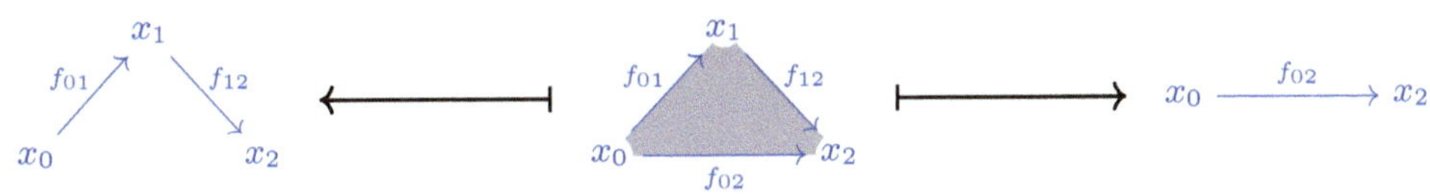

$$X_1 \times_{X_0} X_1 \xleftarrow{\quad (d_2, d_0) \quad} X_2 \xrightarrow{\quad d_1 \quad} X_1$$

More informally, we complete a pair of composable morphisms (the spine of a 2-simplex) to a unique 2-simplex, and then forget everything about that 2-simplex except the unique maximal edge.

More generally, the 1-Segal conditions provide n-fold composition maps[3]

$$\underbrace{X_1 \times_{X_0} \cdots \times_{X_0} X_1}_{\times n} \longrightarrow X_1$$

defined by completing a spine to the corresponding n-simplex, and then forgetting everything except the maximal edge. In dimension 3, this looks like

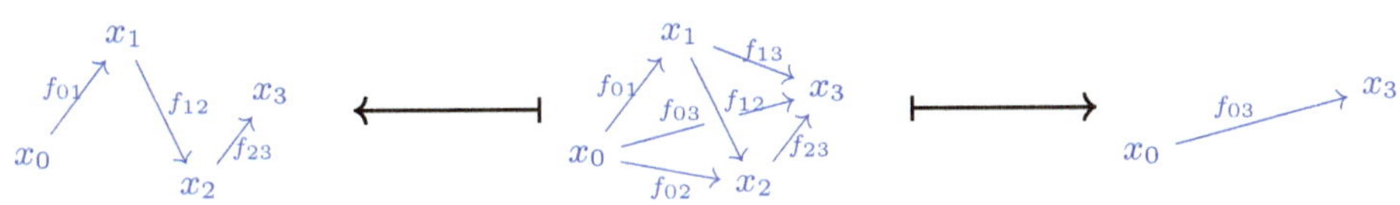

Now that we understand how composition maps arise from the 1-Segal conditions, let us try to untangle associativity. The simplest expression of the associativity condition is that, writing κ for the composition map

$$\kappa\colon X_1 \times_{X_0} X_1 \longrightarrow X_1,$$

we require

$$\kappa \circ (\kappa \times_{X_0} \mathrm{id}) = \kappa \circ (\mathrm{id} \times_{X_0} \kappa). \tag{2.1}$$

[3] In homotopical algebra and higher category theory, these higher compositions allow 1-Segal *spaces* to encode $(\infty, 1)$-categories and A_∞-algebras. This perspective will reappear later on in the present work, as the 2-Segal conditions relate to *coherent* associativity, even in the Set-valued setting.

Since each copy of κ is defined by a span of sets, let's draw what happens in the composition $\kappa \circ (\kappa \times_{X_0} \mathrm{id})$.

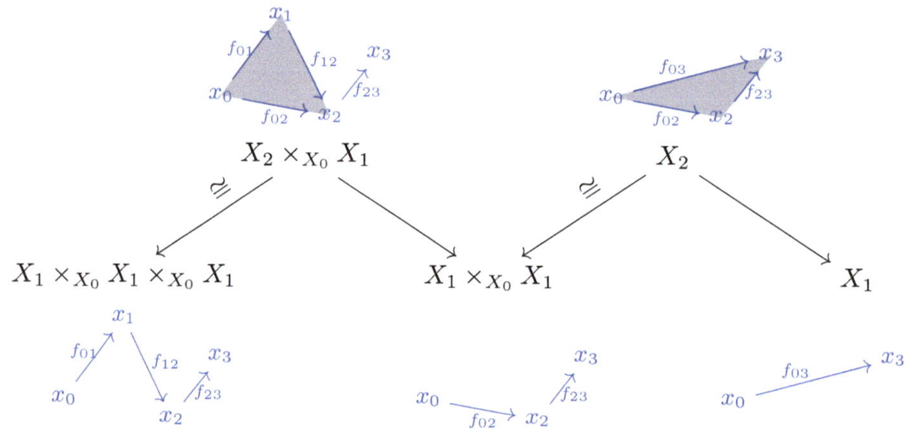

As we examine this picture, we might notice something interesting: we could skip the middle "forgetting" step. Instead, we could first complete the 2-simplex $x_0 \to x_1 \to x_2$, then complete the 2-simplex $x_0 \to x_2 \to x_3$, and then forget anything except the maximal edge. Categorically, we can interpret this by replacing the pair of spans above with a single span given by the pullback

$$X_2 \times_{X_1 \times_{X_0} X_1} (X_2 \times_{X_0} X_1)$$

of the three sets in the middle of the diagram.

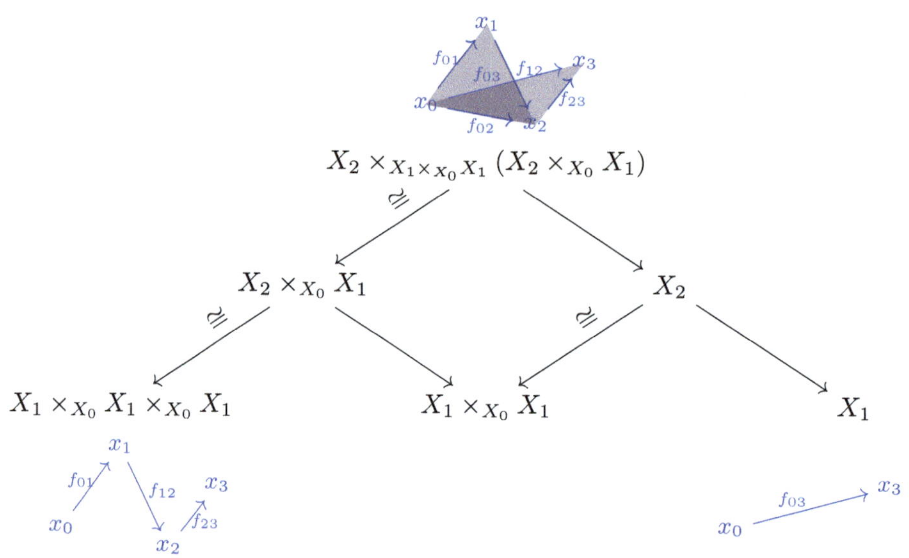

Examining the diagrams above, we see that this pullback is isomorphic to the pullback $X_2^{d_2} \times_{X_1}^{d_1} X_2$, where we indicate with superscripts the maps defining the pullback diagram.

Repeating this procedure for the composite $\kappa \circ (\mathrm{id} \times_{X_0} \kappa)$, we obtain two spans, representing the two composites in our associativity conditions. These are

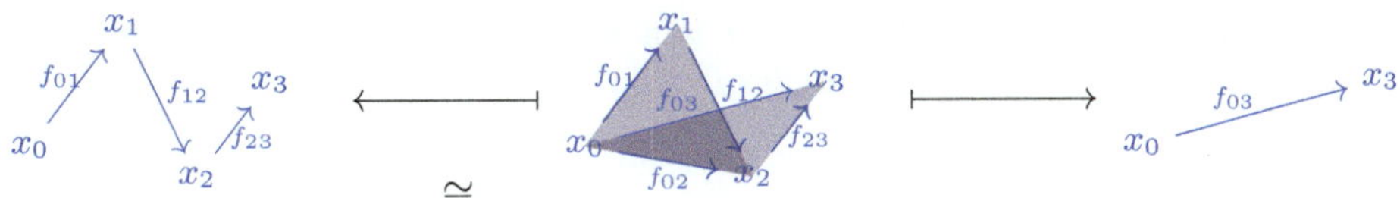

$$X_1 \times_{X_0} X_1 \times_{X_0} X_1 \xleftarrow{\ \cong\ } X_2 {}^{d_2}\!\times^{d_1}_{X_1} X_2 \longrightarrow X_1$$

for $\kappa \circ (\kappa \times_{X_0} \mathrm{id})$, and

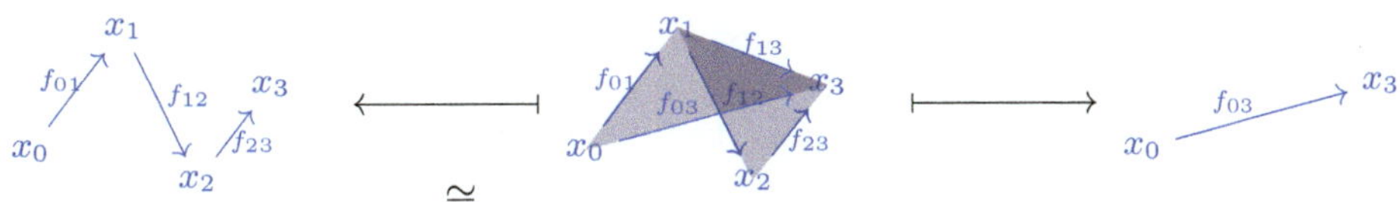

$$X_1 \times_{X_0} X_1 \times_{X_0} X_1 \xleftarrow{\ \cong\ } X_2 {}^{d_0}\!\times^{d_1}_{X_1} X_2 \longrightarrow X_1$$

for $\kappa \circ (\mathrm{id} \times_{X_0} \kappa)$.

So why do the 1-Segal conditions guarantee that the maps defined by these spans are equal? Simply put, the roofs of the two spans we have computed are isomorphic, in a way compatible with the maps defining the spans. That is, each of the "tacos"[4] pictured above can be uniquely completed to a full 3-simplex of X.

More precisely, each of these spans is isomorphic to the span defining the 3-fold composition, and the diagram

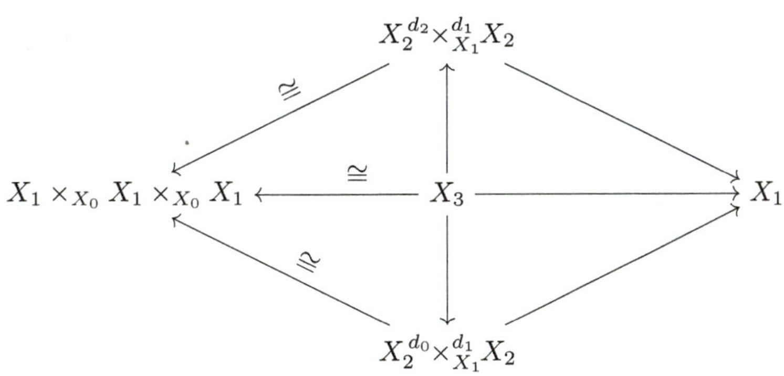

commutes. Thus, this associativity diagram is equivalent to requiring that both legs of the span

$$X_2 {}^{d_2}\!\times^{d_1}_{X_1} X_2 \xleftarrow{(d_1,d_3)} X_3 \xrightarrow{(d_2,d_0)} X_2 {}^{d_0}\!\times^{d_1}_{X_1} X_2$$

are isomorphisms. Equivalently, this amounts to requiring that the squares

$$\begin{array}{ccc} X_3 & \xrightarrow{d_3} & X_2 \\ {\scriptstyle d_1}\downarrow & & \downarrow{\scriptstyle d_1} \\ X_2 & \xrightarrow[d_2]{} & X_1 \end{array} \qquad \text{and} \qquad \begin{array}{ccc} X_3 & \xrightarrow{d_0} & X_2 \\ {\scriptstyle d_2}\downarrow & & \downarrow{\scriptstyle d_1} \\ X_2 & \xrightarrow[d_0]{} & X_1 \end{array} \tag{2.2}$$

are pullback squares.

[4] A wonderfully evocative term introduced by Contreras, Keller, and Mehta in [3].

We now, after a half-dozen pages of exploration, are ready to give a slogan explaining what we want the 2-Segal conditions to be.

SLOGAN: A 2-Segal set $X\colon \Delta^{\mathrm{op}} \to \mathsf{Set}$ is a set X_1 equipped with a composition operation which *may not be well-defined*[5], but *is* associative.

We thus remove the 1-Segal condition entirely, and *define* our n-fold compositions to *be* the spans

$$\underbrace{X_1 \times_{X_0} \cdots \times_{X_0} X_1}_{\times n} \longleftarrow X_n \longrightarrow X_1.$$

We then require that the two squares of (2.2) are pullback.

However, this is not quite sufficient, for a rather subtle reason. If we write κ_n for this n-fold composition operation, then combining the 1-Segal conditions with the pullback squares of (2.2) implies that the relations

$$\kappa_m \circ (\mathrm{id} \times \kappa_n \times \mathrm{id}) \cong \kappa_{n+m-1} \tag{2.3}$$

should hold no matter which input of the m-fold composition we plug the n-fold composition into.[6]

If we forget the 1-Segal conditions, however, the pullback squares (2.2) no longer suffice to prove these additional associativity conditions. We must, therefore, impose explicit associativity conditions for each such equation (2.3).

If we unwind the conditions exactly as we did above, each equation (2.3) yields a square

$$\begin{array}{ccc} X_{n+m-1} & \longrightarrow & X_n \\ \downarrow & & \downarrow \\ X_m & \longrightarrow & X_1 \end{array}$$

which we must require to be pullback. We thus obtain the full *2-Segal conditions.*

DEFINITION 2.2. A simplicial set $X\colon \Delta^{\mathrm{op}} \to \mathsf{Set}$ is called *2-Segal* if, for every $k \geq 1$ and every $0 \leq i < j \leq k$, the square of ordered sets

$$\begin{array}{ccc} \{i,j\} & \longrightarrow & \{i, i+1, \dots, j-1, j\} \\ \downarrow & & \downarrow \\ \{0,1,\dots,i,j,j+1,\dots,k\} & \longrightarrow & [k] \end{array}$$

(when interpreted as a square in Δ) is sent by X to a pullback square. Equivalently, X is *2-Segal* if for any $n, m \geq 1$, and any $0 \leq i < n$, the squares

$$\begin{array}{ccc} X_{n+m-1} & \longrightarrow & X_n \\ \downarrow & & \downarrow{\scriptstyle \{i,i+1\}} \\ X_m & \xrightarrow[\{0,m\}]{} & X_1 \end{array}$$

are pullback.[7]

[5]By this, we mean that the composition may be partially-defined, multiply-defined, or both. That is, the composition is a span.

[6]Notice that these isomorphisms are part of the data of a coherently associative (i.e., A_∞- or E_1-) algebra.

[7]This second description does not actually uniquely specify the squares in question as stated. One must, additionally, require that the underlying maps in Δ are injective.

These pullback conditions are collectively referred to as the *2-Segal* conditions.

REMARK 2.3. There are many sets of pullback conditions which are equivalent to the 2-Segal conditions (we will see at least one other when we discuss the geometric interpretation). The formulation of *decomposition spaces* in [**10**] is another way of describing the same conditions.

REMARK 2.4. As with the 1-Segal conditions, the 2-Segal conditions make sense in a much broader context than that of simplicial sets. Indeed, one can easily impose the 2-Segal conditions on any simplicial object X in a category $\mathcal{C}$ with pullbacks. More generally, if $\mathcal{C}$ is an ∞-category, one can require that the squares in question be ∞-categorical pullback squares. Since, as we are in the process of showing, 2-Segal objects are intimately connected to associativity in higher algebra, it is these ∞-categorical versions which are often of the greatest use.

Since the 2-Segal conditions are meant to encode associativity, and we arrived at them by contemplating the associativity of composition in the nerves of categories, we might expect that standard associative composition laws also give rise to 2-Segal simplicial sets. It is precisely this that the next condition makes precise.

THEOREM 2.5 ([**7**], 2.3.3 and 2.5.3). *If* $X\colon \Delta^{\mathrm{op}}\mathsf{Set}$ *is 1-Segal, then it is 2-Segal.*

PROOF. Suppose that X is 1-Segal, let $n \geq 2$, and $0 \leq i < j \leq n$. We will show that the corresponding 2-Segal square is pullback. The proof is reasonably straightforward, but keeping track of the maps involved can be difficult. As such, we will parallel each commutative diagram with figures depicting generic elements of the corresponding sets in the special case where $n = 3$, $i = 1$ and $j = 3$.

We can use the spine map to extend the associated 2-Segal square to a diagram

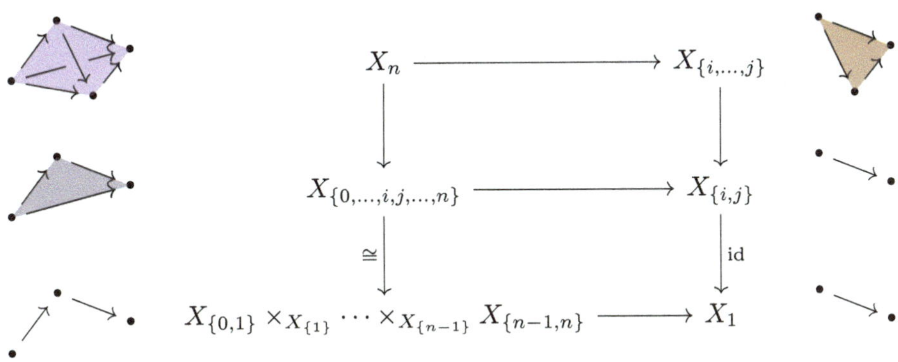

$$\begin{array}{ccc} X_n & \longrightarrow & X_{\{i,\dots,j\}} \\ \downarrow & & \downarrow \\ X_{\{0,\dots,i,j,\dots,n\}} & \longrightarrow & X_{\{i,j\}} \\ \cong\downarrow & & \downarrow \mathrm{id} \\ X_{\{0,1\}}\times_{X_{\{1\}}}\cdots\times_{X_{\{n-1\}}} X_{\{n-1,n\}} & \longrightarrow & X_1 \end{array}$$

The lower square is pullback, since two of its parallel legs are isomorphisms, and thus we can identify the map from X_n to the pullback of the upper square with the map

$$X_n \longrightarrow X_{\{0,1\}}\times_{X_{\{1\}}}\cdots X_{\{i-1,i\}}\times_{X_{\{i\}}} X_{\{i,\dots,j\}}\times_{X_{\{j\}}}\cdots\times_{X_{\{n-1\}}} X_{\{n-1,n\}}.$$

Composing with the spine isomorphism

$$X_{\{i,\ldots,j\}} \longrightarrow X_{\{i,i+1\}} \times_{X_{\{i+1\}}} \cdots \times_{X_{\{j-1\}}} X_{\{j-1,j\}}$$

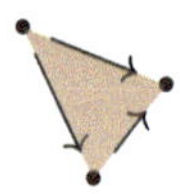

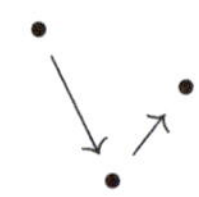

yields precisely the spine map for X_n. Thus, by the 2-out-of-3 property, the map from X_n to the pullback is an isomorphism, proving the proposition. □

2.1. Addendum: unitality. Having now understood how associativity leads us naturally to the 2-Segal conditions, it is natural to ask whether there are similar conditions governing associativity. It turns out that there are. Indeed, the resulting conditions are called the *unitality conditions* in [**7**]. However, as we will see, the story behind unitality becomes a bit stranger.

First, let us unwind the meaning of unitality in our framework. Within the rubric of our "categories with multiply- and/or partially-defined composition," let's require a fairly strict definition. For $x \in X_0$, the element[8] $s_0(x)$ should be the left unit in the sense that, for any $f \in X_1$ with $d_0(f) = x$, the composition $f \circ s_0(x)$ should have precisely one value, f. We can interpret this diagrammatically, as we did with the first associativity condition.

The requirement of strict (left) unitality can be expressed by writing that $\kappa_2 \circ (s_0 \times \mathrm{id}) = \mathrm{id}_{X_1}$. Expressing the left-hand side of this equation as a pair of spans, we obtain

$$\begin{array}{ccccccc} & & X_0 \times_{X_0} X_1 & & & & X_2 & & \\ & \overset{\cong}{\swarrow} & & \overset{s_0 \times id}{\searrow} & & \overset{(d_2, d_0)}{\swarrow} & & \searrow & \\ X_0 \times_{X_0} X_1 & & & & X_1 \times_{X_0} X_1 & & & & X_1 \end{array}$$

The condition then requires that, reformulating this as a single span by the use of a pullback, the result must be isomorphic to the identity span

$$X_1 \xleftarrow{\mathrm{id}} X_1 \xrightarrow{\mathrm{id}} X_1.$$

The right unitality conditions can be analyzed similarly, and we arrive at the conclusion that unitality amounts to the squares

$$\begin{array}{ccc} X_1 & \xrightarrow{s_0} & X_2 \\ {\scriptstyle d_1}\downarrow & & \downarrow{\scriptstyle d_2} \\ X_0 & \xrightarrow[s_0]{} & X_1 \end{array} \qquad \text{and} \qquad \begin{array}{ccc} X_1 & \xrightarrow{s_1} & X_2 \\ {\scriptstyle d_0}\downarrow & & \downarrow{\scriptstyle d_0} \\ X_0 & \xrightarrow[s_0]{} & X_1 \end{array}$$

being pullback.

As we did with associativity, we must also include the considerations of unitality for the n-fold compositions. This yields the following conditions.

---

[8]One might reasonably object at this point that there is no reason that the unit map $X_0 \to X_1$ should not also be multiply- and/or partially-defined like our multiplications. While this is reasonable, we will set it aside until we reinterpret the 2-Segal conditions in terms of algebras.

DEFINITION 2.6. A 2-Segal simplicial set $X\colon \Delta^{\mathrm{op}} \to \mathsf{Set}$ is called *unital* if, for every $n \geq 2$ and every $0 \leq i < n$, it sends the square

$$\begin{array}{ccc} [n-1] & \xleftarrow{\;s_i\;} & [n] \\ \uparrow & & \uparrow \\ \{i\} & \longleftarrow & \{i, i+1\} \end{array}$$

to a pullback square.

Surprisingly, though, it turns out that this condition is redundant. That is, the 2-Segal conditions which guarantee unitality *also* guarantee associativity! This is less strange than it first appears, because the degeneracy maps in a simplicial set already encode a lot of information pertaining to units. Indeed, if one traces carefully through our work on associativity above, we will see that the associativity makes no use of *any* degeneracy maps.

As a result, we can think of a partially-/multiply-defined multiplication which is associative, but not unital, as a 2-Segal *semi-simplicial* set. That is, a functor

$$X\colon \Delta_{\mathrm{inj}}^{\mathrm{op}} \longrightarrow \mathsf{Set}$$

satisfying the 2-Segal pullback conditions, where Δ_{inj} denotes the wide subcategory of Δ which contains only the injective maps. In general, the nerve of a non-unital category is a semi-simplicial set, and semi-simplicial sets satisfying an analogue of the 1-Segal condition are often used to model non-unital or weakly unital higher categories. See, e.g., [**28**] and [**17**] for some discussion of the role of semi-simplicial sets in modeling weak units.

The remarkable result, however, is that imposing unitality consists only of extending to a simplicial structure, and does not require new conditions.

THEOREM 2.7 ([**9**]). *If* $X\colon \Delta^{\mathrm{op}} \to \mathsf{Set}$ *is 2-Segal, then* X *is unital.*

3. The geometric perspective: polygonal decompositions

In addition to the purely algebraic intuition discussed above, there is a purely geometric way of thinking about 2-Segal spaces in terms of triangulations of polygons. In this section, we will first develop this perspective, and then explain how it relates to the previous, algebraic perspective.

The basic idea is to try and define a kind of "state-sum" construction for polygons.[9] What we mean by a "state-sum" is that we will try to assign elements of an algebraic or combinatorial structure to triangles in a triangulation of a surface in a consistent way such that the set of all labelings is independent of the choice of triangulation.[10] In our case, the algebraic input will be a simplicial set $X\colon \Delta^{\mathrm{op}} \to \mathsf{Set}$, and the geometric input will be an $(n+1)$-gon P_{n+1} with vertices labeled $0, 1, \ldots, n$ counterclockwise. The slogan for this section is thus

SLOGAN: A 2-Segal set $X\colon \Delta^{\mathrm{op}} \to \mathsf{Set}$ is a simplicial set such that the "state-sum" of X over any polygon P_{n+1} is independent of the choice of triangulation.

[9] With some additional structure on the inputs, the construction we give here can be formalized into a state-sum construction for punctured surfaces. See [**8**, §V] and [**27**, §3.3] for more details.

[10] I am being deliberately vague in this description, as there are many variations on state sums and lattice models in the literature.

To understand this slogan, we need to explain our "state-sum" construction. Given any triangulation $\mathcal{T}$ of P_{n+1} *with the same vertices as* P_{n+1}, we can think of the labeling of the vertices of P_{n+1} as orienting the edges of $\mathcal{T}$, as pictured below.

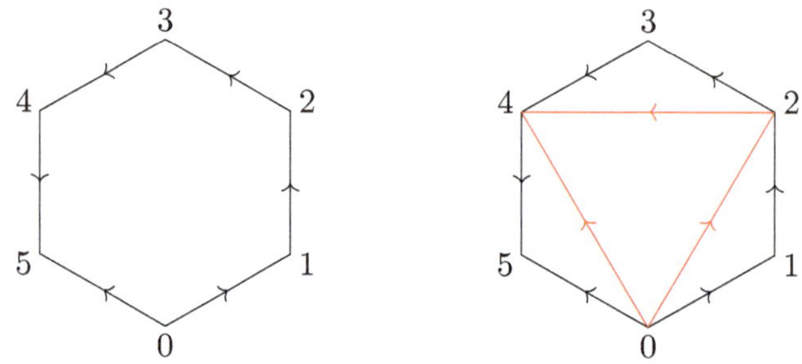

This means that we can identify each triangle of the triangulation as a 2-simplex, and each edge of the triangulation as a 1-simplex. Put another way, we can think of the subsets of $\{0, \ldots, n\}$ defining the 1- and 2-simplices of $\mathcal{T}$ as elements of Δ via the canonical order on $\{0, \ldots, n\}$.

We will then label each 2-simplex (triangle) t of the triangulation by a 2-simplex $\sigma_t \in X_2$ and each 1-simplex (edge) e of the triangulation by a 1-simplex $x_e \in X_1$. We will require our labelings to be compatible in the sense that $d_i(\sigma_t) = x_{d_i(t)}$ for any face map. Such a labeling is completely determined by the 2-simplices of X_2 assigned to the triangles of $\mathcal{T}$, which are then subject to compatibilities. For instance, for the triangulated pentagon above, a generic labelling looks like

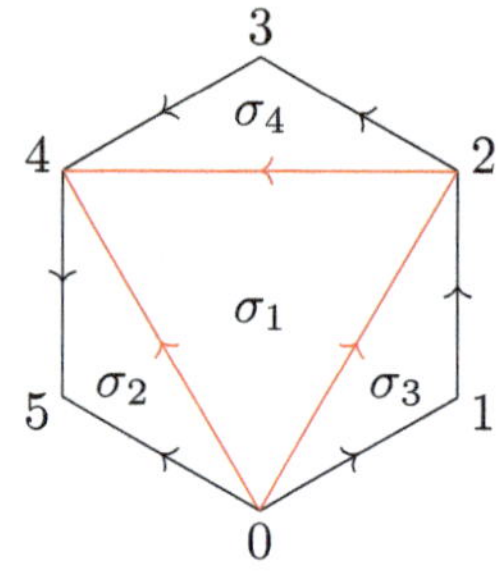

$$d_0(\sigma_1) = d_1(\sigma_4)$$
$$d_1(\sigma_1) = d_2(\sigma_2)$$
$$d_2(\sigma_1) = d_1(\sigma_3).$$

The set of all such labelings, which we will write as $(\mathcal{T}, X)$, can be expressed as a limit over a particular diagram. Again thinking of the subsets of $\{0, \ldots, n\}$ which define the 1- and 2-simplices of $\mathcal{T}$ as elements of Δ, we obtain a diagram in the simplex category which sends 1- and 2-simplices of $\mathcal{T}$ to the corresponding ordered sets in Δ. We can further apply X to obtain a diagram in Set. In the case of the

example above, these diagrams are

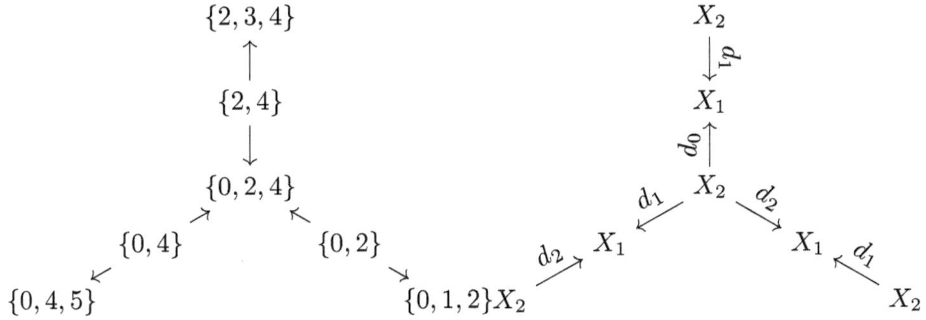

Let us denote the indexing category of this resulting diagram $I_{\mathcal{T}}$, and the functor $X_{\mathcal{T}}$. We can then express our proposed state-sum construction as the limit of this diagram of sets, i.e.,

$$(\mathcal{T}, X) := \lim_{I_{\mathcal{T}}} X_{\mathcal{T}}.$$

However, in general, this construction will depend explicitly on the choice of triangulation, which runs counter to the very idea of a state sum. To rectify this, we notice that, since our original diagram was defined in terms of ordered subsets of $[n] = \{0, \ldots, n\}$, the set X_n automatically fits into a cone over $X_{\mathcal{T}}$, the components of which are induced by the inclusions of the subsets into $[n]$. Thus, there is a unique map

$$f_{\mathcal{T}} \colon X_n \longrightarrow (\mathcal{T}, X)$$

induced by this cone. Following [**7**], we will call $f_{\mathcal{T}}$ the *2-Segal map corresponding to* $\mathcal{T}$. We can then define precisely the condition which will make our state-sum invariant under triangulation.

DEFINITION 3.1. We call a simplicial set $X \colon \Delta^{\mathrm{op}} \to \mathsf{Set}$ *geometrically 2-Segal*[11] if for any $n \geq 2$ and any triangulation $\mathcal{T}$ of the labeled $(n+1)$-gon P_{n+1} with the same vertex set as P_{n+1}, the 2-Segal map

$$f_{\mathcal{T}} \colon X_n \longrightarrow (\mathcal{T}, X)$$

is an isomorphism. Equivalently, this is the requirement that the canonical cone with tip X_n over $X_{\mathcal{T}}$ is a limit cone. In particular, this definition makes our slogan from the beginning of the section precise, in that it means the "state-sum" $(\mathcal{T}, X)$ is independent of the choice of triangulation.

This definition is all well and good, and may even seem reasonable in isolation, but our choice of terminology should raise an important question: what does this state-sum invariance condition have to do with associativity?

One answer to this question lies in an operadic view of associativity. Schematically, suppose that an object Y of some monoidal category $\mathcal{C}$ has a multiplication operation $\mu \colon Y \otimes Y \longrightarrow Y$. We can draw this multiplication as the planar

[11]Warning: this terminology, unlike most of our terminology and notation, is non-standard.

rooted tree

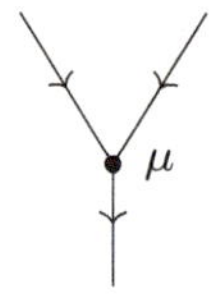

Associativity then says that any planar rooted binary tree with n inputs — interpreted as a composition of $n-1$ copies of μ with itself — must be equal. So, for instance, the associativity condition of (2.1) is expressed by saying the compositions of multiplications assigned to the two planar rooted binary trees with two leaves must be equal. Schematically

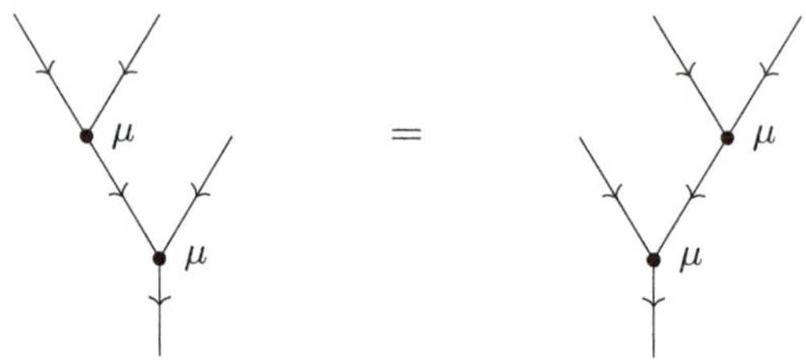

Reinterpreting this in terms of a 2-Segal simplicial set $X\colon \Delta^{\mathrm{op}} \to \mathsf{Set}$ in the algebraic terms of the previous section, this requires that the squares of equation (2.2) be pullback.

Given a triangulation $\mathcal{T}$ of P_{n+1} as above, there is a standard way to extract a planar, rooted, binary tree, which we will call the *dual graph* to the triangulation.[12] To each triangle of $\mathcal{T}$ one assigns a vertex of the tree, and to each edge $\{i,j\}$ of $\mathcal{T}$ one assigns an edge $e_{i,j}$ of the tree. The orientations are chosen so that the edges $\{i,j\}$ and $e_{i,j}$ form an oriented basis.[13] For example:

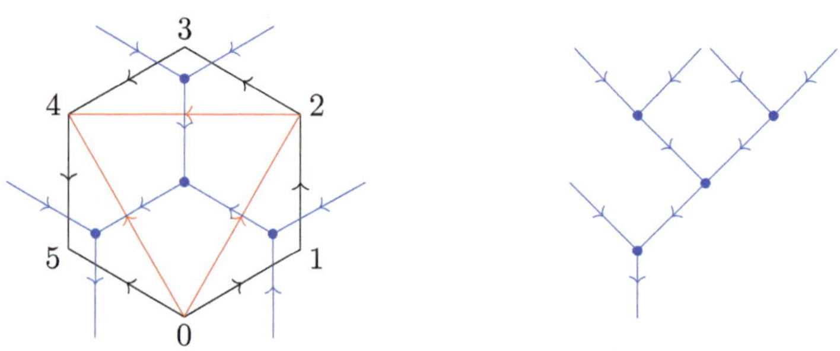

We can thus see that the limits arising from our triangulations are precisely iterated versions of the pullbacks we used to explore our associativity conditions. As such, we should expect that the 2-Segal conditions defining associativity would be equivalent to the geometric 2-Segal conditions.

However, this association — from triangulations, to binary trees, to limits — can also be generalized. By allowing $\mathcal{T}$ to divide P_{n+1} not into triangles, but into $(k+1)$-gons, we obtain planar rooted trees which are not necessarily binary, and

[12]For discussions of this correspondence in the context of Catalan numbers, see, e.g., [**25**, Theorem 1.5.1].

[13]The orientations here are included to better align with the operadic intuition. However, they are essential to neither our state-sum construction nor the formalization which will follow in the next section, and may be safely ignored should the reader so choose.

thereby obtain associativity conditions explicitly involving the n-fold composition operations defined by X. We will call such a subdivision $\mathcal{T}$ a *polygonal subdivision*, and note that for any polygonal subdivision $\mathcal{T}$, the construction of $I_{\mathcal{T}}$, $X_{\mathcal{T}}$, $(\mathcal{T}, X)$, and $f_{\mathcal{T}}$ works precisely as it does for triangulations.

With this in hand, the intuition we have developed connecting the algebraic and geometric 2-Segal conditions can be made precise.

THEOREM 3.2 ([**7**, Proposition 2.3.2]). *For a simplicial set* $X\colon \Delta^{\mathrm{op}} \to \mathsf{Set}$, *the following are equivalent.*

(1) *The simplicial set X is 2-Segal.*
(2) *The simplicial set X is geometrically 2-Segal.*
(3) *For every polygonal subdivision $\mathcal{T}$ of P_{n+1} for $n \geq 2$, the canonical map*

$$f_{\mathcal{T}}\colon X_n \longrightarrow (\mathcal{T}, X)$$

is an isomorphism.

We will provide a proof of this theorem in the next section, as we formalize our geometric intuition.

4. Formalizing the geometric perspective

Now that we have walked through two intuitions behind 2-Segal simplicial sets, let's try to formalize them. To make the geometric intuition formal, we introduce a tool from [**7**, §2.2]: *membrane sets.*[14]

DEFINITION 4.1. Let D and X be simplicial sets. We define the *set of D-membranes in X* to be the set

$$(D, X) := \mathrm{Hom}_{\mathsf{Set}_\Delta}(D, X).$$

Applying the standard result that D is a colimit over its category of simplices, we can alternatively express the membrane set as

$$(D, X) = \mathrm{Hom}_{\mathsf{Set}_\Delta}(D, X) \cong \mathrm{Hom}_{\mathsf{Set}_\Delta}(\mathrm{colim}_{\Delta/D}\, \Delta^n, X) \cong \lim_{(\Delta/D)^{\mathrm{op}}} X_n.$$

REMARK 4.2. It is the latter formula which is of use when X is a simplicial object in a category other than sets, and which allows us to generalize the work in this section beyond simplicial sets.

The notion of membrane sets will be most useful when $D \subset \Delta^n$ is a subset of a standard n-simplex. In this case, the limit expressing the membrane set can be rewritten using the category of *non-degenerate* simplices of X. However, there is another important feature in this case which is key to the 2-Segal conditions.

DEFINITION 4.3. Let X be a simplicial set, and $D \subset \Delta^n$ a simplicial subset. The inclusion $D \to \Delta^n$ induces a map

$$f_D\colon X_n \longrightarrow (D, X)$$

by the contravariant functoriality of $\mathrm{Hom}_{\mathsf{Set}_\Delta}(-, X)$. This map can alternatively be viewed as the universal map induced by the cone with tip X_n over the functor

$$(\Delta/D)^{\mathrm{op}} \longrightarrow \Delta^{\mathrm{op}} \xrightarrow{X} \mathsf{Set}.$$

[14]In [**7**], the authors focus on the higher categorical (space-enriched) case, and thus always speak of membrane *spaces*, rather than membrane sets.

Following [7], we will call f_D the *generalized Segal map* associated to $D \subset \Delta^n$.

EXAMPLES 4.4.

(1) The motivation for the terminology *generalized Segal map* comes from the case where $D = \mathrm{Sp}(\Delta^n) \subset \Delta^n$. The generalized Segal map associated to this inclusion is simply the nth 1-Segal map

$$f_D\colon X_n \longrightarrow X_{\{0,1\}} \times_{X_{\{1\}}} \cdots \times_{X_{\{n-1\}}} X_{\{n-1,n\}}.$$

(2) Let $D \subset \Delta^n$ be a 1-dimensional simplicial set which is a tree and contains every vertex of Δ^n. We will call such a subset a *spanning tree for* Δ^n. Notice that each spine inclusion is, in particular, a spanning tree. If X is a simplicial set such that the generalized Segal maps associated to spanning trees are always isomorphisms, then X is isomorphic to the nerve of a groupoid (this statement is implicit in the discussion of [**16**, §3], but I am unaware of a more direct reference).

For our purposes, the most important examples of generalized Segal maps are those arising from triangulations and polygonal subdivisions. Let $\mathcal{T}$ be a polygonal subdivision of P_{n+1} which has the same vertex set as P_{n+1}. Then each polygon of $\mathcal{T}$ can be uniquely identified with a subset $I \subset [n]$ via the labeling of the vertices. We then define a simplicial subset $\Delta^{\mathcal{T}} \subset \Delta^n$ associated to the polygonal subdivision $\mathcal{T}$ by the formula

$$\Delta^{\mathcal{T}} := \bigcup_{I \in \mathcal{T}} \Delta^I.$$

That is, we define $\Delta^{\mathcal{T}}$ to be the full simplicial subset of Δ^n on the simplices representing polygons of the decomposition $\mathcal{T}$. With this terminology in place, we can reformulate our geometric definition of the 2-Segal conditions.

To make it clear how this connects to the dual graph pictures we drew in the previous section, let us fix some further definitions.

DEFINITION 4.5. Let $\mathcal{T}$ be a polygonal subdivision of P_{n+1} with the same vertex set as P_{n+1}. Write $(\Delta/\Delta^{\mathcal{T}})_{\mathrm{nd}}$ for the category of non-degenerate simplices of $\Delta^{\mathcal{T}}$. Let $I_{\mathcal{T}} \subset (\Delta/\Delta^{\mathcal{T}})_{\mathrm{nd}}$ be the full subcategory on the objects of form Δ^I for I a polygon of $\mathcal{T}$ and the objects $\Delta^{\{i,j\}}$ for $\{i,j\}$ an edge of $\mathcal{T}$. We will call $I_{\mathcal{T}}$ the *dual graph category* of $\mathcal{T}$.

The dual graph category is simply a formalization of the indexing category introduced under the same notation in the previous section.[15] As a simple example, we have

[15]The one slight difference is that, in the previous section, we did not include the objects of $I_{\mathcal{T}}$ which correspond to external edges of the polygon. As these do not alter the limits of the diagrams we consider, this change makes no difference to the overall picture.

where we have drawn a triangulation of the square on the left in black, and the corresponding dual graph category on the right in blue.

REMARK 4.6. Treating the dual graph of the triangulation $\mathcal{T}$ as a topological space, one can prove that the realization $|I_{\mathcal{T}}|$ is homeomorphic to the dual graph. We will not, however, pursue this further here.

PROPOSITION 4.7. *The inclusion*

$$\iota\colon I_{\mathcal{T}} \longrightarrow (\Delta/\Delta^{\mathcal{T}})_{\mathrm{nd}}$$

induces an isomorphism

$$\lim_{I_{\mathcal{T}}} X_m \cong \lim_{(\Delta/\Delta^{\mathcal{T}})^{\mathrm{op}}_{\mathrm{nd}}} X_m$$

PROOF. An explicit levelwise computation in Set shows that the induced map

$$\mathrm{colim}_{I_{\mathcal{T}}} \Delta^m \longrightarrow \Delta^{\mathcal{T}}$$

is an isomorphism of simplicial sets. The proposition then follows from the continuity of the Yoneda embedding by applying $\mathrm{Hom}_{\mathsf{Set}_{\Delta}}(-, X)$. □

Since, as we saw above, $(\Delta^{\mathcal{T}}, X) \cong \lim_{(\Delta/\Delta^{\mathcal{T}})^{\mathrm{op}}_{\mathrm{nd}}} X_m$, Proposition 4.7 shows that we can identify the generalized Segal map $f_{\mathcal{T}}$ associated to $\Delta^{\mathcal{T}}$ with the map

$$X_n \longrightarrow \lim_{I_{\mathcal{T}}} X_m$$

induced by the canonical cone.

We are now in a position to prove Theorem 3.2.

PROOF (OF THEOREM 3.2). Since every triangulation is a polygonal subdivision, we have that (3) $\Longrightarrow$ (2). To see that (3) $\Longrightarrow$ (1), let $n \geq 3$, and $0 \leq i < j \leq n$, and consider the square

$$\begin{array}{ccc} X_n & \longrightarrow & X_{\{i,\ldots,j\}} \\ \downarrow & & \downarrow \\ X_{\{0,\ldots,i,j,\ldots,n\}} & \longrightarrow & X_{\{i,j\}} \end{array}$$

This is pullback if and only if the induced map

$$X_n \longrightarrow X_{\{i,\ldots,j\}} \times_{X_{\{i,j\}}} X_{\{0,\ldots,i,j,\ldots,n\}}$$

is an isomorphism. However, this is simply the generalized Segal map corresponding to the polygonal subdivision which divides P_{n+1} into two polygons along the edge from i to j.

To see that (2) $\Longrightarrow$ (3), choose a polygonal subdivision $\mathcal{T}$ of P_{n+1}, and a refinement to a triangulation $\mathcal{S}$. For each polygon Δ^J of $\mathcal{T}$, there is a corresponding generalized Segal map g_J for the restriction $\mathcal{S} \cap J$ of triangulation $\mathcal{S}$ to J. The generalized Segal map $f_{\mathcal{S}}$ can be identified with the composite

$$X_n \xrightarrow{f_{\mathcal{T}}} \lim_{I_{\mathcal{T}}} X_J \xrightarrow{\{g_J\}} \lim_{I_{\mathcal{T}}} \lim_{I_{\mathcal{S}\cap J}} X_L.$$

However, since (2) holds, the g_J's and $f_{\mathcal{S}}$ are isomorphisms. Thus, by 2-out-of-3, $f_{\mathcal{T}}$ is an isomorphism.

A similar 2-out-of-3 argument, building arbitrary subdivisions out of subdivisions into two polygons, shows that (1) $\Longrightarrow$ (3), completing the proof. □

4.1. Digression: the 4-truncation. It is well known that every 1-Segal simplicial set X is completely determined by the underlying 2-dimensional data of X, and thus is 2-coskeletal. Rather less well-known is that, if a 2-coskeletal simplical set X satisfies the 1-Segal conditions for 2-dimensional and 3-dimensional spines, then it is 1-Segal (see [**22**, Lemma 5.2] for a closely-related proof). In this section, we will establish analogues of these results for 2-Segal simplicial sets.

The geometric discussion above could be taken to suggest that 2-Segal simplicial sets are determined by their 2-dimensional data — i.e., their 2-truncations. However, as we will see, this is not quite the case. The reason for this, loosely speaking, is that the *choice* of isomorphism between the membrane sets associated to the two triangulations of the square is a fundamental part of the 2-Segal data. Since this isomorphism is determined by the set X_3 of 3-simplices and the concomitant maps to X_2, we might adjust our expectations to predict that 2-Segal simplicial sets are determined by their 3-truncations. This is, in fact, the case.

PROPOSITION 4.8. *Every 2-Segal simplicial set X is 3-coskeletal.*

PROOF. This is [**1**, Corollary 1.7]. □

Perhaps surprisingly, it is not true that every 3-coskeletal simplicial set for which the 2-Segal maps

$$X_2 \times_{X_1} X_2 \longleftarrow X_3 \longrightarrow X_2 \times_{X_1} X_2$$

are isomorphisms is 2-Segal. The reason for this is easier to interpret from the algebraic perspective. These 3-dimensional 2-Segal conditions encode the existence of *associators*: isomorphisms encoding the associativity of the 2-fold composition/multiplication. However, true higher-categorical associativity requires that these associators satisfy coherence conditions, which it will turn out are encoded in X_4. As such, the proper converse to Proposition 4.8 is the following.

PROPOSITION 4.9. *Let X be a 3-coskeletal simplicial set. If X satisfies the 2-Segal conditions corresponding to subdivisions of the square and pentagon, then X is 2-Segal.*

PROOF. Fix $n > 4$. We define a poset U whose elements are the simplicial subsets $\Delta^{\mathcal{R}} \subset \Delta^n$ where $\mathcal{R}$ is one of the following types of subdivision of P_{n+1}.

- A triangulation $\mathcal{R}$.
- A subdivision $\mathcal{R}$ consisting of triangles and a single square.
- A subdivision $\mathcal{R}$ consisting of triangles and two squares.
- A subdivision $\mathcal{R}$ consisting of triangles and a single pentagon.

By [**11**, Theorem 6.32], the realization of this poset is the 2-skeleton of the nth associahedron. Since the associahedra are contractible, by cellular approximation, we see that the realization of U is simply-connected.

For any fixed $\Delta^{\mathcal{T}} \in U$, we construct a diagram

$$\begin{array}{ccccc} & & (\operatorname{colim}_{\Delta^{\mathcal{R}} \in U} \Delta^{\mathcal{R}}, X) & & \\ & \swarrow^{A} & \downarrow & \nwarrow & \\ (\Delta^{\mathcal{T}}, X) & \longleftarrow & (\mathrm{sk}_3(\Delta^n), X) & \xleftarrow{\cong} & (\Delta^n, X) \end{array}$$

in which the sequence of morphisms on the bottom are induced by the inclusions

$$\Delta^{\mathcal{T}} \longrightarrow \mathrm{sk}_3(\Delta^n) \longrightarrow \Delta^n$$

and the sequence of morphisms on the top are induced by the morphisms

$$\Delta^{\mathcal{T}} \longrightarrow \operatorname{colim}_{\Delta^{\mathcal{R}} \in U} \Delta^{\mathcal{R}} \longrightarrow \Delta^n$$

which are the component of the defining cocone at $\Delta^{\mathcal{T}}$ and the universal map on the colimit induced by the canonical cone with tip Δ^n, respectively. Note that the bottom-right map is an isomorphism because X is 3-coskeletal.

Once we have shown that A is an isomorphism, constructed the dashed arrow, and shown that the dashed arrow is an isomorphism, this diagram will immediately show that the 2-Segal map

$$f_{\mathcal{T}}\colon (\Delta^n, X) \longrightarrow (\Delta^{\mathcal{T}}, X)$$

associated to $\mathcal{T}$ is an isomorphism. Since $\Delta^{\mathcal{R}} \in U$ for every triangulation $\mathcal{R}$ of P_{n+1}, it will immediately follow that X is 2-Segal, as desired.

We will first show that A is an isomorphism. Since X satisfies the 2-Segal conditions corresponding to subdivisions of the square and pentagon, the functor

$$\begin{aligned} M\colon U^{\mathrm{op}} &\longrightarrow \mathsf{Set} \\ \Delta^{\mathcal{R}} &\longmapsto (\Delta^{\mathcal{R}}, X) \end{aligned}$$

sends every morphism to an equivalence. Thus, since U is simply connected, the limit of this diagram is isomorphic to $(\mathcal{R}, X)$ via the component of the limit cone at $\Delta^{\mathcal{R}}$ for any chosen subdivision $\mathcal{R}$ in U. Note that we have natural isomorphisms

$$\begin{aligned} \lim_{\Delta^{\mathcal{R}} \in U^{\mathrm{op}}} (\Delta^{\mathcal{R}}, X) &= \lim_{\Delta^{\mathcal{R}} \in U^{\mathrm{op}}} \mathrm{Hom}_{\mathsf{Set}_\Delta}(\Delta^{\mathcal{R}}, X) \\ &\cong \mathrm{Hom}_{\mathsf{Set}_\Delta}\left(\operatorname{colim}_{\Delta^{\mathcal{R}} \in U} \Delta^{\mathcal{R}}, X\right) \cong \left(\operatorname{colim}_{\Delta^{\mathcal{R}} \in U} \Delta^{\mathcal{R}}, X\right) \end{aligned}$$

which identify precomposition with this component with map A, proving that A is an isomorphism.

To complete the proof of the proposition, it will thus suffice to show that the 3-skeleton of $\operatorname{colim}_{\Delta^{\mathcal{R}} \in U} \Delta^{\mathcal{R}}$ is isomorphic to $\mathrm{sk}_3(\Delta^n)$, thereby producing the dashed arrow and showing it is an isomorphism. Since every 3-simplex of Δ^n is represented by a square in at least one subdivision of P_{n+1}, it is clear that the map $\operatorname{colim}_{\Delta^{\mathcal{R}} \in U} \Delta^{\mathcal{R}} \to \Delta^n$ defined by universal property surjects onto the 3-skeleton. However, any two subdivisions containing the same square are related by a sequence of moves which fix that square and act on the rest of the subdivision by replacing a pair of triangles forming a square by the other pair of triangles which could form that square (see, for instance, the main theorem of [**18**]). This sequence of moves corresponds to a zig-zag in U consisting of subdivisions containing that same square. Thus, any two 3-simplices which correspond to the same 3-simplex in Δ^n are identified in the colimit, completing the proof. □

This result and its proof may seem very technical, and it certainly relies on a great many combinatorial facts and constructions — associahedra and cellular approximation, for example — which we do not have the space to fully explore here. However, this is to be expected. As we will see when we formalize the algebraic perspective, Proposition 4.9 can be seen as a simplicial set analogue of MacLane's celebrated Coherence Theorem for monoidal categories [**21**, Theorem 5.2], and as such we should expect its proof to be similarly challenging.

The proof of [**1**, Proposition 1.6] can be adapted to show that a 4-coskeletal simplicial set that satisfies the 2-Segal conditions necessary for Proposition 4.9 to hold is, in fact, 3-coskeletal. This allows us to rephrase Proposition 4.9 into a statement about 4-coskeletal simplicial sets, in which one condition involves only statements we can verify on the the 4-truncation.

Proposition 4.10. *Let X be a 4-coskeletal simplicial set. The following are equivalent.*

(1) *X is 2-Segal.*
(2) *X satisfies the 2-Segal conditions corresponding to subdivisions of the square and pentagon.*

Proposition 4.10 will be of use in formalizing the algebraic perspective, since it allows us to identify the category of 2-Segal simplicial sets with the category of 4-truncated simplicial sets satisfying condition (2) of the proposition.

5. Formalizing the algebraic perspective

Now that we have a rigorous geometric perspective on the 2-Segal conditions, we can return to the algebraic implications. We will formalize our previous discussion of the 2-Segal conditions in terms of algebraic structures in span bicategories, as in [**3, 4, 12, 26, 27**].

It is worth pointing out that the formalization presented here will differ from what one might expect from the intuitive discussion above. Using the discussion in Section 2 as a guide, one would be led to define a notion of categories (weakly) enriched in spans, a precise formalization of our notion of categories whose composition and unit morphisms are spans. This is the perspective taken in [**7**, §3.3], where such categories are called μ-categories.

To simplify our formalization, and avoid presenting the somewhat esoteric definition of μ-categories and their concomitant functors, we will need modify our intuitions slightly. If, for a moment, we stop thinking of the elements of X_0 as *objects*, we can reinterpret the spans defining our n-fold compositions as spans

$$X_1 \times \cdots \times X_1 \longleftarrow X_n \longrightarrow X_1$$

in which the left-hand leg simply *happens* to take values in $X_1 \times_{X_0} \cdots \times_{X_0} X_1$. That is, we can think of the span as defining a span from $X_1^{\times n}$ to X_1. Similarly, we can think of the degeneracy map $s_0 \colon X_0 \to X_1$ as defining a span

$$* \longleftarrow X_0 \longrightarrow X_1$$

from the singleton $*$ to X_1.

From this perspective, our discussion in section 2 yields the following structure associated to a 2-Segal simplicial set X:

- A set X_1 equipped with *n-fold multiplication spans* from $X_1^{\times n}$ to X_1.
- A *unit span* from $*$ to X_0.
- Associativity and unitality data for the multiplication and unit spans.

Conveniently, these data seem to specify a (coherently) associative algebra in the category of spans. We will make this observation precise in the remainder of this section. The reader preferring the multivalued category perspective is encouraged to supplement the treatment here with [**7**, §3.3].

As our formalization will rely on the bicategory of spans of sets, we take a moment to define this bicategory before continuing.

Definition 5.1. Let X and Y be sets. A *span* from X to Y is a diagram

$$X \xleftarrow{f_1} F \xrightarrow{f_2} Y$$

in Set. The *category of spans from X to Y* is the slice category $\mathsf{Set}_{/X\times Y}$, interpreted as the category whose objects are spans from X to Y, and whose morphisms are commutative diagrams

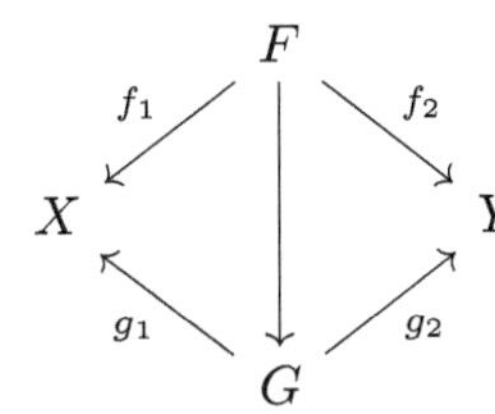

For sets X, Y, and Z, the *composition functor* of spans

$$\circ\colon\ \mathsf{Set}_{/Y\times Z} \times \mathsf{Set}_{/X\times Y} \longrightarrow \mathsf{Set}_{/X\times Z}$$

is given on objects by sending

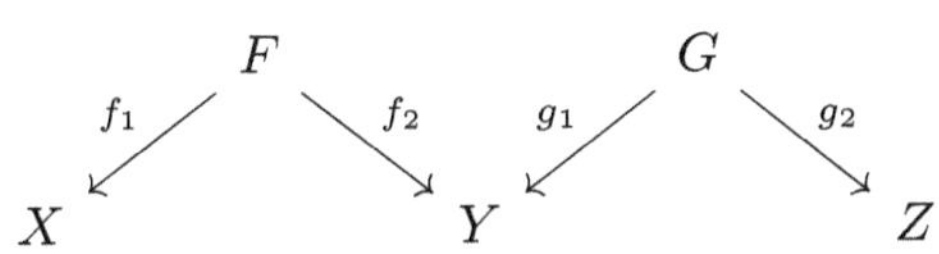

to the pullback span

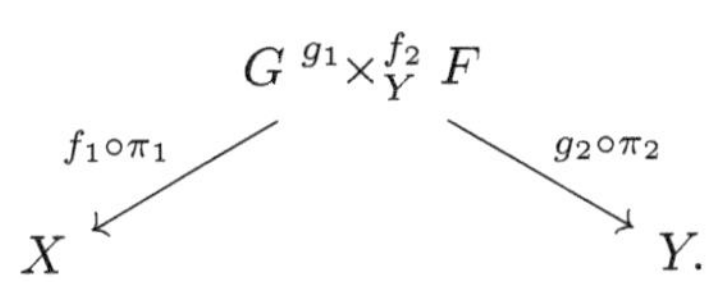

On morphisms, the composition functor is uniquely determined up to unique natural isomorphism by universal property.

Definition 5.2. We define the *bicategory of spans of sets* Span(Set) — which we will typically denote simply by Span — to consist of the following data.

(1) The objects of Span are sets.
(2) The hom-categories $\mathsf{Span}(X,Y)$ are the categories $\mathsf{Set}_{/X\times Y}$ of spans of sets.
(3) The composition is the composition functor

$$\circ\colon\ \mathsf{Set}_{/Y\times Z} \times \mathsf{Set}_{/X\times Y} \longrightarrow \mathsf{Set}_{/X\times Z}.$$

(4) The *identity* on X is the span

$$X \xleftarrow{\mathrm{id}} X \xrightarrow{\mathrm{id}} X.$$

(5) The associators and unitors are the unique natural transformations defined by universal property.

It is a straightforward, if tedious, process, to verify that these data form a bicategory. A proof, in a more general setting, can be found in [**2**]. The bicategory Span is, in fact, a symmetric monoidal bicategory under the Cartesian product of sets, a fact which is also verified in [**2**].

REMARK 5.3. One can similarly define a bicategory $\mathsf{Span}(\mathcal{C})$ of spans in any category $\mathcal{C}$ with pullbacks. Part of the reason that such "span categories" are of great use is that their morphisms generalize many different sensible notions of morphisms derived from those of $\mathcal{C}$. For instance, both $\mathcal{C}$ and $\mathcal{C}^{\mathrm{op}}$ are equivalent to subcategories of $\mathsf{Span}(\mathcal{C})$ — the wide subcategories whose morphisms are spans in which the left-hand and right-hand morphisms, respectively, are identities. Similarly, categories of partially-defined maps can typically be viewed as subcategories of span categories.

Similarly useful is the fact that reversing spans and reading them backwards provides an equivalence between $\mathsf{Span}(\mathcal{C})$ and $\mathsf{Span}^{\mathrm{op}}(\mathcal{C})$. While we will not pursue this further here, this also provides an identification between the algebras in Span we consider in this section, and *coalgebras* in Span.

5.1. Algebras in spans. The aim of this section is to provide a proof that 2-Segal simplicial sets correspond to coherently associative algebras in Span. To this end, we first define such coherently associative algebras and their isomorphisms, following [**6**].[16]

DEFINITION 5.4. An *(associative) algebra* in Span consists of the following data:

(1) A set A_1.
(2) A *multiplication span*

$$A_1 \times A_1 \xleftarrow{(\mu_2,\mu_0)} A_2 \xrightarrow{\mu_1} A_1$$

which we will sometimes briefly denote as μ.
(3) A *unit span*

$$* \longleftarrow A_0 \xrightarrow{\nu_0} A_1$$

which we will sometimes briefly denote as ν.
(4) An isomorphism α in the category $\mathsf{Span}(A_1 \times A_1 \times A_1, A_1)$ from $\mu \circ (\mu \times \mathrm{id}_{A_1})$ to $\mu \circ (\mathrm{id}_{A_1} \times \mu)$, where we are identifying $(A_1 \times A_1) \times A_1$ with $A_1 \times (A_1 \times A_1)$ via the unique isomorphism determined by universal property of products.
(5) Isomorphisms in the category $\mathsf{Span}(A_1, A_1)$

$$\lambda\colon\ \mu \circ (\nu \times \mathrm{id}_{A_1}) \xRightarrow{\simeq} \mathrm{id}_{A_1}$$

and

$$\rho\colon\ \mu \circ (\mathrm{id}_{A_1} \times \nu) \xRightarrow{\simeq} \mathrm{id}_{A_1}\,.$$

Here we are identifying $A_1 \times * \cong A_1 \cong * \times A_1$, again by universal property.

These data must additionally satisfy the following *coherence conditions.*

(1) The *pentagon diagram* (Figure 1) commutes.

[16]In op. cit., Day and Street use the term *pseudomonoids* for what we here call *algebras.* The reason for the terminology change is two-fold: on the one hand to better accord with the ∞-categorical terminology used in [**12, 26**], and on the other to reserve the term *monoid* for associative algebras in Cartesian monoidal structures.

(2) The *triangle diagram* (Figure 2) commutes.

REMARK 5.5. Notice that the isomorphisms α, λ, and ρ are equivalently isomorphisms of sets

$$\alpha\colon\ A_2{}^{\mu_2}\!\times_{A_1}^{\mu_1} A_2 \xrightarrow{\ \cong\ } A_2{}^{\mu_0}\!\times_{A_1}^{\mu_1} A_2$$

$$\lambda\colon\ A_2{}^{\mu_2}\!\times_{A_1}^{\nu_0} A_0 \xrightarrow{\quad\cong\quad} A_1$$

$$\rho\colon\ A_2{}^{\mu_0}\!\times_{A_1}^{\nu_0} A_0 \xrightarrow{\quad\cong\quad} A_1$$

which commute with the morphisms defining the composite spans. For instance, λ must make the diagram

$$\begin{array}{ccccc} & & A_2{}^{\mu_2}\!\times_{A_1}^{\nu_0} A_0 & & \\ & \swarrow^{\mu_0} & \downarrow{\scriptstyle\lambda} & \searrow^{\mu_1} & \\ A_1 & & & & A_1 \\ & \nwarrow_{\mathrm{id}} & & \nearrow_{\mathrm{id}} & \\ & & A_1 & & \end{array}$$

commute.

DEFINITION 5.6. For two associative algebras A and B in Span, a *oplax morphism*[17] from A to B consists of maps of sets $f\colon A_i \to B_i$, for $0 \leq i \leq 2$, which commute with the defining maps of the spans μ and ν, as well as with the morphisms α, λ, and ρ of Remark 5.5. We will denote the 1-category of algebras in spans simply by Alg.

CONSTRUCTION 5.7. Let X be a 2-Segal simplicial set. Define an algebra $A(X)$ in Span to have underlying set X_1, multiplication span

$$X_1 \times X_1 \xleftarrow{(d_2,d_0)} X_2 \xrightarrow{\ d_1\ } X_1,$$

unit span

$$* \longleftarrow X_0 \xrightarrow{\ s_0\ } X_1,$$

associator given by the (invertible) span

$$X_2^{d_2}\!\times_{X_1}^{d_1} X_2 \xleftarrow{(d_1,d_3)} X_3 \xrightarrow{(d_2,d_0)} X_2^{d_0}\!\times_{X_1}^{d_1} X_2$$

consisting of the 2-Segal maps, and unitors λ and ρ given by the inverses of the 2-Segal maps

$$\lambda^{-1}\colon\ X_1 \xrightarrow{(s_0,d_1)} X_2{}^{d_2}\!\times_{X_1}^{s_0} X_0 \tag{5.1}$$

$$\rho^{-1}\colon\ X_1 \xrightarrow{(s_1,d_0)} X_2{}^{d_2}\!\times_{X_1}^{s_0} X_0. \tag{5.2}$$

---

[17] As shown in [**12**], pseudonatural transformations between such algebras whose components are spans correspond to spans of morphisms of simplicial sets satisfying very specific properties. The morphisms defined here correspond to oplax transformations whose components are morphisms of sets, hence the terminology we have chosen.

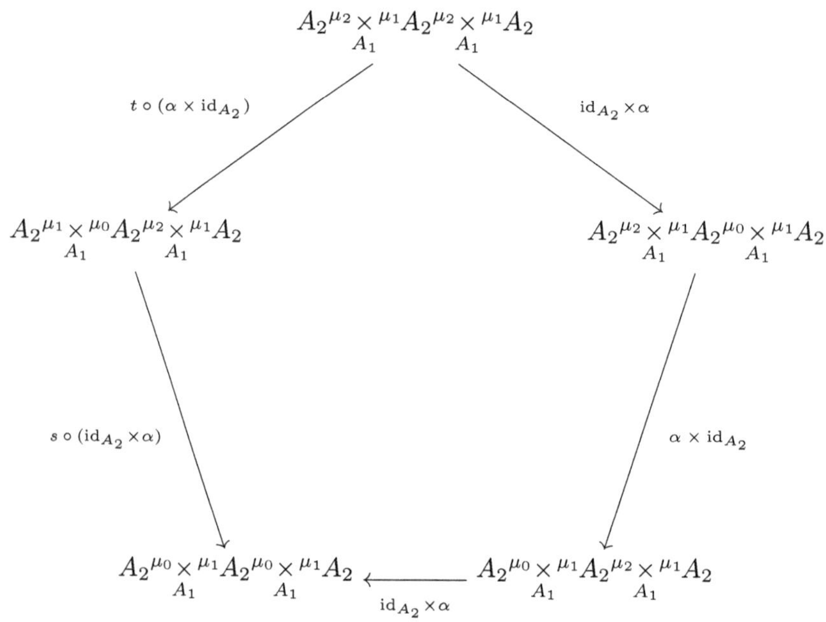

FIGURE 1. The associativity pentagon for algebras in Span, where permutations have been inserted to ease the writing of pull-backs. Here s and t are the permutations $s(x,y,z) = (y,z,x)$ and $t(x,y,z) = y,x,z$, respectively. Each corner of the pentagon corresponds to a composite of three copies of μ, and thus corresponds to a binary tree and a triangulation of the pentagon P_5. To ease notation, we have identified $A_2{}^{(\mu_2,\mu_0)} \times_{A_1}{}^{(\mu_1,\mu_1)} (A_2 \times A_2)$ with $A_2{}^{\mu_1}\times_{A_1}{}^{\mu_0}A_2{}^{\mu_2}\times_{A_1}{}^{\mu_1}A_2$ at the upper-left vertex of the pentagon.

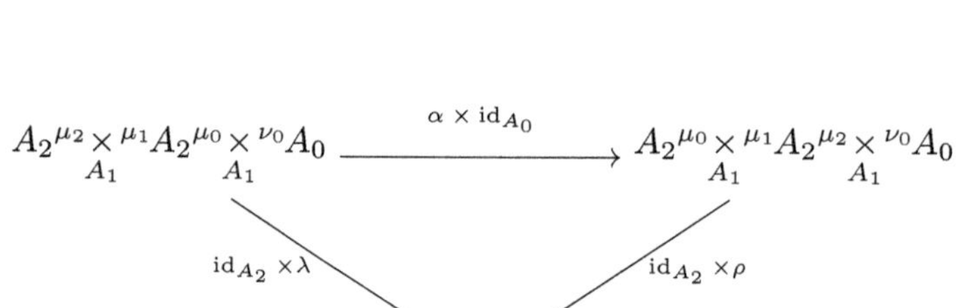

FIGURE 2. The triangle diagram for an algebra in Span. Here, we are implicitly identifying $A_2 \times_{A_1} A_1 \cong A_2$ at the bottom of the diagram.

The pentagon condition follows by placing a copy of X_4 in the center of the pentagon of Figure 1, with the 2-Segal maps for X_4 directed radially outward, using the face maps indicated by the inclusions of subdivisions of the pentagon in the diagram

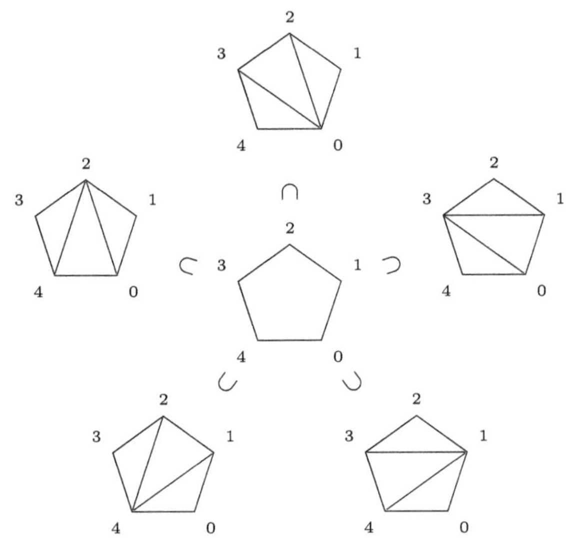

This yields a commutative diagram in which every morphism is an equivalence by the 2-Segal conditions. The triangle condition follows similarly.

It is immediate that given a morphism $f\colon X \to Y$, the components f_0, f_1, and f_2 define an oplax morphism $A(f)\colon A(X) \to A(Y)$ in Alg. As a result we obtain a functor

$$\mathrm{Al}\colon \mathsf{2Seg} \longrightarrow \mathsf{Alg}.$$

The formalization of the algebraic condition then has the following form:

THEOREM 5.8. *The functor*

$$\mathrm{Al}\colon \mathsf{2Seg} \longrightarrow \mathsf{Alg}$$

is an equivalence of categories.

Before we prove this theorem it is worth noting that [**15**, Example 3.6] is a special case which is particularly approachable. A partially-defined map of sets can be viewed as a span whose left leg is injective (as mentioned in Remark 5.3), and so every partial monoid can be viewed as an algebra in Span.

5.2. Essential surjectivity. The construction of the inverse to A is a rather subtle business. In principle, it amounts to proving a kind of coherence result. However, we will make use of Proposition 4.9 to substantially simplify the matter, implicitly shifting the issue of coherence into that result. In particular, we need only construct a 4-truncated, 3-coskeletal simplicial set from an algebra. Even with this simplification, due to length constraints we only sketch the construction which demonstrates essential surjectivity.

Let $A \in \mathsf{Alg}$ be an algebra in Span, and define a 4-truncated simplicial set X_A as follows. The sets of 0-, 1-, and 2-simplices are A_0, A_1, and A_2, respectively. We define A_3 to be the limit of the diagram

$$\alpha\colon A_2{}^{\mu_2}\!\times^{\mu_1}_{A_1} A_2 \xrightarrow{\;\cong\;} A_2{}^{\mu_0}\!\times^{\mu_1}_{A_1} A_2$$

of sets. Since the associator is an isomorphism, this is canonically isomorphic to both pullbacks. The set A_4 is defined to be the limit of the associativity pentagon. The face and degeneracy maps are defined as follows.

Face Maps:

- The face maps $A_1 \to A_0$ are given by $d_1 = p \circ \lambda^{-1}$ and $d_0 = q \circ \rho^{-1}$, where p and q are the projections from the pullback to A_0.
- The face maps $A_2 \to A_1$ are the maps defining the multiplication span, i.e., $d_i = \mu_i$.
- The face maps $A_3 \to A_2$ are the composites of the defining maps of the limit $A_3 \to A_2{}^{\mu_2}\times^{\mu_1}_{A_1} A_2$ and $A_3 \to A_2{}^{\mu_2}\times^{\mu_1}_{A_1} A_2$ with the projections to the four copies of A_2.
- The face maps $A_4 \to A_3$ are the composites of the defining maps of the limit from A_4 to the corners of the pentagon (Figure 1) with the projections of the pullbacks to copies of $A_2{}^{\mu_2}\times^{\mu_1}_{A_1} A_2$ and $A_2{}^{\mu_0}\times^{\mu_1}_{A_1} A_2$. These latter are canonically identified with A_3 by construction.

Degeneracy Maps:

- The degeneracy $s_0\colon A_0 \to A_1$ is identified with ν_0.
- The degeneracy maps s_0 and s_1 from A_1 to A_2 are identified with the composites of λ^{-1} and ρ^{-1}, respectively, with the projections to A_2.
- The definition of A_3 and A_4 yield isomorphisms

$$A_n{}^{\phi_i}\times^{\nu_0}_{A_1} A_0 \xrightarrow{\cong} A_{n-1}$$

for $n = 3, 4$ and $0 \leq i < n$, where ϕ_i is induced by the inclusion of $\{i, i+1\}$ into $[n]$. The isomorphisms are obtained by decomposing A_n into a pullback of A_2's and then applying λ or ρ. The triangle diagram (Figure 2) implies that these are, in fact, unique. We can thus define the degeneracy map s_i to be the composite

$$A_{n-1} \xrightarrow{\cong} A_n{}^{\phi_i}\times^{\nu_0}_{A_1} A_0 \xrightarrow{\mathrm{pr}_1} A_n$$

of the inverse isomorphisms with the projection.

It is time-consuming and notationally heavy to verify that these definitions yield a 4-truncated simplicial set, and we omit such verifications for the sake of brevity.

By construction, this 4-truncated simplicial set satisfies the 2-Segal conditions arising from triangulations of the square and pentagon. Thus, by Proposition 4.10, it uniquely determines a 2-Segal set. By construction, $\mathrm{Al}(X_A) \cong A$, and so the functor Al is essentially surjective.

5.3. Fully faithfulness. Again, we leverage coskeletalness to provide a short proof that the functor Al is fully faithful. In this case, since every 2-Segal simplicial set is 3-coskeletal by Proposition 4.8, we may consider our morphisms of 2-Segal simplicial sets to be morphisms between 3-truncations. This immediately implies that Al is faithful. To see that it is full, we need only note that, given an oplax morphism $(f_0, f_1, f_2)\colon A \to B$ of algebras, the fact that the f_i commute with α, λ and ρ allow us to uniquely extend them to a morphism of 3-truncated simplicial sets, completing the proof.

The one subtlety in the above argument is that, by the lowest-dimensional 2-Segal conditions, f_2 and f_1 uniquely determine maps $A_3 \to B_3$. However, there are two maps so determined, one corresponding to each subdivision of the square. Remembering f_3 amounts to remembering that these maps commute with the associativity isomorphism.

5.4. Coherence, truncation, and coskeletalness. We now arrive at a cluster of ideas which bear profound and interesting interrelations. Though we have defined associative algebras in Span by means of finite data satisfying relations, the most natural way of defining associative algebras in a bicategory is via infinite coherent data. More precisely, one can think of an algebra in a monoidal bicategory $(\mathbb{B}, \otimes, I)$ as consisting of an object A, together with morphisms

$$\mu_n \colon A^{\otimes n} \longrightarrow A$$

for $n \geq 0$, together with 2-isomorphisms

$$\alpha_{m,n} \colon \mu_m \circ \mu_n \Longrightarrow \mu_{m+n-1}$$

and

$$\eta \colon \mu_1 \Longrightarrow \mathrm{id}_A$$

which are *coherent* in the sense that any two parallel 2-morphisms built from these data using tensor product and compositions are equal.[18]

The fact that algebras in the sense of our Definition 5.4 are equivalent to this kind of coherent data is a highly non-trivial theorem — a generalization of MacLane's famed COHERENCE THEOREM for monoidal categories.

Intuitively, however, algebras in Span presented by the infinite data above correspond very naturally to simplicial sets. The roof A_n of the span μ_n from $A_1^{\times n}$ to A_1 is the set of n-simplices, and the associativity conditions amount precisely to the 2-Segal conditions. As such, one can think of Proposition 4.10 as a kind of coherence theorem for algebras in Span — the 4-truncated data which correspond to algebras in the sense of Definition 5.4 can be uniquely extended to the infinite data of an algebra in the sense introduced in this section.

Our proof of Proposition 4.9 is very topological, and the essentially topological character of coherence theorems has long been recognized. In ∞-category theory, one can rephrase MacLane's coherence theorem to state that the space of n-ary multiplications on a monoidal category is contractible, i.e., is a trivial groupoid when equiped with (products and composites of) the associativity and unitality natural isomorphisms. For a treatment of coherence making more explicit use of topological methods see, e.g., [**5**].

Acknowledgments

My thanks go to Julie Bergner, Joachim Kock, and Maru Sarazola, for organizing the wonderful workshop "Higher Segal Spaces and their Applications to Algebraic K-Theory, Hall Algebras, and Combinatorics," and for inviting me to give the talk on which this article is based. I am also very grateful to Philip Hackney for helpful comments on an earlier draft of this article.

[18]Such presentations of algebras are sometimes called *unbiased*, with the implication that they do not have a bias towards binary operations. This is the perspective taken towards monoidal structures in [**19**].

References

[1] J. E. Bergner, A. M. Osorno, V. Ozornova, M. Rovelli, and C. I. Scheimbauer, *2-Segal sets and the Waldhausen construction*, Topology Appl. **235** (2018), 445–484, DOI 10.1016/j.topol.2017.12.009. MR3760213

[2] A. Carboni, G. M. Kelly, R. F. C. Walters, and R. J. Wood, *Cartesian bicategories II*, Theory Appl. Categ. **19** (2007), 93–124. MR3656673

[3] I. Contreras, M. Keller, and R. A. Mehta, *Frobenius objects in the category of spans*, Rev. Math. Phys. **34** (2022), no. 10, Paper No. 2250036, 34, DOI 10.1142/S0129055X22500362. MR4514384

[4] I. Contreras, R. A. Mehta, and W. H. Stern, *Frobenius and commutative pseudomonoids in the bicategory of spans*, J. Geom. Phys. **207** (2025), Paper No. 105309, 31, DOI 10.1016/j.geomphys.2024.105309. MR4798060

[5] P.-L. Curien and G. Laplante-Anfossi, *Topological proofs of categorical coherence* (English, with English and French summaries), Cah. Topol. Géom. Différ. Catég. **65** (2024), no. 4, 357–389. MR4811426

[6] B. Day and R. Street, *Monoidal bicategories and Hopf algebroids*, Adv. Math. **129** (1997), no. 1, 99–157, DOI 10.1006/aima.1997.1649. MR1458415

[7] T. Dyckerhoff and M. Kapranov, *Higher Segal spaces*, Lecture Notes in Mathematics, vol. 2244, Springer, Cham, 2019, DOI 10.1007/978-3-030-27124-4. MR3970975

[8] T. Dyckerhoff and M. Kapranov, *Crossed simplicial groups and structured surfaces*, Stacks and categories in geometry, topology, and algebra, Contemp. Math., vol. 643, Amer. Math. Soc., Providence, RI, 2015, pp. 37–110, DOI 10.1090/conm/643/12896. MR3381470

[9] M. Feller, R. Garner, J. Kock, M. U. Proulx, and M. Weber, *Every 2-Segal space is unital*, Commun. Contemp. Math. **23** (2021), no. 2, Paper No. 2050055, 6, DOI 10.1142/S0219199720500558. MR4201031

[10] I. Gálvez-Carrillo, J. Kock, and A. Tonks, *Decomposition spaces, incidence algebras and Möbius inversion I: Basic theory*, Adv. Math. **331** (2018), 952–1015, DOI 10.1016/j.aim.2018.03.016. MR3804694

[11] S. Garoufalidis, C. D. Hodgson, J. H. Rubinstein, and H. Segerman, *1-efficient triangulations and the index of a cusped hyperbolic 3-manifold*, Geom. Topol. **19** (2015), no. 5, 2619–2689, DOI 10.2140/gt.2015.19.2619. MR3416111

[12] J. Gödicke, *An ∞-category of 2-segal spaces*, 2024.

[13] P. G. Goerss and J. F. Jardine, *Simplicial homotopy theory*, Modern Birkhäuser Classics, Birkhäuser Verlag, Basel, 2009. Reprint of the 1999 edition [MR1711612], DOI 10.1007/978-3-0346-0189-4. MR2840650

[14] A. Grothendieck, *Techniques de construction et théorèmes d'existence en géométrie algébrique III : préschémas quotients*, Séminaire Bourbaki : années 1960/61, exposés 205-222, Séminaire Bourbaki, no. 6, Société mathématique de France, pp. 99–118.

[15] P. Hackney, *The decomposition space perspective*, Workshop on Higher Segal Spaces and their Applications to Algebraic K-Theory, Hall Algebras, and Combinatorics, Contemporary Mathematics, American Mathematical Soc., 2024.

[16] P. Hackney and J. Lynd, *Partial groups as symmetric simplicial sets*, J. Pure Appl. Algebra **229** (2025), no. 2, Paper No. 107864, 22, DOI 10.1016/j.jpaa.2025.107864. MR4850560

[17] Y. Harpaz, *Quasi-unital ∞-categories*, Phd thesis, Hebrew University of Jerusalem, Jerusalem, Israel, October 2012, Available at `https://www.math.univ-paris13.fr/~harpaz/thesis.pdf`.

[18] C. L. Lawson, *Transforming triangulations*, Discrete Math. **3** (1972), 365–372, DOI 10.1016/0012-365X(72)90093-3. MR311491

[19] T. Leinster, *Higher operads, higher categories*, London Mathematical Society Lecture Note Series, vol. 298, Cambridge University Press, Cambridge, 2004, DOI 10.1017/CBO9780511525896. MR2094071

[20] J. Lurie, *Kerodon*, `https://kerodon.net`, 2018.

[21] S. Mac Lane, *Natural associativity and commutativity*, Rice Univ. Stud. **49** (1963), no. 4, 28–46. MR170925

[22] E. Riehl, *On the structure of simplicial categories associated to quasi-categories*, Math. Proc. Cambridge Philos. Soc. **150** (2011), no. 3, 489–504, DOI 10.1017/S0305004111000053. MR2784771

[23] G. Segal, *Classifying spaces and spectral sequences*, Inst. Hautes Études Sci. Publ. Math. **34** (1968), 105–112. MR232393

[24] The Stacks Project Authors, *Stacks Project*, `https://stacks.math.columbia.edu`, 2018.

[25] R. P. Stanley, *Catalan numbers*, Cambridge University Press, New York, 2015, DOI 10.1017/CBO9781139871495. MR3467982

[26] W. H. Stern, *2-Segal objects and algebras in spans*, J. Homotopy Relat. Struct. **16** (2021), no. 2, 297–361, DOI 10.1007/s40062-021-00282-8. MR4266208

[27] W. H. Stern, *Open topological field theories and 2-segal objects*, Phd thesis, Rheinische Friedrich-Wilhelms-Universität Bonn, Bonn, Germany, October 2019, Available at `https://bonndoc.ulb.uni-bonn.de/xmlui/handle/20.500.11811/8081`.

[28] H. L. Tanaka, *Functors (between ∞-categories) that aren't strictly unital*, J. Homotopy Relat. Struct. **13** (2018), no. 2, 273–286, DOI 10.1007/s40062-017-0182-2. MR3802796

Technische Universität München, CIT Department of Mathematics, Garching bei München, Germany

Email address: `walker.stern@tum.de`

Contemporary Mathematics
Volume **838**, 2026
https://doi.org/10.1090/conm/838/16736

The decomposition space perspective

Philip Hackney

ABSTRACT. This paper provides an introduction to decomposition spaces and 2-Segal spaces, unifying the two perspectives. We begin by defining decomposition spaces using the active-inert factorization system on the simplicial category, and show their equivalence to 2-Segal spaces. Key results include the path space criterion, which characterizes decomposition spaces in terms of their upper and lower décalages, and the edgewise subdivision criterion. We also introduce free decomposition spaces arising from outer face complexes, providing a rich source of examples. Formal prerequisites are minimal—readers should have a working knowledge of simplicial methods and basic category theory.

1. Introduction

Decomposition spaces were introduced by Imma Gálvez-Carrillo, Joachim Kock, and Andrew Tonks in a long preprint in 2014 that now comprises six papers [**23–28**]. Their goals were to develop a very general framework for incidence algebras and Möbius inversion that could unify and extend various examples arising in algebraic combinatorics. Further developments along these lines came later, including consideration of antipodes [**10**], incorporation of Schmitt's hereditary species into the picture [**9**], investigation of the universality of Lawvere's Hopf algebra of Möbius intervals [**22**], and the introduction of a new type of species governing other known Hopf algebras [**11**].

The aim of this paper is to give a gentle introduction to decomposition spaces, similar to Stern's introduction to 2-Segal spaces [**46**]. The theory of decomposition spaces encourages us to reduce as much as possible to a calculus of pullback squares, and we try to follow this principle whenever feasible. All told, the prerequisites for this work are mild, and similar to [**46**]: some elementary category theory, and some comfort with simplicial sets (but see Section 1.1).

Decomposition spaces are equivalent to the 2-Segal spaces of Dyckerhoff and Kapranov [**19**]. This equivalence was first observed for *unital* 2-Segal spaces, as described in [**24**, p. 959]:

> The coincidence of the notions was noticed by Mathieu Anel because two of the basic results are the same: specifically, the

2020 *Mathematics Subject Classification.* Primary 18N50; Secondary 18F20, 55U10, 18N60.
Key words and phrases. Decomposition space, 2-Segal space.
This work was supported by a grant from the Simons Foundation (#850849) and by the Louisiana Board of Regents through the Board of Regents Support fund LEQSF(2024-27)-RD-A-31.

> characterisation in terms of decalage and Segal spaces ... and the result that the Waldhausen $S_\bullet$-construction of a stable ∞-category is a decomposition space ...

Later on it was realized by Feller, Garner, Kock, Proulx, and Weber [**21**] that, in fact, the unitality condition is automatic, providing the missing ingredient to see that 2-Segal spaces and decomposition spaces are the same. We will give an account of the equivalence in Section 4, following [**21**, **24**].

The next three sections (§5, §6, §7) are all about relationships between decomposition spaces and Segal spaces. The first of these is simple: decomposition spaces are meant to be a generalization of Segal spaces. We establish that each Segal space is indeed a decomposition space, and that Segal spaces are characterized among decomposition spaces by checking whether a single square is a pullback. In Section 6 we turn to the *path space criterion*, which says that a simplicial space is a decomposition space if and only if both of its décalages/path spaces are Segal spaces. Finally, in Section 7 we prove the closely related edgewise subdivision criterion, which says that X is a decomposition space if and only if $\operatorname{sd} X$ is a Segal space.

The last section of the paper describes a source of examples of decomposition spaces, namely the simplicial sets associated to outer face complexes.

There is fundamental material about decomposition spaces that is not covered in depth here, but can be found in the original sources and in other papers in this volume.

1.1. Ways to read this paper. In following the original sources [**24**–**26**], we have elected mostly to phrase everything within the $(\infty, 1)$-category of simplicial *spaces*.[1] But knowing about $(\infty, 1)$-categories and their limits and such is not a prerequisite for this paper. Readers may safely replace "spaces" with "sets" and replace any $(\infty, 1)$-categorical pullbacks/limits with ordinary pullbacks/limits. In more detail, any of the following are valid ways to read this paper:

(1) Focus solely on the category $\mathsf{sSet} := \mathsf{Fun}(\Delta^{\mathrm{op}}, \mathsf{Set})$ of simplicial sets. This approach requires basic familiarity with simplicial sets as combinatorial objects (but nothing of their homotopy theory), as well as basic notions from category theory such as functors and pullbacks.
(2) All statements before Section 8 also hold, using the same arguments, for simplicial objects in a 1-category C.[2] Particular examples may not make sense in this generality.
(3) Work in the $(\infty, 1)$-category $\mathsf{sSpace} := \mathsf{Fun}(\Delta^{\mathrm{op}}, \mathsf{Space})$ of space-valued presheaves on Δ. Here Space denotes the $(\infty, 1)$-category of spaces (or ∞-groupoids). See [**24**] for details. We emphasize that any pullbacks or limits that appear are necessarily $(\infty, 1)$-categorical pullbacks or limits (see Convention 3.2).
(4) One could also work with simplicial objects in an arbitrary $(\infty, 1)$-category C.

---

[1] For concreteness, '$(\infty, 1)$-category' may be interpreted to mean 'quasi-category' / '∞-category.' That said, as in the original sources (see [**24**, 1.3]), we work in a model-independent, synthetic fashion, and use nothing specific to quasi-categories.

[2] Typically, one would require that C appearing in (2) and (4) has finite limits. This assumption can be dropped for what we do in this paper, by instead asking that certain diagrams arising from simplicial objects *are* limit diagrams.

We will altogether avoid discussing anything from the perspective of Quillen model categories. A number of sources, especially on the 2-Segal side of things, give model-categorical accounts of some of what appears below; see, for instance, [**4**, **19**, **20**].

Acknowledgments

This paper originated from my talks at the workshop "Higher Segal Spaces and their Applications to Algebraic K-Theory, Hall Algebras, and Combinatorics" held at the Banff International Research Station (BIRS) in January 2024. I thank the workshop organizers (Julie Bergner, Joachim Kock, Maru Sarazola) for cultivating a vibrant and stimulating atmosphere and giving me the opportunity to speak, and BIRS for providing an ideal venue for the event. I am grateful to the other participants for their contributions and discussions. Special thanks go to Julie Bergner, Matt Feller, Joachim Kock, Justin Lynd, Felix Naß, and Walker Stern.

2. The active-inert factorization system on the simplicial category

The *simplicial category* Δ has objects $[n] = \{0 < 1 < \cdots < n\}$ for $n \geq 0$, and morphisms are the (weakly) order preserving maps (if $i \leq j$, then $\alpha(i) \leq \alpha(j)$). Let us recall the standard generating maps for this category. For each $i = 0, 1, \dots, n$, there is a *coface map* $\delta^i \colon [n-1] \to [n]$ that is injective and does not have i in its image, and a *codegeneracy map* $\sigma^i \colon [n+1] \to [n]$ that is surjective with the preimage of i having cardinality two. In formulas:

$$\delta^i(k) = \begin{cases} k & k \leq i-1 \\ k+1 & k \geq i \end{cases} \qquad \sigma^i(k) = \begin{cases} k & k \leq i \\ k-1 & k \geq i+1. \end{cases}$$

A *simplicial object* in category C is a functor $X \colon \Delta^{\mathrm{op}} \to C$ that is written as $[n] \mapsto X_n$ and $(\alpha \colon [n] \to [m]) \mapsto (\alpha^* \colon X_m \to X_n)$. As usual, we write $d_i \colon X_n \to X_{n-1}$ for the image of δ^i, and $s_i \colon X_n \to X_{n+1}$ for the image of σ^i; these maps are called *face* and *degeneracy* operators and lead to the usual picture of a simplicial object X:

$$X_0 \underset{d_0}{\overset{d_1}{\leftleftarrows}} \xrightarrow{s_0} X_1 \;\underset{d_0}{\overset{d_2}{\Lleftarrow}}\; X_2 \;\Lleftarrow\; X_3 \cdots . \tag{1}$$

These basic maps define a simplicial object if and only if they satisfy the usual *simplicial identities*

$$\begin{aligned} d_i d_j &= d_{j-1} d_i \text{ if } i < j \\ s_i s_j &= s_{j+1} s_i \text{ if } i \leq j \\ d_i s_j &= \begin{cases} s_{j-1} d_i & \text{if } i < j \\ \mathrm{id} & \text{if } i = j \text{ or } i = j+1 \\ s_j d_{i-1} & \text{if } i > j+1. \end{cases} \end{aligned}$$

Vertical reflection of the picture (1) yields the following contrivance, which allows us to infer some statements by duality.

DEFINITION 2.1. If X is a simplicial object, then its *opposite* X^{op} has $(X^{\mathrm{op}})_n = X_n$ together with the following face and degeneracy operators:

$$\begin{aligned} \left(d_k^{\mathrm{op}} \colon X_n^{\mathrm{op}} \to X_{n-1}^{\mathrm{op}}\right) &= \left(d_{n-k} \colon X_n \to X_{n-1}\right) \\ \left(s_k^{\mathrm{op}} \colon X_n^{\mathrm{op}} \to X_{n+1}^{\mathrm{op}}\right) &= \left(s_{n-k} \colon X_n \to X_{n+1}\right). \end{aligned}$$

More formally, there is an endofunctor $\Delta \to \Delta$ which sends $f\colon [n] \to [m]$ to $g\colon [n] \to [m]$ with $g(i) = m - f(n-i)$. The functor $X^{\mathrm{op}}\colon \Delta^{\mathrm{op}} \to \mathsf{Space}$ is given by precomposing X with this endofunctor.

We say that a morphism $\alpha\colon [n] \to [m]$ in Δ is *active* if it is endpoint preserving, that is, if $\alpha(0) = 0$ and $\alpha(n) = m$. Active maps are denoted by barred arrows $\alpha\colon [n] \rightarrow\!\!\!\!\!\!\mapsto [m]$. We write $\Delta_{\mathrm{act}} \subset \Delta$ for the subcategory containing the same objects as Δ, but with only the active maps; we call such a subcategory *wide*. Notice that, for each n, there is a *unique* active map $[1] \rightarrow\!\!\!\!\!\!\mapsto [n]$.

A morphism $\alpha\colon [n] \to [m]$ in Δ is *inert* if it is distance preserving, that is, if $\alpha(i+1) = \alpha(i) + 1$ for all $i \in [n]$. Alternatively, we have $\alpha(\text{-}) = (\text{-}) + c$, where the constant c is $\alpha(0)$. Inert maps are written with tailed arrows $\alpha\colon [n] \rightarrowtail [m]$, and the wide subcategory of inert maps is denoted by $\Delta_{\mathrm{int}} \subset \Delta$.

The only generating morphisms of Δ that are inert are the *outer coface maps* $\delta^0, \delta^n\colon [n{-}1] \to [n]$. The codegeneracy maps and the inner coface maps $\delta^i\colon [n{-}1] \to [n]$ for $0 < i < n$ are active. The outer face maps in a simplicial set are important enough in what follows that we give them special names: we write

$$\begin{aligned} d_\perp &:= d_0\colon X_n \to X_{n-1} \\ d_\top &:= d_n\colon X_n \to X_{n-1} \end{aligned}$$

for the bottom and top face maps. (We also have bottom and top degeneracy maps $s_\perp = s_0$ and $s_\top = s_n\colon X_n \to X_{n+1}$.)

Given an arbitrary map $\alpha\colon [n] \to [m]$ in Δ, there is a unique factorization of α as an active map followed by an inert map. Namely, setting $p = \alpha(n) - \alpha(0)$, the map α factors as

$$[n] \xrightarrow[\alpha(\text{-})-\alpha(0)]{} [p] \xrightarrow[(\text{-})+\alpha(0)]{} [m], \qquad \text{composite } \alpha.$$

The first map applies α and then subtracts the constant $\alpha(0)$, while second map adds the constant $\alpha(0)$. It is immediate that the first map is active and the second map is inert, and that the composite of the two is equal to α. A bit more thought shows that this active-inert factorization of α is unique, and we arrive at the following.

LEMMA 2.2. *The pair of subcategories* $(\Delta_{\mathrm{act}}, \Delta_{\mathrm{int}})$ *constitutes a factorization system on* Δ. □

An *orthogonal factorization system*[3] on a category C consists of a pair of wide subcategories (L, R) such that every map in C factors as $r \circ \ell$ for some $r \in R$ and $\ell \in L$, this factorization is unique up to unique isomorphism, and $L \cap R$ contains all isomorphisms. A *strict factorization system* is similar, except that we only require one condition on the wide subcategories: every map factors *uniquely* as $r \circ \ell$. In general, the two notions are not directly comparable (though there is an enveloping construction for strict factorization systems [**30**, 2.1], [**43**, 3.1]), but in our case there is no ambiguity, as the only isomorphisms in Δ are identities. Generalizing

[3]The name 'orthogonal' comes from the following lifting property: for every commutative square $\begin{smallmatrix} \cdot & \longrightarrow & \cdot \\ {\scriptstyle\ell}\downarrow & \nearrow & \downarrow{\scriptstyle r} \\ \cdot & \longrightarrow & \cdot \end{smallmatrix}$ with $\ell \in L$ and $r \in R$, there exists a unique dotted map making both triangles commute.

this factorization system to other shape categories is an essential component of Segalic approaches to higher structures [**1**, **12**, **33**]; see Definition 5.1 below.

In the next section, we make use of a special property of Δ, which is that active-inert pushouts exist.

LEMMA 2.3. *Suppose* $\alpha\colon [n] \twoheadrightarrow [m]$ *is active and* $\iota\colon [n] \rightarrowtail [k]$ *is inert. Then there is a pushout square*

$$\begin{array}{ccc} [n] & \xrightarrow{\alpha} & [m] \\ {\scriptstyle\iota}\downarrow & \ulcorner & \downarrow{\scriptstyle\theta} \\ [k] & \xrightarrow[\phi]{} & [p] \end{array} \tag{2}$$

in Δ, *with* θ *inert and* ϕ *active.* □

For brevity we leave the proof as an exercise for the reader; see [**24**, 2.6] for this fact, or [**15**, §3] for a discussion of *all* pushout squares that exist in Δ. Concrete formulas for θ and ϕ allow one to see that (2) is also a pullback square.

3. Decomposition spaces and 2-Segal spaces

With active-inert pushouts in hand, we are ready to define decomposition spaces [**24**, §3]. While categories encode associative and unital *composition* of morphisms, decomposition spaces shift attention to possible *decompositions* of morphisms, requiring them to be associative and unital in an appropriate sense. This idea can be used to encode a number of coalgebras that come from decomposition of combinatorial structures into smaller pieces (see Example 3.5) via a general incidence coalgebra construction [**17**, **23**, **24**].

DEFINITION 3.1 (Decomposition space). A simplicial space X is a *decomposition space* if it sends every active-inert pushout square to a pullback square.

In more detail, given a pushout square on the left below (as in (2) of Lemma 2.3), the associated square on the right is a pullback square of spaces:

$$\begin{array}{ccc} [n] & \xrightarrow{\alpha} & [m] \\ {\scriptstyle\iota}\downarrow & \ulcorner & \downarrow{\scriptstyle\theta} \\ [k] & \xrightarrow[\phi]{} & [p] \end{array} \quad \rightsquigarrow \quad \begin{array}{ccc} X_p & \xrightarrow{\phi^*} & X_k \\ {\scriptstyle\theta^*}\downarrow & \lrcorner & \downarrow{\scriptstyle\iota^*} \\ X_m & \xrightarrow[\alpha^*]{} & X_n. \end{array}$$

CONVENTION 3.2. In this definition, and in what follows, 'pullback' means 'pullback in the $(\infty,1)$-category of spaces' or 'homotopy pullback.' For a reader working with simplicial sets rather than simplicial spaces, these are just ordinary pullbacks.

A simplicial set which is a decomposition space is called a *discrete decomposition space*.

DIVERSION 3.3 (Variations). What happens if one replaces the active-inert factorization system on Δ in Definition 3.1 with a factorization system on some other indexing category? This question is broad and vague, but special cases have been considered in [**1**, Remark 3.15], [**7**, p. 635], [**16**], and [**31**].[4]

[4]It does not seem to be the case that cyclic 2-Segal objects (see [**5**, **18**]) are of this form, owing to the lack of a compatible orthogonal factorization systems on Connes' cyclic category Λ

If C is a category, then the nerve of C is a (discrete) decomposition space. We return to this example in Section 5. A rich source of combinatorial examples are the free decomposition spaces, which we discuss in Section 8. We give additional examples in a moment, but first we give an equivalent characterization, to be discussed in the next section. The following appears in [**21**].

DEFINITION 3.4. A simplicial space X is said to be *upper 2-Segal* (respectively *lower 2-Segal*) if the square below left (respectively below right) is a pullback for all $0 < i < n$:

$$(\heartsuit) \qquad \begin{array}{ccc} X_{n+1} & \xrightarrow{d_{i+1}} & X_n \\ {\scriptstyle d_\perp}\downarrow & \lrcorner & \downarrow{\scriptstyle d_\perp} \\ X_n & \xrightarrow[d_i]{} & X_{n-1} \end{array} \qquad\qquad \begin{array}{ccc} X_{n+1} & \xrightarrow{d_{i}} & X_n \\ {\scriptstyle d_\top}\downarrow & \lrcorner & \downarrow{\scriptstyle d_\top} \\ X_n & \xrightarrow[d_i]{} & X_{n-1}. \end{array}$$

A *2-Segal space* is a simplicial space which is both upper and lower 2-Segal.

As a simplicial space X is lower 2-Segal if and only if X^{op} is upper 2-Segal, many results below address only the upper 2-Segal situation, with the understanding that there is always a dual version. This definition of 2-Segal space is not exactly what appeared in [**46**], but it turns out to be equivalent. We leave this proof to the reader once we have introduced the relevant tools; see Exercise 4.5. Also note that upper and lower 2-Segal spaces are independently of interest: Garner, Kock, and Weber [**29**] showed that unary operadic categories are the same thing as upper 2-Segal sets (see also [**32**]), while Carawan [**8**] gives examples of lower 2-Segal objects coming from the $S_\bullet$-construction applied to a cofibration category.

EXAMPLE 3.5 (Simplicial groupoid of rooted forests). A fundamental example of [**24**] concerns the simplicial groupoid H that underlies the Butcher–Connes–Kreimer Hopf algebra [**14**]. For $n \geq 0$, H_{n+1} is the groupoid of forests together with n (compatible) admissible cuts. For example, the following forest is an object of H_1:

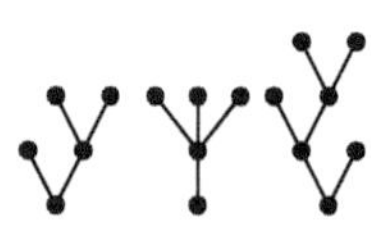

which is in turn the image under d_1 of the following forest with a cut in H_2:

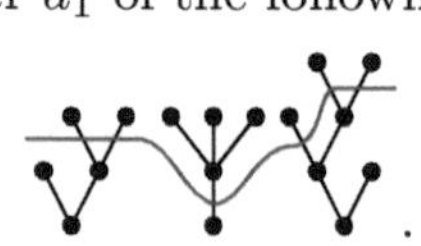

The outer faces clip above and below the cut, producing

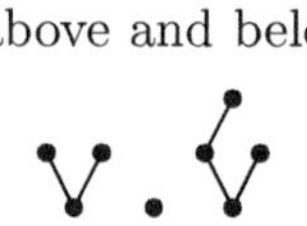 and 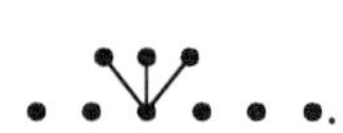

This example is a 2-Segal space, but not a Segal space in the sense of Definition 5.1 below, as there are many 2-simplices having the above forests as their outer faces. As an illustration, the tree with two cuts in the upper left of Figure 1 (which is an

[**13**]. Indeed, if the right class of a factorization system on Λ contains Δ_{int}, then it contains the positive part of the generalized Reedy factorization system on Λ (see [**2**]), while if the left class of a factorization system contains Δ_{act}, then it contains all of Λ.

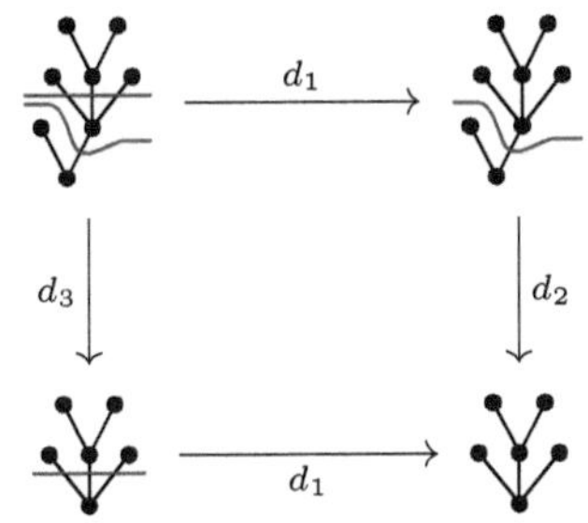

FIGURE 1. Pullback in rooted forests

element of H_3) is uniquely determined by the remainder of the diagram, giving an indication of why the right square of ($\heartsuit$) is a pullback in the case $n = 2$.

EXAMPLE 3.6 (Partial monoids). In [**44**, §2], a *partial monoid* is defined to be a set M together with an identity element $1 \in M$ and a partially defined multiplication function $M \times M \twoheadrightarrow M$. They are unital in the sense that $1 \cdot x = x = x \cdot 1$ are defined and equal for all $x \in M$, and are associative in the sense that if one of $x \cdot (y \cdot z)$ or $(x \cdot y) \cdot z$ is defined, then so is the other, and they are equal. There is an associated simplicial set where the n-simplices are the length n words $(x_1, \ldots, x_n)$ such that the multiplication of these n elements is defined for some parenthesization (and hence for all possible parenthesizations). It was shown in [**3**, Example 2.1] that this simplicial set is always 2-Segal.

EXERCISE 3.7. Show that every partial monoid is 2-Segal using only Definition 3.4.

We give an extension to a kind of partial category. Let Par denote the category whose objects are sets and whose morphisms are partial functions, considered as a monoidal category using the cartesian product of sets.

EXERCISE 3.8. Let C be a small category enriched in Par. Show that if the set $C(x, x)$ is nonempty for every object x of C, then C determines a simplicial set that is 2-Segal.

REMARK 3.9. Each partial monoid may be regarded as a monoid with a zero element, by adjoining a new absorbing element 0 to M and declaring that all undefined multiplications take 0 as their value. Likewise, for any monoid with a zero element different from 1, we obtain a partial monoid by discarding 0. In a similar manner, the Par-enriched categories from the preceding exercise correspond to categories with (nonidentity) zero morphisms [**40**, §1.7].

The examples of partial monoids and Par-enriched categories suggest a useful perspective: simplicial sets that are 2-Segal (or equivalently, discrete decomposition spaces) can be understood as encoding multivalued "associative" composition laws. See [**19**, §3.3] and [**46**] for precise statements. The preceding examples cover the case of composition having at most one value, and the single-valued composition laws correspond to the Segal objects (Section 5).

We mention, in passing, a class of maps that plays an important role in the theory of decomposition spaces. These maps do not make a significant appearance in this paper, but reappear elsewhere in this volume [**17**, **23**]. In particular, [**17**] explains why a culf map between decomposition spaces induces a coalgebra homomorphism between the associated incidence coalgebras (originally from [**24**, Lemma 8.2]).

DEFINITION 3.10 (Culf maps). A map $f\colon X \to Y$ between simplicial spaces is *culf* if it is cartesian on active maps. Concretely, we require that the following squares are pullbacks, for $0 < i < n$ and $0 \leq j \leq n$:

$$\begin{array}{ccc} X_n & \xrightarrow{d_i} & X_{n-1} \\ {\scriptstyle f}\downarrow & \lrcorner & \downarrow{\scriptstyle f} \\ Y_n & \xrightarrow[d_i]{} & Y_{n-1} \end{array} \qquad \begin{array}{ccc} X_n & \xrightarrow{s_j} & X_{n+1} \\ {\scriptstyle f}\downarrow & \lrcorner & \downarrow{\scriptstyle f} \\ Y_n & \xrightarrow[s_j]{} & Y_{n+1}. \end{array}$$

4. 2-Segal spaces are decomposition spaces

In this section we discuss the following fundamental result; as mentioned in the introduction, it was known early on that decomposition spaces were the same thing as unital 2-Segal spaces, and later on it was discovered that the unitality condition is automatic [**21**].

THEOREM 4.1. *A simplicial space X is a decomposition space if and only if it is both upper and lower 2-Segal.*

It is immediate that every decomposition space is 2-Segal, since the squares in (♡) of Definition 3.4 are active-inert pullback squares. The converse is more interesting. If X is a decomposition space, then the following basic squares (for all $0 \leq i \leq n$) come from active-inert pushout squares in Δ, so must be pullbacks:

$$\begin{array}{ccc} X_{n+1} & \xrightarrow{s_{i+1}} & X_{n+2} \\ {\scriptstyle d_\bot}\downarrow & \lrcorner & \downarrow{\scriptstyle d_\bot} \\ X_n & \xrightarrow[s_i]{} & X_{n+1} \end{array} \qquad \begin{array}{ccc} X_{n+1} & \xrightarrow{s_i} & X_{n+2} \\ {\scriptstyle d_\top}\downarrow & \lrcorner & \downarrow{\scriptstyle d_\top} \\ X_n & \xrightarrow[s_i]{} & X_{n+1}. \end{array}$$

However, the 2-Segal condition makes no mention of any squares involving degeneracies. The main effort of the proof of Theorem 4.1 is addressing this issue. Most proofs in the rest of the paper rely on two basic tools. Here is the first.

PASTING LAW FOR PULLBACKS. *Suppose we have a diagram*

$$\begin{array}{ccccc} \bullet & \longrightarrow & \bullet & \longrightarrow & \bullet \\ \downarrow & & \downarrow & \lrcorner & \downarrow \\ \bullet & \longrightarrow & \bullet & \longrightarrow & \bullet \end{array}$$

with right square a pullback. Then the left square is a pullback if and only if the outer (composite) square is a pullback.

For a 1-category C, such as that of sets, this law is traditionally left as an exercise, but see [**6**, Proposition 2.5.9] for a proof. In an $(\infty, 1)$-category C (such as that of spaces), this law is sometimes called the *prism lemma*, since the diagram above should be specified by a functor $\Delta^2 \times \Delta^1 \to C$ (see [**37**, Lemma 4.4.2.1] or [**38**, Tag 03FZ]).

As an illustration of the utility of the pasting law, we establish the following characterization of the (upper) 2-Segal condition from [**24**, Lemma 3.6].

PROPOSITION 4.2. *The following are equivalent for a simplicial space X.*

(1) *The simplicial space X is upper 2-Segal.*
(2) *For each $n \geq 2$, the left square of $(\heartsuit)$ is a pullback for some $0 < i < n$.*

PROOF. It is immediate that 1 implies 2. To prove the converse, it is helpful to add a third equivalent condition:

(3) For each $n \geq 2$, the square

$$\begin{tikzcd} X_{n+1} \arrow[r, "d_2^{n-1}"] \arrow[d, "d_\perp"'] & X_2 \arrow[d, "d_\perp"] \\ X_n \arrow[r, "d_1^{n-1}"'] & X_1 \end{tikzcd}$$

is a pullback.

The following diagram is used for the remaining implications:

$$\begin{tikzcd} X_{n+2} \arrow[r, "d_{i+1}"] \arrow[d, "d_\perp"'] & X_{n+1} \arrow[r, "d_2^{n-1}"] \arrow[d, "d_\perp"'] \arrow[dr, phantom, "\lrcorner", very near start] & X_2 \arrow[d, "d_\perp"] \\ X_{n+1} \arrow[r, "d_i"'] & X_n \arrow[r, "d_1^{n-1}"'] & X_1. \end{tikzcd}$$

To prove that 2 implies 3, we assume inductively that the right square above is a pullback, and that for some $0 < i < n+1$ the left square is a pullback, which implies that the outer rectangle is also a pullback. But $d_2^{n-1} d_{i+1} = d_2^n$ and $d_1^{n-1} d_i = d_1^n$, so this outer rectangle is the case of 3 in the next level. To show that 3 implies 1, let $0 < i < n+1$ be arbitrary. Then the outer rectangle and the right square in the diagram are pullbacks, hence so is the left square. Thus X is upper 2-Segal. □

LEMMA 4.3. *Suppose X is an upper 2-Segal space. If $n > 0$ and $0 \leq i \leq n$, then the square*

$$\begin{tikzcd} X_{n+1} \arrow[r, "s_{i+1}"] \arrow[d, "d_\perp"'] \arrow[dr, phantom, "\lrcorner", very near start] & X_{n+2} \arrow[d, "d_\perp"] \\ X_n \arrow[r, "s_i"'] & X_{n+1} \end{tikzcd}$$

is a pullback.

PROOF. This proof is an application of the pasting law. We can choose $j \in \{i, i+1\}$ with $0 < j < n+1$, and then we have the following diagram:

$$\begin{tikzcd} X_{n+1} \arrow[r, "s_{i+1}"] \arrow[d, "d_\perp"'] \arrow[rr, bend left, "\mathrm{id}"] & X_{n+2} \arrow[r, "d_{j+1}"] \arrow[d, "d_\perp"'] \arrow[dr, phantom, "\lrcorner", very near start] & X_{n+1} \arrow[d, "d_\perp"] \\ X_n \arrow[r, "s_i"'] \arrow[rr, bend right, "\mathrm{id}"'] & X_{n+1} \arrow[r, "d_j"'] & X_n. \end{tikzcd}$$

The right square is a pullback since X is upper 2-Segal, and the outer square is a pullback. Hence the left square is a pullback. □

Unfortunately, the proof of the lemma does not cover the $n = 0$ case, for which we need our second basic tool.

RETRACT STABILITY. *Retracts of pullbacks are pullbacks.*

Being a pullback is a property of an object of $\mathsf{Fun}(\Delta^1 \times \Delta^1, C)$, the category of commutative squares in C, and this property is stable under retracts. See [**21**, Appendix A] for homotopy pullbacks in model categories; for quasi-categories use the duals of [**37**, Lemma 5.1.6.3] or [**38**, Tag 05E6]. The following is proved in [**21**].

LEMMA 4.4. *If X is an upper 2-Segal space, then the square*

$$\begin{array}{ccc} X_1 & \xrightarrow{s_1} & X_2 \\ {\scriptstyle d_0}\downarrow & \lrcorner & \downarrow{\scriptstyle d_0} \\ X_0 & \xrightarrow[s_0]{} & X_1 \end{array}$$

is a pullback.

PROOF. The indicated square is a retract of

$$\begin{array}{ccc} X_2 & \xrightarrow{s_2} & X_3 \\ {\scriptstyle d_\perp}\downarrow & \lrcorner & \downarrow{\scriptstyle d_\perp} \\ X_1 & \xrightarrow[s_1]{} & X_2 \end{array}$$

which is known to be a pullback by Lemma 4.3. This property is witnessed by the maps of squares

$$S = \begin{array}{c|c} s_1 & s_1 \\ \hline s_0 & s_0 \end{array} \qquad \text{and} \qquad D = \begin{array}{c|c} d_1 & d_1 \\ \hline d_0 & d_0 \end{array}$$

with $DS = \mathrm{id}$. More explicitly, maps between squares are cubes, and the diagram

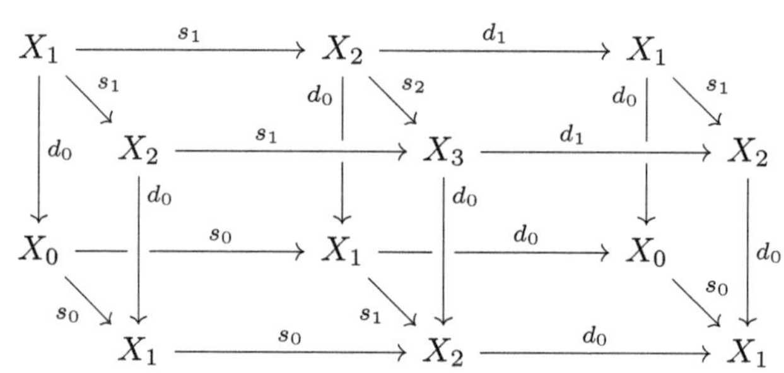

is commutative. ∎

Dually, if X is a lower 2-Segal space then for $n \geq 0$ and $0 \leq i \leq n$, the square

$$\begin{array}{ccc} X_{n+1} & \xrightarrow{s_i} & X_{n+2} \\ {\scriptstyle d_\top}\downarrow & \lrcorner & \downarrow{\scriptstyle d_\top} \\ X_n & \xrightarrow[s_i]{} & X_{n+1} \end{array}$$

is a pullback: apply Lemma 4.3 and Lemma 4.4 to X^{op}. To prove Theorem 4.1, one can now follow [**24**, 3.4], or reason about the situation directly as follows.

PROOF SKETCH FOR THEOREM 4.1. First consider active-inert pushout squares

$$\begin{array}{ccc} [n] & \xrightarrow{\alpha} & [m] \\ {\scriptstyle \delta^k}\downarrow & \ulcorner & \downarrow{\scriptstyle \theta} \\ [n+1] & \xrightarrow[\phi]{} & [m+1] \end{array}$$

where k is either 0 or $n+1$ (implying $\theta = \delta^\ell$ for ℓ either 0 or $m+1$). By factoring α into degeneracies and inner faces, we see that

$$\begin{array}{ccc} X_{m+1} & \xrightarrow{\phi^*} & X_{n+1} \\ {\scriptstyle d_\ell}\downarrow & \lrcorner & \downarrow{\scriptstyle d_k} \\ X_m & \xrightarrow[\alpha^*]{} & X_n \end{array} \tag{3}$$

is a pullback using the pasting law and our known pullback squares. Then, for an arbitrary active-inert pushout square as in (2) of Lemma 2.3, use induction on $k-n$ along with the squares (3) and the pasting law. □

EXERCISE 4.5. Recall that one characterization of 2-Segal spaces given in [**46**] is that X sends the pushout diagram of finite ordered sets

$$\begin{array}{ccc} \{i,j\} & \longrightarrow & \{i, i+1, \dots, j\} \\ \downarrow & & \downarrow \\ \{0, 1, \dots, i, j, j+1, \dots, n\} & \longrightarrow & [n] \end{array} \tag{4}$$

to a pullback for each $0 \leq i < j \leq n$. Prove this condition is equivalent to the definition of 2-Segal space from Definition 3.4.

The squares (4) correspond to those squares appearing in (2) of Lemma 2.3 where $n = 1$ and α is injective. As in [**19**, Proposition 2.3.2], one only needs to require these squares be sent to pullbacks when $i = 0$ or $j = n$.

5. Segal spaces are decomposition spaces

In [**46**], we saw that the nerve of a category is a 2-Segal set, under the characterization from Exercise 4.5, a fact that is true more generally: Segal spaces [**42, 45**] are always decomposition spaces. In this brief section we use the pasting law to establish this fact, and to characterize the Segal spaces among decomposition spaces.

DEFINITION 5.1. A simplicial space X is a *Segal space* if the following equivalent conditions hold.

(1) For each $n \geq 1$, the following square is a pullback.

$$\begin{array}{ccc} X_{n+1} & \xrightarrow{d_\bot} & X_n \\ {\scriptstyle d_\top}\downarrow & \lrcorner & \downarrow{\scriptstyle d_\top} \\ X_n & \xrightarrow[d_\bot]{} & X_{n-1} \end{array}$$

(2) For each $n \geq 2$, the map

$$X_n \xrightarrow[(d_\top^{n-i} d_\bot^{i-1})_{i=1}^n]{\simeq} X_1 \times_{X_0} X_1 \times_{X_0} \cdots \times_{X_0} X_1$$

is an equivalence.

EXERCISE 5.2. Prove that 1 and 2 in Definition 5.1 are equivalent. (For a hint or solution, see [**24**, Lemma 2.10].)

PROPOSITION 5.3. *Every Segal space is a decomposition space.*

PROOF. We show that every Segal space is upper 2-Segal. Since the opposite of a Segal space is a Segal space, and the opposite of an upper 2-Segal space is a lower 2-Segal space, this argument suffices. By Proposition 4.2, we need only show that the left square in the diagram

$$\begin{array}{ccccc} X_{n+1} & \xrightarrow{d_n} & X_n & \xrightarrow{d_\top} & X_{n-1} \\ {\scriptstyle d_\bot}\downarrow & & {\scriptstyle d_\bot}\downarrow & \lrcorner & \downarrow{\scriptstyle d_\bot} \\ X_n & \xrightarrow[d_{n-1}]{} & X_{n-1} & \xrightarrow[d_\top]{} & X_{n-2} \end{array}$$

is a pullback. The right square is a pullback since X is Segal. The outer rectangle is equal to the outer rectangle of the diagram

$$\begin{array}{ccccc} X_{n+1} & \xrightarrow{d_\top} & X_n & \xrightarrow{d_\top} & X_{n-1} \\ {\scriptstyle d_\bot}\downarrow & \lrcorner & {\scriptstyle d_\bot}\downarrow & \lrcorner & \downarrow{\scriptstyle d_\bot} \\ X_n & \xrightarrow[d_\top]{} & X_{n-1} & \xrightarrow[d_\top]{} & X_{n-2}. \end{array}$$

Since both squares are pullbacks, so is the outer rectangle, which implies that the left square in first diagram is a pullback, as desired. □

PROPOSITION 5.4. *A simplicial space X is a Segal space if and only if it is a decomposition space and the following square is a pullback:*

$$\begin{array}{ccc} X_2 & \xrightarrow{d_0} & X_1 \\ {\scriptstyle d_2}\downarrow & & \downarrow{\scriptstyle d_1} \\ X_1 & \xrightarrow[d_0]{} & X_0. \end{array}$$

PROOF. The forward direction is immediate from Proposition 5.3. For the reverse direction, one shows via induction that the squares in 1 of Definition 5.1 are pullbacks; the case $n = 1$ is an assumption. For the inductive step, use the two rectangles from the proof of Proposition 5.3 in a different way, assuming that both squares in the first rectangle and the right square in the second rectangle are pullbacks, in order to deduce that the left square in the second rectangle is a pullback. □

6. The path space criterion

We now turn to the path space criterion, which is originally due to Dyckerhoff and Kapranov [**19**, Theorem 6.3.2]. Our presentation separates the lower and upper 2-Segal conditions, following Poguntke [**41**], who also gave higher-dimensional versions. A version for decomposition spaces appeared as [**24**, Theorem 4.10].

THEOREM 6.1. *A simplicial space X is upper 2-Segal if and only if its upper décalage* $\mathrm{dec}_\top X$ *is Segal.*

The upper décalage $\mathrm{dec}_\top X$ is given by precomposition of $X\colon \Delta^{\mathrm{op}} \to \mathsf{Space}$ with the endofunctor $\Delta \to \Delta$ that sends $[n]$ to $[n]\star[0] \cong [n+1]$.[5] This endofunctor takes $f\colon [n] \to [m]$ to $g\colon [n+1] \to [m+1]$ satisfying $g(n+1) = m+1$ and $g(k) = f(k)$ for $0 \leq k \leq n$.

More explicitly, we have $(\mathrm{dec}_\top X)_n = X_{n+1}$ and the face and degeneracy operators are given for $0 \leq k \leq n$ by

$$\big(d_k\colon (\mathrm{dec}_\top X)_n \to (\mathrm{dec}_\top X)_{n-1}\big) = \big(d_k\colon X_{n+1} \to X_n\big)$$
$$\big(s_k\colon (\mathrm{dec}_\top X)_n \to (\mathrm{dec}_\top X)_{n+1}\big) = \big(s_k\colon X_{n+1} \to X_{n+2}\big).$$

Notably, we have forgotten about the top face $d_\top = d_{n+1}\colon X_{n+1} \to X_n$ and top degeneracy $s_\top = s_n\colon X_n \to X_{n+1}$. The upper décalage is also called the *final path space* and denoted by $P^{\triangleright}X$.

PROOF OF THEOREM 6.1. For $n \geq 2$, the following two squares are the same:

$$\begin{array}{ccc} (\mathrm{dec}_\top X)_n & \xrightarrow{d_\top} & (\mathrm{dec}_\top X)_{n-1} \\ {\scriptstyle d_\bot}\downarrow & & \downarrow{\scriptstyle d_\bot} \\ (\mathrm{dec}_\top X)_{n-1} & \xrightarrow[d_\top]{} & (\mathrm{dec}_\top X)_{n-2} \end{array} \qquad \begin{array}{ccc} X_{n+1} & \xrightarrow{d_n} & X_n \\ {\scriptstyle d_\bot}\downarrow & & \downarrow{\scriptstyle d_\bot} \\ X_n & \xrightarrow[d_{n-1}]{} & X_{n-1}. \end{array}$$

Proposition 4.2 says that the right square is a pullback for all $n \geq 2$ if and only if X is upper 2-Segal. □

Similarly, there is a lower décalage $\mathrm{dec}_\bot X$ (or initial path space $P^{\triangleleft}X$) with $(\mathrm{dec}_\bot X)_n = X_{n+1}$ obtained by forgetting the bottom face and degeneracy operators (and shifting indices by one). It can be formally obtained using precomposition with the endofunctor $[n] \mapsto [0] \star [n]$ of Δ, sending $f\colon [n] \to [m]$ to $g\colon [n+1] \to [m+1]$ satisfying $g(0) = 0$ and $g(k) = f(k-1)+1$ for $1 \leq k \leq n+1$. In other words, the face and degeneracy operators are as follows:

$$\big(d_k\colon (\mathrm{dec}_\bot X)_n \to (\mathrm{dec}_\bot X)_{n-1}\big) = \big(d_{k+1}\colon X_{n+1} \to X_n\big)$$
$$\big(s_k\colon (\mathrm{dec}_\bot X)_n \to (\mathrm{dec}_\bot X)_{n+1}\big) = \big(s_{k+1}\colon X_{n+1} \to X_{n+2}\big).$$

One checks that $(\mathrm{dec}_\top X)^{\mathrm{op}} = \mathrm{dec}_\bot(X^{\mathrm{op}})$, so that applying Theorem 6.1 to X^{op} we obtain the following.

COROLLARY 6.2. *A simplicial space X is lower 2-Segal if and only if its lower décalage* $\mathrm{dec}_\bot X$ *is Segal.* □

Using Theorem 4.1, we conclude the original form of the path space criterion.

COROLLARY 6.3. *A simplicial space X is a decomposition space if and only if both* $\mathrm{dec}_\top X$ *and* $\mathrm{dec}_\bot X$ *are Segal.* □

When X is a decomposition space, it turns out that the maps $d_\top\colon \mathrm{dec}_\top X \to X$ and $d_\bot\colon \mathrm{dec}_\bot X \to X$ are culf (in the sense of Definition 3.10); see [**24**, Theorem 4.10].

---

[5] Here $\star$ denotes the *join* of categories [**38**, Tag 0160], where the set of objects of $C \star D$ is the disjoint union of the sets of objects of C and D, the categories C and D are full subcategories of $C \star D$, and, for $c \in C$ and $d \in D$, there are no morphisms $d \to c$ and exactly one morphism $c \to d$.

7. The edgewise subdivision criterion

In this section we explain the following theorem from [**4**]. The proof presented here utilizes the exact same techniques that we have been using since Section 4. This proof relies on the path space criterion (Corollary 6.3), which is appropriate as there is a close connection between the edgewise subdivision and the upper/lower décalage pair.

THEOREM 7.1. *A simplicial space X is a decomposition space if and only if its edgewise subdivision* $\operatorname{sd} X$ *is Segal.*

As with the décalage functors, the edgewise subdivision is the result of precomposition with an endofunctor of Δ, namely the one that sends $[n]$ to $[n]^{\mathrm{op}} \star [n] \cong [2n+1]$. A map $f\colon [n] \to [m]$ is sent to $g\colon [2n+1] \to [2m+1]$ with

$$g(t) = \begin{cases} m - f(n-t) & t \leq n \\ m + 1 + f(t-n-1) & n+1 \leq t. \end{cases}$$

If we write $[2n+1]$ in the following way

$$\begin{array}{l} 0 \longrightarrow 1 \longrightarrow \cdots \longrightarrow n \\ \uparrow \\ 0' \longleftarrow 1' \longleftarrow \cdots \longleftarrow n', \end{array}$$

then g can be interpreted as $t' \mapsto f(t)'$ and $t \mapsto f(t)$. If X is a simplicial space, then the simplicial space $Z = \operatorname{sd} X$ has $Z_n = X_{2n+1}$ and

$$\begin{aligned} \big(d_i\colon Z_n \to Z_{n-1}\big) &= \big(d_{n-i} d_{n+i+1}\colon X_{2n+1} \to X_{2n-1}\big) \\ \big(s_i\colon Z_n \to Z_{n+1}\big) &= \big(s_{n-i} s_{n+i+1}\colon X_{2n+1} \to X_{2n+3}\big). \end{aligned}$$

One can check that $\operatorname{sd} X = \operatorname{sd}(X^{\mathrm{op}})$. If X is the nerve of a category C, then $\operatorname{sd} X$ is (isomorphic to) the nerve of the *twisted arrow category* of C, whose objects are the morphisms of C, and where a morphism $f \to g$ is given by a factorization of g as below:

$$\begin{array}{ccc} c' & \xrightarrow{k} & d' \\ {\scriptstyle f}\uparrow & & \uparrow{\scriptstyle g} \\ c & \xleftarrow[h]{} & d. \end{array}$$

One manifestation of the connection between déclage and edgewise subdivision is that iterated bottom and top degeneracies yield simplicial maps

$$\operatorname{dec}_\bot X \longrightarrow \operatorname{sd} X \longleftarrow (\operatorname{dec}_\top X)^{\mathrm{op}}.$$

For the map on the left, recall that $\operatorname{dec}_\bot X$ is given by precomposition with the endofunctor $[n] \mapsto [0] \star [n]$. The maps $[n]^{\mathrm{op}} \star [n] \to [0] \star [n]$ collapsing the first component to a point gives a natural transformation of functors, and hence the indicated map. It is specified by

$$\operatorname{dec}_\bot X_n = X_{n+1} \xrightarrow{s_\bot^n} X_{2n+1} = \operatorname{sd} X_n. \tag{5}$$

Likewise, the leftward arrow is given by

$$\operatorname{sd} X_n = X_{2n+1} \xleftarrow{s_\top^n} X_{n+1} = (\operatorname{dec}_\top X)^{\mathrm{op}}_n,$$

which arises from the natural transformation $[n]^{\mathrm{op}} \star [n] \to [n]^{\mathrm{op}} \star [0]$.

LEMMA 7.2. *If X is a decomposition space, then $\operatorname{sd} X$ is Segal.*

PROOF. Let $Z = \operatorname{sd} X$. For $n \geq 1$, the following two squares are the same:

$$\begin{array}{ccc} Z_{n+1} & \xrightarrow{d_{n+1}} & Z_n \\ {\scriptstyle d_0}\downarrow & & \downarrow{\scriptstyle d_0} \\ Z_n & \xrightarrow[d_n]{} & Z_{n-1} \end{array} \qquad \begin{array}{ccc} X_{2n+3} & \xrightarrow{d_0 d_{2n+3}} & X_{2n+1} \\ {\scriptstyle d_{n+1}d_{n+2}}\downarrow & & \downarrow{\scriptstyle d_n d_{n+1}} \\ X_{2n+1} & \xrightarrow[d_0 d_{2n+1}]{} & X_{2n-1}. \end{array}$$

The result follows since the right-hand square is an active-inert pullback square. □

LEMMA 7.3. *Let X be a simplicial space and write $Y = \operatorname{dec}_\perp X$, $Z = \operatorname{sd} X$. For each $n \geq 1$, the square below left is a retract of the square below right:*

$$\begin{array}{ccc} Y_{n+1} & \xrightarrow{d_\perp} & Y_n \\ {\scriptstyle d_\top}\downarrow & & \downarrow{\scriptstyle d_\top} \\ Y_n & \xrightarrow[d_\perp]{} & Y_{n-1} \end{array} \qquad \begin{array}{ccc} Z_{n+1} & \xrightarrow{d_\perp} & Z_n \\ {\scriptstyle d_\top}\downarrow & & \downarrow{\scriptstyle d_\top} \\ Z_n & \xrightarrow[d_\perp]{} & Z_{n-1}. \end{array}$$

Consequently, if $\operatorname{sd} X$ *is Segal, so is* $\operatorname{dec}_\perp X$.

PROOF. The conclusion follows since pullbacks are stable under retracts. The map from left to right is given by the simplicial map $Y \to Z$ from (5). We explicitly give the retraction from right to left (which does *not* come from a simplicial map $Z \to Y$). Utilizing the simplicial identities, all faces in the following cube commute:

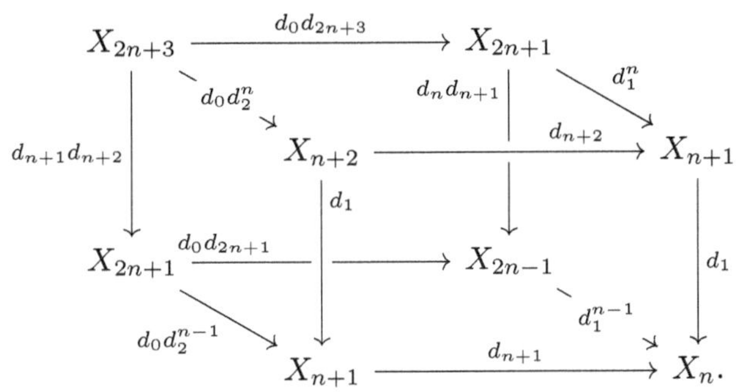

In terms of the face maps of Y and Z, the front and back faces of the cube are as follows:

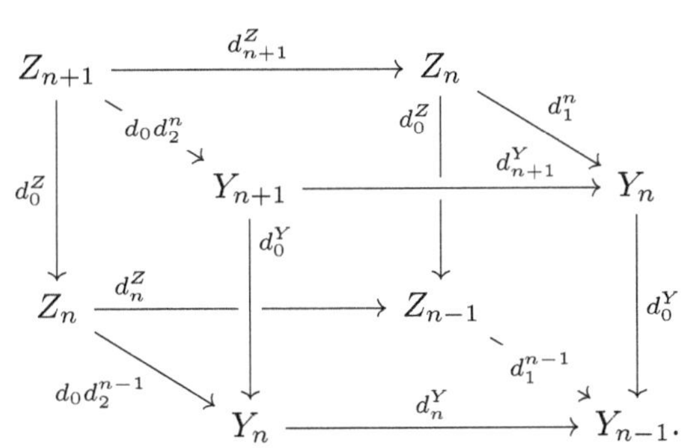

We thus see that the Y square is indeed a retract of the Z square since $d_1^n s_0^n = \mathrm{id}_{X_{n+1}} = d_0 d_2^{n-1} s_0^n$ for all $n \geq 1$. □

LEMMA 7.4. *If* $\operatorname{sd} X$ *is Segal, so is the upper décalage* $\operatorname{dec}_\top X$.

PROOF. We have $\operatorname{sd} X = \operatorname{sd}(X^{\text{op}})$ for any simplicial space X. By Lemma 7.3 we conclude that $\operatorname{dec}_\perp(X^{\text{op}}) = (\operatorname{dec}_\top X)^{\text{op}}$ is Segal. □

PROOF OF THEOREM 7.1. The forward direction is Lemma 7.2. If $\operatorname{sd} X$ is Segal, then the preceding two lemmas show that $\operatorname{dec}_\perp X$ and $\operatorname{dec}_\top X$ are Segal as well, so the path space criterion implies that X is a decomposition space. □

REMARK 7.5. If X is a discrete decomposition space, then the three simplicial sets $\operatorname{sd} X$, $\operatorname{dec}_\perp X$, and $\operatorname{dec}_\top X$ are (nerves of) categories. We have inclusions of simplicial sets $\operatorname{dec}_\perp X \hookrightarrow \operatorname{sd} X$ and $(\operatorname{dec}_\top X)^{\text{op}} \hookrightarrow \operatorname{sd} X$. Identifying each of these subcategories with its image, the category $\operatorname{sd} X$ admits two (strict) factorization systems $(\operatorname{dec}_\perp X, (\operatorname{dec}_\top X)^{\text{op}})$ and $((\operatorname{dec}_\top X)^{\text{op}}, \operatorname{dec}_\perp X)$; see [**32**] for one explanation.

There is also an interesting story about culf maps (Definition 3.10) and edgewise subdivision, which gives the following theorem from [**36**] (see [**34**] for a generalization).

THEOREM 7.6. *If X is a discrete decomposition space, then there is an equivalence of categories between the category of culf maps over X and the category of presheaves on* $\operatorname{sd} X$.

8. Free decomposition spaces

We conclude by briefly explaining a class of examples. For simplicity of presentation, in this section we restrict ourselves to the category sSet of simplicial sets. This section is based on [**35**], which addresses the more general case of simplicial spaces.

DEFINITION 8.1. An *outer face complex* is a presheaf on the category of inert maps, that is, a functor $A\colon \Delta_{\text{int}}^{\text{op}} \to \mathsf{Set}$.

As the top and bottom coface maps in Δ generate the subcategory Δ_{int}, an outer face complex is given by a sequence of sets and maps as displayed (compare to (1)):

$$A_0 \underset{d_\perp}{\overset{d_\top}{\leftleftarrows}} A_1 \underset{d_\perp}{\overset{d_\top}{\leftleftarrows}} A_2 \underset{d_\perp}{\overset{d_\top}{\leftleftarrows}} A_3 \cdots \qquad d_\top d_\perp = d_\perp d_\top. \tag{6}$$

Let $j\colon \Delta_{\text{int}} \to \Delta$ be the inclusion of the category of inert maps. An outer face complex then determines a simplicial set by left Kan extension along j, which we denote by $X = j_! A$. A surprising fact is that this simplicial set is always 2-Segal.

THEOREM 8.2. *If A is an outer face complex, then $X = j_! A$ is a (discrete) decomposition space.*

Rather than proving this theorem formally (which is already done in [**35**]), we instead sketch out the idea about why it might be true. We begin by unpacking the structure of $X = j_! A$, which we calculate as follows:

$$\begin{aligned} X_k = j_! A_k &\cong \operatorname*{colim}_{([k]\downarrow j)^{\text{op}}} A_m \\ &\cong \sum_{[k] \twoheadrightarrow [m]} A_m \\ &\cong \sum_{\ell_1, \dots, \ell_k \in \mathbb{N}} A_m \qquad \text{where } m = \ell_1 + \cdots + \ell_k. \end{aligned}$$

The first identification is the standard pointwise colimit formula for the left Kan extension along $j^{\mathrm{op}}\colon \Delta_{\mathrm{int}}^{\mathrm{op}} \to \Delta^{\mathrm{op}}$ (see [**39**, X.3]), along with an isomorphism of categories $(j^{\mathrm{op}} \downarrow [k]) \cong ([k] \downarrow j)^{\mathrm{op}}$. The second identification uses a cofinality argument: the discrete subcategory of $[k] \downarrow j$ on the set of active maps $[k] \twoheadrightarrow [m]$ is an initial subcategory. (By the unique factorization in Section 2, each active map is an initial object in a connected component of $[k] \downarrow j$.) Taking opposites makes the inclusion a final functor, so the colimit may instead be computed on this discrete subcategory [**39**, IX.3]. The last identification simply uses the following bijection:

$$\begin{aligned}\{[k] \twoheadrightarrow [m]\} &\cong \{(\ell_1, \dots, \ell_k) \in \mathbb{N}^k \mid \textstyle\sum \ell_i = m\}\\ \alpha &\mapsto (\alpha(1) - \alpha(0), \alpha(2) - \alpha(1), \dots, \alpha(k) - \alpha(k-1))\\ (t \mapsto \textstyle\sum_{i=1}^t \ell_i) &\leftarrow\!\shortmid (\ell_1, \dots, \ell_k).\end{aligned}$$

We now give an indication of the face maps. Writing an element of X_k as $(a, \ell_1, \dots, \ell_k)$ where $a \in A_m$ (for $m = \sum \ell_i$), the inner face maps are given by adding adjacent natural numbers:

$$d_i(a, \ell_1, \dots, \ell_k) = (a, \ell_1, \dots, \ell_i + \ell_{i+1}, \dots, \ell_k)$$

for $i = 1, \dots, k-1$. The outer face maps discard an the outer element of the list of natural numbers, and use the structure maps from our original outer face complex A:

$$\begin{aligned}d_0(a, \ell_1, \dots, \ell_k) &= (d_\perp^{\ell_1} a, \ell_2, \dots, \ell_k)\\ d_k(a, \ell_1, \dots, \ell_k) &= (d_\top^{\ell_k} a, \ell_1, \dots, \ell_{k-1}).\end{aligned}$$

Degeneracies are given by inserting a zero into the list of natural numbers. We leave it to the reader to trace through the identifications above to verify that these formulas are correct.

Formulas for the face maps at hand, it is straightforward to check that X is a 2-Segal set. For instance the square

$$\begin{array}{ccc} X_3 & \xrightarrow{d_2} & X_2 \\ {\scriptstyle d_0}\downarrow & \lrcorner & \downarrow{\scriptstyle d_0} \\ X_2 & \xrightarrow[d_1]{} & X_1 \end{array}$$

is a pullback since if we have $(a, \ell_1, \ell_2), (b, m_1, m_2) \in X_2$ (with $a \in A_{\ell_1+\ell_2}$ and $b \in A_{m_1+m_2}$) such that $d_1(a, \ell_1, \ell_2) = (a, \ell_1 + \ell_2)$ is equal to $d_0(b, m_1, m_2) = (d_\perp^{m_1} b, m_2)$ then the only element $x = (b', k_1, k_2, k_3) \in X_3$ with $d_0 x = (a, \ell_1, \ell_2)$ and $d_2 x = (b, m_1, m_2)$ is $x = (b, m_1, \ell_1, \ell_2)$. All other pullback squares are verified in a similar manner.

The most basic example is to apply this construction to the terminal outer face complex, where each $A_m = *$ is a one-element set. In this case,

$$X_k = \sum_{\ell_1, \dots, \ell_k \in \mathbb{N}} * \cong \mathbb{N} \times \overset{k}{\cdots} \times \mathbb{N}.$$

This simplicial set X is isomorphic to $B\mathbb{N}$, the nerve of the monoid of natural numbers. This example generalizes to other free monoids:

EXAMPLE 8.3. Fix an alphabet $A_1 \in \mathsf{Set}$, and let A_m denote the set of length m words in A_1. The face maps $d_\perp, d_\top \colon A_m \to A_{m-1}$ discard the first or last letters of a word. Let $X = j_! A$. The set of vertices $X_0 = A_0$ is a one-point set containing only the empty word. The set of edges $X_1 = \sum_{m\geq 0} A_m$ is the set of all finite words $(a_1 \dots a_m)$. We have

$$X_2 = \sum_{\ell_1,\ell_2} A_{\ell_1+\ell_2} = \sum_{\ell_1+\ell_2} A_{\ell_1} \times A_{\ell_2} = \{(a_1 \dots a_{\ell_1} | a_{\ell_1+1} \dots a_{\ell_1+\ell_2})\}$$

is the set of all words together with a cut point. The outer face maps forget one side of the cut, while the inner face map forgets the cut point. This pattern continues, with X_3 consisting of words with two cut points, and so on. This example is Segal, since it is the nerve of the free monoid on A_1. However, we can tweak it by fixing a positive integer N and considering a sub-outer face complex A' consisting only of words of length at most N:

$$\begin{array}{lllllllll} A: & A_0 \overset{d_\top}{\underset{d_\bot}{\leftleftarrows}} A_1 \overset{d_\top}{\underset{d_\bot}{\leftleftarrows}} A_2 \cdots A_{N-1} \overset{d_\top}{\underset{d_\bot}{\leftleftarrows}} A_N \overset{d_\top}{\underset{d_\bot}{\leftleftarrows}} A_{N+1} \cdots \\ A': & A_0 \overset{d_\top}{\underset{d_\bot}{\leftleftarrows}} A_1 \overset{d_\top}{\underset{d_\bot}{\leftleftarrows}} A_2 \cdots A_{N-1} \overset{d_\top}{\underset{d_\bot}{\leftleftarrows}} A_N \leftleftarrows \varnothing \cdots . \end{array}$$

Then $X' = j_! A'$ is the decomposition space of words of length at most N, which is not Segal.

Notice that the decomposition space of words of length at most N also falls under the setting of Example 3.6, by letting M be the subset of the free monoid on A_1 with length bounded by N.

EXAMPLE 8.4. Let G be a directed graph, such as the following.

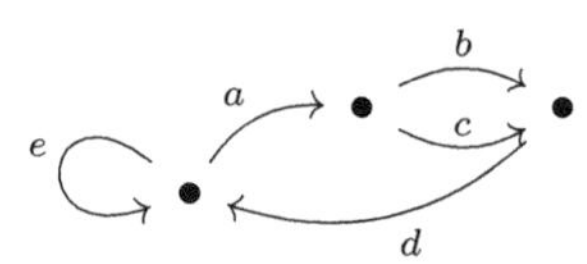

The associated outer face complex is the one where A_n consists of length n paths in the graph. Functionally, this is a modification of Example 8.3 where the words are restricted to only allow certain length 2-subwords; for instance ac and ee are acceptable subwords but bc and eb are not.

We leave many other interesting examples to [**35**], and conclude with a statement of the main theorem of that paper.

THEOREM 8.5. *The category of outer face complexes is equivalent to the category of culf maps over* $B\mathbb{N}$.

The theorem says, roughly, that outer face complexes may be identified with decomposition spaces X equipped with a well-behaved notion of (integer-valued) edge length. One can classify the culf maps over $B\mathbb{N}$ whose domains are Segal; these are precisely those categories which are free on a directed graph [**47**, §1]. The corresponding outer face complex is exactly the one from Example 8.4 (see [**35**, 4.1]).

References

[1] C. Berger, *Moment categories and operads*, Theory Appl. Categ. **38** (2022), Paper No. 39, 1485–1537. MR4541944

[2] C. Berger and I. Moerdijk, *On an extension of the notion of Reedy category*, Math. Z. **269** (2011), no. 3-4, 977–1004, DOI 10.1007/s00209-010-0770-x. MR2860274

[3] J. E. Bergner, A. M. Osorno, V. Ozornova, M. Rovelli, and C. I. Scheimbauer, *2-Segal sets and the Waldhausen construction*, Topology Appl. **235** (2018), 445–484, DOI 10.1016/j.topol.2017.12.009. MR3760213

[4] J. E. Bergner, A. M. Osorno, V. Ozornova, M. Rovelli, and C. I. Scheimbauer, *The edgewise subdivision criterion for 2-Segal objects*, Proc. Amer. Math. Soc. **148** (2020), no. 1, 71–82, DOI 10.1090/proc/14679. MR4042831

[5] J. E. Bergner and W. H. Stern, *Cyclic Segal spaces*, this volume, `arXiv:2409.11945 [math.AT]`.

[6] F. Borceux, *Handbook of categorical algebra. 1*, Encyclopedia of Mathematics and its Applications, vol. 50, Cambridge University Press, Cambridge, 1994. Basic category theory. MR1291599

[7] S. Burkin, *Twisted arrow categories, operads and Segal conditions*, Theory Appl. Categ. **38** (2022), Paper No. 16, 595–660. MR4403280

[8] T. Carawan, *2-Segal maps associated to a category with cofibrations*, `arXiv:2405.11561 [math.AT]`.

[9] L. Carlier, *Hereditary species as monoidal decomposition spaces, comodule bialgebras, and operadic categories*, Int. Math. Res. Not. IMRN **8** (2022), 5745–5780, DOI 10.1093/imrn/rnaa260. MR4410767

[10] L. Carlier and J. Kock, *Antipodes of monoidal decomposition spaces*, Commun. Contemp. Math. **22** (2020), no. 2, 1850081, 15, DOI 10.1142/S0219199718500815. MR4077086

[11] A. Cebrian and W. Forero, *Directed hereditary species and decomposition spaces*, `arXiv:2211.07721 [math.CO]`.

[12] H. Chu and R. Haugseng, *Homotopy-coherent algebra via Segal conditions*, Adv. Math. **385** (2021), Paper No. 107733, 95, DOI 10.1016/j.aim.2021.107733. MR4256131

[13] A. Connes, *Cohomologie cyclique et foncteurs* Ext^n (French, with English summary), C. R. Acad. Sci. Paris Sér. I Math. **296** (1983), no. 23, 953–958. MR777584

[14] A. Connes and D. Kreimer, *Hopf algebras, renormalization and noncommutative geometry*, Comm. Math. Phys. **199** (1998), no. 1, 203–242, DOI 10.1007/s002200050499. MR1660199

[15] C. Constantin, T. Fritz, P. Perrone, and B. T. Shapiro, *Weak cartesian properties of simplicial sets*, J. Homotopy Relat. Struct. **18** (2023), no. 4, 477–520, DOI 10.1007/s40062-023-00334-1. MR4671178

[16] I. Contreras, R. A. Mehta, and W. H. Stern, *Frobenius and commutative pseudomonoids in the bicategory of spans*, J. Geom. Phys. **207** (2025), Paper No. 105309, 31, DOI 10.1016/j.geomphys.2024.105309. MR4798060

[17] B. Cooper and M. B. Young, *Hall algebras via 2-Segal spaces*, this volume, `arXiv:2409.19384 [math.CT]`.

[18] T. Dyckerhoff and M. Kapranov, *Triangulated surfaces in triangulated categories*, J. Eur. Math. Soc. (JEMS) **20** (2018), no. 6, 1473–1524, DOI 10.4171/JEMS/791. MR3801819

[19] T. Dyckerhoff and M. Kapranov, *Higher Segal spaces*, Lecture Notes in Mathematics, vol. 2244, Springer, Cham, 2019, DOI 10.1007/978-3-030-27124-4. MR3970975

[20] M. Feller, *Quasi-2-Segal sets*, Tunis. J. Math. **5** (2023), no. 2, 327–367, DOI 10.2140/tunis.2023.5.327. MR4596737

[21] M. Feller, R. Garner, J. Kock, M. U. Proulx, and M. Weber, *Every 2-Segal space is unital*, Commun. Contemp. Math. **23** (2021), no. 2, Paper No. 2050055, 6, DOI 10.1142/S0219199720500558. MR4201031

[22] W. Forero, *The Gálvez-Carrillo-Kock-Tonks conjecture for locally discrete decomposition spaces*, Commun. Contemp. Math. **26** (2024), no. 4, Paper No. 2350011, 54, DOI 10.1142/S0219199723500116. MR4730614

[23] I. Gálvez-Carrillo, J. Kock, and A. Tonks, *Decomposition spaces in combinatorics*, this volume, `arXiv:1612.09225 [math.CO]`.

[24] I. Gálvez-Carrillo, J. Kock, and A. Tonks, *Decomposition spaces, incidence algebras and Möbius inversion I: Basic theory*, Adv. Math. **331** (2018), 952–1015, DOI 10.1016/j.aim.2018.03.016. MR3804694

[25] I. Gálvez-Carrillo, J. Kock, and A. Tonks, *Decomposition spaces, incidence algebras and Möbius inversion II: Completeness, length filtration, and finiteness*, Adv. Math. **333** (2018), 1242–1292, DOI 10.1016/j.aim.2018.03.017. MR3818099

[26] I. Gálvez-Carrillo, J. Kock, and A. Tonks, *Decomposition spaces, incidence algebras and Möbius inversion III: The decomposition space of Möbius intervals*, Adv. Math. **334** (2018), 544–584, DOI 10.1016/j.aim.2018.03.018. MR3828744

[27] I. Gálvez-Carrillo, J. Kock, and A. Tonks, *Homotopy linear algebra*, Proc. Roy. Soc. Edinburgh Sect. A **148** (2018), no. 2, 293–325, DOI 10.1017/S0308210517000208. MR3777576

[28] I. Gálvez-Carrillo, J. Kock, and A. Tonks, *Decomposition spaces and restriction species*, Int. Math. Res. Not. IMRN **21** (2020), 7558–7616, DOI 10.1093/imrn/rny089. MR4176832

[29] R. Garner, J. Kock, and M. Weber, *Operadic categories and décalage*, Adv. Math. **377** (2021), Paper No. 107440, 23, DOI 10.1016/j.aim.2020.107440. MR4186001

[30] M. Grandis, *Weak subobjects and the epi-monic completion of a category*, J. Pure Appl. Algebra **154** (2000), no. 1-3, 193–212, DOI 10.1016/S0022-4049(99)00191-7. Category theory and its applications (Montreal, QC, 1997). MR1787597

[31] J. Gödicke, *An ∞-category of 2-Segal spaces*, `arXiv:2407.13357 [math.AT]`.

[32] P. Hackney, *Operadic categories and 2-Segal sets*, `arXiv:2312.00756 [math.CT]`.

[33] P. Hackney, *Segal conditions for generalized operads*, Higher structures in topology, geometry, and physics, Contemp. Math., vol. 802, Amer. Math. Soc., Providence, RI, 2024, pp. 161–194, DOI 10.1090/conm/802/16077. MR4773880

[34] P. Hackney and J. Kock, *Culf maps and edgewise subdivision*, Trans. Amer. Math. Soc. **379** (2026), no. 4, 2955–3007, DOI 10.1090/tran/9577, with an appendix coauthored with Jan Steinebrunner.

[35] P. Hackney and J. Kock, *Free decomposition spaces*, Collect. Math. **76** (2025), no. 3, 575–599, DOI 10.1007/s13348-024-00446-8. MR4950314

[36] J. Kock and D. I. Spivak, *Decomposition-space slices are toposes*, Proc. Amer. Math. Soc. **148** (2020), no. 6, 2317–2329, DOI 10.1090/proc/14834. MR4080877

[37] J. Lurie, *Higher topos theory*, Annals of Mathematics Studies, vol. 170, Princeton University Press, Princeton, NJ, 2009, DOI 10.1515/9781400830558. MR2522659

[38] J. Lurie, *Kerodon*, `https://kerodon.net`, 2024.

[39] S. Mac Lane, *Categories for the working mathematician*, 2nd ed., Graduate Texts in Mathematics, vol. 5, Springer-Verlag, New York, 1998. MR1712872

[40] B. Pareigis, *Categories and functors*, Pure and Applied Mathematics, Vol. 39, Academic Press, New York-London, 1970. Translated from the German. MR265428

[41] T. Poguntke, *Higher Segal structures in algebraic K-theory*, `arXiv:1709.06510 [math.AT]`.

[42] C. Rezk, *A model for the homotopy theory of homotopy theory*, Trans. Amer. Math. Soc. **353** (2001), no. 3, 973–1007, DOI 10.1090/S0002-9947-00-02653-2. MR1804411

[43] R. Rosebrugh and R. J. Wood, *Distributive laws and factorization*, J. Pure Appl. Algebra **175** (2002), no. 1-3, 327–353, DOI 10.1016/S0022-4049(02)00140-8. Special volume celebrating the 70th birthday of Professor Max Kelly. MR1941618

[44] G. Segal, *Configuration-spaces and iterated loop-spaces*, Invent. Math. **21** (1973), 213–221, DOI 10.1007/BF01390197. MR331377

[45] G. Segal, *Categories and cohomology theories*, Topology **13** (1974), 293–312, DOI 10.1016/0040-9383(74)90022-6. MR353298

[46] W. H. Stern, *Perspectives on the 2-Segal conditions*, this volume.

[47] R. Street, *Categorical structures*, Handbook of algebra, Vol. 1, Handb. Algebr., vol. 1, Elsevier/North-Holland, Amsterdam, 1996, pp. 529–577, DOI 10.1016/S1570-7954(96)80019-2. MR1421811

DEPARTMENT OF MATHEMATICS, UNIVERSITY OF LOUISIANA AT LAFAYETTE
Email address: `philip@phck.net`

Contemporary Mathematics
Volume **838**, 2026
https://doi.org/10.1090/conm/838/16737

Waldhausen's $S_\bullet$-construction

Viktoriya Ozornova and Martina Rovelli

ABSTRACT. We survey various versions of Waldhausen's $S_\bullet$-construction and the role they play in defining K-theory, and we discuss their 2-Segality properties.

Introduction

A prominent invariant for a ring (spectrum) R is its K-theory space, in fact a spectrum, $K(R)$. Various constructions are available to access $K(R)$, most notably:

- Quillen's *+-construction* [**Qui72**], defined for every exact category;
- Quillen's *Q-construction* [**Qui73**], defined for every exact category;
- Segal's *group completion* [**Seg74**], defined for every symmetric monoidal category;
- Grayson's *$S^{-1}S$-construction* [**Gra76**], defined for every category of finitely generated projective modules over a ring;
- Waldhausen's *$S_\bullet$-construction* [**Wal85**], defined for every category with cofibrations, which are nowadays referred to as Waldhausen categories.

The perspective of the $S_\bullet$-construction was prone to generalizations for a very large variety of type of input structure, and it is the focus of this note.

From any ring R, one can consider the exact category $\mathcal{E}_R$ of finitely generated projective R-modules. The structure of an exact category for $\mathcal{E}_R$ takes into account the behavior of its class of monomorphisms and its class of epimorphisms, as well as the role of the zero module. For instance, key properties are the facts that monomorphisms (resp. epimorphisms) can be composed, that pushout (resp. pullback) of a monomorphism (resp. an epimorphism) exists and is still a monomorphism (resp. an epimorphism). These are essential to study the short exact sequences and the bicartesian squares, and to develop the homological algebra in $\mathcal{E}_R$.

Further, given $\mathcal{E}$ any exact (∞-)category, or even something more general such as a Waldhausen (∞-)category, its K-theory space $K(\mathcal{E})$ can be obtained through the realization of a certain simplicial space: Waldhausen's *$S_\bullet$-construction* $S_\bullet(\mathcal{E})$.

A lot of the *structure* of $\mathcal{E}$, such as information about the bicartesian squares in $\mathcal{E}$, is naturally retained in the simplicial space $S_\bullet(\mathcal{E})$. Similarly, a lot of the relevant *properties* of $\mathcal{E}$, such as the fact that bicartesian squares are determined by the span of which they are a colimit cone, are manifested in an interesting way in $S_\bullet(\mathcal{E})$: they are encoded in the fact that $S_\bullet(\mathcal{E})$ is a (lower) 2-*Segal space*.

Whether $S_\bullet(\mathcal{E})$ is a 2-Segal space or just a lower 2-Segal space depends on the structure on $\mathcal{E}$, and we will survey in the remainder of the paper the contributions to date towards this perspective. The simplicial space $S_\bullet(\mathcal{E})$ is a 2-Segal space when $\mathcal{E}$ is an exact category, or more generally it is *pointed* and *stable* in an appropriate sense. Instead, the simplicial space $S_\bullet(\mathcal{E})$ is a priori only a lower 2-Segal space when $\mathcal{E}$ is a Waldhausen category, or more generally it is pointed and satisfies just one half of the stability requirements, which we may refer to as *semi-stable.*

The notion of a 2-Segal space was discovered independently in [**DK19**] (cf. the contribution [**Ste24**]) and in [**GCKT18**] under the name of *decomposition space* (cf. the contribution [**Hac24**]), and the $S_\bullet$-construction $S_\bullet(\mathcal{E})$ was identified as a main example by both sets of authors. It was later shown in [**BOO$^+$18, BOO$^+$21a**] (cf. the contribution [**Rov24**]) that in fact the $S_\bullet$-construction defines an equivalence between 2-Segal spaces and certain double Segal spaces (i.e., bisimplicial spaces which are Segal in each variable) with an appropriate enhancement. Hence, every 2-Segal space does indeed arise as an appropriate $S_\bullet$-construction.

In most cases of interest for $\mathcal{E}$, the K-theory space $K(\mathcal{E})$ can be delooped infinitely many times, so it is in fact an Ω-spectrum. An n-fold delooping for $K(\mathcal{E})$ is given by the realization of an appropriate iterated version of the $S_\bullet$-construction: the multisimplicial space $S^{(n)}_{\bullet,\dots,\bullet}(\mathcal{E})$.

Again, a lot of the structure and properties of $\mathcal{E}$ are manifested through a 2-Segality property of the multisimplicial space $S^{(n)}_{\bullet,\dots,\bullet}(\mathcal{E})$ in each simplicial variable individually. For instance, it was proved (cf. [**Pen17**]) that when $\mathcal{E}$ is an exact category and $n=2$ the bisimplicial space $S^{(2)}_{\bullet,\bullet}(\mathcal{E})$ is 2-Segal in each variable.

In this note we discuss various flavors of $S_\bullet$-constructions, how they can be used to define K-theory, their 2-Segality properties, and how this is remembering features of the original input.

How to read this paper. The paper hopes to accomplish two goals: 1) describe to a novice reader the classical instance of Waldhausen's $S_\bullet$-construction, and 2) compile for a more experienced reader a literature review for many generalizations of the original $S_\bullet$. With this in mind, we suggest different reading strategies.

The following table can be regarded as a 2-dimensional table of contents:

structure of $\mathcal{E}$	$S_\bullet(\mathcal{E})$	2-Segality	$S^{(n)}_{\bullet,\dots,\bullet}(\mathcal{E})$
$\mathcal{E}=\mathcal{E}_R$ (e.g. category of modules over R)	§1.1	-	§2.1
$\mathcal{E}$ is stable pointed (e.g. exact category)	§1.2.1	§1.3.1	-
$\mathcal{E}$ is semistable pointed (e.g. Waldhausen category)	§1.2.2	§1.3.2	-
$\mathcal{E}$ is a Σ-space	§1.2.3	-	§2.2

The topics are covered by columns. A more experienced reader can follow the content in order, spending less time on the first row overall. However, we encourage novice readers to start with the first row.

Acknowledgments. We are grateful to Julie Bergner, Joachim Kock, Mark Penney, and Maru Sarazola for organizing a wonderful workshop at the Banff International Research Station: *Higher Segal Spaces and their Applications to Algebraic*

K-theory, Hall Algebras, and Combinatorics (24w5266, originally 20w5173). We are also thankful to Claudia Scheimbauer for creating and letting us use the picture macros from our previous papers and to Lennart Meier for helpful discussions on the history of the subject. MR acknowledges the support from the National Science Foundation under Grant No. DMS-2203915. We thank the referee and Joachim Kock for their helpful comments and suggestions.

1. The $S_\bullet$-construction and K-theory space

We review Waldhausen's $S_\bullet$-construction and some of its generalizations.

1.1. Classical $S_\bullet$-construction. We discuss in detail what Waldhausen's $S_\bullet$-construction is and how it is used to define K-theory for the case of an exact category of modules over a ring. More details can be found in Waldhausen's original source [**Wal85**, §1.3].

1.1.1. *The sequence construction.* Let R be a unital associative commutative ring and let $\mathcal{E}_R$ be the category of modules over R or a full subcategory thereof.

The easiest case to keep in mind throughout this section is the case $\mathcal{E}_R$ being the category of all R-modules, which is a primary example of an *abelian* category. However, this choice leads to a trivial K-theory (cf. Remark 1.12). Alternatively, one can consider the case of R being a local regular ring and $\mathcal{E}_R$ being the category of finitely generated R-modules, which is also an abelian category. Overall, what is considered the most interesting case is that of R being any ring and $\mathcal{E}_R$ being the category of finitely generated projective R-modules, which fails to be an abelian category, but can be given the structure of an *exact* category. In fact, as we will mention later, all the constructions presented in this section work with minor adjustments for any exact or abelian category $\mathcal{E}$.

When drawing diagrams valued in $\mathcal{E}_R$, we follow the convention that arrows drawn horizontally, resp. vertically, as

$$\boxed{A_0 \rightarrowtail A_1} \qquad \text{resp.} \qquad \boxed{\begin{array}{c} A_0 \\ \downarrow\!\!\downarrow \\ A_1 \end{array}}$$

correspond to an injective (resp. surjective) morphism in $\mathcal{E}_R$, that displayed squares of the form

$$\boxed{\begin{array}{ccc} A_{00} & \rightarrowtail & A_{10} \\ \downarrow\!\!\downarrow & & \downarrow\!\!\downarrow \\ A_{10} & \rightarrowtail & A_{11} \end{array}}$$

are bicartesian squares (i.e., squares which are both cartesian and cocartesian squares in $\mathcal{E}$), and that objects denoted Z_i are R-trivial modules, i.e., a zero module.

Some first relevant information about the ring R which is naturally stored in $\mathcal{E}_R$ is the understanding of sequences of subobjects. For instance, if $\mathcal{E}_\mathbb{Z}$ is the category of finitely generated projective abelian groups, for all $n \geq 0$ there exists a sequence of proper submodules of the object $\mathbb{Z}$ of length n:

$$2^n\mathbb{Z} \subsetneq 2^{n-1}\mathbb{Z} \subsetneq \ldots \subsetneq 4\mathbb{Z} \subsetneq 2\mathbb{Z}.$$

Instead, if $\mathcal{E}_{\mathbb{Z}/2}$ is the category of finitely generated $\mathbb{Z}/2$-modules, the length of a chain of subobjects of a given object is bounded by the rank of this object.

So one could consider:

CONSTRUCTION 1.1. For $k \geq 0$, let $\mathrm{seq}_k(\mathcal{E}_R)$ denote the set of all sequences of subobjects in $\mathcal{E}_R$ of length k (intended to be the singleton of the empty sequence of subobjects if $k = 0$). For instance, for $k = 3$ the generic element is of the form

$$\boxed{A_1 \rightarrowtail A_2 \rightarrowtail A_3}$$

For $0 < i \leq k$, let the i-th *face map* $d_i\colon \mathrm{seq}_k(\mathcal{E}_R) \to \mathrm{seq}_{k-1}(\mathcal{E}_R)$ be given by skipping the i-th term. For instance, if $k = 3$ this gives:

$$\boxed{A_1 \rightarrowtail A_2 \rightarrowtail A_3} \overset{d_i}{\longmapsto} \begin{cases} \boxed{A_2 \rightarrowtail A_3}, & i = 1 \\ \boxed{A_1 \rightarrowtail A_3}, & i = 2 \\ \boxed{A_1 \rightarrowtail A_2}, & i = 3 \end{cases}$$

Let the 0-th *face map* $d_0\colon \mathrm{seq}_k(\mathcal{E}_R) \to \mathrm{seq}_{k-1}(\mathcal{E}_R)$ be given by taking quotients modulo the first term. For instance, if $k = 3$ this gives:

$$\boxed{A_1 \rightarrowtail A_2 \rightarrowtail A_3} \overset{d_0}{\longmapsto} \boxed{A_2/A_1 \rightarrowtail A_3/A_1}$$

For $0 \leq i \leq k$, let the i-th *degeneracy map* $s_i\colon \mathrm{seq}_k(\mathcal{E}_R) \to \mathrm{seq}_{k+1}(\mathcal{E}_R)$ be given by doubling the i-th term for $i > 0$ and adding a zero object 0 for $i = 0$. For instance, if $k = 2$ this gives

$$\boxed{A_1 \rightarrowtail A_2} \overset{s_i}{\longmapsto} \begin{cases} \boxed{0 \rightarrowtail A_1 \rightarrowtail A_2}, & i = 0 \\ \boxed{A_1 \rightarrowtail A_1 \rightarrowtail A_2}, & i = 1 \\ \boxed{A_1 \rightarrowtail A_2 \rightarrowtail A_2}, & i = 2 \end{cases}$$

As discussed in [**Wal78**, §4], the construction $\mathrm{seq}_k(\mathcal{E})$ is however not natural in the variable k:

REMARK 1.2. The assignment $[k] \mapsto \mathrm{seq}_k(\mathcal{E}_R)$ does *not* define a simplicial set. Given σ in $\mathrm{seq}_3(\mathcal{E}_R)$

$$\boxed{A_1 \rightarrowtail A_2 \rightarrowtail A_3}$$

many simplicial identities hold, such as

$$d_1 d_2(\sigma) = A_3 = d_1 d_1(\sigma).$$

However, any expression involving the face d_0 is problematic. For instance, when comparing d_0d_0 to d_0d_1, the best that one can say is that there is an isomorphism, yet generally not an equality, of the form

$$(1) \qquad d_0d_0(\sigma) = (A_3/A_1)/(A_2/A_1) \cong A_3/A_2 = d_0d_1(\sigma).$$

Overall, the construction $\mathrm{seq}_k(\mathcal{E}_R)$ suffers from the following shortcomings:

(1) The assignment $[k] \mapsto \mathrm{seq}_k(\mathcal{E}_R)$ does not define a simplicial set of sequences in $\mathcal{E}_R$, as discussed in Remark 1.2.
(2) The definition of $\mathrm{seq}_k(\mathcal{E}_R)$ breaks the natural symmetry present in the structure of $\mathcal{E}_R$, in that it prioritizes subobjects over quotients.
(3) The *set* of sequences $\mathrm{seq}_k(\mathcal{E}_R)$ is not so meaningful in itself, in that it is not capable of recognizing when two sequences are *isomorphic*, and should instead be replaced with the *groupoid* or *space* of sequences.

A way to address at least some of these issues is to consider the following variant (which can be seen as a variant of [**Wal85**, §1] and as an instance of [**Wal78**, §4], taking in $\mathcal{E}_R$ cofibrations to be monomorphisms and weak equivalences to be isomorphisms):

CONSTRUCTION 1.3. For $k \geq 0$, let $\mathrm{Seq}_k(\mathcal{E}_R)$ denote the groupoid of all sequences of k subobjects in $\mathcal{E}_R$ (intended to be the singleton groupoid if $k = 0$); that is, the maximal groupoid in the category of functors from $[k-1]$ to $\mathcal{E}_R$ spanned by the set of objects $\mathrm{seq}_k(\mathcal{E}_R)$. For $0 \leq i \leq k$, let the i-th *face map* $d_i\colon \mathrm{Seq}_k(\mathcal{E}_R) \to \mathrm{Seq}_{k-1}(\mathcal{E}_R)$ and the i-th *degeneracy map* $s_i\colon \mathrm{Seq}_k(\mathcal{E}_R) \to \mathrm{Seq}_{k+1}(\mathcal{E}_R)$ be defined by extending the formulas from Construction 1.1 in the obvious way.

When assessing the functoriality of the construction $\mathrm{Seq}_k(\mathcal{E}_R)$ with respect to the parameter k, the best that one can say is that the construction is *pseudo-functorial* in k (cf. e.g. [**JY21**, §4.1]). More precisely, by making appropriately coherent the isomorphism witnessing the failure of the simplicial identities for $\mathrm{Seq}_k(\mathcal{E}_R)$, e.g. the one from (1), one could show:

PROPOSITION 1.4. *The assignment* $[k] \mapsto \mathrm{Seq}_k(\mathcal{E}_R)$ *defines a pseudo simplicial groupoid.*

The construction $\mathrm{Seq}_k(\mathcal{E}_R)$ addresses (3), but it only partially addresses (1) (in that it involves a higher categorical metatheory) and it does not address (2).

1.1.2. *The $S_\bullet$-construction.* An attempt at solving the issues (1), (3) and (2) going in a different direction is the following (cf. [**Wal85**, §1.4]):

Denote by $\mathrm{Ar}[k] = [k]^{[1]}$ the *arrow category* of $[k]$; that is, the category of functors $[1] \to [k]$ and natural transformations.

CONSTRUCTION 1.5. For $k \geq 0$, let $s_k(\mathcal{E}_R)$ denote the set of *exact functors* $\mathrm{Ar}[k] \to \mathcal{E}_R$. That is, of functors $\mathrm{Ar}[k] \to \mathcal{E}_R$ which:

- send the objects of the form (i,i) to a zero object in $\mathcal{E}_R$,
- send the morphisms of the form $(i,j) \to (i, j+\ell)$ to monomorphisms in $\mathcal{E}_R$,
- send the morphisms $(i,j) \to (i+k, j)$ to epimorphisms in $\mathcal{E}_R$,
- send the commutative squares involving two maps of the form $(i,j) \to (i+k,j)$ and two maps of the form $(i,j) \to (i,j+\ell)$ to bicartesian squares in $\mathcal{E}_R$.

For instance, a functor $\mathrm{Ar}[3] \to \mathcal{E}_R$ defining an object of $s_3(\mathcal{E}_R)$ is of the form

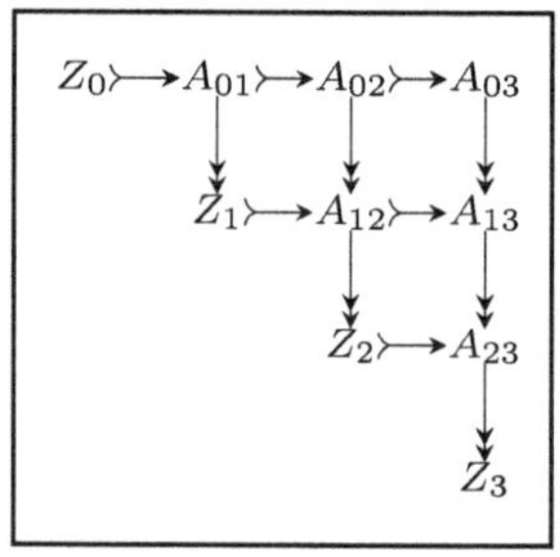

with the convention that all objects featuring on the diagonal are zero objects, all horizontal morphisms are monomorphisms, all vertical morphisms are epimorphisms, and all squares are bicartesian squares. For $0 \leq i \leq k$, let the i-th *face map* $d_i \colon s_k(\mathcal{E}_R) \to s_{k-1}(\mathcal{E}_R)$ be given by skipping the i-th row and the i-th column. For instance, if $k = 3$ the face maps d_2 and d_3 act as:

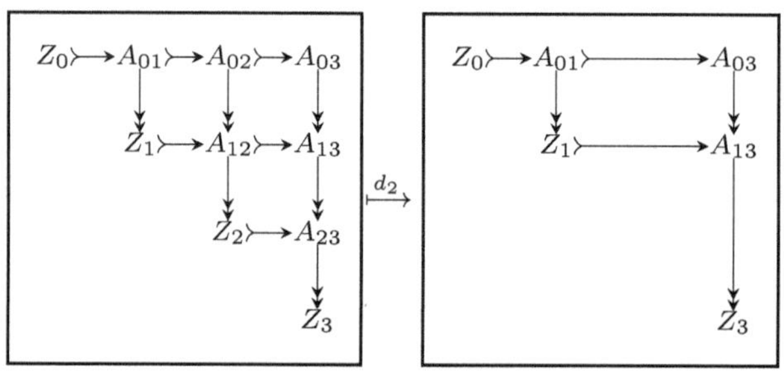

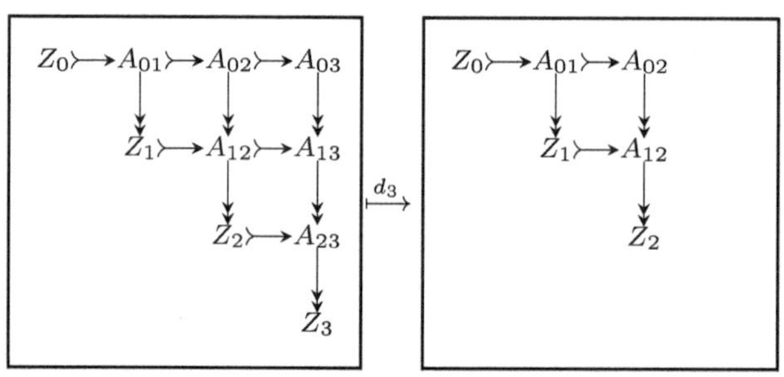

For $0 \leq i \leq k$, let the i-th *degeneracy map* $s_i \colon s_k(\mathcal{E}_R) \to s_{k+1}(\mathcal{E}_R)$ be given by doubling the i-th term. For instance, if $k = 2$ the degeneracy maps s_0 and s_1 are as follows

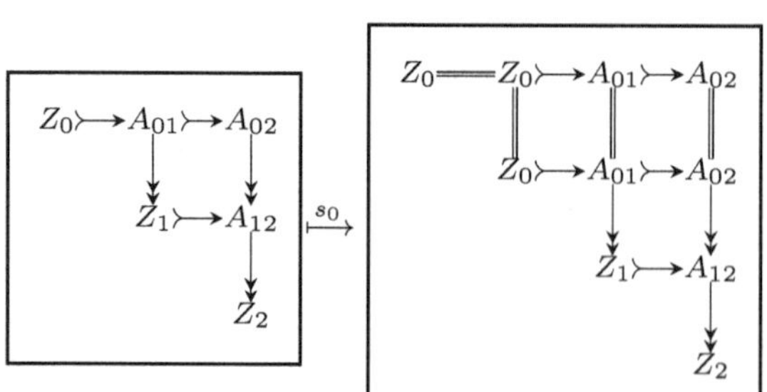

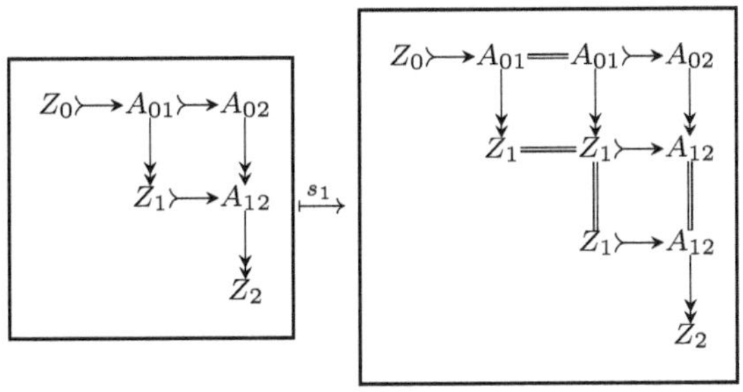

One can then check that the construction $s_k(\mathcal{E})$ is natural in k:

PROPOSITION 1.6. *The assignment* $[k] \mapsto s_k(\mathcal{E}_R)$ *defines a simplicial set* $s_\bullet(\mathcal{E}_R)$.

The construction $s_k(\mathcal{E}_R)$ solves (1) without requiring higher categorical meta-theory, and it solves (2), but does not fix (3). So one can then mix the two adjustments $\mathrm{Seq}_k(\mathcal{E}_R)$ and $s_k(\mathcal{E}_R)$, culminating with Waldhausen's original $S_\bullet$-construction (cf. [**Wal85**, §1.3] in the special case where weak equivalences are chosen to be isomorphisms):

CONSTRUCTION 1.7. For $k \geq 0$, let $S_k(\mathcal{E}_R)$ denote the groupoid of exact functors from $\mathrm{Ar}[k]$ to $\mathcal{E}_R$; that is, the maximal groupoid in the category of functors $\mathrm{Ar}[k]$ to $\mathcal{E}_R$ spanned by the set of objects $s_k(\mathcal{E}_R)$. For $0 \leq i \leq k$, let the i-th *face map* $d_i \colon S_k(\mathcal{E}_R) \to S_{k-1}(\mathcal{E}_R)$ and the i-th *degeneracy map* $s_i \colon S_k(\mathcal{E}_R) \to S_{k+1}(\mathcal{E}_R)$ be defined by extending the formulas from Construction 1.5 in the obvious way.

One can then check that this produces a simplicial object (cf. [**Wal85**, §1.3]), which we refer to as the *$S_\bullet$-construction* of $\mathcal{E}_R$:

PROPOSITION 1.8. *The assignment* $[k] \mapsto S_k(\mathcal{E}_R)$ *defines a simplicial groupoid* $S_\bullet(\mathcal{E}_R)$.

The following proposition explains how the construction $S_\bullet(\mathcal{E}_R)$ can really be understood as an appropriate correction of $\mathrm{Seq}_\bullet(\mathcal{E}_R)$, in that (cf. [**DK19**, Lemma 2.4.9]):

PROPOSITION 1.9. *For $k \geq 0$, the canonical projection on "the 0-th row" is an equivalence of groupoids*

$$S_k(\mathcal{E}_R) \xrightarrow{\simeq} \mathrm{Seq}_k(\mathcal{E}_R).$$

IDEA. For instance, for $k = 3$ this map of groupoids can be depicted as

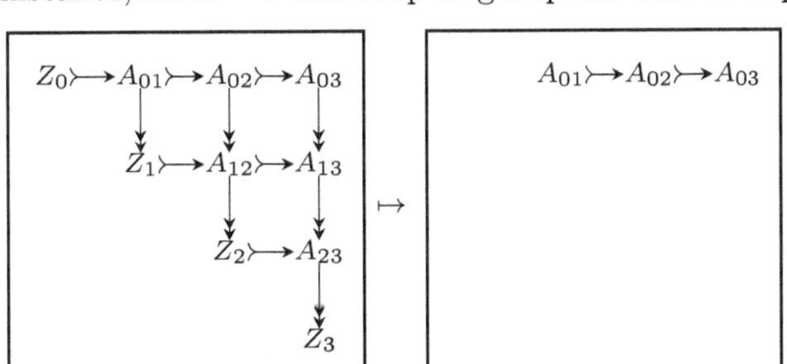

By putting the zero module in the diagonal positions of the diagram and completing all the relevant spans to pushout squares in $\mathcal{E}_R$, which can be done is an essentially

unique way, one can consider the assignment

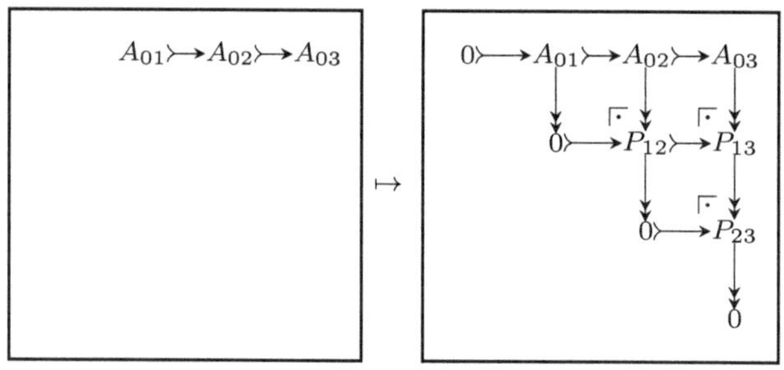

and verify it defines an inverse equivalence of groupoids

$$\mathrm{Seq}_k(\mathcal{E}_R) \xrightarrow{\simeq} S_k(\mathcal{E}_R).$$

for the canonical projection. The statement for general k can be proven through an appropriate enhancement of this argument. □

We can take a closer look at how, in addition to the sequences from Proposition 1.9, many other relevant "homological" features of R are retained in $S_\bullet(\mathcal{E}_R)$:

REMARK 1.10. We have that:

(0) If $\mathrm{Z}(\mathcal{E}_R)$ denotes the groupoid of zero objects in $\mathcal{E}_R$, there is an isomorphism of groupoids

$$S_0(\mathcal{E}_R) \cong \mathrm{Z}(\mathcal{E}_R).$$

(1) If $\mathrm{Ob}(\mathcal{E}_R)$ denotes the core of $\mathcal{E}_R$, i.e., the maximal groupoid inside $\mathcal{E}_R$, as a special case of Proposition 1.9 there is an equivalence of groupoids

$$S_1(\mathcal{E}_R) \simeq \mathrm{Ob}(\mathcal{E}_R).$$

(2) If $\mathrm{ExSeq}(\mathcal{E}_R)$ denotes the groupoid of short exact sequences in $\mathcal{E}_R$, there is an equivalence of groupoids

$$S_2(\mathcal{E}_R) \simeq \mathrm{ExSeq}(\mathcal{E}_R),$$

which can be depicted as

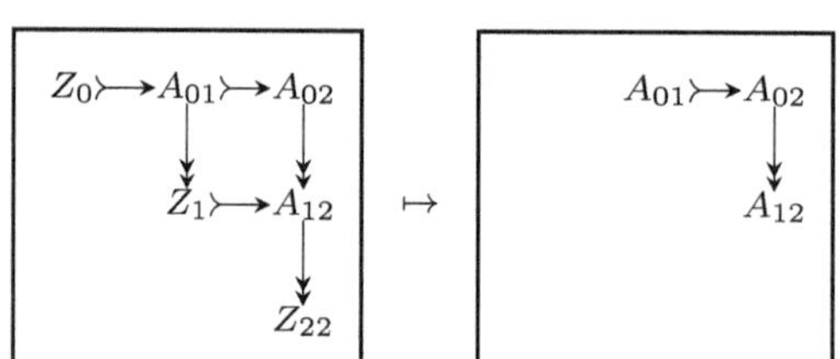

(3) If $\mathrm{BiCartSq}(\mathcal{E}_R)$ denotes the groupoid of bicartesian squares involving two (admissible) monomorphisms and two (admissible) epimorphisms in $\mathcal{E}_R$, there is an equivalence of groupoids

$$S_3(\mathcal{E}_R) \simeq \mathrm{BiCartSq}(\mathcal{E}_R),$$

which can be depicted as

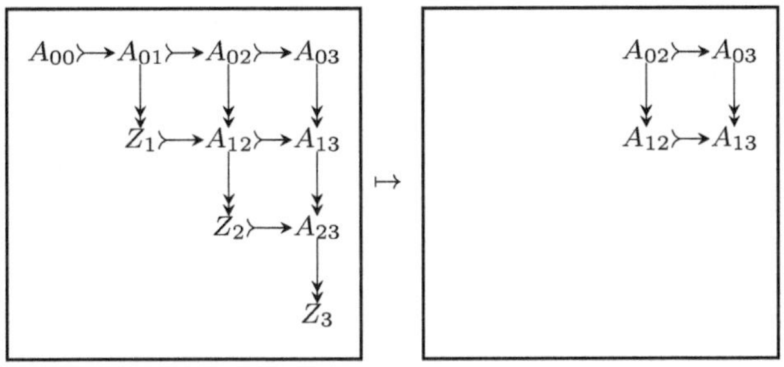

with inverse equivalence given by

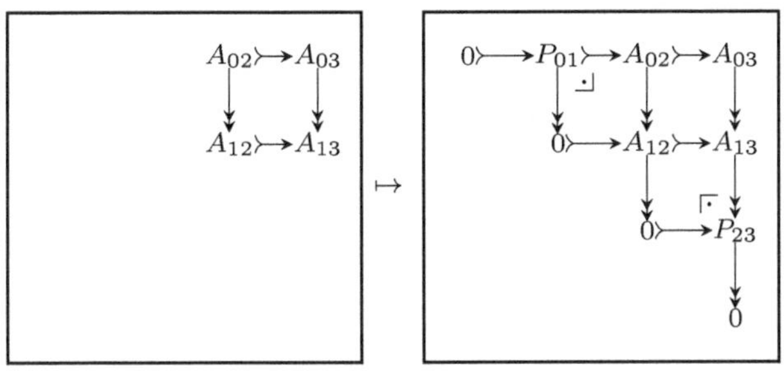

From $S_\bullet(\mathcal{E}_R)$, one can define the K-theory space $K(\mathcal{E}_R)$ (cf. [**Wal85**, §1.3]), by further applying Real: $\mathcal{G}pd^{\Delta^{\mathrm{op}}} \to \mathcal{S}^{\Delta^{\mathrm{op}}} \to \mathcal{S}$, the realization of simplicial groupoids through simplicial spaces (cf. [**Rie14**, §8.2]):

CONSTRUCTION 1.11. The *K-theory space* of $\mathcal{E}_R$ is

$$K(\mathcal{E}_R) := \Omega(\mathrm{Real}(S_\bullet(\mathcal{E}_R))).$$

REMARK 1.12. If $\mathcal{E}_R$ is the category of modules over R, then $K(\mathcal{E}_R)$ is contractible (see e.g. [**Wei13**, §II.2]). Instead, if $\mathcal{E}_R$ is the category of finitely generated projective modules over R, then $K(\mathcal{E}_R)$ is typically interesting, and it is what one refers to as the K-theory space $K(R)$ of the ring R. The homotopy groups of $K(R)$ have been fully computed for the case of R being a finite field (see [**Qui72**, Theorem 8]), and are only partially known for $R = \mathbb{Z}$ (see [**Wei05**]).

We will see later in Section 2 how the space $K(\mathcal{E}_R)$ can be delooped infinitely many times, so it actually comes from an Ω-spectrum.

The formalism in this section applies with only minor adjustments to the case of exact categories. All constructions were carried out using the exact category structure of $\mathcal{E}_R$, where the admissible monomorphisms (resp. admissible epimorphisms) are the module maps that are injective (resp. surjective), and the zero objects are the trivial modules.

More generally, the datum of an *exact category* $\mathcal{E}$ in the sense of Quillen consists of a category endowed with two classes of distinguished morphisms, called *admissible monomorphisms* and *admissible epimorphisms*, and a distinguished class of objects. We refer the reader to [**Qui73**, §2] for a complete definition, and we just mention that the axioms imposed on $\mathcal{E}$ guarantee that *zero objects* exist, that admissible monomorphisms (resp. admissible epimorphisms) can be composed, that pushout (resp. pullback) of an admissible monomorphism (resp. an admissible epimorphism) exists and is still an admissible monomorphism (resp. an admissible epimorphism). This is in particular enough structure to talk about admissible *short exact sequences* and *bicartesian squares.* As a consequence of the axioms, the property of pointedness and a version of stability hold for $\mathcal{E}$. An *exact functor* is a functor that preserves the admissible short exact sequences.

REMARK 1.13. By the Freyd–Mitchell embedding theorem [**Wei94**, §1.6] any abelian category is a subcategory of a category of R-modules of some ring R (however, the embedding is noncanonical). Moreover, by the Gabriel–Quillen embedding theorem [**Qui73**, §2] any exact category is a full subcategory of an abelian category. In total, any exact category is an exact sub-category of a category of R-modules of some ring R. In virtue of these theorems, understanding the contents of this section in the case of categories of modules over a ring is not far from the general case.

1.2. Generalized $S_\bullet$-construction. We now mention several generalizations of Waldhausen's construction that have been studied in the literature, culminating with a detailed description in Section 1.2.3 of a rather general one. This is intended as a guide to navigate the literature on generalizations and see how they fit together, rather than a detailed account of these. The reader who wishes to postpone this discussion can jump directly to Section 2.1 to see how to iterate the $S_\bullet$-construction.

1.2.1. *Pointed stable contexts.* We collect here many extensions of Waldhausen's $S_\bullet$-construction to a generalized input structure $\mathcal{E}$ that maintains a version of the properties of "stability" and "pointedness" (or more generally "augmentation"). The structure of $\mathcal{E}$ needs to include a class of objects, a class of distinguished objects, two classes of morphisms (horizontal and vertical) and a class of squares. The properties required of $\mathcal{E}$ roughly encode the following, to be understood in a strict or weak sense, depending on the context:

◇ **pasting**: the fact that the horizontal morphisms can be composed, vertical morphisms can be composed, and squares can be composed both horizontally and vertically;
◇ **pointedness** (or more generally **augmentation**): the fact that there is a distinguished class of objects (typically the class of zero objects, or a generalization thereof) with the property that every object receives uniquely a horizontal morphism from an object in this distinguished class, and every object has a unique vertical morphism to an object in this distinguished class;
◇ **stability**: the fact that every square is uniquely determined by the span in its boundary and also uniquely determined by the cospan in its boundary, through the depicted assignments:

(2)
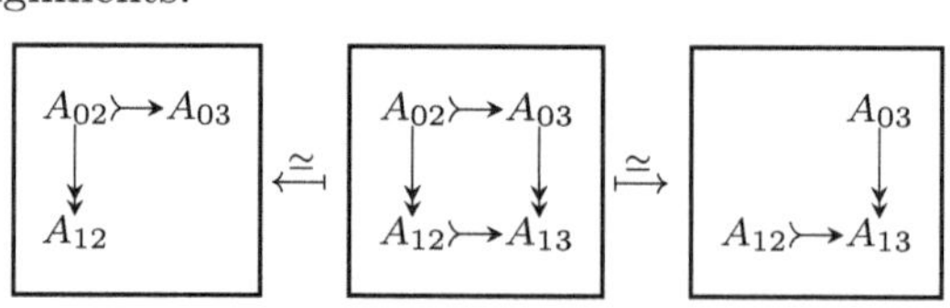

Categorical structures that follow this scheme, roughly in the chronological order in which they were considered, include the case of $\mathcal{E}$ being:

- an *abelian category*; a key example is the abelian category of abelian groups, or more generally the abelian category of modules over a ring R.
- a *(proto)exact category*, as in [**Qui73**, §1.2] and [**DK19**, §2.4]; a key example is the exact category of finitely generated abelian groups, or more generally the

exact category of finitely generated modules over a ring R, or the proto-exact category of finite pointed sets;

- a *stable ∞-category*, as in [**Lur06**]; a key example is the stable ∞-category of spectra, or more generally the stable ∞-category of (compact) modules over a ring spectrum.
- an *exact ∞-category*, as in [**DK19**, §7.2] (cf. also [**Bar15**]); a key example is the exact ∞-category of connective spectra.
- a *stable pointed double category* (or more generally a *stable augmented double category*), as in [**BOO$^+$18**, §3]; a key example is the path construction of a reduced 2-Segal set (or more generally any 2-Segal set); in fact it is shown in [**BOO$^+$18**, Theorem 6.1] that all examples are of this kind.
- a *stable pointed double Segal space* (or more generally a *stable augmented double Segal space*), as in [**BOO$^+$21a**, §2.5]; a key example is the path construction of a reduced 2-Segal space (or more generally any 2-Segal space); in fact it is shown in [**BOO$^+$21a**, Theorem 6.1] that all examples are of this kind.
- a *CGW category* as in [**CZ22**, §2], (cf. also an *ACGW category* as in [**CZ22**, §5], and an *ECGW category* as in [**SS21**, Part 1]); a key example is the category of reduced schemes of finite type.
- an *isostable squares category* as in [**CS24**, §3]; a key example is the squares category of polytopes from [**CS24**, Examples 2.14].

These different frameworks can be organized as follows:

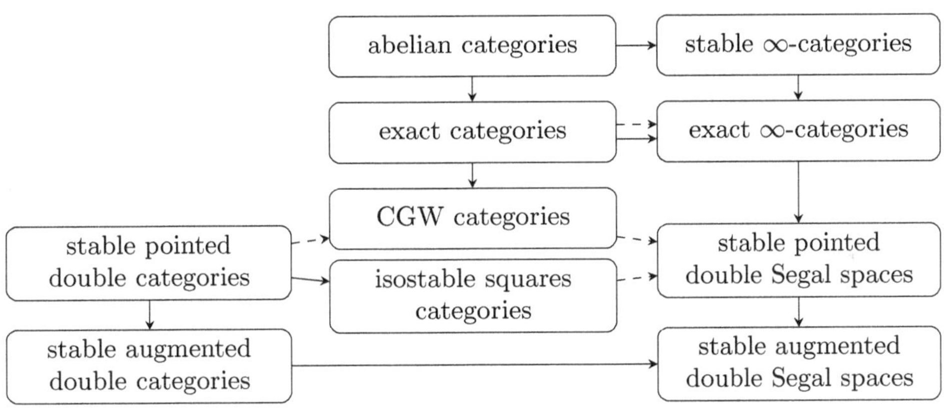

In this diagram, solid arrows represent established constructions, while dashed arrows indicate an additional construction that we believe should exist, but has not yet been fully studied (see below for a list of references).

Some existing comparisons between these frameworks include: [**Qui73**, §2] for how exact categories recover abelian categories, [**DK19**, Example 7.2.3] for how exact ∞-categories recover exact categories, [**Nee90**] for how exact ∞-categories possibly recover exact categories through the derived ∞-category of chain complexes, [**Lur18**, §1.3.2] for how stable ∞-categories recover abelian categories, [**CZ22**, Example 3.1] for how CGW categories recover exact categories, [**CS24**, Proposition 3.11] for how isostable squares categories recover stable pointed double categories, [**BOO$^+$21b**] for how stable augmented double Segal spaces recover exact categories and stable ∞-categories through appropriate nerve constructions.

The study of how CGW categories recover pointed stable double categories and of how pointed stable double Segal spaces recover CGW categories is subject of ongoing work by some workshop participants (cf. Bergner's contribution in [**BLLM24**, pp. 29-30]) A comment on how CGW categories possibly relate to isostable square categories is in [**CS24**, Remark 3.27].

An appropriate $S_\bullet$-construction was considered for each of these situations. This was done in [**Wal85**, §1.3] for exact categories (as a special case of Waldhausen categories, there referred to as categories with cofibrations and weak equivalences), in [**Bar15**, Recollection 5.8] (as a special case of Waldhausen ∞-categories) and [**DK19**, §2.4] for exact ∞-categories, in [**BOO$^+$18**, §3] for pointed or augmented stable double categories, in [**BOO$^+$21a**, §2, §7] for pointed or augmented stable double Segal spaces, and in [**CS24**, §2.2] for isostable squares categories. Many are recovered as a special case of the $S_\bullet$-construction for Σ-spaces, which will be recalled in Section 1.2.3.

Results establishing the compatibility between the various $S_\bullet$-constructions include [**BOO$^+$21b**, §2] for the compatibility of the one for stable augmented double categories with the one for stable augmented double Segal spaces, [**BOO$^+$21b**, §3] for the compatibility of the one for exact categories with the one for stable augmented double Segal spaces and [**CS24**, Proposition 3.6] for the compatibility of the one for stable augmented double categories with the one for isostable squares categories.

The versions of $S_\bullet$-constructions considered for different frameworks, however, do not always agree in a strict sense. For instance, the $S_\bullet$-construction for abelian categories and the one for stable ∞-categories only agree up to equivalence upon taking the loop space of their geometric realizations, essentially as an instance of the Gillet–Waldhausen theorem (see [**TT90**, Theorem 1.11.7] for an exact statement).

1.2.2. *Pointed semi-stable context.* One could also generalize in a different direction. In many cases of interest the left half of (2), asking for a square to be determined by the span contained in its boundary, holds. However, right half of (2), asking for a square to be determined by the cospan contained in its boundary, does not. This is the case, for instance, when the class of distinguished squares are certain pushout squares in a given category, which are not necessarily pullbacks. Categorical structures that fit this framework, roughly in the chronological order in which they were considered, include the case of $\mathcal{W}$ being:

- a *Waldhausen category* as in [**Wei13**, Definition II.9.1.1] (originally referred to as *category with cofibrations and weak equivalences* in [**Wal85**, §1.2]); a key example is the Waldhausen category of pointed CW-complexes, or more generally the category of retractive spaces over a CW-complex X (see [**Wal85**, §1.1]);
- a *Waldhausen ∞-category*, as in [**Bar16**, Definition 2.7]; this notion offers a common framework to talk about stable ∞-categories and Waldhausen categories;
- a *proto-Waldhausen squares category*, as in [**CS24**, §2.3].

An appropriate $S_\bullet$-construction was considered for several of these situations. This was done in [**Wal85**, §1.3] for $\mathcal{W}$ being a Waldhausen category, in [**Bar16**, §5] (through a slightly different though equivalent formalism) for $\mathcal{W}$ being a Waldhausen ∞-category, and in [**CS24**, §2.2] for $\mathcal{W}$ being a proto-Waldhausen squares category. An $S_\bullet$-construction for stable augmented double Segal spaces (in fact for any Σ-space) is given in Section 1.2.3.

Sometimes, however, the versions of $S_\bullet$-construction considered for different framework do not agree in a strict sense. For instance, as explained in [**CS24**, Remark 2.34], the $S_\bullet$-construction for proto-Waldhausen categories and the one for Waldhausen categories only agree up to equivalence upon taking geometric realization.

1.2.3. *General context.* We describe here a general framework for $S_\bullet$-construction recovering many of the instances discussed.

Let Σ denote the category from [**BOO$^+$21a**, Definition 2.9], obtained by freely adding a terminal object $[-1]$ to the category $\Delta \times \Delta$, and consider the category $\mathcal{S}^{\Sigma^{\text{op}}}$ of Σ-spaces. More explicitly, a Σ-space X consists of a bisimplicial space $X_{\bullet,\bullet}$ together with an augmentation space X_{-1} and an augmentation map $X_{-1} \to X_{0,0}$ subject to appropriate relations.

We briefly recall how the framework of Σ-spaces recovers the framework of exact categories through an appropriate nerve construction, as done in [**BOO$^+$21b**, §2]. Nerve constructions that have been studied for other frameworks from Sections 1.2.1 and 1.2.2 include [**BOO$^+$21b**, §3] for stable ∞-categories and [**BOO$^+$21b**, §4] for exact ∞-categories.

If $\mathcal{S}$ denotes the category of spaces and $\mathcal{E}xCat$ denotes the category of exact categories and exact functors, we recall the exact nerve construction $N^{\text{ex}}\colon \mathcal{E}xCat \to \mathcal{S}^{\Sigma^{\text{op}}}$ from [**BOO$^+$21b**, Definition 2.2]. Given an exact category $\mathcal{E}$, let

$$N^{\text{ex}}_{-1}(\mathcal{E}) := \mathrm{Z}(\mathcal{E})$$

denote the groupoid of zero objects in $\mathcal{E}$, and for $a, b \geq 0$ we let

$$N^{\text{ex}}_{a,b}(\mathcal{E}) := \mathrm{Hom}^{\text{ex}}([a] \times [b], \mathcal{E})$$

denote the groupoid of *exact functors* from $[a]\times[b]$ to $\mathcal{E}$; that is, functors $[a]\times[b] \to \mathcal{E}$ which:

- send the morphisms of the form $(i,j) \to (i, j+\ell)$ to admissible monomorphisms in $\mathcal{E}$,
- send the morphisms $(i,j) \to (i+k, j)$ to admissible epimorphisms in $\mathcal{E}$,
- send the commutative squares involving two maps of the form $(i,j) \to (i+k, j)$ and two maps of the form $(i,j) \to (i, j+\ell)$ to bicartesian squares in $\mathcal{E}$.

For instance, a functor $[2] \times [3] \to \mathcal{E}$ defining an object of $N^{\text{ex}}_{2,3}(\mathcal{E})$ is of the form

$$\begin{array}{ccccccc}
A_{00} & \rightarrowtail & A_{01} & \rightarrowtail & A_{02} & \rightarrowtail & A_{03} \\
\downarrow & & \downarrow & & \downarrow & & \downarrow \\
A_{10} & \rightarrowtail & A_{11} & \rightarrowtail & A_{12} & \rightarrowtail & A_{13} \\
\downarrow & & \downarrow & & \downarrow & & \downarrow \\
A_{20} & \rightarrowtail & A_{21} & \rightarrowtail & A_{22} & \rightarrowtail & A_{23}
\end{array}$$

with the convention that all horizontal morphisms are admissible monomorphisms, all vertical morphisms are admissible epimorphisms, and all squares are bicartesian squares.

We now turn to the $S_\bullet$-construction in the context of Σ-spaces. Let $p\colon \Sigma \to \Delta$ denote the ordinal sum given by $[-1] \mapsto [0]$ and $[a,b] \mapsto [a+1+b]$, as considered in [**BOO$^+$21a**, Definition 2.15] (see [**Rov24**] in this volume). If $\mathcal{S}$ denotes the category of spaces, there is an induced adjunction

$$\mathcal{P} = p^*\colon \mathcal{S}^{\Delta^{\text{op}}} \leftrightarrows \mathcal{S}^{\Sigma^{\text{op}}} : p_* = S_\bullet$$

between the category $\mathcal{S}^{\Delta^{\text{op}}}$ of simplicial spaces and the category $\mathcal{S}^{\Sigma^{\text{op}}}$ of Σ-spaces.

So we obtain the description of the $S_\bullet$-construction:

CONSTRUCTION 1.14. For $k \geq 0$, and X a Σ-space, we let

$$S_k(X) := \mathrm{Map}(\mathcal{P}\Delta[k], X)$$

denote the space of maps of Σ-spaces $\mathcal{P}\Delta[k] \to X$. For $0 \leq i \leq n$, the i-th *face map* $d_i\colon S_k(X) \to S_{k-1}(X)$ and the i-th *degeneracy map* $s_i\colon S_k(X) \to S_{k+1}(X)$ are induced by the cosimplicial structure of $\mathcal{P}\Delta[\bullet]$, and can be described in a similar fashion to those of Construction 1.5.

In total we get the *$S_\bullet$-construction* of any Σ-space X:

PROPOSITION 1.15. *If X is a Σ-space, the assignment $[k] \mapsto S_k(X)$ defines a simplicial space $S_\bullet(X)$.*

The $S_\bullet$-construction for exact categories and the one for Σ-spaces are compatible through the exact nerve $N^{\mathrm{ex}}\colon \mathcal{E}x\mathcal{C}at \to \mathcal{S}^{\Sigma^{\mathrm{op}}}$:

REMARK 1.16. It is shown in [**BOO$^+$21b**, Theorem 2.18] that there is a commutative diagram of categories

$$\begin{array}{ccc} \mathcal{E}x\mathcal{C}at & \xrightarrow{S_\bullet} & \mathcal{G}pd^{\Delta^{\mathrm{op}}} \\ {\scriptstyle N^{\mathrm{ex}}}\downarrow & & \downarrow{\scriptstyle N_*} \\ \mathcal{S}^{\Sigma^{\mathrm{op}}} & \xrightarrow[S_\bullet]{} & \mathcal{S}^{\Delta^{\mathrm{op}}} \end{array}$$

Indeed, for $n \geq 0$ and $\mathcal{E}$ an exact category, there is a natural bijection

$$S_k(N^{\mathrm{ex}}\mathcal{E}) \cong \mathrm{Map}(\mathcal{P}\Delta[k], N^{\mathrm{ex}}\mathcal{E}) \cong N(\mathrm{Hom}^{\mathrm{ex}}(\mathrm{Ar}[k], \mathcal{E})) \cong N(S_k(\mathcal{E})) = (N_*S)_k(\mathcal{E}).$$

From $S_\bullet(X)$, one can define the K-theory space $K(X)$ for a very general type of input X, by further applying $\mathrm{Real}\colon \mathcal{G}pd^{\Delta^{\mathrm{op}}} \to \mathcal{S}^{\Delta^{\mathrm{op}}} \to \mathcal{S}$, the realization of simplicial groupoids through simplicial spaces (cf. [**Rie14**, §8.2]). If X is a Σ-space,

$$K(X) := \Omega(\mathrm{Real}(S_\bullet(X))).$$

1.3. 2-Segality properties. Recall the definition of *2-Segal space* from [**DK19**] (see [**Ste24**] in this volume), a.k.a. *decomposition space in* [**GCKT18**] (see [**Hac24**] in this volume), and its generalization *lower 2-Segal space* from [**Pog17**, Definition 2.2].

We discuss how the properties of the input $\mathcal{E}$ are reflected in the 2-Segality property of its $S_\bullet$-construction. We will compose any existing $S_\bullet$-construction valued in $\mathcal{G}pd^{\Delta^{\mathrm{op}}}$ from previous sections with the levelwise nerve functor $N_*\colon \mathcal{G}pd^{\Delta^{\mathrm{op}}} \to \mathcal{S}^{\Delta^{\mathrm{op}}}$, keeping the same notation.

1.3.1. *The 2-Segal case.* We start by treating the stable pointed case, namely the case of $\mathcal{E}$ being one of the structures from Section 1.2.1.

THEOREM 1.17. *If $\mathcal{E}$ is any of the structures discussed in Section 1.2.1, the corresponding $S_\bullet$-construction $S_\bullet\mathcal{E}$ is a 2-Segal space.*

This is shown in [**GCKT18**, Theorem 10.10] for $\mathcal{E}$ being an abelian category, in [**DK19**, §2.4] for $\mathcal{E}$ being a (proto-)exact category, in [**BOO$^+$18**, Theorem 4.8] for $\mathcal{E}$ being a stable augmented double category, in [**BOO$^+$21a**, Theorem 5.1] for $\mathcal{E}$ being a stable augmented double Segal space, in [**CS24**, Theorem B] for $\mathcal{E}$ being an isostable squares category.

We briefly illustrate how to treat one of the lowest-dimensional 2-Segal maps. The following sketch can be interpreted in any of the settings discussed in Section 1.2.1, provided one considers the corresponding $S_\bullet$-construction, uses the properties of stability and augmentation of the given context, and interprets the pictures appropriately.

IDEA. The 2-Segal map

$$(d_2, d_0)\colon S_3(\mathcal{E}) \to S_2(\mathcal{E}) \times_{S_1(\mathcal{E})} S_2(\mathcal{E})$$

can be depicted as

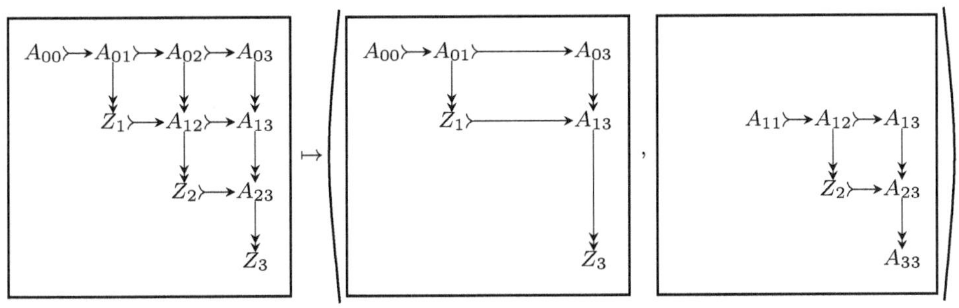

Using the structure of $\mathcal{E}$ and properties of stability and augmentation, one can consider the assignment

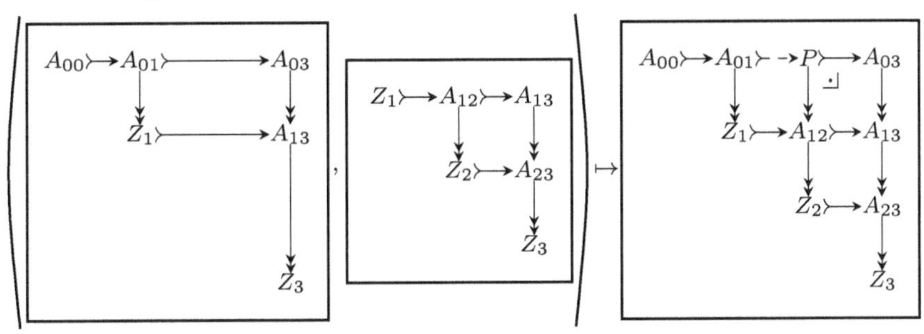

and verify that it defines an inverse equivalence of groupoids

$$S_2(\mathcal{E}) \times_{S_1(\mathcal{E})} S_2(\mathcal{E}) \xrightarrow{\simeq} S_3(\mathcal{E}).$$

The other map

$$(d_1, d_3)\colon S_3(\mathcal{E}) \to S_2(\mathcal{E}) \times_{S_1(\mathcal{E})} S_2(\mathcal{E})$$

can be treated analogously. □

REMARK 1.18. One can also see that $S_\bullet(\mathcal{E})$ is a 2-Segal space by observing that its edgewise subdivision $\operatorname{esd}(S_\bullet(\mathcal{E}))$ is a Segal space (which is closely related to Quillen's Q-construction $Q(\mathcal{E})$ from [**Qui73**, §I.2]), hence by [**BOO**$^+$**20**, Theorem 2.11] the $S_\bullet$-construction $S_\bullet(\mathcal{E})$ is a 2-Segal space.

It was shown in [**BOO**$^+$**21a**] (and [**BOO**$^+$**18**] for the discrete version of the statement, see also [**Rov24**] in this volume) that in fact all 2-Segal spaces can be obtained through the $S_\bullet$-construction from an augmented stable double Segal space:

THEOREM 1.19. *The $S_\bullet$-construction is an equivalence between stable augmented double Segal spaces and 2-Segal spaces.*

1.3.2. *The lower 2-Segal case.* We now briefly discuss one of the pointed semi-stable cases, namely the case of $\mathcal{W}$ being one of those from Section 1.2.2. While one half of the proof of Theorem 1.17 (and of the argument we sketched in Section 1.2.2) can be reproduced at this further level of generality, the other half cannot.

In fact, lacking one half of the stability conditions, the $S_\bullet$-construction from Section 1.2.2 does not generally define a 2-Segal space. To see this, let us consider the following explicit example (obtained by adjusting the one from [**BOO$^+$20**, Example 4.3]) in the case of the $S_\bullet$-construction of a Waldhausen category $\mathcal{W}$.

REMARK 1.20. Let $\mathcal{W}$ denote the Waldhausen category of finite pointed CW-complexes and pointed cellular maps, with cellular embeddings as cofibrations and homotopy equivalences as weak equivalences (a special case of [**Wal85**, §1.1–1.2] with X being a point). Then we can see that the 2-Segal map

$$(d_2, d_0)\colon S_3(\mathcal{W}) \to S_2(\mathcal{W}) \times_{S_1(\mathcal{W})} S_2(\mathcal{W})$$

is not injective on connected components, hence in particular it is not an equivalence. Indeed, if P is a finite 2-dimensional CW-complex which is not contractible but whose suspension is contractible (for example, the classifying space of the perfect group from [**Hat02**, Example 2.38]), the following two objects of $S_3(\mathcal{W})$ live in different connected components but map to the same connected component in $S_2(\mathcal{W}) \times_{S_1(\mathcal{W})} S_2(\mathcal{W})$ under the map (d_2, d_0):

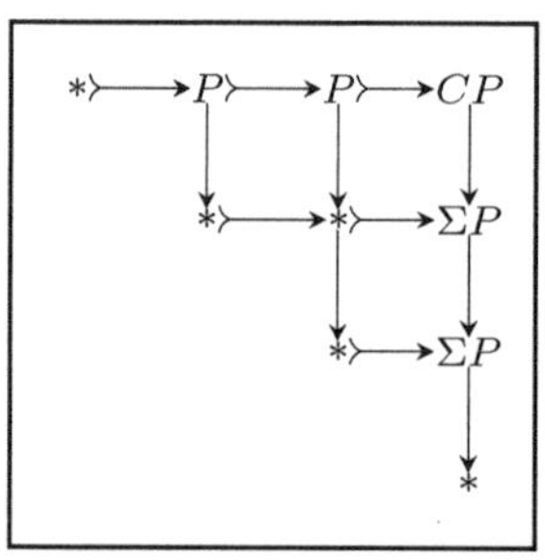

and

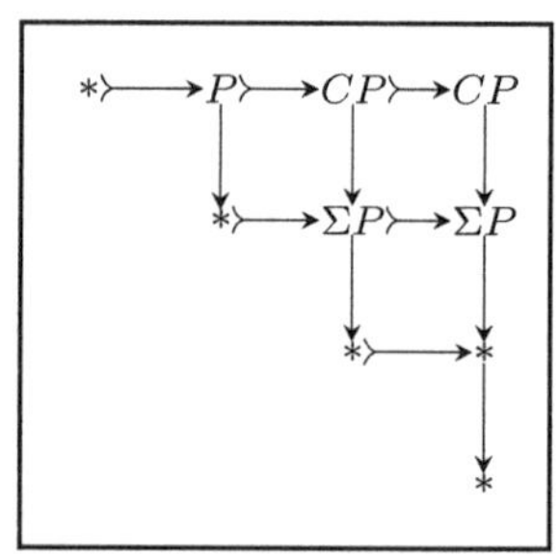

In particular, the simplicial space $S_\bullet(\mathcal{W})$ is *not* a 2-Segal space.

However, one half of the proof of Theorem 1.17 remains valid. It is shown in [**Car24**, §7, §8] that:

THEOREM 1.21. *If $\mathcal{W}$ is a Waldhausen category, the $S_\bullet$-construction $S_\bullet\mathcal{W}$ is a lower 2-Segal space.*

It is possible that if $\mathcal{W}$ is any of the structures discussed in Section 1.2.2 the $S_\bullet$-construction $S_\bullet\mathcal{W}$ is a lower 2-Segal space.

2. Iterated $S_\bullet$-construction and K-theory spectrum

We discuss how to deloop K-theory through an appropriate iteration of Waldhausen's construction under appropriate assumptions.

The iteration of the $S_\bullet$-construction was originally done in Waldhausen [**Wal85**, §1.3] in the formulation of Remark 2.4, and was later explicitly described through the "all-at-once construction" of Construction 2.2 in [**BM11**, Construction 2.2].

2.1. Classical iterated $S_\bullet$-construction. We recall the classical way to deloop the K-theory space of $\mathcal{E}_R$ and more generally of an exact category $\mathcal{E}$.

REMARK 2.1. If $\mathcal{E}$ is an exact category, there is an exact category $[\mathrm{Ar}[k], \mathcal{E}]$. The objects are the exact functors from $\mathrm{Ar}[k]$ to $\mathcal{E}$, and the exact structure is inherited from that of $\mathcal{E}$ (and is essentially a special case of [**Wal85**, §1.3] and [**Pen17**, Lemma 3.15]). More precisely, the ordinary $S_\bullet$-construction can be lifted to a functor

$$\widetilde{S}_\bullet := [\mathrm{Ar}[\bullet], -] \colon \mathcal{E}x\mathcal{C}at \to \mathcal{E}x\mathcal{C}at^{\Delta^{\mathrm{op}}}.$$

Using this, and inspired by [**BM11**, Construction 2.2] (cf. also [**Pen17**, (3.6)] for the case $n = 2$), we can define the *iterated* $S_\bullet$-construction – the multisimplicial groupoid $S^{(n)}_{\bullet,\dots,\bullet}(\mathcal{E}_R)$ – as follows:

CONSTRUCTION 2.2. For $n \geq 0$ and $k_1, \dots, k_n \geq 0$, let $S^{(n)}_{k_1,\dots,k_n}(\mathcal{E})$ denote the groupoid

$$S^{(n)}_{k_1,\dots,k_n}(\mathcal{E}) := \mathrm{Hom}^{\mathrm{ex}}(\mathrm{Ar}[k_1] \times \cdots \times \mathrm{Ar}[k_n], \mathcal{E})$$

of *exact functors* from $\mathrm{Ar}[k_1] \times \cdots \times \mathrm{Ar}[k_n]$ to $\mathcal{E}$. That is, of functors from $\mathrm{Ar}[k_1] \times \cdots \times \mathrm{Ar}[k_n]$ to $\mathcal{E}$ which:

- send the objects of the form $((i_1, i_1), \dots, (i_n, i_n))$ to a zero object in $\mathcal{E}$,
- send the morphisms of the form
$$((i_1, j_1), \dots, (i_n, j_n)) \to ((i_1, j_1 + \ell_1), \dots, (i_n, j_n + \ell_n))$$
to monomorphisms in $\mathcal{E}$,
- send the morphisms of the form
$$((i_1, j_1), \dots, (i_n, j_n)) \to ((i_1 + k_1, j_1), \dots, (i_n + k_n, j_n))$$
to epimorphisms in $\mathcal{E}$,
- send the commutative squares involving two maps of the form
$$((i_1, j_1), \dots, (i_n, j_n)) \to ((i_1 + k_1, j_1), \dots, (i_n + k_n, j_n))$$
and two maps of the form
$$((i_1, j_1), \dots, (i_n, j_n)) \to ((i_1, j_1 + \ell_1), \dots, (i_n, j_n + \ell_n))$$
to bicartesian squares in $\mathcal{E}$.

So in total we get a multisimplicial groupoid:

PROPOSITION 2.3. *The assignment* $[k_1, \dots, k_n] \mapsto S^{(n)}_{k_1,\dots,k_n}(\mathcal{E})$ *defines an n-fold simplicial groupoid* $S^{(n)}_{\bullet,\dots,\bullet}(\mathcal{E})$.

One can understand $S^{(n)}_{\bullet,\dots,\bullet}$ as an analog n-fold iteration of $S_\bullet$, which justifies the name "iterated" $S_\bullet$-construction (reconciling with the original viewpoint from [**Wal85**, §1.3], [**Wei13**, §IV.8.5.5] and [**Pen17**, Corollary 3.18] for $n = 2$). Let us explain this more in detail for the case $n = 2$.

REMARK 2.4. If $\mathrm{Ob} \colon \mathcal{E}x\mathcal{C}at \to \mathcal{G}pd$ denotes the core functor, there is a commutative diagram of categories:

$$\begin{array}{ccccc}
\mathcal{E}x\mathcal{C}at & \xrightarrow{\widetilde{S}_\bullet} & \mathcal{E}x\mathcal{C}at^{\Delta^{\mathrm{op}}} & \xrightarrow{(\widetilde{S}_\bullet)_*} & (\mathcal{E}x\mathcal{C}at^{\Delta^{\mathrm{op}}})^{\Delta^{\mathrm{op}}} \\
& \searrow^{S^{(2)}_{\bullet,\bullet}} & & & \downarrow{\scriptstyle ((\mathrm{Ob})_*)_*} \\
& & & & ((\mathcal{G}pd^{\Delta^{\mathrm{op}}})^{\Delta^{\mathrm{op}}}
\end{array}$$

Indeed, given an exact category $\mathcal{E}$ and $a, b \geq 0$, there are natural bijections for $a, b \geq 0$:

$$\begin{aligned} \mathrm{Ob}((\widetilde{S}_a)_*(\widetilde{S}_b(\mathcal{E}))) &\cong \mathrm{Ob}((\widetilde{S}_a)_*([\mathrm{Ar}[b], \mathcal{E}])) \\ &\cong \mathrm{Ob}([\mathrm{Ar}[a], [\mathrm{Ar}[b], \mathcal{E}]]) \\ &\cong \mathrm{Hom}^{\mathrm{ex}}(\mathrm{Ar}[a], [\mathrm{Ar}[b], \mathcal{E}]) \\ &\cong \mathrm{Hom}^{\mathrm{ex}}(\mathrm{Ar}[a] \times \mathrm{Ar}[b], \mathcal{E}) = S^{(2)}_{a,b}(\mathcal{E}). \end{aligned}$$

Denote by $\mathrm{Real}^{(n)}\colon \mathcal{G}pd^{\Delta^{\mathrm{op}} \times \cdots \times \Delta^{\mathrm{op}}} \to \mathcal{S}^{\Delta^{\mathrm{op}} \times \cdots \times \Delta^{\mathrm{op}}} \to \mathcal{S}$ the realization of n-fold simplicial groupoids through n-fold simplicial spaces. It is a special case of [**Wal85**, §1.3] (cf. [**Wei13**, §IV.8.5.5]) that the K-theory space $K(\mathcal{E})$ can be delooped infinitely many times as follows:

THEOREM 2.5. *Given an exact category $\mathcal{E}$, for $n \geq 1$, there is an equivalence of spaces*

$$K(\mathcal{E}) \simeq \Omega^n \mathrm{Real}^{(n)}(S^{(n)}_{\bullet,\ldots,\bullet}(\mathcal{E})).$$

2.2. Generalized iterated $S_\bullet$-construction. In this last section we provide a more general version of the iterated construction that works for any Σ-space. This might perhaps allow for an enhancement of Waldhausen's proof of Theorem 2.5, delooping the $S_\bullet$-construction in more general situations.

Let $p^{(n)} = (p, \ldots, p)\colon \Sigma \to \Delta \times \cdots \times \Delta$ be given by $\sigma \mapsto (p(\sigma), \ldots, p(\sigma))$. There is an adjunction

$$\mathcal{P}^{(n)} = p^{(n),*}\colon \mathcal{S}^{\Delta^{\mathrm{op}} \times \cdots \times \Delta^{\mathrm{op}}} \leftrightarrows \mathcal{S}^{\Sigma^{\mathrm{op}}} : p^{(n)}_* = S^{(n)}_{\bullet,\ldots,\bullet}$$

between the category $\mathcal{S}^{\Delta^{\mathrm{op}} \times \cdots \times \Delta^{\mathrm{op}}}$ of n-fold simplicial spaces and the category $\mathcal{S}^{\Sigma^{\mathrm{op}}}$ of Σ-spaces.

We can understand the functor $\mathcal{P}^{(n)}$ on the representable functor $\Delta[k_1, \ldots k_n]$:

PROPOSITION 2.6. *For $n \geq 1$ and $k_1, \ldots, k_n \geq 0$, there is an isomorphism of Σ-spaces:*

$$\mathcal{P}^{(n)}\Delta[k_1, \ldots k_n] \cong \mathcal{P}\Delta[k_1] \times \cdots \times \mathcal{P}\Delta[k_n] \cong N^{\mathrm{ex}}\,\mathrm{Ar}[k_1] \times \ldots N^{\mathrm{ex}}\,\mathrm{Ar}[k_n].$$

PROOF. Let us do the case $n = 2$, the general case working similarly. For α in Σ, and $k_1, k_2 \geq 0$, there are natural isomorphisms of spaces

$$\begin{aligned} (\mathcal{P}^{(2)}\Delta[k_1, k_2])_\alpha &\cong \mathrm{Map}(\Sigma[\alpha], p^{(2),*}\Delta[k_1, k_2]) \\ &\cong \mathrm{Map}(p^{(2)}_!\Sigma[\alpha], \Delta[k_1, k_2]) \\ &\cong \mathrm{Map}(\Delta[p(\alpha), p(\alpha)], \Delta[k_1, k_2]) \\ &\cong \mathrm{Map}(\Delta[p(\alpha)], \Delta[k_1]) \times \mathrm{Map}(\Delta[p(\alpha)], \Delta[k_2]) \\ &\cong \mathrm{Map}(\Sigma[\alpha], \mathcal{P}\Delta[k_1]) \times \mathrm{Map}(\Sigma[\alpha], \mathcal{P}\Delta[k_2]) \\ &\cong (\mathcal{P}\Delta[k_1])_\alpha \times (\mathcal{P}\Delta[k_2])_\alpha, \end{aligned}$$

as desired. □

We can then describe the functor $S^{(n)}_{\bullet,\ldots,\bullet}\colon \mathcal{S}^{\Sigma^{\mathrm{op}}} \to \mathcal{S}^{\Delta^{\mathrm{op}} \times \cdots \times \Delta^{\mathrm{op}}}$ explicitly:

CONSTRUCTION 2.7. Given a Σ-space X, $n \geq 0$ and $k_1, \ldots, k_n \geq 0$, we set

$$S^{(n)}_{k_1,\ldots,k_n}(X) = \mathrm{Map}(\mathcal{P}\Delta[k_1] \times \cdots \times \mathcal{P}\Delta[k_n], X)$$

to be the space of maps of Σ-spaces $\mathcal{P}\Delta[k_1] \times \cdots \times \mathcal{P}\Delta[k_n] \to X$.

One can check that the construction above defines a multisimplicial space:

PROPOSITION 2.8. *Given a Σ-space X, the assignment*

$$[k_1, \dots, k_n] \mapsto S^{(n)}_{k_1,\dots,k_n}(X)$$

defines an n-fold simplicial space $S^{(n)}_{\bullet,\dots,\bullet}(X)$.

This construction recovers the one for exact categories through the exact nerve $N^{\mathrm{ex}}\colon \mathcal{E}x\mathcal{C}at \to \mathcal{S}^{\Sigma^{\mathrm{op}}}$ from [**BOO$^+$21b**, Definition 2.2] as follows:

REMARK 2.9. There is a commutative diagram of categories

$$\begin{array}{ccc} \mathcal{E}x\mathcal{C}at & \xrightarrow{S^{(2)}_{\bullet,\bullet}} & \mathcal{G}pd^{\Delta^{\mathrm{op}}\times\Delta^{\mathrm{op}}} \\ {\scriptstyle N^{\mathrm{ex}}}\downarrow & & \downarrow{\scriptstyle N_*} \\ \mathcal{S}^{\Sigma^{\mathrm{op}}} & \xrightarrow[S^{(2)}_{\bullet,\bullet}]{} & \mathcal{S}^{\Delta^{\mathrm{op}}\times\Delta^{\mathrm{op}}} \end{array}$$

Indeed, for $k_1, k_2 \geq 0$ and $\mathcal{E}$ an exact category, there is a natural bijection

$$\begin{aligned} S^{(2)}_{k_1,k_2}(N^{\mathrm{ex}}\mathcal{E}) &\cong \mathrm{Map}(\mathcal{P}\Delta[k_1] \times \mathcal{P}\Delta[k_2], N^{\mathrm{ex}}\mathcal{E}) \\ &\cong N\,\mathrm{Hom}^{\mathrm{ex}}(\mathrm{Ar}[k_1] \times \mathrm{Ar}[k_2], \mathcal{E}) \\ &\cong NS^{(2)}_{k_1,k_2}(\mathcal{E}) \cong (N_*S^{(2)})_{k_1,k_2}(\mathcal{E}). \end{aligned}$$

It is possible to recognize the functor $S^{(n)}_{\bullet,\dots,\bullet}$ as an appropriate iteration of the $S_\bullet$-construction in the context of Σ-spaces, by observing that the $S_\bullet$-construction can be lifted to a functor

$$\widetilde{S}_\bullet := (-)^{\mathcal{P}\Delta[\bullet]}\colon \mathcal{S}^{\Sigma^{\mathrm{op}}} \to (\mathcal{S}^{\Sigma^{\mathrm{op}}})^{\Delta^{\mathrm{op}}}.$$

and replacing the category $\mathcal{E}x\mathcal{C}at$ with the category $\mathcal{S}^{\Sigma^{\mathrm{op}}}$ in Remark 2.4:

REMARK 2.10. If $\mathrm{ev}_{0,0}\colon \mathcal{S}^{\Sigma^{\mathrm{op}}} \to \mathcal{S}$ denotes the evaluation functor at $(0,0)$, there is a commutative diagram of categories:

$$\begin{array}{ccccc} \mathcal{S}^{\Sigma^{\mathrm{op}}} & \xrightarrow{\widetilde{S}_\bullet} & (\mathcal{S}^{\Sigma^{\mathrm{op}}})^{\Delta^{\mathrm{op}}} & \xrightarrow{(\widetilde{S}_\bullet)_*} & ((\mathcal{S}^{\Sigma^{\mathrm{op}}})^{\Delta^{\mathrm{op}}})^{\Delta^{\mathrm{op}}} \\ & \searrow^{S^{(2)}_{\bullet,\bullet}} & & & \downarrow{\scriptstyle (((\mathrm{ev}_{0,0})_*)_*} \\ & & & & ((\mathcal{S}^{\Delta^{\mathrm{op}}})^{\Delta^{\mathrm{op}}} \end{array}$$

Indeed, given a stable augmented double Segal space X and $a, b \geq 0$, there are natural equivalences for $a, b \geq 0$:

$$\begin{aligned} \mathrm{ev}_{0,0}((\widetilde{S}_a)_*(\widetilde{S}_b(X))) &\cong \mathrm{ev}_{0,0}((\widetilde{S}_a)_*([\mathcal{P}\Delta[b], X])) \\ &\cong \mathrm{ev}_{0,0}([\mathcal{P}\Delta[a], [\mathcal{P}\Delta[b], X]]) \\ &\cong \mathrm{Map}(\mathcal{P}\Delta[a], [\mathcal{P}\Delta[b], X]) \\ &\cong \mathrm{Map}(\mathcal{P}\Delta[a] \times \mathcal{P}\Delta[b], X) = S^{(2)}_{a,b}(X). \end{aligned}$$

One *could* then hope for a delooping in larger generality, so that $K(X)$ would come from an Ω-spectrum:

QUESTION 2.11. *For $n \geq 1$, under which conditions on a Σ-space X does one have equivalence of spaces*

$$K(X) \simeq \Omega^n \mathrm{Real}^{(n)}(S^{(n)}_{\bullet,\dots,\bullet}(X))?$$

This type of statement was proven in [**Wal85**, §1.3] in the context of Waldhausen categories, and in [**SS21**, §7] in the context of ECGW categories, and it might hold in larger generality.

References

[Bar15] C. Barwick, *On exact ∞-categories and the theorem of the heart*, Compos. Math. **151** (2015), no. 11, 2160–2186, DOI 10.1112/S0010437X15007447. MR3427577

[Bar16] C. Barwick, *On the algebraic K-theory of higher categories*, J. Topol. **9** (2016), no. 1, 245–347, DOI 10.1112/jtopol/jtv042. MR3465850

[BLLM24] *Homotopical algebra and higher structures*, Oberwolfach Rep. **21** (2024), no. 3, 2251–2307, DOI 10.4171/owr/2024/39. Abstracts from the workshop held August 26–30, 2024; Organized by Michael Batanin, Andrey Lazarev, Muriel Livernet and Martin Markl. MR4865595

[BM11] A. J. Blumberg and M. A. Mandell, *Derived Koszul duality and involutions in the algebraic K-theory of spaces*, J. Topol. **4** (2011), no. 2, 327–342, DOI 10.1112/jtopol/jtr003. MR2805994

[BOO+18] J. E. Bergner, A. M. Osorno, V. Ozornova, M. Rovelli, and C. I. Scheimbauer, *2-Segal sets and the Waldhausen construction*, Topology Appl. **235** (2018), 445–484, DOI 10.1016/j.topol.2017.12.009. MR3760213

[BOO+20] J. E. Bergner, A. M. Osorno, V. Ozornova, M. Rovelli, and C. I. Scheimbauer, *The edgewise subdivision criterion for 2-Segal objects*, Proc. Amer. Math. Soc. **148** (2020), no. 1, 71–82, DOI 10.1090/proc/14679. MR4042831

[BOO+21a] J. E. Bergner, A. M. Osorno, V. Ozornova, M. Rovelli, and C. I. Scheimbauer, *2-Segal objects and the Waldhausen construction*, Algebr. Geom. Topol. **21** (2021), no. 3, 1267–1326, DOI 10.2140/agt.2021.21.1267. MR4299667

[BOO+21b] J. E. Bergner, A. M. Osorno, V. Ozornova, M. Rovelli, and C. I. Scheimbauer, *Comparison of Waldhausen constructions*, Ann. K-Theory **6** (2021), no. 1, 97–136, DOI 10.2140/akt.2021.6.97. MR4283095

[Car24] T. N. Carawan, *2-Segal maps associated to a category with cofibrations*, `arXiv:2405.11561v1` (2024).

[CS24] M. E. Calle and M. Sarazola, *Squares K-theory and 2-Segal spaces*, `arXiv:2409.16428v1` (2024).

[CZ22] J. A. Campbell and I. Zakharevich, *Dévissage and localization for the Grothendieck spectrum of varieties*. part A, Adv. Math. **411** (2022), no. part A, Paper No. 108710, 80, DOI 10.1016/j.aim.2022.108710. MR4512395

[DK19] T. Dyckerhoff and M. Kapranov, *Higher Segal spaces*, Lecture Notes in Mathematics, vol. 2244, Springer, Cham, 2019, DOI 10.1007/978-3-030-27124-4. MR3970975

[GCKT18] I. Gálvez-Carrillo, J. Kock, and A. Tonks, *Decomposition spaces, incidence algebras and Möbius inversion I: Basic theory*, Adv. Math. **331** (2018), 952–1015, DOI 10.1016/j.aim.2018.03.016. MR3804694

[Gra76] D. Grayson, *Higher algebraic K-theory. II (after Daniel Quillen)*, Algebraic K-theory (Proc. Conf., Northwestern Univ., Evanston, Ill., 1976), Lecture Notes in Math., Vol. 551, Springer, Berlin-New York, 1976, pp. 217–240. MR574096

[Hac24] P. Hackney, *The decomposition space perspective*, `arXiv:2409.19061v1` (2024), this volume.

[Hat02] A. Hatcher, *Algebraic topology*, Cambridge University Press, Cambridge, 2002. MR1867354

[JY21] N. Johnson and D. Yau, *2-dimensional categories*, Oxford University Press, Oxford, 2021, DOI 10.1093/oso/9780198871378.001.0001. MR4261588

[Lur06] J. Lurie, *Stable infinity categories*, `arXiv:math/0608228v5` (2006).

[Lur18] J. Lurie, *Higher algebra*, preprint available at `http://www.math.harvard.edu/~lurie/papers/HA.pdf`, retrieved in December 2024, 2018.

[Nee90] A. Neeman, *The derived category of an exact category*, J. Algebra **135** (1990), no. 2, 388–394, DOI 10.1016/0021-8693(90)90296-Z. MR1080854

[Pen17] M. D. Penney, *The universal Hall bialgebra of a double 2-Segal space*, `arXiv:1711.10194v1` (2017).

[Pog17] T. Poguntke, *Higher Segal structures in algebraic K-theory*, `arXiv:1709.06510v1` (2017).

[Qui72] D. Quillen, *On the cohomology and K-theory of the general linear groups over a finite field*, Ann. of Math. (2) **96** (1972), 552–586, DOI 10.2307/1970825. MR315016

[Qui73] D. Quillen, *Higher algebraic K-theory. I*, Algebraic K-theory, I: Higher K-theories (Proc. Conf., Battelle Memorial Inst., Seattle, Wash., 1972), Lecture Notes in Math., Vol. 341, Springer, Berlin-New York, 1973, pp. 85–147. MR338129

[Rie14] E. Riehl, *Categorical homotopy theory*, New Mathematical Monographs, vol. 24, Cambridge University Press, Cambridge, 2014, DOI 10.1017/CBO9781107261457. MR3221774

[Rov24] M. Rovelli, *The $S_\bullet$-construction as an equivalence between 2-Segal spaces and stable augmented double Segal spaces*, 2024, this volume.

[Seg74] G. Segal, *Categories and cohomology theories*, Topology **13** (1974), 293–312, DOI 10.1016/0040-9383(74)90022-6. MR353298

[SS21] M. Sarazola and B. Shapiro, *A Gillet–Waldhausen theorem for chain complexes of sets*, arXiv:2107.07701v2 (2021).

[Ste24] W. H. Stern, *The 2-Segal space perspective*, 2024, this volume.

[TT90] R. W. Thomason and T. Trobaugh, *Higher algebraic K-theory of schemes and of derived categories*, The Grothendieck Festschrift, Vol. III, Progr. Math., vol. 88, Birkhäuser Boston, Boston, MA, 1990, pp. 247–435, DOI 10.1007/978-0-8176-4576-2_10. MR1106918

[Wal78] F. Waldhausen, *Algebraic K-theory of topological spaces. I*, Algebraic and geometric topology (Proc. Sympos. Pure Math., Stanford Univ., Stanford, Calif., 1976), Proc. Sympos. Pure Math., XXXII, Amer. Math. Soc., Providence, RI, 1978, pp. 35–60. MR520492

[Wal85] F. Waldhausen, *Algebraic K-theory of spaces*, Algebraic and geometric topology (New Brunswick, N.J., 1983), Lecture Notes in Math., vol. 1126, Springer, Berlin, 1985, pp. 318–419, DOI 10.1007/BFb0074449. MR802796

[Wei94] C. A. Weibel, *An introduction to homological algebra*, Cambridge Studies in Advanced Mathematics, vol. 38, Cambridge University Press, Cambridge, 1994, DOI 10.1017/CBO9781139644136. MR1269324

[Wei05] C. Weibel, *Algebraic K-theory of rings of integers in local and global fields*, Handbook of K-theory. Vol. 1, 2, Springer, Berlin, 2005, pp. 139–190, DOI 10.1007/3-540-27855-9_5. MR2181823

[Wei13] C. A. Weibel, *The K-book*, Graduate Studies in Mathematics, vol. 145, American Mathematical Society, Providence, RI, 2013. An introduction to algebraic K-theory, DOI 10.1090/gsm/145. MR3076731

Max Planck Institute for Mathematics, Bonn, Germany
Email address: viktoriya.ozornova@mpim-bonn.mpg.de

Department of Mathematics and Statistics, University of Massachusetts Amherst, Amherst
Email address: mrovelli@umass.edu

Contemporary Mathematics
Volume **838**, 2026
https://doi.org/10.1090/conm/838/16744

The $S_\bullet$-construction as an equivalence between 2-Segal spaces and stable augmented double Segal spaces

Martina Rovelli

ABSTRACT. This contribution is a streamlined exposition of a result by Bergner–Osorno–Ozornova–Rovelli–Scheimbauer. We discuss how a generalized version of Waldhausen's $S_\bullet$-construction describes a correspondence between 2-Segal spaces and certain double Segal spaces that satisfy further conditions of stability and augmentation.

Introduction

As the equivalent notions of a *2-Segal space* [**DK19**] and of a *decomposition space* [**GCKT18**] arose, the output of various flavors of *Waldhausen's $S_\bullet$-construction* [**Wal85**] was identified as a prominent example. For instance, this was proven for the $S_\bullet$-construction of abelian categories in [**GCKT18**, §10], of (proto-)exact categories in [**DK19**, §2.4], and of exact and stable ∞-categories in [**DK19**, §7]; see [**OR24**] in this volume for a more comprehensive survey.

Two natural questions arose:

(1) Do all examples of 2-Segal spaces arise as the $S_\bullet$-construction of a suitably generalized input?
(2) Does the $S_\bullet$-construction remember all the structure of the original input?

In other words, it is natural to wonder to which degree the $S_\bullet$-construction is surjective and to which degree it is injective.

These questions were addressed by one of the teams of *Women in Topology II* – Bergner–Osorno–Ozornova–Rovelli–Scheimbauer – in [**BOO**$^+$**18**, **BOO**$^+$**21a**]. The team identified an appropriate class of double Segal spaces with further structure, called *stable augmented double Segal spaces*, to which the $S_\bullet$-construction can be generalized, as well as an inverse construction: the *path construction* $\mathcal{P}$ (closely related to the total décalage construction from [**Ill71**]).

In this note, we outline the main ingredients in the proof of a soft version of the correspondence between 2-Segal spaces and stable augmented double Segal spaces, which we state as Theorem 5.4:

THEOREM (Soft version). *The path construction $\mathcal{P}$ and (a variant of) the $S_\bullet$-construction define inverse bijections*

$$\mathcal{P}\colon 2\mathrm{SegSp}/_{\simeq} \;\cong\; \mathrm{saDblSegSp}/_{\simeq}\colon S_\bullet \circ \widetilde{(-)}$$

We acknowledge the support from the National Science Foundation under Grant No. DMS-2203915.

between 2-Segal spaces up to equivalence and stable augmented double Segal spaces up to equivalence.

We also state as Theorem 5.5 the stronger version of the result, which studies the correspondence between 2-Segal spaces and stable augmented double Segal spaces at the level of ∞-categories:

THEOREM (Strong version). *The path construction $\mathcal{P}$ and (a variant of) the $S_\bullet$-construction define inverse equivalences of ∞-categories*

$$\boldsymbol{\mathcal{P}}\colon \mathbf{2SegSp} \simeq \mathbf{saDblSegSp}\colon \boldsymbol{S_\bullet} \circ \widetilde{(-)}$$

between the ∞-category of 2-Segal spaces and the ∞-category of stable augmented double Segal spaces.

This note is written with the intention of being as elementary as possible, focusing on conveying intuition rather than describing the most general framework. The information presented is accurate; however, certain technical details have been excluded to favor exposition, referring the reader to the original source(s) for the complete proofs.

Acknowledgments. We are grateful to Julie Bergner, Joachim Kock, Mark Penney, and Maru Sarazola for organizing a wonderful workshop at the Banff International Research Station: *Higher Segal Spaces and their Applications to Algebraic K-Theory, Hall Algebras, and Combinatorics* (24w5266, originally 20w5173). We are also thankful to Viktoriya Ozornova for assistance in writing this note, to Claudia Scheimbauer for creating and letting us use the picture macros from our previous papers, and to Joachim Kock and the referee for useful feedback.

1. 2-Segal spaces

In this section we recall the notion of a 2-*Segal space*, which was introduced in [**DK19**] and [**GCKT18**] independently, and note the properties which will be used in the remainder of the paper. We also discuss various structures one can consider on the class of 2-Segal spaces.

Let Δ denote the simplex category, and $\mathcal{S}$ the category of simplicial sets, which we refer to as *spaces*. For the notion of homotopy pullback with respect to the Kan–Quillen homotopy theory of simplicial sets we refer the reader to e.g. [**Hov99**].

NOTATION 1.1. A *simplicial space* is a functor $X\colon \Delta^{\mathrm{op}} \to \mathcal{S}$. Explicitly, it consists of a space X_n of n-*simplices* for $n \geq 0$, together with the following structure maps:

- the i-th *face map* $d_i\colon X_n \to X_{n-1}$ for $n > 0$ and $0 \leq i \leq n$; and
- the i-th *degeneracy map* $s_i\colon X_n \to X_{n+1}$ for $n \geq 0$ and $0 \leq i \leq n$.

These structure maps are subject to the usual simplicial identities.

In this note we take the following as the definition of a 2-*Segal space*. It was shown in [**BOO$^+$21a**, Proposition 1.18] that it is equivalent to the original definition from [**DK19**, Definition 2.3.1].

DEFINITION 1.2. A simplicial space X is a 2-*Segal space* if, for all $n \geq 2$, the 2-*Segal maps* induced respectively by the commutative squares of spaces

$$\begin{array}{ccc} X_{n+1} & \longrightarrow & X_{\{0,2,\ldots,n+1\}} \\ \downarrow & & \downarrow \\ X_{\{0,1,2\}} & \longrightarrow & X_{\{0,2\}} \end{array} \quad \text{and} \quad \begin{array}{ccc} X_{n+1} & \longrightarrow & X_{\{0,\ldots,n-2,n\}} \\ \downarrow & & \downarrow \\ X_{\{n-2,n-1,n\}} & \longrightarrow & X_{\{n-2,n\}} \end{array}$$

are equivalences of spaces

$$X_{n+1} \xrightarrow{\simeq} X_{\{0,1,2\}} \times^h_{X_{\{0,2\}}} X_{\{0,2,\ldots,n+1\}}$$

$$\text{and } X_{n+1} \xrightarrow{\simeq} X_{\{n-2,n-1,n\}} \times^h_{X_{\{n-2,n\}}} X_{\{0,\ldots,n-2,n\}}.$$

REMARK 1.3. It was shown in [**FGK$^+$21**] that every 2-Segal space X is *unital* in the sense of [**DK19**, Definition 2.5.2] (equivalently, every 2-Segal space X is a *decomposition space* in the sense of [**GCKT18**, Definition 3.1]). This guarantees in particular that there are equivalences of spaces

$$(s_1, d_0)\colon X_1 \xrightarrow{\simeq} X_2 \times^h_{X_1} X_0 \quad \text{and} \quad (s_0, d_1)\colon X_1 \xrightarrow{\simeq} X_2 \times^h_{X_1} X_0$$

induced, respectively, by the commutative squares

$$\begin{array}{ccc} X_1 & \xrightarrow{d_0} & X_0 \\ {\scriptstyle s_1}\downarrow & & \downarrow{\scriptstyle s_0} \\ X_2 & \xrightarrow[d_0]{} & X_1 \end{array} \quad \text{and} \quad \begin{array}{ccc} X_1 & \xrightarrow{d_1} & X_0 \\ {\scriptstyle s_0}\downarrow & & \downarrow{\scriptstyle s_0} \\ X_2 & \xrightarrow[d_2]{} & X_1. \end{array}$$

The main goal is to explain that there is a correspondence between the notion of a 2-Segal space and that of a certain double Segal space, from which a variety of examples will arise.

To make this correspondence precise, we shall identify the appropriate categorical structure that 2-Segal spaces assemble into. As a first approximation, one can consider an ordinary category:

DEFINITION 1.4. Let 2SegSp be the full subcategory of the category of simplicial spaces spanned by the 2-Segal spaces.

To encode the homotopy theory of 2-Segal spaces, we need a notion of equivalence:

DEFINITION 1.5. Let X and Y be 2-Segal spaces. A map of simplicial spaces $f\colon X \to Y$ is an *equivalence* of 2-Segal spaces if for all $n \geq 0$ it defines an equivalence of spaces $f_n\colon X_n \xrightarrow{\simeq} Y_n$.

With this notion of equivalence between 2-Segal spaces, one can consider the collection of equivalence classes of 2-Segal spaces:

REMARK 1.6. Consider the relation on the class of 2-Segal spaces given by declaring that X is related to Y if there exists an equivalence $f\colon X \xrightarrow{\simeq} Y$ of 2-Segal spaces. It follows from the definitions that this relation is reflexive and transitive, though a priori possibly not symmetric. Nevertheless, one can consider the set $\mathrm{2SegSp}/_{\simeq}$ of 2-Segal spaces up to the equivalence relation it generates.

We will see in Theorem 5.4 that the set $\mathrm{2SegSp}/_{\simeq}$ of equivalence classes can be used to phrase an elementary version of the classification theorem. However, to state the most comprehensive version of the result, which we will state as Theorem 5.5, one should consider the ∞-category of 2-Segal spaces, in which equivalences are formally inverted.

One way to incarnate the notion of an *∞-category* is in the form of a simplicial category. Recall that a *simplicial category* is a category enriched over the category of simplicial sets. It consists in particular of a collection of objects, and a mapping space between any two objects, together with identity and composition simplicial operators that satisfy associativity and unitality axioms.

Given a category with a class of equivalences, one can construct its simplicial localization in the form of a simplicial category. Various implementations of this procedure are available, and one of them is the *hammock localization* from [**DK80a**] (cf. also [**Hin16**]).

By simplicially localizing the category of 2-Segal spaces at the class of equivalences of 2-Segal spaces, we obtain a simplicial category in which all equivalences of 2-Segal spaces are universally inverted in an appropriate weak sense:

DEFINITION 1.7. Denote by **2SegSp** the simplicial localization of the category of 2-Segal spaces at the equivalences of 2-Segal spaces. It is a simplicial category in which:

- The objects are the 2-Segal spaces;
- Given 2-Segal spaces X and Y, the mapping space of maps from a X to Y is given by the space of hammocks from X to Y.
- Composition and identity operators are given by concatenating hammocks, and taking the trivial hammock, respectively.

2. Stable augmented double Segal spaces

In this section we discuss the notion of a *stable augmented double Segal space* from [**BOO$^+$21a**], and note the properties which will be used in the remainder of the paper. We also discuss various structures one can consider on the class of stable augmented double Segal spaces.

Denote by Σ the category obtained by freely adding a terminal object to $\Delta\times\Delta$, which could be visualized as:

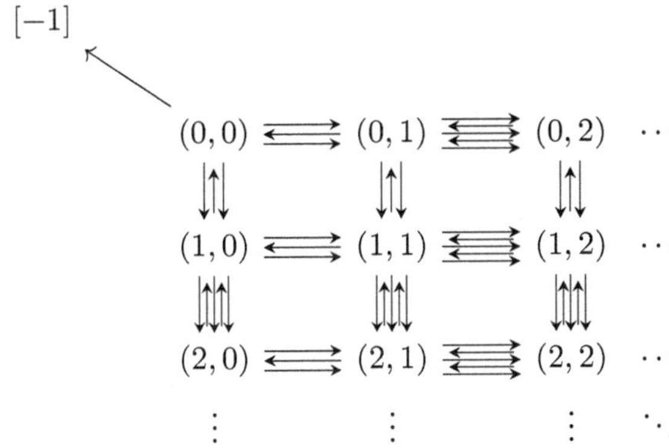

NOTATION 2.1. A *preaugmented bisimplicial space* is a functor $\mathcal{D}\colon \Sigma^{\mathrm{op}} \to \mathcal{S}$. Precisely, it consists of a space $\mathcal{D}_{a,b}$ of *(a,b)-bisimplices* for $a,b\geq 0$ and an *augmentation space* $\mathcal{D}_{-1}$, together with the following structure maps (for $0\leq i\leq n$ in each case):

- the *vertical i-th face map* $d_i^v\colon \mathcal{D}_{n,\ell}\to\mathcal{D}_{n-1,\ell}$ for $\ell\geq 0$ and $n>0$,
- the *vertical i-th degeneracy map* $s_i^v\colon \mathcal{D}_{n,\ell}\to\mathcal{D}_{n+1,\ell}$ for $\ell\geq 0$ and $n\geq 0$,
- the *horizontal i-th face map* $d_i^h\colon \mathcal{D}_{\ell,n}\to\mathcal{D}_{\ell,n-1}$ for $\ell\geq 0$ and $n>0$,
- the *horizontal i-th degeneracy map* $s_i^h\colon \mathcal{D}_{\ell,n}\to\mathcal{D}_{\ell,n+1}$ for $\ell\geq 0$ and $n\geq 0$,
- the *augmentation map* $\varepsilon\colon \mathcal{D}_{-1}\to\mathcal{D}_{0,0}$.

Face and degeneracy maps are subject to the usual bisimplicial identities, and the augmentation is not subject to any relation. For further reference, we fix the notation for:

- the *vertical iterated top face map* $d_\top^v := d_n^v\circ d_{n-1}^v\circ\cdots\circ d_{k+1}^v\colon \mathcal{D}_{n,\ell}\to\mathcal{D}_{k,\ell}$ for $\ell\geq 0$ and $n\geq k\geq 0$;
- the *horizontal iterated top face map* $d_\top^h := d_n^h\circ d_{n-1}^h\circ\cdots\circ d_{k+1}^h\colon \mathcal{D}_{\ell,n}\to\mathcal{D}_{\ell,k}$ for $\ell\geq 0$ and $n\geq k\geq 0$;
- the *vertical iterated bottom face map* $d_\perp^v := d_0^v\circ d_0^v\circ\cdots\circ d_0^v\colon \mathcal{D}_{n,\ell}\to\mathcal{D}_{k,\ell}$ for $\ell\geq 0$ and $n\geq k\geq 0$;
- the *horizontal iterated bottom face map* $d_\perp^h := d_0^h\circ d_0^h\circ\cdots\circ d_0^h\colon \mathcal{D}_{\ell,n}\to\mathcal{D}_{\ell,k}$ for $\ell\geq 0$ and $n\geq k\geq 0$;

REMARK 2.2. If $\mathcal{D}$ is a preaugmented bisimplicial space, the spaces in lowest degree are interpreted as follows.

- $\mathcal{D}_{0,0}$ is a space of objects, which we generically depict as

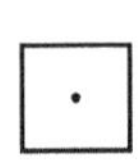

- $\mathcal{D}_{-1}$ is a subspace of distinguished objects (cf. Proposition 2.4), which we generically depict as

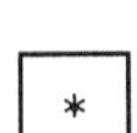

- $\mathcal{D}_{1,0}$ and $\mathcal{D}_{0,1}$ are, respectively, a space of vertical morphisms and a space of horizontal morphisms, which we generically depict as

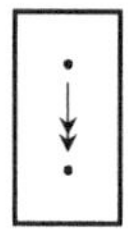

and

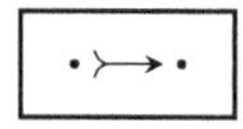

- $\mathcal{D}_{1,1}$ is a space of squares, which we picture as

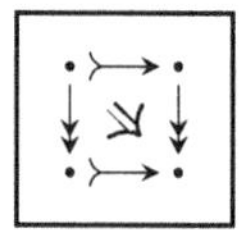

- $\mathcal{D}_{1,2}$ and $\mathcal{D}_{2,1}$ are, respectively, spaces of horizontally composable pairs of squares with specified composites and vertically composable pairs of squares with specified composites, which we picture as

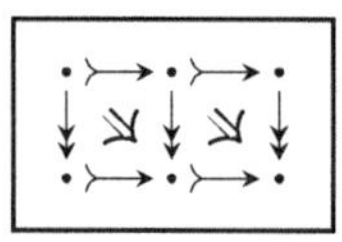

and

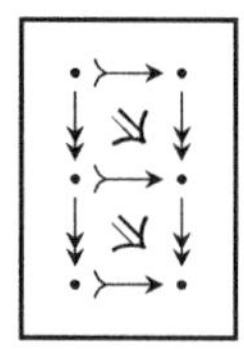

Under this interpretation, we think of:

- the augmentation map $\varepsilon\colon \mathcal{D}_{-1} \to \mathcal{D}_{0,0}$ as a (homotopy-)inclusion of $\mathcal{D}_{-1}$ into $\mathcal{D}_{0,0}$;
- the face map $d_1^v\colon \mathcal{D}_{1,0} \to \mathcal{D}_{0,0}$ (resp. $d_0^v\colon \mathcal{D}_{1,0} \to \mathcal{D}_{0,0}$) as the source (resp. target) map for vertical morphisms;
- the face map $d_1^h\colon \mathcal{D}_{0,1} \to \mathcal{D}_{0,0}$ (resp. $d_0^h\colon \mathcal{D}_{0,1} \to \mathcal{D}_{0,0}$) as the source (resp. target) map for horizontal morphisms;
- the face map $d_1^v\colon \mathcal{D}_{1,1} \to \mathcal{D}_{0,1}$ (resp. $d_0^v\colon \mathcal{D}_{1,1} \to \mathcal{D}_{0,1}$) as the horizontal source (resp. horizontal target) map for squares;
- the face map $d_1^h\colon \mathcal{D}_{1,1} \to \mathcal{D}_{1,0}$ (resp. $d_0^h\colon \mathcal{D}_{1,1} \to \mathcal{D}_{1,0}$) as the vertical source (resp. vertical target) map for squares;
- the degeneracy map $s_0^v\colon \mathcal{D}_{0,0} \to \mathcal{D}_{1,0}$ (resp. $s_0^h\colon \mathcal{D}_{0,0} \to \mathcal{D}_{0,1}$) as the vertical identity map for objects (resp. horizontal identity map for objects).

In order to enforce the desired behavior from the various pieces of the structure, we impose the following further conditions:

Definition 2.3. A preaugmented bisimplicial space $\mathcal{D}$ is a *stable augmented double Segal space* if the following hold:

(1) *Double Segal condition*: For all $k, \ell, m \geq 0$, the maps induced, respectively, by the commutative squares of spaces

$$\begin{array}{ccc} \mathcal{D}_{k+\ell,m} & \xrightarrow{d_\perp^v} & \mathcal{D}_{k,m} \\ {\scriptstyle d_\top^v}\downarrow & & \downarrow{\scriptstyle d_\top^v} \\ \mathcal{D}_{\ell,m} & \xrightarrow[d_\perp^v]{} & \mathcal{D}_{0,m} \end{array} \quad \text{and} \quad \begin{array}{ccc} \mathcal{D}_{m,k+\ell} & \xrightarrow{d_\top^h} & \mathcal{D}_{m,k} \\ {\scriptstyle d_\perp^h}\downarrow & & \downarrow{\scriptstyle d_\perp^h} \\ \mathcal{D}_{m,\ell} & \xrightarrow[d_\top^h]{} & \mathcal{D}_{m,0} \end{array}$$

are equivalences of spaces

$$(d^v_\top, d^v_\bot)\colon \mathcal{D}_{k+\ell,m} \xrightarrow{\simeq} \mathcal{D}_{k,m} \times^h_{\mathcal{D}_{0,m}} \mathcal{D}_{\ell,m}$$

$$\text{and} \quad (d^h_\top, d^h_\bot)\colon \mathcal{D}_{m,k+\ell} \xrightarrow{\simeq} \mathcal{D}_{m,k} \times^h_{\mathcal{D}_{m,0}} \mathcal{D}_{m,\ell}.$$

(2) *Stability*: The maps induced, respectively, by the commutative squares of spaces

$$\begin{array}{ccc} \mathcal{D}_{1,1} & \xrightarrow{d^v_1} & \mathcal{D}_{0,1} \\ {\scriptstyle d^h_1}\downarrow & & \downarrow{\scriptstyle d^h_1} \\ \mathcal{D}_{1,0} & \xrightarrow[d^h_1]{} & \mathcal{D}_{0,0} \end{array} \quad \text{and} \quad \begin{array}{ccc} \mathcal{D}_{1,1} & \xrightarrow{d^v_0} & \mathcal{D}_{1,0} \\ {\scriptstyle d^h_0}\downarrow & & \downarrow{\scriptstyle d^h_0} \\ \mathcal{D}_{0,1} & \xrightarrow[d^v_0]{} & \mathcal{D}_{0,0} \end{array}$$

are equivalences of spaces

$$(d^h_1, d^v_1)\colon \mathcal{D}_{1,1} \xrightarrow{\simeq} \mathcal{D}_{1,0} \times^h_{\mathcal{D}_{0,0}} \mathcal{D}_{0,1} \quad \text{and} \quad (d^v_0, d^h_0)\colon \mathcal{D}_{1,1} \xrightarrow{\simeq} \mathcal{D}_{0,1} \times^h_{\mathcal{D}_{0,0}} \mathcal{D}_{1,0}.$$

(3) *Augmentation*: The maps d^v_1 and d^h_0 induce, respectively, equivalences of spaces

$$d^v_1\colon \mathcal{D}_{1,0} \times^h_{\mathcal{D}_{0,0}} \mathcal{D}_{-1} \xrightarrow{\simeq} \mathcal{D}_{0,0} \quad \text{and} \quad d^h_0\colon \mathcal{D}_{-1} \times^h_{\mathcal{D}_{0,0}} \mathcal{D}_{0,1} \xrightarrow{\simeq} \mathcal{D}_{0,0},$$

where the homotopy pullbacks are given by

$$\begin{array}{ccc} \mathcal{D}_{1,0} \times^h_{\mathcal{D}_{0,0}} \mathcal{D}_{-1} & \longrightarrow & \mathcal{D}_{1,0} \\ \downarrow & \lrcorner & \downarrow{\scriptstyle d^v_0} \\ \mathcal{D}_{-1} & \xrightarrow[\varepsilon]{} & \mathcal{D}_{0,0} \end{array} \quad \text{and} \quad \begin{array}{ccc} \mathcal{D}_{0,1} \times^h_{\mathcal{D}_{0,0}} \mathcal{D}_{-1} & \longrightarrow & \mathcal{D}_{0,1} \\ \downarrow & \lrcorner & \downarrow{\scriptstyle d^h_1} \\ \mathcal{D}_{-1} & \xrightarrow[\varepsilon]{} & \mathcal{D}_{0,0}. \end{array}$$

While the structure map $\varepsilon\colon \mathcal{D}_{-1} \to \mathcal{D}_{0,0}$ may not be injective, if $\mathcal{D}$ is a stable augmented double Segal spaces it is at least a retract of $\mathcal{D}_{0,0}$ up to homotopy, as now detailed:

PROPOSITION 2.4. *If $\mathcal{D}$ is a stable augmented double Segal space, the augmentation map factors as a retract map followed by an equivalence of spaces, as follows:*

$$\begin{array}{ccc} \mathcal{D}_{-1} & \xrightarrow{\varepsilon} & \mathcal{D}_{0,0}. \\ & \searrow & \uparrow{\scriptstyle \simeq} \\ & & \mathcal{D}_{0,1} \times^h_{\mathcal{D}_{0,0}} \mathcal{D}_{-1} \end{array}$$

PROOF. Consider the homotopy pullback

$$\begin{array}{ccc} \mathcal{D}_{1,0} \times^h_{\mathcal{D}_{0,0}} \mathcal{D}_{-1} & \xrightarrow{\pi_1} & \mathcal{D}_{1,0} \\ \downarrow & \lrcorner & \downarrow{\scriptstyle d^v_0} \\ \mathcal{D}_{-1} & \xrightarrow[\varepsilon]{} & \mathcal{D}_{0,0}. \end{array}$$

We observe that there is a commutative diagram of spaces

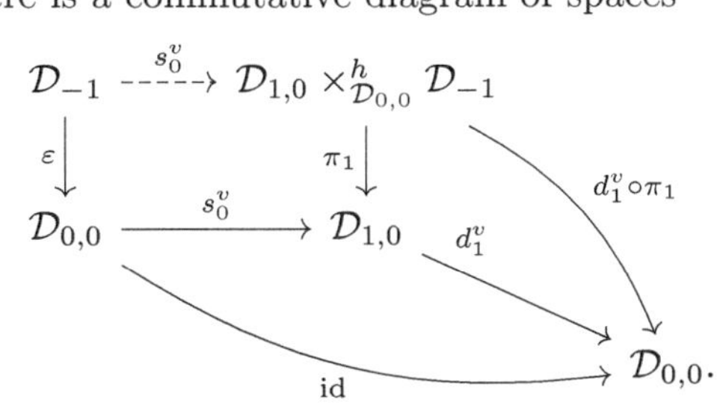

Here,

- the top dashed map $s_0^v \colon \mathcal{D}_{-1} \to \mathcal{D}_{0,1} \times^h_{\mathcal{D}_{0,0}} \mathcal{D}_{-1}$ is a retract because it admits the retraction $d_0^v \colon \mathcal{D}_{0,1} \times^h_{\mathcal{D}_{0,0}} \mathcal{D}_{-1} \to \mathcal{D}_{-1}$,
- the tilted map $d_1^v \circ \pi_1 \colon \mathcal{D}_{1,0} \times^h_{\mathcal{D}_{0,0}} \mathcal{D}_{-1} \xrightarrow{\simeq} \mathcal{D}_{0,0}$ is an equivalence of spaces because of the augmentation condition, and
- the total composite is the augmentation $\varepsilon \colon \mathcal{D}_{-1} \to \mathcal{D}_{0,0}$.

The claim then follows. □

REMARK 2.5. Relying on the intuition laid out in Remark 2.2, the axioms are designed to encode, respectively, the following properties:

(1) The double Segal condition demands that the underlying bisimplicial space $\mathcal{D}|_{\Delta\times\Delta}$ of $\mathcal{D}$ is a double Segal space in the usual sense. This encodes the fact that horizontal morphisms can be composed, vertical morphisms can be composed, and squares can be composed horizontally and vertically, the composition law is associative and unital in an appropriate sense.

(2) The stability conditions demand that each square is determined in an appropriate sense by the span contained in its boundary and also by the cospan contained in its boundary. Indeed, the stability conditions give equivalences of spaces

$$(d_1^v, d_1^h) \colon \mathcal{D}_{1,1} \xrightarrow{\simeq} \mathcal{D}_{1,0} \times^h_{\mathcal{D}_{0,0}} \mathcal{D}_{0,1} \quad \text{and} \quad (d_0^h, d_0^v) \colon \mathcal{D}_{1,1} \xrightarrow{\simeq} \mathcal{D}_{0,1} \times^h_{\mathcal{D}_{0,0}} \mathcal{D}_{1,0},$$

which can be depicted as

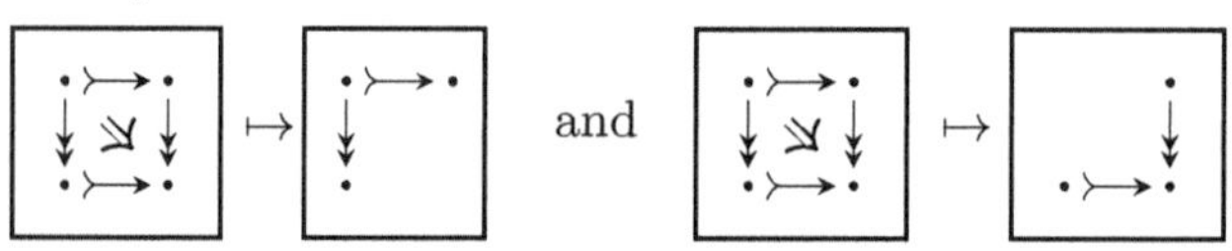

(3) The augmentation condition essentially demands that every object has a unique vertical morphism to an object in the augmentation space, and every object has a unique horizontal morphism from an object in the augmentation space. Indeed, the two augmentation conditions give equivalences of spaces

$$d_1^v \colon \mathcal{D}_{1,0} \times^h_{\mathcal{D}_{0,0}} \mathcal{D}_{-1} \xrightarrow{\simeq} \mathcal{D}_{0,0} \quad \text{and} \quad d_0^h \colon \mathcal{D}_{-1} \times^h_{\mathcal{D}_{0,0}} \mathcal{D}_{0,1} \xrightarrow{\simeq} \mathcal{D}_{0,0},$$

which can be depicted as

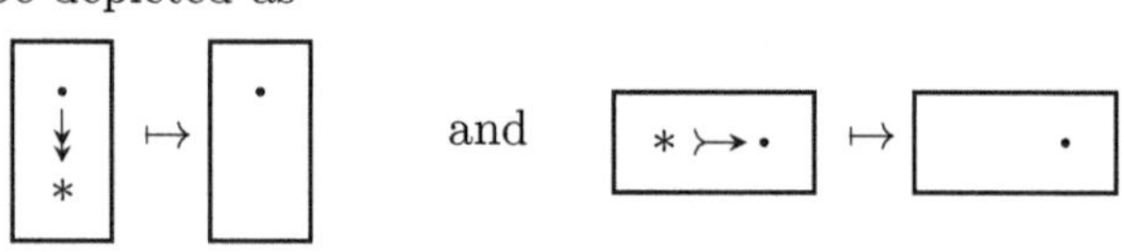

The definition is general enough to recover a variety of examples of interest, typically through a nerve construction. Let us discuss the construction from [**BOO$^+$21b**, §2] in the case where $\mathcal{E}$ is the category of abelian groups, or more generally an *abelian category* (in the sense of [**ML98**]), or more generally an *exact category* (in the sense of Quillen [**Qui73**, §2]). When thinking of an abelian category as an exact category, all monos are *admissible* and all epis are *admissible.*

CONSTRUCTION 2.6. Given an exact category $\mathcal{E}$, the preaugmented bisimplicial space $N^{\mathrm{ex}}\mathcal{E}$ is defined as follows. The space $N^{\mathrm{ex}}_{-1}\mathcal{E}$ is set to be the nerve of the groupoid of zero objects in $\mathcal{E}$, and for $a, b \geq 0$ we set $N^{\mathrm{ex}}_{a,b}\mathcal{E}$ to be the nerve of the groupoid of functors $[a] \times [b] \to \mathcal{E}$ such that

- all the restrictions $[0] \times [1] \to \mathcal{E}$ are admissible monomorphisms in $\mathcal{E}$,
- all the restrictions $[1] \times [0] \to \mathcal{E}$ are admissible epimorphisms in $\mathcal{E}$,
- all the restrictions $[1] \times [1] \to \mathcal{E}$ are bicartesian squares in $\mathcal{E}$.

The structure maps are the obvious ones.

PROPOSITION 2.7 ([**BOO⁺21b**, §2]). *Given an exact category $\mathcal{E}$, the preaugmented bisimplicial space $N^{\mathrm{ex}}\mathcal{E}$ is a stable augmented double Segal space.*

PROOF IDEA. The Segal maps

$$(d_2^v, d_0^v)\colon N_{2,1}^{\mathrm{ex}}\mathcal{E} \to N_{1,1}^{\mathrm{ex}}\mathcal{E} \times^h_{N_{0,1}^{\mathrm{ex}}\mathcal{E}} N_{1,1}^{\mathrm{ex}}\mathcal{E}$$

$$\text{and} \quad (d_2^h, d_0^h)\colon N_{1,2}^{\mathrm{ex}}\mathcal{E} \to N_{1,1}^{\mathrm{ex}}\mathcal{E} \times^h_{N_{1,0}^{\mathrm{ex}}\mathcal{E}} N_{1,1}^{\mathrm{ex}}\mathcal{E}$$

define equivalences of spaces because the classes of pushout squares and of pullback squares in $\mathcal{E}$ are closed under pasting.

The stability maps

$$(d_1^v, d_1^h)\colon N_{1,1}^{\mathrm{ex}}\mathcal{E} \to N_{0,1}^{\mathrm{ex}}\mathcal{E} \times^h_{N_{0,0}^{\mathrm{ex}}\mathcal{E}} N_{1,0}^{\mathrm{ex}}\mathcal{E}$$

$$\text{and} \quad (d_0^h, d_0^v)\colon N_{1,1}^{\mathrm{ex}}\mathcal{E} \to N_{1,0}^{\mathrm{ex}}\mathcal{E} \times^h_{N_{0,0}^{\mathrm{ex}}\mathcal{E}} N_{0,1}^{\mathrm{ex}}\mathcal{E}$$

define equivalences of spaces because every pushout (resp. pullback) square in $\mathcal{E}$ is determined up to isomorphism by the span (resp. cospan) contained in its boundary.

The augmentation maps

$$d_1^v\colon N_{1,0}^{\mathrm{ex}}\mathcal{E} \times^h_{N_{0,0}^{\mathrm{ex}}\mathcal{E}} N_{-1}^{\mathrm{ex}}\mathcal{E} \xrightarrow{\simeq} N_{0,0}^{\mathrm{ex}}\mathcal{E}$$

$$\text{and} \quad d_0^h\colon N_{0,1}^{\mathrm{ex}}\mathcal{E} \times^h_{N_{0,0}^{\mathrm{ex}}\mathcal{E}} N_{-1}^{\mathrm{ex}}\mathcal{E} \xrightarrow{\simeq} N_{0,0}^{\mathrm{ex}}\mathcal{E}$$

define equivalences of spaces because any zero object in $\mathcal{E}$ is initial in the category of admissible monomorphisms and terminal in the category of admissible epimorphisms of $\mathcal{E}$. □

One can mimic the same nerve construction and corresponding argument in various other contexts:

REMARK 2.8. Stable augmented double Segal spaces arise via an appropriate nerve constructions in many situations:

- Given a (proto-)exact (∞-)category (in the sense of [**Bar16**, **DK19**]), a nerve construction in the form of a stable augmented double Segal space is provided in [**BOO⁺21b**, §2, §4].
- Given a stable ∞-category (in the sense of [**Lur06**]), a nerve construction resulting in a stable augmented double Segal space is provided in [**BOO⁺21b**, §3].
- Given an appropriate flavor of CGW-categories (in the sense of [**CZ22**]), a nerve construction resulting in a stable augmented double Segal space should exist, and is currently the subject of further investigation by some of the workshop participants (cf. Bergner's contribution in [**BLLM24**]).
- Warning: Given a Waldhausen category (in the sense of [**Wal85**]), the canonical choice for a nerve construction results in a preaugmented bisimplicial space which is not generally a stable augmented double Segal space. For instance, it generally only satisfies one of the stability conditions.
- Given a 2-Segal space (as recalled in Section 1), we will define a path construction which will result in a stable augmented double Segal space in Section 3.

Again, in order to state the main result we shall consider a categorical structure for the class of stable augmented double Segal spaces. As a first attempt, one can consider an ordinary category:

DEFINITION 2.9. Let saDblSegSp be the full subcategory of the category of preaugmented bisimplicial spaces spanned by the stable augmented double Segal spaces.

To encode the homotopy theory of stable augmented double Segal spaces, we need a notion of equivalence:

DEFINITION 2.10. Let $\mathcal{D}$ and $\mathcal{E}$ be stable augmented double Segal spaces. A map of preaugmented bisimplicial spaces $\varphi\colon \mathcal{D} \to \mathcal{E}$ is an *equivalence* of stable augmented double Segal spaces if for all $a, b \geq 0$ it defines an equivalence of spaces $\varphi_{a,b}\colon \mathcal{D}_{a,b} \xrightarrow{\simeq} \mathcal{E}_{a,b}$, as well as an equivalence of spaces $\varphi_{-1}\colon \mathcal{D}_{-1} \xrightarrow{\simeq} \mathcal{E}_{-1}$.

With this notion of equivalence between stable augmented double Segal spaces, one can consider the collection of equivalence classes of stable augmented double Segal spaces:

REMARK 2.11. Consider the relation on the class of stable augmented double Segal spaces given by declaring that $\mathcal{D}$ is related to $\mathcal{E}$ if there exists an equivalence $\varphi\colon \mathcal{D} \xrightarrow{\simeq} \mathcal{E}$. It follows from the definitions that this relation is reflexive and transitive, though a priori possibly not symmetric. Nevertheless, one can consider the collection saDblSegSp$/_{\simeq}$ of stable augmented double Segal spaces up to the equivalence relation $\simeq$ it generates.

By simplicially localizing the category of stable augmented double Segal spaces at the class of equivalences of stable augmented double Segal spaces, we obtain a simplicial category in which all equivalences of stable augmented double Segal spaces are universally inverted in an appropriate weak sense:

DEFINITION 2.12. Denote by **saDblSegSp** the simplicial localization of the category of stable augmented double Segal spaces at the equivalences of stable augmented double Segal spaces. It is a simplicial category in which:

- The objects are the stable augmented double Segal spaces;
- Given stable augmented double Segal spaces $\mathcal{D}$ and $\mathcal{E}$, the mapping space of maps from $\mathcal{D}$ to $\mathcal{E}$ is given by the space of hammocks from $\mathcal{D}$ to $\mathcal{E}$.
- Composition and identity operators are given by concatenating hammocks, and taking the trivial hammock, respectively.

3. The path construction

In this section we introduce the *path construction* $\mathcal{P}$ of a 2-Segal space from [**BOO**$^+$**21a**], and show it is a stable augmented double Segal space.

Let $p\colon \Sigma \to \Delta$ denote the ordinal sum given by $[-1] \mapsto [0]$ and $[a, b] \mapsto [a+1+b]$, as considered in [**BOO**$^+$**21a**, Definition 2.15]. If $\mathcal{S}$ denotes the category of spaces, there is an induced functor

$$\mathcal{P} := p^*\colon \mathcal{S}^{\Delta^{\mathrm{op}}} \to \mathcal{S}^{\Sigma^{\mathrm{op}}}$$

between the category $\mathcal{S}^{\Delta^{\mathrm{op}}}$ of simplicial spaces and the category $\mathcal{S}^{\Sigma^{\mathrm{op}}}$ of preaugmented bisimplicial spaces.

CONSTRUCTION 3.1. Given a simplicial space X, we define the *path construction* of X, which is a preaugmented bisimplicial space $\mathcal{P}X$. It consists of the spaces

$$\mathcal{P}_{-1}X := X_0 \quad \text{and} \quad \mathcal{P}_{a,b}X := X_{a+1+b} \quad \text{for } a, b \geq 0,$$

together with the following structure maps:

- the vertical i-th face map $d_i^v\colon \mathcal{P}_{n,\ell}X \to \mathcal{P}_{n-1,\ell}X$ is given by $d_i\colon X_{n+1+\ell} \to X_{n+\ell}$ for $\ell \geq 0$, $n > 0$ and $0 \leq i \leq n$,
- the horizontal i-th face map $d_i^h\colon \mathcal{P}_{\ell,n}X \to \mathcal{P}_{\ell,n-1}X$ is given by $d_{i+1+\ell}\colon X_{\ell+1+n} \to X_{\ell+n}$ for $\ell \geq 0$, $n > 0$ and $0 \leq i \leq n$,
- the vertical i-th degeneracy map $s_i^v\colon \mathcal{P}_{n,\ell}X \to \mathcal{P}_{n+1,\ell}X$ is given by $s_i\colon X_{n+1+\ell} \to X_{n+2+\ell}$ for $\ell \geq 0$ and $n \geq 0$,
- the horizontal i-th degeneracy map $s_i^h\colon \mathcal{P}_{\ell,n}X \to \mathcal{P}_{\ell,n+1}X$ is given by $s_{i+1+\ell}\colon X_{\ell+1+n} \to X_{\ell+2+n}$ for $\ell \geq 0$, $n \geq 0$ and $0 \leq i \leq n$,
- the augmentation map $\varepsilon\colon \mathcal{P}_{-1}X \to \mathcal{P}_{0,0}X$ is given by $s_0\colon X_0 \to X_1$.

To provide more intuition, let us spell out some of the basic data of $\mathcal{P}X$:

REMARK 3.2. Let X be a simplicial space.

- The space of objects $\mathcal{P}_{0,0}X$ is given by the space of 1-simplices of X.
- The augmentation space $\mathcal{P}_{-1}X$ is given by the 0-simplices of X, regarded as degenerate 1-simplices.
- The spaces of vertical and horizontal morphisms $\mathcal{P}_{1,0}X$ and $\mathcal{P}_{0,1}X$ are both given by the space of 2-simplices of X, but the structure maps act differently. Indeed, a 2-simplex σ defines at the same time a vertical morphism and a horizontal morphism which can be depicted as follows:

$$\boxed{\begin{array}{c} d_1\sigma \\ \big\downarrow \\ d_0\sigma \end{array}} \qquad \text{and} \qquad \boxed{d_2\sigma \rightarrowtail d_1\sigma}$$

- The space of squares $\mathcal{P}_{1,1}X$, and the spaces of pairs of composable vertical and horizontal morphisms $\mathcal{P}_{0,2}X$ and $\mathcal{P}_{2,0}X$ are all three given by the space of 3-simplices of X, but the structure maps act differently. Indeed, a 3-simplex τ defines at the same time a pair of composable vertical morphisms, a square, and a pair of composable horizontal morphisms which can be depicted as follows:

$$\boxed{\begin{array}{c} d_{\{0,3\}}\tau \\ \big\downarrow \\ d_{\{1,3\}}\tau \\ \big\downarrow \\ d_{\{2,3\}}\tau \end{array}} \qquad \text{and} \qquad \boxed{\begin{array}{ccc} d_{\{0,2\}}\tau & \rightarrowtail & d_{\{0,3\}}\tau \\ \big\downarrow & \searrow & \big\downarrow \\ d_{\{1,2\}}\tau & \rightarrowtail & d_{\{1,3\}}\tau \end{array}} \qquad \text{and}$$

$$\boxed{d_{\{0,1\}}\tau \rightarrowtail d_{\{0,2\}}\tau \rightarrowtail d_{\{0,3\}}\tau}$$

PROPOSITION 3.3. *If X is a 2-Segal space, the preaugmented bisimplicial space $\mathcal{P}X$ is a stable augmented double Segal space.*

PROOF. Since X is a 2-Segal space, by Definition 1.2 and Remark 1.3 for all $n \geq 2$ there are homotopy pullback squares of spaces

$$\begin{array}{ccc} X_3 & \xrightarrow{d_0} & X_2 \\ {\scriptstyle d_2}\downarrow & \lrcorner & \downarrow{\scriptstyle d_1} \\ X_1 & \xrightarrow[d_0]{} & X_1 \end{array} \quad \text{and} \quad \begin{array}{ccc} X_3 & \xrightarrow{d_3} & X_2 \\ {\scriptstyle d_1}\downarrow & \lrcorner & \downarrow{\scriptstyle d_1} \\ X_2 & \xrightarrow[d_2]{} & X_1 \end{array} \quad \text{and} \quad \begin{array}{ccc} X_1 & \xrightarrow{s_0} & X_2 \\ {\scriptstyle d_1}\downarrow & \lrcorner & \downarrow{\scriptstyle d_2} \\ X_0 & \xrightarrow[s_0]{} & X_1. \end{array}$$

By definition of $\mathcal{P}X$, these can be rewritten as homotopy pullback squares of spaces

$$\begin{array}{ccc} \mathcal{P}_{2,0}X & \xrightarrow{d_0^v} & \mathcal{P}_{1,0}X \\ {\scriptstyle d_2^v}\downarrow & \lrcorner & \downarrow{\scriptstyle d_1^v} \\ \mathcal{P}_{1,0}X & \xrightarrow[d_0^v]{} & \mathcal{P}_{0,0}X \end{array} \quad \text{and} \quad \begin{array}{ccc} \mathcal{P}_{1,1}X & \xrightarrow{d_1^h} & \mathcal{P}_{1,0}X \\ {\scriptstyle d_1^v}\downarrow & \lrcorner & \downarrow{\scriptstyle d_1^v} \\ \mathcal{P}_{0,1}X & \xrightarrow[d_1^h]{} & \mathcal{P}_{0,0}X \end{array} \quad \text{and} \quad \begin{array}{ccc} \mathcal{P}_{0,0}X & \longrightarrow & \mathcal{P}_{0,1}X \\ \downarrow & \lrcorner & \downarrow{\scriptstyle d_1^h} \\ \mathcal{P}_{-1}X & \xrightarrow[\varepsilon]{} & \mathcal{P}_{0,0}X. \end{array}$$

These imply, respectively, the first smallest instances of the Double Segal, stability and (after thinking about it a little bit) augmentation conditions from Definition 2.3 for $\mathcal{P}X$. The dual conditions and the general double Segal property can be proven with a similar reasoning. □

We can unpack the path construction $\mathcal{P}\Delta[n]$ of $\Delta[n]$ for $n \leq 4$, and let the reader imagine that of $\Delta[n]$ for general n.

EXAMPLE 3.4. Following the conventions established in Remark 2.2, the path constructions

$$\mathcal{P}\Delta[0] \quad , \quad \mathcal{P}\Delta[1] \quad , \quad \mathcal{P}\Delta[2] \quad , \quad \mathcal{P}\Delta[3] \quad , \quad \mathcal{P}\Delta[4]$$

are the stable augmented double Segal spaces generated by the following data, respectively:

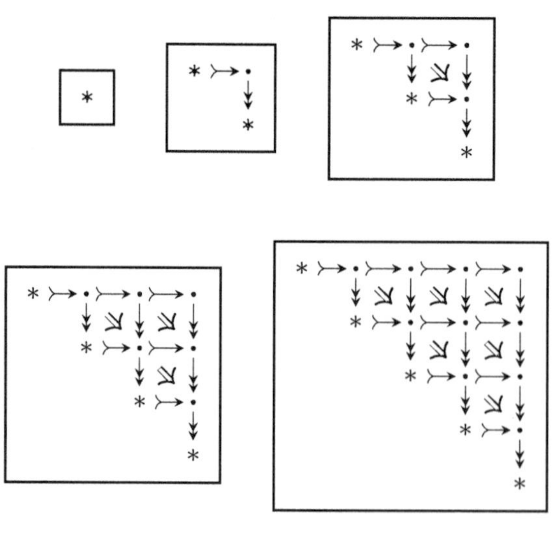

PROPOSITION 3.5. *The path construction induces a functor*

$$\mathcal{P}\colon \text{2SegSp} \to \text{saDblSegSp}.$$

This functor is compatible with the homotopy theory of the categories involved:

PROPOSITION 3.6. *If $f\colon X \xrightarrow{\simeq} Y$ is an equivalence of 2-Segal spaces, then*

$$\mathcal{P}f\colon \mathcal{P}X \xrightarrow{\simeq} \mathcal{P}Y$$

is an equivalence of stable augmented double Segal spaces.

PROOF. Since $f\colon X \xrightarrow{\simeq} Y$ is an equivalence of 2-Segal spaces, for all $a, b \geq 0$ it induces equivalences of spaces

$$f_{a+1+b}\colon X_{a+1+b} \xrightarrow{\simeq} Y_{a+1+b} \quad \text{and} \quad f_0\colon X_0 \xrightarrow{\simeq} Y_0.$$

and equivalences of spaces

$$\mathcal{P}_{a,b}f\colon \mathcal{P}_{a,b}X \xrightarrow{\simeq} \mathcal{P}_{a,b}Y \quad \text{and} \quad \mathcal{P}_{-1}f\colon \mathcal{P}X_{-1} \xrightarrow{\simeq} \mathcal{P}_{-1}Y.$$

So $\mathcal{P}f\colon \mathcal{P}X \xrightarrow{\simeq} \mathcal{P}Y$ is an equivalence of stable augmented double Segal spaces, as desired. □

Thanks to Proposition 3.6, we obtain that the path construction descends to equivalence classes:

PROPOSITION 3.7. *The path construction defines a function*

$$\mathcal{P}\colon \text{2SegSp}/_{\simeq} \to \text{saDblSegSp}/_{\simeq}.$$

Moreover, it follows from [**BK12**, §1.5] that the path construction also induces a functor at the level of simplicial localizations:

PROPOSITION 3.8. *The path construction induces a simplicial functor*

$$\boldsymbol{\mathcal{P}}\colon \textbf{2SegSp} \to \textbf{saDblSegSp}.$$

4. The $S_\bullet$-construction

In this section we introduce the $S_\bullet$-*construction* of a stable augmented double Segal space from [**BOO⁺21a**], and show it is a 2-Segal space under the mild technical assumption of injective fibrancy that we come to shortly.

If $\mathcal{S}$ denotes the category of spaces, there is an induced adjunction

$$\mathcal{P} = p^*\colon \mathcal{S}^{\Delta^{\text{op}}} \leftrightarrows \mathcal{S}^{\Sigma^{\text{op}}} : p_* =: S_\bullet$$

between the category $\mathcal{S}^{\Delta^{\text{op}}}$ of simplicial spaces and the category $\mathcal{S}^{\Sigma^{\text{op}}}$ of Σ-spaces. In order to spell out the $S_\bullet$-construction, we should introduce some further structure for the category of preaugmented bisimplicial spaces.

NOTATION 4.1. Given a preaugmented bisimplicial space $\mathcal{A}$ and a space S, the *tensor* $\mathcal{A} \boxtimes S$ of $\mathcal{A}$ with S is the preaugmented bisimplicial space which consists of the spaces

$$(\mathcal{A} \boxtimes S)_{-1} := \mathcal{A}_{-1} \times S \quad \text{and } (\mathcal{A} \boxtimes S)_{a,b} := \mathcal{A}_{a,b} \times S \quad \text{for } a, b \geq 0.$$

The structures maps are the obvious ones.

NOTATION 4.2. Given preaugmented bisimplicial spaces $\mathcal{A}$ and $\mathcal{B}$, the *mapping space* $\mathrm{Map}(\mathcal{A},\mathcal{B})$ from $\mathcal{A}$ to $\mathcal{B}$ is the simplicial set given, for $\ell \geq 0$, by the set

$$\mathrm{Map}_\ell(\mathcal{A},\mathcal{B}) := \mathrm{Hom}_{\mathrm{saDblSegSp}}(\mathcal{A} \boxtimes \Delta[\ell], \mathcal{B}).$$

The structures maps are the obvious ones.

We can now define the $S_\bullet$-construction of a preaugmented bisimplicial space. We regard the standard simplex $\Delta[n]$ as a discrete simplicial space.

CONSTRUCTION 4.3. Given a preaugmented bisimplicial space $\mathcal{D}$, we define the *$S_\bullet$-construction* of X, which is a simplicial space $S_\bullet\mathcal{D}$. It consists, for $n \geq 0$, of the space

$$S_n\mathcal{D} := \mathrm{Map}(\mathcal{P}\Delta[n], \mathcal{D}).$$

The structure maps are the obvious ones, induced by the cosimplicial structure of $n \mapsto \Delta[n] \mapsto \mathcal{P}\Delta[n]$.

We'd like to say that the $S_\bullet$-construction of a stable augmented double Segal space is a 2-Segal space, but this fails in such generality. The problem can be resolved by incorporating a technical condition, which we acknowledge for the sake of accuracy; however, we encourage readers who are not acquainted with it to disregard this detail in order to focus on the primary narrative.

We refer the reader to [**Hir03**, §6.11] for the notion of an *injectively fibrant* preaugmented bisimplicial space (i.e., a fibrant object in the injective model structure of Σ-spaces). What one needs to know is just that for every preaugmented simplicial space $\mathcal{D}$ there exists an injectively fibrant preaugmented bisimplicial space $\widetilde{\mathcal{D}}$ and a natural map $\mathcal{D} \to \widetilde{\mathcal{D}}$ which induces equivalences of spaces $\mathcal{D}_{-1} \to \widetilde{\mathcal{D}}_{-1}$ and $\mathcal{D}_{a,b} \to \widetilde{\mathcal{D}}_{a,b}$ for $a, b \geq 0$. Moreover, if $\mathcal{D}$ is a stable augmented bisimplicial space, so is $\widetilde{\mathcal{D}}$. To give an idea of the flavor of the injective fibrancy condition, we can mention that the spaces $\widetilde{\mathcal{D}}_{a,b}$ and $\widetilde{\mathcal{D}}_{-1}$ are always Kan complexes.

PROPOSITION 4.4. *If $\mathcal{D}$ is an injectively fibrant stable augmented double Segal space, then the simplicial space $S_\bullet\mathcal{D}$ is a 2-Segal space.*

Proving the proposition amounts to analyzing the 2-Segal maps of $S_\bullet\mathcal{D}$:

LEMMA 4.5. *Let $\mathcal{D}$ be an injectively fibrant stable augmented double Segal space. For all $n \geq 0$ the second canonical map from Definition 1.2 is an equivalence of spaces*

$$S_{n+1}\mathcal{D} \xrightarrow{\simeq} S_n\mathcal{D} \times^h_{S_1\mathcal{D}} S_2\mathcal{D}.$$

When $n = 2$, the map

$$S_3\mathcal{D} \xrightarrow{\simeq} S_2\mathcal{D} \times^h_{S_1\mathcal{D}} S_2\mathcal{D}$$

is induced by the commutative square of spaces

$$\begin{array}{ccc} S_3\mathcal{D} & \xrightarrow{d_1} & S_2\mathcal{D} \\ {\scriptstyle d_3}\downarrow & & \downarrow{\scriptstyle d_2} \\ S_2\mathcal{D} & \xrightarrow[d_1]{} & S_1\mathcal{D}, \end{array} \tag{1}$$

and it can be depicted as

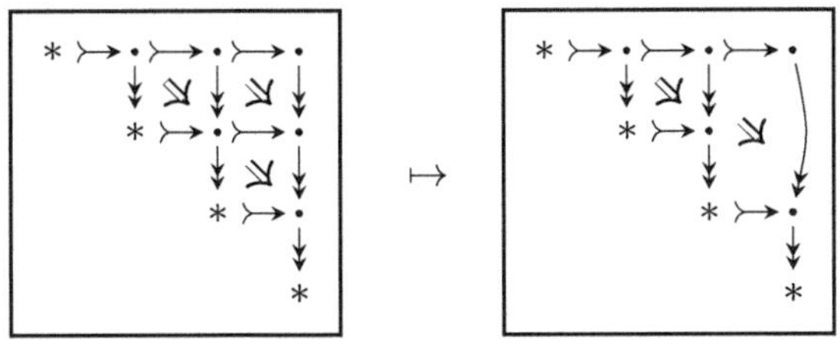

Here, the right-hand side is understood as the shape formed by gluing together the shapes corresponding to the components of the pullback. Specifically, the contribution from the copy of $S_2\mathcal{D}$ in the top right corner of the commutative square (1) is seen in the picture as the one containing the "rectangular" square, while the copy of $S_2\mathcal{D}$ from the bottom left corner of the commutative square is seen in the part of the picture around the "regularly sized" square. These two parts are glued together along a copy of $S_1\mathcal{D}$, which includes the second bullet in the first row of the picture.

While this graphical method does not clearly indicate which composites should be included, the ambiguity does not affect the argument or the underlying intuition when the double Segal property holds for $\mathcal{D}$.

We illustrate the idea of the proof in the case $n = 2$. An analogous (though much more intricate) argument could be provided for the case of arbitrary n using [**BOO$^+$21a**, Lemma 5.12] and an appropriate generalization of [**DK19**, Proposition 2.4.8].

Proof Idea of Lemma 4.5. There are equivalences of spaces of the form

$$S_{n+1}\mathcal{D} \xrightarrow{\simeq} S_n\mathcal{D} \times^h_{S_1\mathcal{D}} S_2\mathcal{D},$$

which can be constructed as follows:

- First, there is an equivalence of spaces

$$S_{n+1}\mathcal{D} \xrightarrow{\simeq} S_n\mathcal{D} \times^h_{S_1\mathcal{D}} S_2\mathcal{D} \times^h_{\mathcal{D}_{1,1}\times_{\mathcal{D}_{0,0}}\mathcal{D}_{0,2}} \mathcal{D}_{1,2} =: F,$$

  which essentially forgets the bottom square and one of the vertical morphisms, and can be depicted as

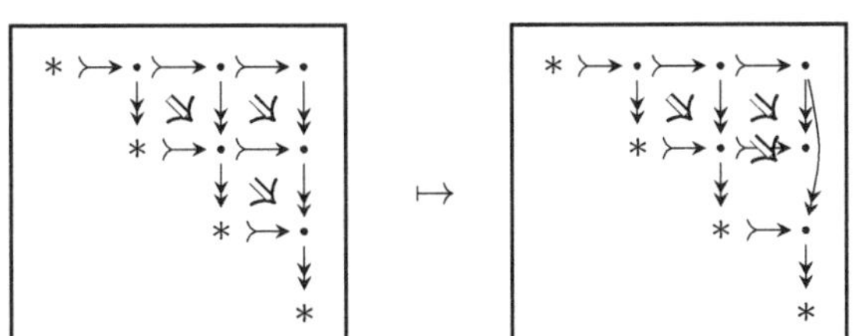

  Roughly speaking, an inverse equivalence for the depicted assignment would be given by
  - first adding the lowest square using stability,
  - then adding the vertical composite of the two squares on the right using the double Segal property,
  - and then observing that the two "tall squares" (the one we just added and the one that existed all along) need to agree by stability because they share the same span in their boundaries.

To see that $S_n\mathcal{D} \to F$ is indeed an equivalence of spaces, it is possible to factor it as the composite of equivalences. Precisely, a (homotopy) pullback along the equivalence

$$\mathcal{D}_{1,1} \xrightarrow{\simeq} \mathcal{D}_{1,1} \times^h_{\mathcal{D}_{1,0}\times^h_{\mathcal{D}_{0,0}}\mathcal{D}_{0,1}} \mathcal{D}_{1,1},$$

which encodes the fact that a span can be uniquely filled to a square because of stability, followed by a (homotopy) pullback along the double Segal map

$$\mathcal{D}_{2,1} \xrightarrow{\simeq} \mathcal{D}_{1,1} \times^h_{\mathcal{D}_{0,1}} \mathcal{D}_{1,1},$$

which forgets a vertical composite of squares, and followed by a pullback of the stability map

$$\mathcal{D}_{1,1} \xrightarrow{\simeq} \mathcal{D}_{1,0} \times^h_{\mathcal{D}_{0,0}} \mathcal{D}_{0,1},$$

which forgets the lowest square.

- Next, there is an equivalence of spaces

$$F \to S_n\mathcal{D} \times^h_{S_1\mathcal{D}} S_2\mathcal{D},$$

which essentially forgets the top right square and the cospan contained in its boundary, and can be depicted as

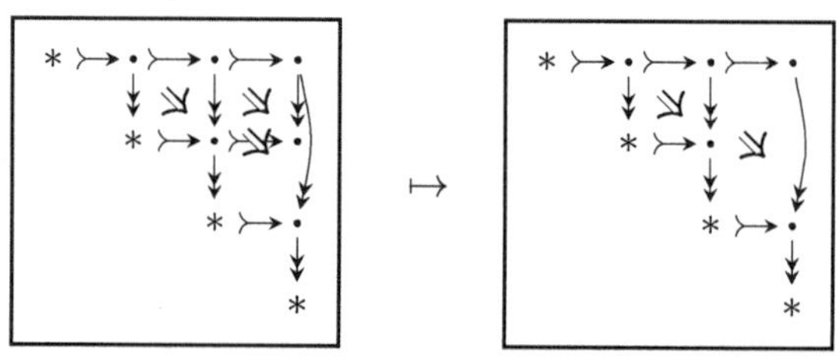

To see that this in indeed an equivalence of spaces, we notice that it can be realized as a pullback of the stability map

$$\mathcal{D}_{1,1} \xrightarrow{\simeq} \mathcal{D}_{1,0} \times^h_{\mathcal{D}_{0,0}} \mathcal{D}_{0,1},$$

which forgets the top right square.

The fact that $\mathcal{D}$ is injectively fibrant allows one to realize all the relevant homotopy pullbacks through strict pullbacks and achieve the desired factorizations. The asserted result then follows. □

We can now prove the proposition:

PROOF OF PROPOSITION 4.4. One kind of 2-Segal map for $S_\bullet\mathcal{D}$ was discussed as Lemma 4.5, and the other one can be treated analogously. □

PROPOSITION 4.6. *There is a functor*

$$S_\bullet\colon \text{saDblSegSp} \to \text{2SegSp}.$$

This functor is compatible with the homotopy theory of the categories involved:

PROPOSITION 4.7. *Let $\mathcal{D}$ and $\mathcal{E}$ be injectively fibrant stable augmented double Segal spaces. If $\varphi\colon \mathcal{D} \xrightarrow{\simeq} \mathcal{E}$ is an equivalence of stable augmented double Segal spaces, then*

$$S_\bullet\varphi\colon S_\bullet\mathcal{D} \xrightarrow{\simeq} S_\bullet\mathcal{E}$$

is an equivalence of 2-Segal spaces.

The proof of the proposition relies on a technical lemma:

LEMMA 4.8. *Let $\mathcal{D}$ be an injectively fibrant stable augmented double Segal space. For all $n \geq 0$ there is an equivalence of spaces*

$$\operatorname{Map}(\mathcal{P}\Delta[n+1], \mathcal{D}) \xrightarrow{\simeq} \mathcal{D}_{0,n}$$

given essentially by "selecting the first row". For instance, when $n = 3$ it can be depicted as

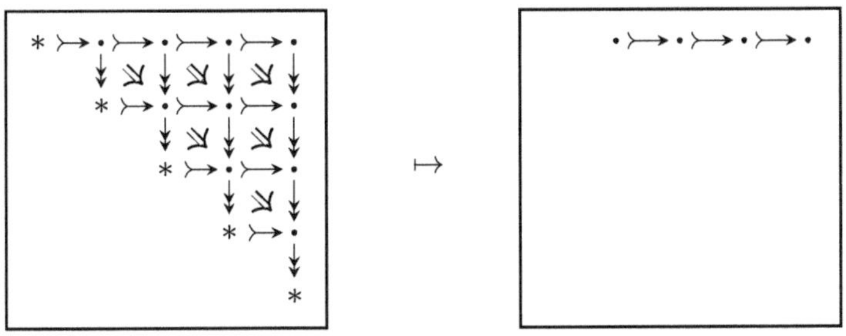

We illustrate the idea in the case $n = 1$. An analogous (though much more intricate) argument can be provided for the case of arbitrary n by adjusting the filtration from [**BOO$^+$21a**, Lemma 5.12].

PROOF IDEA OF LEMMA 4.8 FOR $n = 1$. There is a tower of equivalences of spaces of the form

$$\operatorname{Map}(\mathcal{P}\Delta[2], \mathcal{D}) \xrightarrow{\simeq} F^{(1)} \xrightarrow{\simeq} F^{(2)} \xrightarrow{\simeq} \mathcal{D}_{0,1},$$

which can be constructed as follows.

- First, there is an equivalence of spaces

$$\operatorname{Map}(\mathcal{P}\Delta[2], \mathcal{D}) \xrightarrow{\simeq} \mathcal{D}_{1,1} \times^h_{\mathcal{D}_{0,0}} \mathcal{D}_{-1} =: F^{(1)},$$

  which can be depicted as

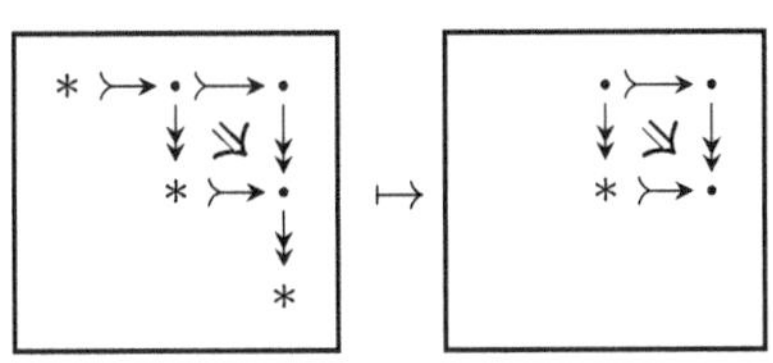

  and can be understood to be an equivalence of spaces using the double Segal and augmentation property for $\mathcal{D}$.
- Next, there is an equivalence of spaces of the form

$$F^{(1)} \xrightarrow{\simeq} \mathcal{D}_{1,0} \times^h_{\mathcal{D}_{0,0}} \mathcal{D}_{0,1} \times^h_{\mathcal{D}_{0,0}} \mathcal{D}_{-1} =: F^{(2)},$$

  which can be depicted as

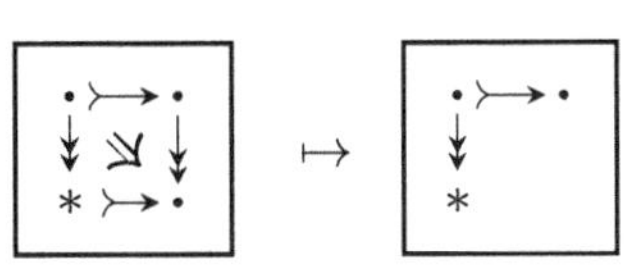

  and can be understood to be an equivalence of spaces using the property of stability for $\mathcal{D}$.

- Finally, there is an equivalence of spaces of the form

$$F^{(2)} \xrightarrow{\simeq} \mathcal{D}_{0,1},$$

which can be depicted as

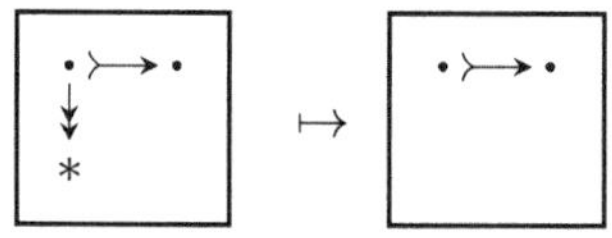

and can be understood to be an equivalence of spaces using the property of augmentation for $\mathcal{D}$.

The desired result then follows. □

We can now prove the proposition.

PROOF OF PROPOSITION 4.7. Since $\varphi \colon \mathcal{D} \to \mathcal{E}$ is an equivalence of stable augmented double Segal spaces, for all $n \geq 0$ there is an equivalence of spaces

$$\varphi_{0,n} \colon \mathcal{D}_{0,n} \xrightarrow{\simeq} \mathcal{E}_{0,n}.$$

By Lemma 4.8, we can deduce that there is also an equivalence of spaces

$$\varphi_* \colon \operatorname{Map}(\mathcal{P}\Delta[n], \mathcal{D}) \xrightarrow{\simeq} \operatorname{Map}(\mathcal{P}\Delta[n], \mathcal{E}),$$

which is by definition the map

$$S_n\varphi \colon S_n\mathcal{D} \xrightarrow{\simeq} S_n\mathcal{E}.$$

We then obtain an equivalence of 2-Segal spaces $S_\bullet\varphi \colon S_\bullet\mathcal{D} \xrightarrow{\simeq} S_\bullet\mathcal{E}$, as desired. □

Thanks to Proposition 4.7, we obtain that the path construction descends to equivalence classes in the following sense:

PROPOSITION 4.9. *There is a function*

$$S_\bullet \circ \widetilde{(-)} \colon \mathrm{saDblSegSp}/_{\simeq} \;\to\; \mathrm{2SegSp}/_{\simeq}.$$

Moreover, it follows from [**BK12**, §1.5] that the path construction also induces a functor at the level of simplicial localizations, so we obtain:

PROPOSITION 4.10. *The $S_\bullet$-construction induces a simplicial functor*

$$\boldsymbol{S_\bullet} \circ \widetilde{(-)} \colon \mathbf{saDblSegSp} \to \mathbf{2SegSp}.$$

5. The equivalence

We can now explain how the path construction $\mathcal{P}$ and the $S_\bullet$-construction define inverse equivalences. We start by analyzing the first composite:

PROPOSITION 5.1. *If X is a 2-Segal space, then there is an equivalence of 2-Segal spaces*

$$\eta^h \colon X \xrightarrow{\simeq} S_\bullet\widetilde{\mathcal{P}X}.$$

PROOF. There is a canonical simplicial map

$$\eta^h\colon X \xrightarrow{\eta} S_\bullet\mathcal{P}X \to S_\bullet\widetilde{\mathcal{P}X}.$$

Now, for all $n \geq 0$, using the naturality of mapping spaces from Notation 4.2 we see that there are commutative diagrams of spaces

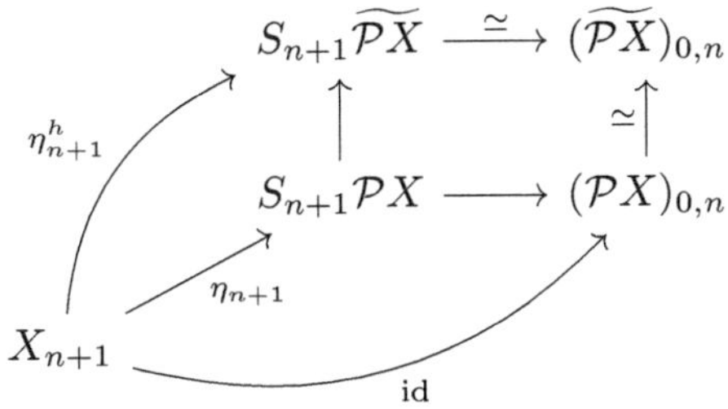

and

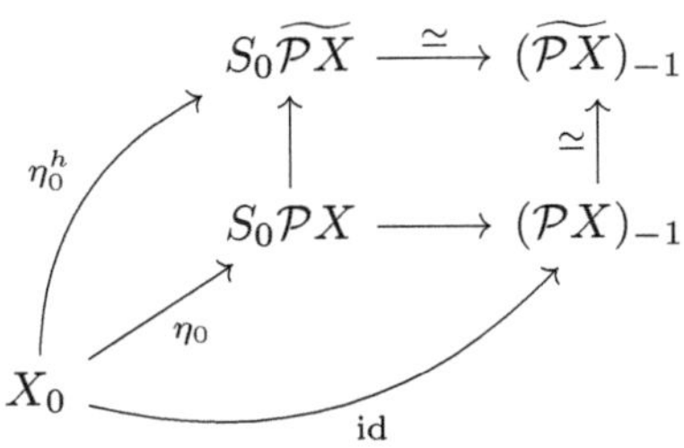

where the top horizontal map in the first diagram is an equivalence as an instance of Lemma 4.8. By two-out-of-three, we then obtain equivalences of spaces

$$\eta^h_{n+1}\colon X \xrightarrow{\simeq} S_{n+1}\widetilde{\mathcal{P}X} \quad \text{and} \quad \eta^h_0\colon X_0 \xrightarrow{\simeq} S_0\widetilde{\mathcal{P}X},$$

and an equivalence of 2-Segal spaces $\eta^h\colon X \to S_\bullet\widetilde{\mathcal{P}X}$, as desired. □

Next, we address the other composite:

PROPOSITION 5.2. *If $\mathcal{D}$ is an injectively fibrant stable augmented double Segal space, then there is an equivalence of stable augmented double Segal spaces*

$$\epsilon\colon \mathcal{P}S_\bullet\mathcal{D} \xrightarrow{\simeq} \mathcal{D}.$$

The proof of the proposition requires an auxiliary lemma:

LEMMA 5.3. *Let $\mathcal{D}$ be an injectively fibrant stable augmented double Segal space. There is an equivalence of spaces*

$$\operatorname{Map}(\mathcal{P}\Delta[3],\mathcal{D}) \xrightarrow{\simeq} \mathcal{D}_{1,1},$$

which can be pictured as

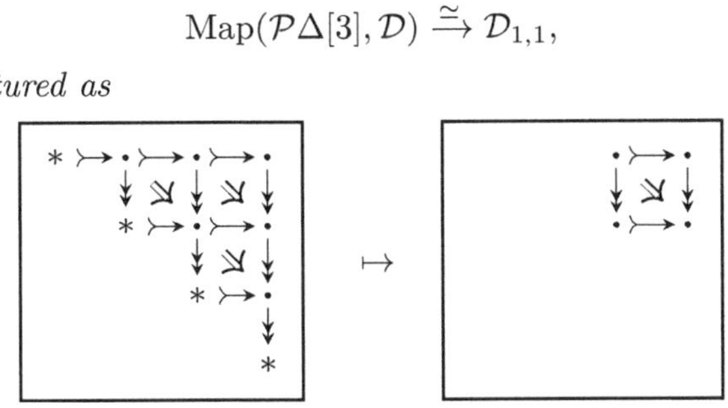

We illustrate the idea, and refer the reader to [**BOO**[+]**21a**, Lemma 6.6] for a rigorous proof.

PROOF IDEA OF LEMMA 5.3. There is a tower of equivalences of spaces of the form:

$$\mathrm{Map}(\mathcal{P}\Delta[3],\mathcal{D}) \xrightarrow{\simeq} F^{(1)} \xrightarrow{\simeq} F^{(2)} \xrightarrow{\simeq} \mathcal{D}_{1,1},$$

which can be constructed as follows:

- First, there is an equivalence of spaces

$$\mathrm{Map}(\mathcal{P}\Delta[3],\mathcal{D}) \xrightarrow{\simeq} \mathcal{D}_{2,1} \times^h_{\mathcal{D}_{1,1}} \mathcal{D}_{1,2} \times^h_{\mathcal{D}_{0,0}} \mathcal{D}_{-1} \times^h_{\mathcal{D}_{0,0}} \mathcal{D}_{-1} =: F^{(1)},$$

  which can be depicted as

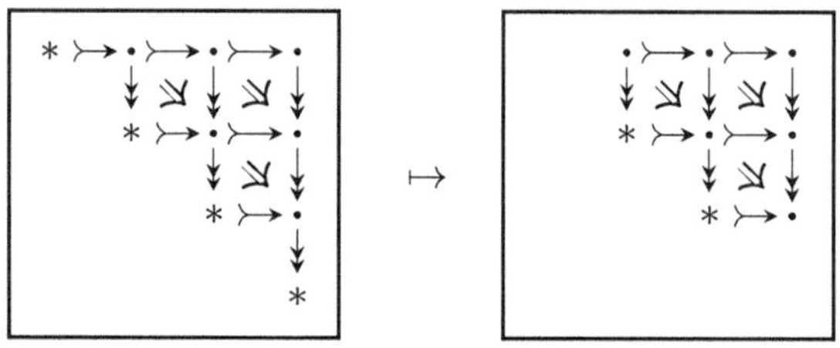

  and can be understood to be an equivalence of spaces using the double Segal and augmentation property for $\mathcal{D}$.
- Next, there is an equivalence of spaces

$$F^{(1)} \xrightarrow{\simeq} \mathcal{D}_{1,1} \times^h_{\mathcal{D}_{0,0}} \mathcal{D}_{0,1} \times^h_{\mathcal{D}_{0,0}} \mathcal{D}_{-1} \times^h_{\mathcal{D}_{0,0}} \mathcal{D}_{0,1} \times^h_{\mathcal{D}_{0,0}} \mathcal{D}_{-1} =: F^{(2)},$$

  which can be depicted as

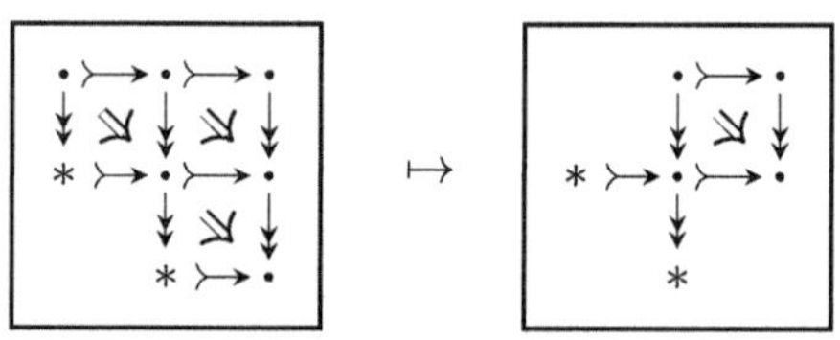

  and can be understood to be an equivalence of spaces using the property of double Segal and stability properties for $\mathcal{D}$.
- Finally, there is an equivalence of spaces

$$F^{(2)} \xrightarrow{\simeq} \mathcal{D}_{1,1},$$

  which can be depicted as

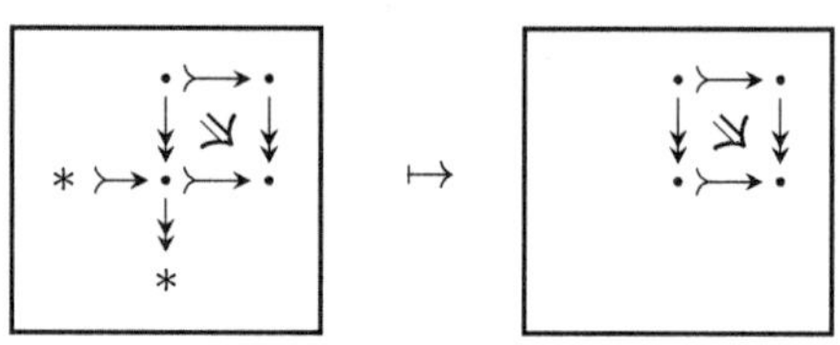

  and can be understood to be an equivalence of spaces using the double Segal and augmentation properties for $\mathcal{D}$.

The desired result then follows. □

We can now prove the proposition:

PROOF OF PROPOSITION 5.2. By Proposition 4.4 and Construction 3.1, both $\mathcal{P}S_\bullet\mathcal{D}$ and $\mathcal{D}$ are stable augmented double Segal spaces. For all $a, b \geq 0$, we consider the map of spaces

$$\mathcal{P}_{a,b}(S_\bullet\mathcal{D}) = S_{a+1+b}\mathcal{D} \to \mathcal{D}_{a,b}$$

which "selects the internal grid of size $[a] \times [b]$". For instance, when $a = 1$ and $b = 2$ it can be depicted as

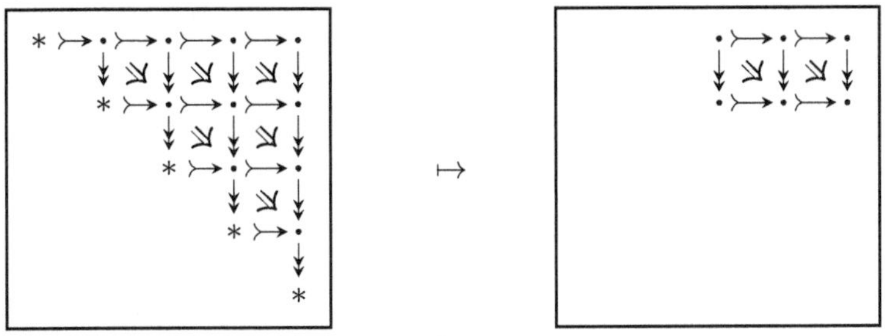

This map can be checked to be natural in $a, b \geq 0$, and it hence defines a map of stable augmented double Segal spaces

$$\epsilon\colon \mathcal{P}S_\bullet\mathcal{D} \to \mathcal{D}.$$

In order to show that it is an equivalence of stable augmented double Segal spaces, it suffices to check that it induces equivalences of spaces ϵ_{-1}, $\epsilon_{0,0}$, $\epsilon_{0,1}$, $\epsilon_{1,0}$, and $\epsilon_{1,1}$, which we now analyze.

The map $\epsilon_{\mathcal{D},-1}$ is an isomorphism of spaces

$$\epsilon_{-1}\colon (\mathcal{P}S_\bullet\mathcal{D})_{-1} = S_0\mathcal{D} = \mathrm{Map}(\mathcal{P}\Delta[0], \mathcal{D}) \cong \mathcal{D}_{-1}.$$

The map $\epsilon_{0,0}$ can be identified with an instance of the equivalence of spaces from Lemma 4.8:

$$\epsilon_{0,0}\colon (\mathcal{P}S_\bullet\mathcal{D})_{0,0} = S_1\mathcal{D} = \mathrm{Map}(\mathcal{P}\Delta[1], \mathcal{D}) \simeq \mathcal{D}_{0,0}.$$

The map $\epsilon_{0,1}$, resp. $\epsilon_{1,0}$, can be identified with an instance of the equivalence of spaces from Lemma 4.8, resp. of an appropriate dual statement of Lemma 4.8:

$$\epsilon_{1,0}\colon (\mathcal{P}S_\bullet\mathcal{D})_{1,0} = S_2\mathcal{D} \cong \mathrm{Map}(\mathcal{P}\Delta[2], \mathcal{D}) \simeq \mathcal{D}_{1,0},$$

resp.

$$\epsilon_{0,1}\colon (\mathcal{P}S_\bullet\mathcal{D})_{0,1} = S_2\mathcal{D} = \mathrm{Map}(\mathcal{P}\Delta[2], \mathcal{D}) \simeq \mathcal{D}_{0,1}.$$

Finally $\epsilon_{1,1}$ can be identified with an instance of the equivalence of spaces from Lemma 5.3:

$$\epsilon_{1,1}\colon (\mathcal{P}S_\bullet\mathcal{D})_{1,1} = S_3\mathcal{D} \cong \mathrm{Map}(\mathcal{P}\Delta[3], \mathcal{D}) \simeq \mathcal{D}_{1,1}.$$

This concludes the proof. □

We can finally establish the desired correspondence:

THEOREM 5.4. *The path construction and $S_\bullet$-construction – modulo injectively fibrant replacement – induce inverse bijections*

$$\mathcal{P}\colon 2\mathrm{SegSp}/_\simeq \;\cong\; \mathrm{saDblSegSp}/_\simeq\colon S_\bullet \circ \widetilde{(-)}.$$

PROOF. The statement follows from Propositions 5.1 and 5.2. □

The stronger version of the statement, which is the main result from [**BOO$^+$21a**], asserts that the correspondence from Theorem 5.4 can actually be upgraded to an equivalence of ∞-categories. In this context, an equivalence of ∞-categories can be understood as an equivalence of simplicial categories. Recall from [**DK80b**, §2.4] that an equivalence of simplicial categories (a.k.a. a *Dwyer–Kan equivalence*) is a simplicial functor which is essentially surjective and induces equivalences at the level of mapping spaces.

THEOREM 5.5 ([**BOO+21a**]). *The path and the $S_\bullet$-construction (modulo injectively fibrant replacement) induce homotopy inverse Dwyer–Kan equivalences*

$$\mathcal{P}\colon \mathbf{2SegSp} \simeq \mathbf{saDblSegSp}\colon S_\bullet \circ \widetilde{(-)}.$$

We briefly comment on the machinery used to prove the theorem. We refer the reader to standard model-categorical references ([**Hir03**, **Hov99**]) for the model categorical terminology.

PROOF STRATEGY. It is described in [**DK19**, §5.3] how to put model structures on the category $\mathcal{S}^{\Delta^{\mathrm{op}}}$ of simplicial spaces in which the fibrant-cofibrant objects are the injectively fibrant 2-Segal spaces. The underlying ∞-category is then equivalent as a simplicial category to the one mentioned in Definition 1.7.

Similarly, it is described in [**BOO+21a**, §4.2] how to put model structures on the category $\mathcal{S}^{\Sigma^{\mathrm{op}}}$ of preaugmented bisimplicial spaces in which the fibrant-cofibrant objects are the injectively fibrant stable augmented double Segal spaces. The underlying ∞-category is then equivalent as a simplicial category to the one mentioned in Definition 2.12.

It is shown in [**BOO+21a**, Theorem 6.1] that the path construction $\mathcal{P}$ and the $S_\bullet$-construction form a Quillen equivalence

$$\mathcal{P} = p^*\colon \mathcal{S}^{\Delta^{\mathrm{op}}} \leftrightarrows \mathcal{S}^{\Sigma^{\mathrm{op}}}\colon p_* =: S_\bullet$$

with respect to these model structures. Using the results from [**MG16**], one then deduces that the path construction $\mathcal{P}$ and the $S_\bullet$-construction induce inverse Dwyer–Kan equivalences of simplicial categories, as desired. □

References

[Bar16] C. Barwick, *On the algebraic K-theory of higher categories*, J. Topol. **9** (2016), no. 1, 245–347, DOI 10.1112/jtopol/jtv042. MR3465850

[BK12] C. Barwick and D. M. Kan, *A characterization of simplicial localization functors and a discussion of DK equivalences*, Indag. Math. (N.S.) **23** (2012), no. 1-2, 69–79, DOI 10.1016/j.indag.2011.10.001. MR2877402

[BLLM24] *Homotopical algebra and higher structures*, Oberwolfach Rep. **21** (2024), no. 3, 2251–2307, DOI 10.4171/owr/2024/39. Abstracts from the workshop held August 26–30, 2024; Organized by Michael Batanin, Andrey Lazarev, Muriel Livernet and Martin Markl. MR4865595

[BOO+18] J. E. Bergner, A. M. Osorno, V. Ozornova, M. Rovelli, and C. I. Scheimbauer, *2-Segal sets and the Waldhausen construction*, Topology Appl. **235** (2018), 445–484, DOI 10.1016/j.topol.2017.12.009. MR3760213

[BOO+21a] J. E. Bergner, A. M. Osorno, V. Ozornova, M. Rovelli, and C. I. Scheimbauer, *2-Segal objects and the Waldhausen construction*, Algebr. Geom. Topol. **21** (2021), no. 3, 1267–1326, DOI 10.2140/agt.2021.21.1267. MR4299667

[BOO+21b] J. E. Bergner, A. M. Osorno, V. Ozornova, M. Rovelli, and C. I. Scheimbauer, *Comparison of Waldhausen constructions*, Ann. K-Theory **6** (2021), no. 1, 97–136, DOI 10.2140/akt.2021.6.97. MR4283095

[CZ22] J. A. Campbell and I. Zakharevich, *Dévissage and localization for the Grothendieck spectrum of varieties*. part A, Adv. Math. **411** (2022), no. part A, Paper No. 108710, 80, DOI 10.1016/j.aim.2022.108710. MR4512395

[DK80a] W. G. Dwyer and D. M. Kan, *Calculating simplicial localizations*, J. Pure Appl. Algebra **18** (1980), no. 1, 17–35, DOI 10.1016/0022-4049(80)90113-9. MR578563

[DK80b] W. G. Dwyer and D. M. Kan, *Function complexes in homotopical algebra*, Topology **19** (1980), no. 4, 427–440, DOI 10.1016/0040-9383(80)90025-7. MR584566

[DK19] T. Dyckerhoff and M. Kapranov, *Higher Segal spaces*, Lecture Notes in Mathematics, vol. 2244, Springer, Cham, 2019, DOI 10.1007/978-3-030-27124-4. MR3970975

[FGK+21] M. Feller, R. Garner, J. Kock, M. U. Proulx, and M. Weber, *Every 2-Segal space is unital*, Commun. Contemp. Math. **23** (2021), no. 2, Paper No. 2050055, 6, DOI 10.1142/S0219199720500558. MR4201031

[GCKT18] I. Gálvez-Carrillo, J. Kock, and A. Tonks, *Decomposition spaces, incidence algebras and Möbius inversion I: Basic theory*, Adv. Math. **331** (2018), 952–1015, DOI 10.1016/j.aim.2018.03.016. MR3804694

[Hin16] V. Hinich, *Dwyer-Kan localization revisited*, Homology Homotopy Appl. **18** (2016), no. 1, 27–48, DOI 10.4310/HHA.2016.v18.n1.a3. MR3460765

[Hir03] P. S. Hirschhorn, *Model categories and their localizations*, Mathematical Surveys and Monographs, vol. 99, American Mathematical Society, Providence, RI, 2003, DOI 10.1090/surv/099. MR1944041

[Hov99] M. Hovey, *Model categories*, Mathematical Surveys and Monographs, vol. 63, American Mathematical Society, Providence, RI, 1999. MR1650134

[Ill71] L. Illusie, *Complexe cotangent et déformations. I* (French), Lecture Notes in Mathematics, Vol. 239, Springer-Verlag, Berlin-New York, 1971. MR491680

[Lur06] J. Lurie, *Stable infinity categories*, `arXiv:math/0608228v5` (2006).

[MG16] A. Mazel-Gee, *Quillen adjunctions induce adjunctions of quasicategories*, New York J. Math. **22** (2016), 57–93. MR3484677

[ML98] S. Mac Lane, *Categories for the working mathematician*, 2nd ed., Graduate Texts in Mathematics, vol. 5, Springer-Verlag, New York, 1998. MR1712872

[OR24] V. Ozornova and M. Rovelli, *Waldhausen's $S_\bullet$-construction*, 2024, this volume.

[Qui73] D. Quillen, *Higher algebraic K-theory. I*, Algebraic K-theory, I: Higher K-theories (Proc. Conf., Battelle Memorial Inst., Seattle, Wash., 1972), Lecture Notes in Math., Vol. 341, Springer, Berlin-New York, 1973, pp. 85–147. MR338129

[Wal85] F. Waldhausen, *Algebraic K-theory of spaces*, Algebraic and geometric topology (New Brunswick, N.J., 1983), Lecture Notes in Math., vol. 1126, Springer, Berlin, 1985, pp. 318–419, DOI 10.1007/BFb0074449. MR802796

Department of Mathematics and Statistics, University of Massachusetts Amherst, Amherst

Email address: `mrovelli@umass.edu`

Contemporary Mathematics
Volume **838**, 2026
https://doi.org/10.1090/conm/838/16738

Hall algebras via 2-Segal spaces

Benjamin Cooper and Matthew B. Young

ABSTRACT. This is an introduction to Hall algebras from the perspective of 2-Segal spaces or, equivalently, decomposition spaces. We explain how linearizations of the 2-Segal space arising as the Waldhausen $\mathcal{S}_\bullet$-construction of a proto-exact category recover various previously known Hall algebras. We use the 2-Segal perspective to study functoriality of the Hall algebra construction and explain how relative variants of 2-Segal spaces lead naturally to representations of Hall algebras.

Introduction

The goal of this survey paper is to explain how Hall algebras, an important class of algebras in representation theory, arise naturally from the perspective of 2-Segal spaces, as introduced by Dyckerhoff and Kapranov [**DK19**], or decomposition spaces, as introduced by Gálvez-Carrillo, Kock and Tonks [**GCKT18a**]. The notions of 2-Segal spaces and decomposition spaces are equivalent, something unsurprising considering that, in each case, the authors made their discoveries through a careful study of the associativity relation which is satisfied by Hall algebras and incidence algebras, respectively.

More precisely, the connection between Hall algebras and 2-Segal spaces occurs when the 2-Segal space arises from Waldhausen's $\mathcal{S}_\bullet$-construction [**Wal85**]. For a suitable category $\mathcal{C}$, the $\mathcal{S}_\bullet$-construction $\mathcal{S}_\bullet(\mathcal{C})$ is a simplicial object whose n-simplices correspond to composition series of length n in $\mathcal{C}$. The homotopy groups of the geometric realization of $\mathcal{S}_\bullet(\mathcal{C})$ agree with the algebraic K-theory of $\mathcal{C}$:

$$K_n(\mathcal{C}) \simeq \pi_{n+1}|\mathcal{S}_\bullet(\mathcal{C})|.$$

For Waldhausen, an important goal was to show that a broad collection of categories $\mathcal{C}$, now known as Waldhausen categories, could be considered suitable, thereby extending Quillen's [**Qui73**] definition of algebraic K-theory to a wider setting. For

2020 *Mathematics Subject Classification.* Primary 54C40, 14E20; Secondary 46E25, 20C20.
Key words and phrases. Higher Segal spaces, Hall algebras and their representations.
The authors thank the organizers of Higher Segal Spaces and their Applications to Algebraic K-Theory, Hall Algebras, and Combinatorics (24w5266), consisting of Julie Bergner, Joachim Kock, Maru Sarazola and (from the planned 2020 workshop) Mark Penney. The authors thank the anonymous referees and Joachim Kock for catching inaccuracies in a pervious version of this paper and suggesting many improvements to the exposition. The authors also thank the workshop participants and BIRS staff. M. B. Y. is partially supported by National Science Foundation grant DMS-2302363 and a Simons Foundation Travel Support for Mathematicians grant (Award ID 853541).

the purposes of the present paper, the crucial observation is that for a slightly different class of categories $\mathcal{C}$ the simplicial object $\mathcal{S}_\bullet(\mathcal{C})$ is a 2-Segal space. Waldhausen's $\mathcal{S}_\bullet$-construction, seen as a 2-Segal space, contains in its shadow the Hall algebra.

The story of Hall algebras can be told in a way which allows these 2-Segal spaces to appear organically. The Hall algebra of a suitable category $\mathcal{C}$ is a ring which is spanned by the isomorphism classes of objects in $\mathcal{C}$ together with a product that is determined by the combinatorics of extensions between these objects. The first Hall algebra, discovered independently by Steinitz [**Ste01**], Delsarte [**Del48**] and Hall [**Hal59**], arises when $\mathcal{C}$ is the category of finite abelian p-groups (p prime). Steinitz and Hall interpreted this Hall algebra in terms of the ring of symmetric functions. Perhaps the most common definition of Hall algebras found in the literature today was introduced by Ringel [**Rin90a**] and concerns abelian categories $\mathcal{C}$ with finiteness properties similar to those of the category of finite abelian p-groups. Ringel's main examples arose from quivers Q and their abelian categories of finite-dimensional representations $\mathsf{rep}_{\mathbb{F}_q}(Q)$ over a finite field $\mathbb{F}_q$. It was known from Gabriel's theorem [**Gab72**] that, for connected quivers Q, the category $\mathsf{rep}_{\mathbb{F}_q}(Q)$ has finitely many isomorphism classes of indecomposable objects if and only if the underlying graph of Q is a Dynkin diagram from the list A_n, D_n or E_6, E_7, E_8. In each case, Ringel [**Rin90b**] identified the Hall algebra of $\mathcal{C} = \mathsf{rep}_{\mathbb{F}_q}(Q)$ with the positive part of the quantum group of the corresponding complex simple Lie algebra $\mathfrak{g}_Q$:

$$H(\mathsf{rep}_{\mathbb{F}_q}(Q)) \simeq U^+_{\sqrt{q}}(\mathfrak{g}_Q).$$

This construction established the Hall algebra as a central object in geometric representation theory. Later, Green [**Gre95**] introduced a coproduct on the Hall algebra and extended Ringel's isomorphism to one of bialgebras. Moreover, taking Q to be the Jordan (one loop) quiver recovers the original Hall algebra of Steinitz. In his study of quantum groups, Lusztig [**Lus10**] built magnificently on these ideas, viewing the Hall algebra as a convolution algebra of locally constant functions on the moduli stack of objects of $\mathcal{C}$. These ideas were also central to the construction of representations of quantum affine algebras by Nakajima [**Nak01**].

The convolution algebra incarnation of the Hall algebra sees only very coarsely the structure of the moduli stacks involved in its definition. Barwick [**Bar16**] proved that the Waldhausen $\mathcal{S}_\bullet$-construction $\mathcal{S}_\bullet(\mathcal{C})$ occurs as the Goodwillie derivative of the functor which associates to $\mathcal{C}$ its moduli space of objects. From the perspective of this paper, the central observation of Dyckerhoff–Kapranov and Gálvez-Carrillo–Kock–Tonks is that the Waldhausen $\mathcal{S}_\bullet$-construction, seen as a 2-Segal space, suffices to encode the Hall algebra. Lusztig's geometric definition factors through the $\mathcal{S}_\bullet$-construction, as evidenced by Section 2, which produces the Hall algebra of Ringel as a linearization of the 2-Segal space.

That the ideas of Steinitz, 123 years ago, have a home within an additive linearization of the Goodwillie derivative of a most complicated object (the moduli stack of objects of a category) that we can associate to a simple object (the Jordan quiver) is a kind of historical coincidence which lends to the subject its own beauty and depth.

What is included in this article. Section 1 contains brief accounts of 2-Segal spaces and the Waldhausen $\mathcal{S}_\bullet$-construction. We work in the generality of proto-exact categories, a non-additive generalization of exact categories which allows for the realization of Hall algebras of so-called $\mathbb{F}_1$-linear categories. Section 2 constructs from a 2-Segal space an associative algebra object in spans. Various linearizations of this algebra object are then discussed and it is explained, via examples, how this recovers a number of Hall algebras which appear in the literature. This includes Ringel's Hall algebra of $\mathbb{F}_q$-linear abelian categories, their $\mathbb{F}_1$-analogues introduced by Szczesny [**Szc12**] and the cohomological and motivic Hall algebras of Kontsevich–Soibelman [**KS08, KS11**] and Joyce [**Joy07**]. We aim to give a variety of examples, ranging from elementary, involving only combinatorial data or Lie algebras, to more advanced, involving various forms of quantum groups. The examples, in particular those that require more background in representation theory, are not required to understand the main thread of the paper. The remainder of the paper is concerned with using 2-Segal spaces as an organizational tool for the study of Hall algebras. Section 3 contains a discussion of the functoriality of the Hall algebra construction while Section 4 explains the role of relative variants of 2-Segal spaces in the representation theory of Hall algebras, two topics for which this abstraction yields constructive approaches.

What is not included in this article. Being a narrow discussion of a subject containing many technicalities, this article inevitably gives few proofs and glosses over some technical details. Readers are referred to the references and excellent surveys by Schiffmann [**Sch12b, Sch18**], who focuses on representation-theoretic aspects of finitary Hall algebras of quivers and curves, and Dyckerhoff [**Dyc18**], who takes the 2-Segal perspective we follow in this article. Adornments of Hall algebras, such as coproducts, and the crucial role of the Euler pairing in the theory can be found elsewhere, including as Schiffmann's survey [**Sch12b**]. The role of Hall algebras in geometric representation theory is discussed only indirectly; see a second survey of Schiffmann [**Sch12a**] for an introduction in this direction. We do not discuss higher-categorical generalizations of Hall algebras, such as Hall monoidal categories or algebra objects in the $(\infty, 2)$-category of spans; see the work of Dyckerhoff–Kapranov [**DK19**]. Finally, we do not discuss other ideas which interface the 2-Segal/decomposition space perspectives of the articles [**DK19, GCKT18a**].

1. Recollections on 2-Segal spaces

We recall the definition of 2-Segal spaces in a form convenient for applications to Hall algebras. The theory of 2-Segal spaces is treated in detail by Dyckerhoff and Kapranov [**DK19**]. See the paper by Stern [**Ste24**] in this volume for a brief introduction.

1.1. 2-Segal spaces. Let Δ be the category of non-empty finite ordinals and their monotone maps. We often use without mention the canonical equivalence of Δ with its full subcategory on the standard ordinals $[n] = \{0, \dots, n\}$, $n \in \mathbb{Z}_{\geq 0}$.

Let $\mathscr{S}$ be a Quillen model category. We refer to homotopy pullbacks and pushouts simply as pullbacks and pushouts, respectively, and similarly for cartesian and cocartesian diagrams. We write pt for the final object of $\mathscr{S}$. The primary example to keep in mind is the model category of small groupoids, in which weak equivalences are equivalences of categories and cofibrations are functors which are injective on objects. With minor changes, $\mathscr{S}$ can also be taken to be an ∞-category with finite limits.

As a warm up and for later reference, we recall the definition of a 1-Segal space.

DEFINITION 1.1.1 ([**Rez01**]). A simplicial object $X_\bullet : \Delta^{\mathrm{op}} \to \mathscr{S}$ is 1-*Segal* if, for every $n \geq 2$ and $0 \leq i \leq n$, the square

$$\begin{array}{ccc} X_n & \longrightarrow & X_{\{i,\ldots,n\}} \\ \downarrow & & \downarrow \\ X_{\{0,\ldots,i\}} & \longrightarrow & X_{\{i\}} \end{array}$$

is cartesian.

In the above diagram, we have written X_n for $X_{[n]}$ and all morphisms are induced by the obvious inclusions of ordinals. Similar comments apply in what follows.

DEFINITION 1.1.2 ([**DK19**]). A simplicial object $X_\bullet : \Delta^{\mathrm{op}} \to \mathscr{S}$ is 2-*Segal* if, for every $n \geq 3$ and $0 \leq i < j \leq n$, the square

$$\begin{array}{ccc} X_n & \longrightarrow & X_{\{i,\ldots,j\}} \\ \downarrow & & \downarrow \\ X_{\{0,\ldots,i,j,\ldots,n\}} & \longrightarrow & X_{\{i,j\}} \end{array}$$

is cartesian.

DEFINITION 1.1.3 ([**DK19**]). A 2-Segal simplicial object $X_\bullet : \Delta^{\mathrm{op}} \to \mathscr{S}$ is *unital* if, for every $n \geq 2$ and $0 \leq i \leq n-1$, the square

$$\begin{array}{ccc} X_{n-1} & \longrightarrow & X_n \\ \downarrow & & \downarrow \\ X_{\{i\}} & \longrightarrow & X_{\{i,i+1\}} \end{array}$$

is cartesian.

The following result asserts that unitality is in fact implied by the 2-Segal condition.

THEOREM 1.1.4 ([**FGK$^+$21**]). *Every* 2-*Segal space is unital.*

1.2. Proto-exact categories. In order to treat various flavours of Hall algebras which appear in the literature, we work with the following non-additive generalization of Quillen's exact categories [**Qui73, Büh10**].

DEFINITION 1.2.1 ([**DK19**, §2.4]). A *proto-exact category* $\mathcal{C}$ is a pointed category, whose zero object is denoted by 0, with two classes of morphisms, $\mathfrak{I}$ and $\mathfrak{D}$, called *inflations* and *deflations* and denoted by $\rightarrowtail$ and $\twoheadrightarrow$, respectively, such that the following properties hold:

(i) For each object $U \in \mathcal{C}$, the morphism $0 \to U$ is in $\mathfrak{I}$ and the morphism $U \to 0$ is in $\mathfrak{D}$.
(ii) Each class $\mathfrak{I}$ and $\mathfrak{D}$ is closed under composition and contains all isomorphisms.

(iii) A commutative square of the form

$$
\begin{array}{ccc}
U & \rightarrowtail & V \\
\downarrow\!\!\!\!\downarrow & & \downarrow\!\!\!\!\downarrow \\
W & \rightarrowtail & X
\end{array}
\tag{1.1}
$$

is cartesian if and only if it is cocartesian.

(iv) Any diagram of the form

$$
\begin{array}{ccc}
 & & V \\
 & & \downarrow\!\!\!\!\downarrow \\
W & \rightarrowtail & X
\end{array}
$$

can be completed to a bicartesian square of the form (1.1).

(v) Any diagram of the form

$$
\begin{array}{ccc}
U & \rightarrowtail & V \\
\downarrow\!\!\!\!\downarrow & & \\
W & &
\end{array}
$$

can be completed to a bicartesian square of the form (1.1).

A bicartesian square of the form (1.1) with $W = 0$ is called a *conflation* or *short exact sequence* and is denoted by $U \rightarrowtail V \twoheadrightarrow X$.

Examples of proto-exact categories include abelian categories, where $\mathfrak{I}$ and $\mathfrak{D}$ are the monomorphisms and epimorphisms, respectively, and exact categories, where $\mathfrak{I}$ and $\mathfrak{D}$ are the admissible monomorphisms and admissible epimorphisms, respectively. A non-exact example is as follows.

EXAMPLE 1.2.2 ([**Szc12**]). The *category* $\mathsf{vect}_{\mathbb{F}_1}$ *of finite-dimensional vector spaces over* $\mathbb{F}_1$ is defined as follows. An object of $\mathsf{vect}_{\mathbb{F}_1}$ is a pair $(V, *)$ consisting of a finite set V with a basepoint $* \in V$. A morphism $f : (V, *) \to (W, *)$ in $\mathsf{vect}_{\mathbb{F}_1}$ is a map of pointed sets for which the restriction of f to $V \setminus f^{-1}(*)$ is injective. The category $\mathsf{vect}_{\mathbb{F}_1}$ admits a proto-exact structure in which inflations (resp. deflations) are injective (resp. surjective) morphisms. Note that $\mathsf{vect}_{\mathbb{F}_1}$ is not pre-additive and so, in particular, is not exact.

1.3. The Waldhausen $\mathcal{S}_\bullet$-construction. The *Waldhausen $\mathcal{S}_\bullet$-construction* of a proto-exact category $\mathcal{C}$ is the simplicial groupoid $\mathcal{S}_\bullet(\mathcal{C})$ whose n-simplices $\mathcal{S}_n(\mathcal{C})$

consist of all diagrams in $\mathcal{C}$ of the form

$$\begin{array}{ccccccccc}
0 & \rightarrowtail & A_{\{0,1\}} & \rightarrowtail & \cdots & \rightarrowtail & A_{\{0,n-1\}} & \rightarrowtail & A_{\{0,n\}} \\
 & & \downarrow & & & & \downarrow & & \downarrow \\
 & & 0 & \rightarrowtail & \cdots & \rightarrowtail & A_{\{1,n-1\}} & \rightarrowtail & A_{\{1,n\}} \\
 & & & & & & \downarrow & & \downarrow \\
 & & & & \ddots & & \vdots & & \vdots \\
 & & & & & & \downarrow & & \downarrow \\
 & & & & & & 0 & \rightarrowtail & A_{\{n-1,n\}} \\
 & & & & & & & & \downarrow \\
 & & & & & & & & 0
\end{array}$$

such that each sub-rectangle is bicartesian. An isomorphism $A_{\{\bullet,\bullet\}} \xrightarrow{\sim} B_{\{\bullet,\bullet\}}$ in $\mathcal{S}_n(\mathcal{C})$ is the data of isomorphisms $A_{\{i,j\}} \xrightarrow{\sim} B_{\{i,j\}}$, $0 \leq i \leq j \leq n$, which make the obvious diagrams commute. The degeneracy map $s_i : \mathcal{S}_n(\mathcal{C}) \to \mathcal{S}_{n+1}(\mathcal{C})$ inserts a row and column of identity morphisms after the ith row and i^{th} column. The face map $\partial_i : \mathcal{S}_n(\mathcal{C}) \to \mathcal{S}_{n-1}(\mathcal{C})$ deletes the ith row and i^{th} column and then composes the adjacent morphisms. See [**Wal85**] for the introduction of $\mathcal{S}_\bullet(\mathcal{C})$ and [**DK19, OR24**] for expositions in the context of 2-Segal spaces.

We make explicit the simplices in low degrees. The 0-simplices $\mathcal{S}_0(\mathcal{C}) = \{0\}$ form the trivial groupoid while $\mathcal{S}_1(\mathcal{C}) \simeq \mathcal{C}^{\sim}$ is the maximal groupoid of $\mathcal{C}$. The 2-simplices form the groupoid of conflations in $\mathcal{C}$:

$$\mathcal{S}_2(\mathcal{C}) = \left\{ \begin{array}{ccccc}
0 & \rightarrowtail & A_{\{0,1\}} & \rightarrowtail & A_{\{0,2\}} \\
 & & \downarrow & & \downarrow \\
 & & 0 & \rightarrowtail & A_{\{1,2\}} \\
 & & & & \downarrow \\
 & & & & 0
\end{array} \right\} \simeq \{U \rightarrowtail V \twoheadrightarrow W\}. \tag{1.2}$$

With respect to these equivalences, the face maps $\partial_0, \partial_1, \partial_2 : \mathcal{S}_2(\mathcal{C}) \to \mathcal{S}_1(\mathcal{C})$ send a conflation $U \rightarrowtail V \twoheadrightarrow W$ to W, V and U, respectively.

The simplicial groupoid $\mathcal{S}_\bullet(\mathcal{C})$ was originally defined by Waldhausen [**Wal85**] in the context of algebraic K-theory, where for a Waldhausen category $\mathcal{C}$—for example, an exact category—it was proved that there is an isomorphism of groups

$$K_i(\mathcal{C}) \simeq \pi_{i+1}|\mathcal{S}_\bullet(\mathcal{C})|. \qquad i \geq 0 \tag{1.3}$$

Here $|\mathcal{S}_\bullet(\mathcal{C})|$ denotes the geometric realization of the nerve of $\mathcal{S}_\bullet(\mathcal{C})$.

THEOREM 1.3.1 ([**DK19**, Proposition 2.4.8]). *Let $\mathcal{C}$ be a proto-exact category. The simplicial groupoid $\mathcal{S}_\bullet(\mathcal{C})$ is 2-Segal.*

The proof of Theorem 1.3.1 is elementary, effectively reducing to the Third Isomorphism Theorem for $\mathcal{C}$. The proof extends with only minor changes to the setting in which $\mathcal{C}$ is an exact ∞-category in the sense of Dyckerhoff and Kapranov [**DK19**, §7.2]; see [**DK19**, Theorem 7.3.3].

2. Hall algebras via 2-Segal spaces

Throughout this section $X_\bullet : \Delta^{\mathrm{op}} \to \mathscr{S}$ denotes a 2-Segal space.

2.1. An algebra in spans of groupoids. In this section we take $\mathscr{S}$ to be the model category of small groupoids. This suffices for all explicit examples below. Modifications (and strengthenings) of the content of this section when $\mathscr{S}$ is an ∞-category can be found in [**Ste21**].

Let $\mathsf{Span}(\mathscr{S})$ be the category of spans in $\mathscr{S}$. An object of $\mathsf{Span}(\mathscr{S})$ is an object of $\mathscr{S}$. A morphism $A \to B$ in $\mathsf{Span}(\mathscr{S})$ is an equivalence class of diagrams $A \leftarrow V \to B$ in $\mathscr{S}$, with two diagrams $A \leftarrow V \to B$ and $A \leftarrow V' \to B$ being equivalent if there exists an equivalence $V \to V'$ making the obvious diagrams commute. A monoidal structure on $\mathsf{Span}(\mathscr{S})$ is induced by cartesian product.

Let $m : X_1 \times X_1 \to X_1$ be the following morphism in $\mathsf{Span}(\mathscr{S})$:

$$X_{\{0,1\}} \times X_{\{1,2\}} \xleftarrow{(\partial_2,\partial_0)} X_2 \xrightarrow{\partial_1} X_{\{0,2\}}. \tag{2.1}$$

The composition $m \circ (m \times \mathrm{id}_{X_1})$ corresponds to the outside span in the diagram

(2.2)

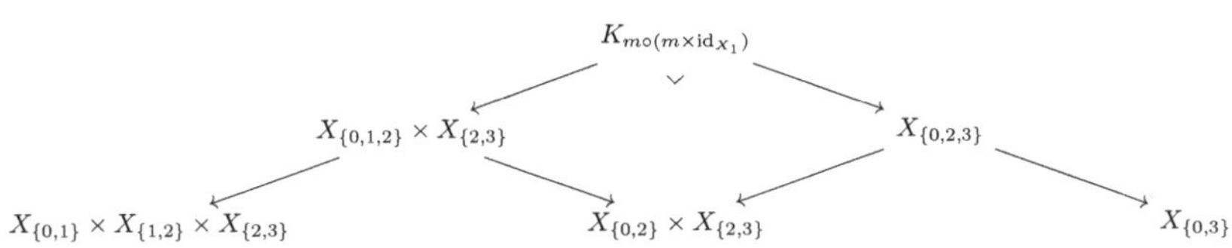

The symbol ⌟ indicates that the enclosing square is a pullback. The apex $K_{m\circ(m\times \mathrm{id}_{X_1})}$ is the middle term in the sequence

$$X_3 \xrightarrow{f} K_{m\circ(m\times \mathrm{id}_{X_1})} \xrightarrow{g} X_{\{0,1,2\}} \times_{X_{\{0,2\}}} X_{\{0,2,3\}},$$

where f is induced by the canonical maps $X_3 \to X_{\{0,1,2\}} \times X_{\{2,3\}}$ and $X_3 \to X_{\{0,2,3\}}$. The map g is clearly an equivalence and the composition $g \circ f$ is an equivalence by the 2-Segal property of $X_\bullet$; take $n = 3$, $i = 0$, $j = 2$ in Definition 1.1.2. The 2-out-of-3 property of equivalences therefore implies that f is an equivalence. Hence, the diagram (2.2) is equivalent to

(2.3)

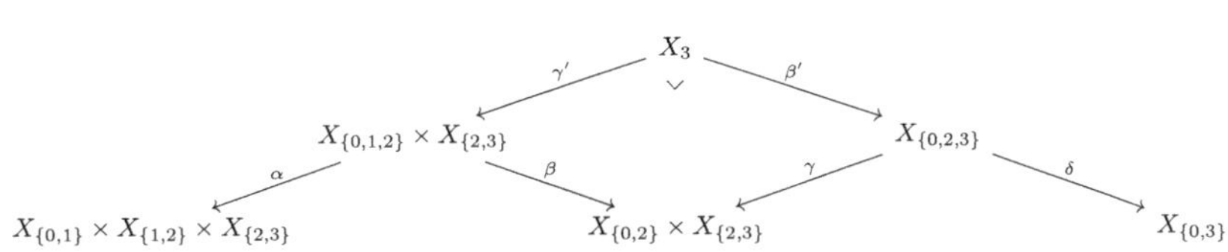

It follows that $m \circ (m \times \mathrm{id}_{X_1})$ is equal to the span

(2.4)

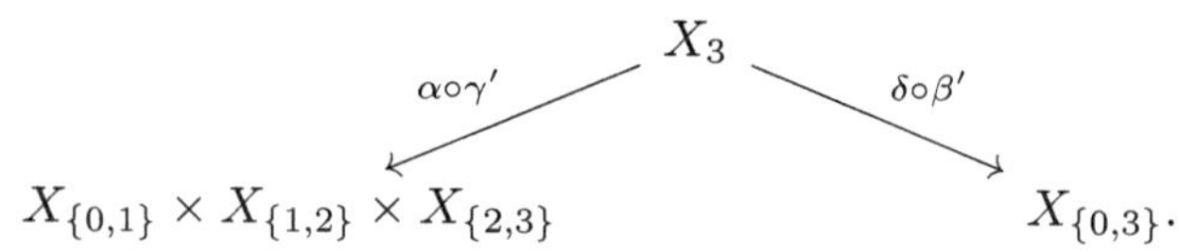

Similarly, the composition $m \circ (\mathrm{id}_{X_1} \times m)$ is given by the outside span in the diagram

(2.5)

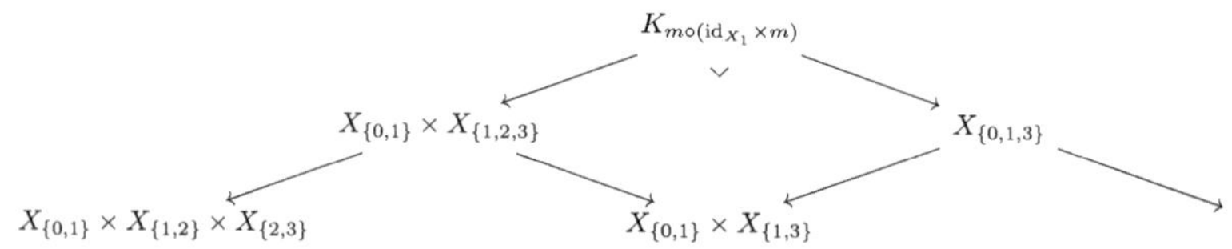

Arguing as above, this time using the case $n = 3$, $i = 1$, $j = 3$ of Definition 1.1.2, the outside span is again seen to equal (2.4).

Summarizing, the pair (X_1, m) is an associative algebra object in $\mathsf{Span}(\mathscr{S})$, associativity resulting from the first non-trivial 2-Segal conditions, that is, the conditions corresponding to $n = 3$ in Definition 1.1.2.

2.2. Linearization via theories with transfer. Following [**DK19**, §8.1], we construct algebras from a 2-Segal space $X_\bullet$. When $\mathscr{S}$ is the category of small groupoids, this can be seen as linearization of the algebra object (X_1, m) in $\mathsf{Span}(\mathscr{S})$ constructed in Section 2.1.

DEFINITION 2.2.1. Subclasses Sm and P of morphisms in $\mathscr{S}$ are called *smooth* and *proper*, respectively, if the following properties hold:

(1) Each subclass contains all weak equivalences and is closed under composition and cartesian product. In particular, Sm and P may be seen as subcategories of $\mathscr{S}$ which contain all objects.
(2) For each cartesian[1] diagram

(2.6)

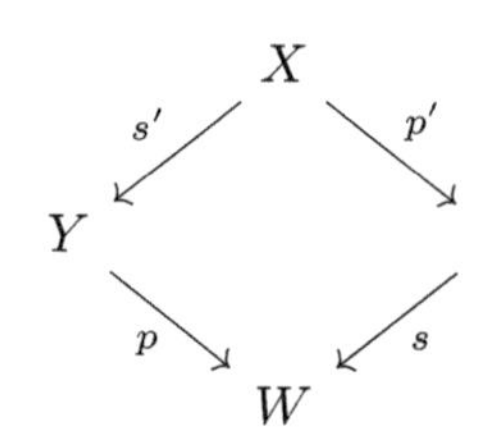

in $\mathscr{S}$ with $s \in \mathsf{Sm}$ and $p \in \mathsf{P}$, we have $s' \in \mathsf{Sm}$ and $p' \in \mathsf{P}$.

Let $\mathcal{V}$ be a monoidal category with monoidal unit $\mathbb{I}_{\mathcal{V}}$.

DEFINITION 2.2.2. A *theory with transfer on $\mathscr{S}$ valued in $\mathcal{V}$* is the data of a covariant functor $(-)_* : \mathsf{P} \to \mathcal{V}$ and a contravariant functor $(-)^* : \mathsf{Sm} \to \mathcal{V}$ with the following properties:

(1) Both $(-)_*$ and $(-)^*$ send weak equivalences in $\mathscr{S}$ to isomorphisms in $\mathcal{V}$.
(2) $X_* = X^*$ for all objects $X \in \mathscr{S}$; denote this common object by $\ell(X)$.

---

[1]Per our conventions for $\mathscr{S}$ outlined in Section 1.1, we abbreviate homotopy cartesian to cartesian.

(3) Monoidal data: there are isomorphisms $\ell(\mathrm{pt}) \xrightarrow{\sim} \mathbb{I}_{\mathcal{V}}$ and

$$\{m_{X,Y} : \ell(X) \otimes \ell(Y) \xrightarrow{\sim} \ell(X \times Y)\}_{X,Y \in \mathscr{S}}$$

which are natural with respect to $(-)^*$ and $(-)_*$ and satisfy the standard associativity and unit coherence conditions.

(4) Given a cartesian diagram (2.6) with $s \in \mathsf{Sm}$ and $p \in \mathsf{P}$, the Beck–Chevalley-type equality

$$s^* \circ p_* = p'_* \circ s'^*$$

of morphisms $\ell(Y) \to \ell(Z)$ holds.

For ease of notation, we denote a theory with transfer simply by ℓ.

Recall again the span (2.1) which defines the multiplication m on X_1. Consider also the span

$$\begin{array}{ccc} & X_0 & \\ {}^{\mathrm{can}}\swarrow & & \searrow^{s_0} \\ \mathrm{pt} & & X_1, \end{array} \tag{2.7}$$

where can is the canonical map.

THEOREM 2.2.3 ([**DK19**, Proposition 8.1.5]). *Let ℓ be a theory with transfer on $\mathscr{S}$ valued in $\mathcal{V}$ and $X_\bullet : \Delta^{\mathrm{op}} \to \mathscr{S}$ a 2-Segal space. If $(\partial_2, \partial_0) \in \mathsf{Sm}$ and $\partial_1 \in \mathsf{P}$, then $H(X_\bullet; \ell) := \ell(X_1)$ is an associative algebra object in $\mathcal{V}$ with multiplication*

$$\ell(X_1) \otimes \ell(X_1) \simeq \ell(X_1 \times X_1) \xrightarrow{\partial_{1*} \circ (\partial_2, \partial_0)^*} \ell(X_1)$$

induced by the span (2.1). *If, moreover,* can $\in \mathsf{Sm}$ *and $s_0 \in \mathsf{P}$, then the morphism*

$$\mathbb{I}_{\mathcal{V}} \xrightarrow{\sim} \ell(\mathrm{pt}) \xrightarrow{s_{0*} \circ \mathrm{can}^*} \ell(X_1)$$

induced by the span (2.7) *is a unit for the algebra $H(X_\bullet; \ell)$.*

PROOF. Write $\mathfrak{m}$ for the multiplication $\ell(X_1) \otimes \ell(X_1) \to \ell(X_1)$. Keeping the notation of diagram (2.3) and using the monoidal coherence data, the composition $\mathfrak{m} \circ (\mathfrak{m} \otimes \mathrm{id}_{\ell(X_1)})$ is equal to

$$\delta_* \circ \gamma^* \circ \beta_* \circ \alpha^* = \delta_* \circ \beta'_* \circ \gamma'^* \circ \alpha^* = (\delta \circ \beta')_* \circ (\alpha \circ \gamma')^*.$$

The first equality is the Beck–Chevalley property and the second is functoriality of $(-)_*$ and $(-)^*$. The final expression is the result of push-pull along the span (2.4). Similarly, the composition $\mathfrak{m} \circ (\mathrm{id}_{\ell(X_1)} \otimes \mathfrak{m})$, which results from push-pull along the span (2.5), is equal to the composition $(\delta \circ \beta')_* \circ (\alpha \circ \gamma')^*$. This establishes the associativity of $\mathfrak{m}$.

By Theorem 1.1.4, the 2-Segal space $X_\bullet$ is unital. Using this, it is straightforward to verify unitality of $\ell(X_1)$. □

2.3. Hall algebras of finitary categories. We specialize the results of Section 2.2 to recover the Hall algebras formulated by Ringel [**Rin90a**].

Given an essentially small groupoid $\mathcal{A}$, let $\ell(\mathcal{A}) = \mathrm{Fun}_c(\pi_0\mathcal{A}, \mathbb{C})$ be the vector space of finitely supported complex-valued functions on the set $\pi_0\mathcal{A}$ of isomorphism classes of objects of $\mathcal{A}$.

DEFINITION 2.3.1 ([**DK19**, §8.2]).

(1) A morphism of groupoids $f : \mathcal{A} \to \mathcal{B}$ is called *weakly proper* if the induced map $\pi_0 f : \pi_0 \mathcal{A} \to \pi_0 \mathcal{B}$ is finite-to-one.
(2) A morphism of groupoids $f : \mathcal{A} \to \mathcal{B}$ is called *locally proper* for each $a \in \mathcal{A}$, the induced group homomorphism $f : \mathrm{Aut}_{\mathcal{A}}(a) \to \mathrm{Aut}_{\mathcal{B}}(f(a))$ has finite kernel and cokernel.

If $f : \mathcal{A} \to \mathcal{B}$ is weakly proper, then there is an obvious pullback morphism $f^* : \ell(\mathcal{B}) \to \ell(\mathcal{A})$. If $f : \mathcal{A} \to \mathcal{B}$ is locally proper, then there is a pushforward $f_* : \ell(\mathcal{A}) \to \ell(\mathcal{B})$ given by

$$(f_*\phi)(b) = \int_{Rf^{-1}(b)} \phi_{|Rf^{-1}(b)},$$

where $Rf^{-1}(b)$ denotes the homotopy fibre of f over $b \in \mathcal{B}$ and the integral is defined as follows. Let $\mathcal{G}$ be a small groupoid with the property that each group $\mathrm{Aut}_{\mathcal{G}}(x)$, $x \in \mathcal{G}$, is finite. Then the integral of a function $\phi \in \ell(\mathcal{G})$ is defined by

$$\int_{\mathcal{G}} \phi = \sum_{x \in \pi_0 \mathcal{G}} \frac{1}{|\mathrm{Aut}_{\mathcal{G}}(x)|} \phi(x).$$

With the above definitions, taking P to be the collection of weakly proper morphisms and Sm the collection of locally proper morphisms, ℓ forms a $\mathsf{Vect}_{\mathbb{C}}$-valued theory with transfer on the model category of small groupoids [**DK19**, Lemma 8.2.7].

DEFINITION 2.3.2. A proto-exact category $\mathcal{C}$ is called *finitary* if it is essentially small and, for each $U, V \in \mathcal{C}$, the sets $\mathrm{Hom}_{\mathcal{C}}(U,V)$ and $\mathrm{Ext}^1_{\mathcal{C}}(U,V)$ are finite.

Let $X_\bullet = \mathcal{S}_\bullet(\mathcal{C})$ be the Waldhausen construction of a finitary proto-exact category $\mathcal{C}$, seen as a 2-Segal simplicial groupoid. The span (2.1) takes the form

$$\begin{array}{ccccc} & & \mathcal{S}_2(\mathcal{C}) & & \\ & \overset{(\partial_2,\partial_0)}{\swarrow} & & \overset{\partial_1}{\searrow} & \\ \mathcal{S}_1(\mathcal{C}) \times \mathcal{S}_1(\mathcal{C}) & & & & \mathcal{S}_1(\mathcal{C}) \end{array}, \qquad \begin{array}{ccccc} & & U \rightarrowtail V \twoheadrightarrow W & & \\ & \swarrow & & \searrow & \\ (U,W) & & & & V. \end{array}$$

LEMMA 2.3.3.

(1) *The morphism (∂_2, ∂_0) is weakly proper, that is, $(\partial_2, \partial_0) \in \mathsf{Sm}$.*
(2) *The morphism ∂_1 is locally proper, that is, $\partial_1 \in \mathsf{P}$.*

PROOF.

(1) This follows from the assumption that for each pair $U, W \in \mathcal{C}$, the set $\mathrm{Ext}^1_{\mathcal{C}}(W,U)$ is finite.
(2) This follows from the assumption that for each pair $U, W \in \mathcal{C}$, the set $\mathrm{Hom}_{\mathcal{C}}(W,U)$ is finite. □

It follows from Theorem 2.2.3 that there is a unital associative algebra

$$H(\mathcal{C}) := H(\mathcal{S}_\bullet(\mathcal{C}); \ell).$$

The algebra $H(\mathcal{C})$ is the *Hall algebra of* $\mathcal{C}$, as originally defined by Ringel [**Rin90a**]. The vector space $H(\mathcal{C})$ has a canonical basis $\{1_U\}_{U \in \pi_0 \mathcal{C}}$ labelled by isomorphism classes of objects of $\mathcal{C}$. In this basis, the multiplication reads

$$1_U \cdot 1_W = \sum_{V \in \pi_0 \mathcal{C}} c^V_{U,W} 1_V.$$

Here $c^V_{U,W}$ is the cardinality of the set

$$\{U' \subset V \mid U' \simeq U,\ V/U' \simeq W\}.$$

Alternatively, we can write

$$c^V_{U,W} = \frac{|\mathrm{Aut}_{\mathcal{C}}(V)|}{|\mathrm{Aut}_{\mathcal{C}}(U)||\mathrm{Aut}_{\mathcal{C}}(W)|} \frac{|\mathrm{Ext}^1_{\mathcal{C}}(W,U)_V|}{|\mathrm{Hom}_{\mathcal{C}}(W,U)|},$$

where $\mathrm{Ext}^1_{\mathcal{C}}(W,U)_V \subset \mathrm{Ext}^1_{\mathcal{C}}(W,U)$ denotes the subset of extensions (seen as isomorphism classes of conflations) whose middle term is isomorphic to V. See the original work of Riedtmann [**Rie94**, §4] or the exposition of Schiffmann [**Sch12b**, §1.3] for proofs. In this form, it is apparent that the finitary assumption on $\mathcal{C}$ ensures that $c^V_{U,W}$ is finite and, for fixed $U, W \in \pi_0\mathcal{C}$, is non-zero for only finitely many $V \in \pi_0\mathcal{C}$. The unit of the algebra $H(\mathcal{C})$ is 1_0.

We work out various instances of the above construction. Let $\mathbb{F}_q$ be a finite field of cardinality q.

EXAMPLE 2.3.4. Let $\mathsf{vect}_{\mathbb{F}_q}$ be the finitary abelian category of finite-dimensional vector spaces over $\mathbb{F}_q$. The Hall algebra

$$H(\mathsf{vect}_{\mathbb{F}_q}) \simeq \bigoplus_{n \in \mathbb{Z}_{\geq 0}} \mathbb{C} \cdot 1_n$$

has multiplication determined by

$$1_n \cdot 1_m = \begin{bmatrix} n+m \\ n \end{bmatrix}_q 1_{n+m},$$

where our conventions for quantum integers read

$$[n]_q = \frac{q^n - 1}{q-1}, \qquad [n]_q! = \prod_{i=1}^{n} [i]_q \qquad \begin{bmatrix} n+m \\ n \end{bmatrix}_q = \frac{[n+m]_q!}{[n]_q![m]_q!}.$$

The structure constant $\left[\begin{smallmatrix} n+m \\ n \end{smallmatrix}\right]_q$ is the number of $\mathbb{F}_q$-rational points of the Grassmannian $\mathrm{Gr}(n, n+m)$.

EXAMPLE 2.3.5. Let $\mathbb{F}_q[x]\text{-}\mathsf{mod}^{\mathsf{nil}}$ be the finitary abelian category of finite-dimensional nilpotent $\mathbb{F}_q[x]$-modules. Jordan canonical form induces a bijection from the set of isomorphism classes of objects of $\mathbb{F}_q[x]\text{-}\mathsf{mod}^{\mathsf{nil}}$ to the set $\mathcal{P}$ of partitions, so that we may identify

$$H(\mathbb{F}_q[x]\text{-}\mathsf{mod}^{\mathsf{nil}}) \simeq \bigoplus_{\lambda \in \mathcal{P}} \mathbb{C} \cdot 1_\lambda.$$

The structure constants $c^\nu_{\lambda,\mu}$ are rather complicated and do not admit a closed-form expression. However, studying these coefficients when μ is of the form (1^r) allows one to prove that $H(\mathbb{F}_q[x]\text{-}\mathsf{mod}^{\mathsf{nil}})$ is a polynomial algebra on the variables $1_{(1^r)}$, $r \geq 1$. Using this, it is not difficult to show that $H(\mathbb{F}_q[x]\text{-}\mathsf{mod}^{\mathsf{nil}})$ is isomorphic to Macdonald's ring of symmetric functions [**Mac95**], an important algebra in combinatorics.

Taking q to be a prime p, the algebra $H(\mathbb{F}_p[x]\text{-}\mathsf{mod}^{\mathsf{nil}})$ is isomorphic to the original Hall algebra, introduced by Steinitz [**Ste01**] in the context of the combinatorics of the lattice of finite abelian p-groups.

EXAMPLE 2.3.6 ([**Rin90b, Gre95**]). Let Q be an acyclic quiver, that is, a finite oriented multigraph without cycles. Let $\mathsf{rep}_{\mathbb{F}_q}(Q)$ be the category of finite-dimensional representations of Q over $\mathbb{F}_q$. Concretely, a representation of Q is a finite-dimensional vector space V_i over $\mathbb{F}_q$ for each vertex i of Q and a $\mathbb{F}_q$-linear map $V_i \to V_j$ for each arrow $i \to j$ of Q. The category $\mathcal{C}$ is finitary abelian. Denote by S_i the 1-dimensional representation which is supported only at the vertex i. Any simple object of $\mathcal{C}$ is isomorphic to a unique S_i. See [**DW17**] for an introduction to the representation theory of quivers.

Let $\mathfrak{g}_Q$ be the complex generalized Kac–Moody algebra whose generalized Dynkin diagram is the underlying multigraph of Q. There is a bijection between the vertices of Q and the set Π^+ of positive simple roots of $\mathfrak{g}_Q$. Let $U_v(\mathfrak{g}_Q)$ be the De Concini–Kac quantum group at quantum parameter $v \in \mathbb{C}^\times$. The algebra $U_v(\mathfrak{g}_Q)$ has a standard presentation by generators and relations. The Chevalley generators are E_i, F_i, $i \in \Pi^+$ and the group-like Cartan generators are $K_\alpha^{\pm 1}$, α a simple coroot. Let $U_v^+(\mathfrak{g}_Q)$ be the subalgebra of $U_v(\mathfrak{g}_Q)$ generated by E_i; its only relations are the quantum Serre relations. For background on Kac–Moody algebras, the reader is referred to [**Kac90**] or, from the perspective of Hall algebras, [**Sch12b**, Appendix A].

It was proved by Ringel [**Rin90b**] and Green [**Gre95**] that the assignment $E_i \mapsto 1_{S_i}$ extends to an algebra embedding

$$U^+_{\sqrt{q}}(\mathfrak{g}_Q) \hookrightarrow H(\mathsf{rep}_{\mathbb{F}_q}(Q))$$

which is an isomorphism if and only if Q is of finite representation type, that is, Q is an orientation of a simply laced Dynkin diagram. The algebra $H(\mathsf{rep}_{\mathbb{F}_q}(Q))$ itself is in fact a quantization of a q-dependent Borcherds algebra [**SVdB01**]. For a general discussion of the relationship between Hall algebras of finitary exact categories and quantum groups, see the work of Berenstein–Greenstein [**BG16**].

More generally, if Q has cycles (including loops), one can consider the category of representations of Q with nilpotency conditions on morphisms associated to cycles. For example, the category $\mathsf{rep}^{\mathsf{nil}}_{\mathbb{F}_q}(Q_{\mathrm{Jor}})$ of nilpotent representations of the Jordan quiver Q_{Jor} (a single vertex with a loop) is equivalent to the category $\mathbb{F}_q[x]\text{-}\mathsf{mod}^{\mathsf{nil}}$ considered in Example 2.3.5. In this way, one can expand the class of generalized quantum Kac–Moody algebras realized by Hall algebras [**KS06**].

The theory of Hall algebras of quivers is rich and interesting. A detailed introduction can be found in [**Sch12b**].

EXAMPLE 2.3.7 ([**Kap97**]). Let X be a smooth projective irreducible curve over a finite field $\mathbb{F}_q$. The category $\mathsf{coh}(X)$ of coherent sheaves on X is finitary abelian and hence has an associated Hall algebra $H(X) := H(\mathsf{coh}(X))$.

The case of $X = \mathbb{P}^1$, the projective line over $\mathbb{F}_q$, was studied by Kapranov [**Kap97**]. See [**BK01**] for an expository account. To explain the main result, note that the full subcategory $\mathsf{tor}(\mathbb{P}^1) \subset \mathsf{coh}(\mathbb{P}^1)$ of torsion sheaves is abelian and closed under extensions. Using functoriality of the Hall algebra construction, as discussed in Section 3 below, this translates to define a subalgebra $H(\mathsf{tor}(\mathbb{P}^1))$ of $H(\mathbb{P}^1)$. Consideration of the support of a torsion sheaf leads to a decomposition

$$\mathsf{tor}(\mathbb{P}^1) \simeq \prod_{x \in \mathbb{P}^1} \mathsf{tor}(\mathbb{P}^1)_x$$

labelled by the closed points of $\mathbb{P}^1$ and hence an isomorphism of algebras

$$H(\mathsf{tor}(\mathbb{P}^1)) \simeq \bigotimes_{x\in\mathbb{P}^1} H(\mathsf{tor}(\mathbb{P}^1)_x).$$

The abelian equivalence $\mathsf{tor}(\mathbb{P}^1)_x \simeq \mathsf{rep}^{\mathsf{nil}}_{\mathbb{F}_{q^{\deg x}}}(Q_{\mathrm{Jor}})$ yields, via Example 2.3.5, a description of $H(\mathsf{tor}(\mathbb{P}^1))$ in terms of algebras of symmetric functions. Similarly, the subcategory $\mathsf{vect}(\mathbb{P}^1) \subset \mathsf{coh}(\mathbb{P}^1)$ of vector bundles and their morphisms of constant rank is exact and closed under extensions. The resulting subalgebra $H(\mathsf{vect}(\mathbb{P}^1))$ of $H(\mathbb{P}^1)$ can be interpreted as the algebra of unramified automorphic forms for the groups GL_n, $n \geq 0$, with parabolic induction as the product. Finally, that the pair $(\mathsf{tor}(\mathbb{P}^1), \mathsf{vect}(\mathbb{P}^1))$ is a torsion theory for $\mathsf{coh}(\mathbb{P}^1)$ translates to the statement that $H(\mathbb{P}^1)$ is a semi-direct product of $H(\mathsf{tor}(\mathbb{P}^1))$ and $H(\mathsf{vect}(\mathbb{P}^1))$ with respect to the Hecke action of the former on the latter. By taking a subalgebra $\tilde{H}(\mathsf{tor}(\mathbb{P}^1))$ of $H(\mathsf{tor}(\mathbb{P}^1))$ obtained as a suitable average of $H(\mathsf{tor}(\mathbb{P}^1)_x)$, $x \in \mathbb{P}^1$, it is proved that there is an algebra isomorphism

$$\tilde{H}(\mathsf{tor}(\mathbb{P}^1)) \rtimes H(\mathsf{vect}(\mathbb{P}^1)) \simeq U^{D,+}_{\sqrt{q}}(\widehat{\mathfrak{sl}_2(\mathbb{C})})$$

and hence a non-surjective algebra embedding

$$U^{D,+}_{\sqrt{q}}(\widehat{\mathfrak{sl}_2(\mathbb{C})}) \hookrightarrow H(\mathbb{P}^1).$$

Here $U^{D,+}_{\sqrt{q}}(\widehat{\mathfrak{sl}_2(\mathbb{C})})$ denotes the positive part of the quantum affine algebra $U_v(\widehat{\mathfrak{sl}_2(\mathbb{C})})$ in Drinfeld's new realization [**Dri87**], specialized to $v = \sqrt{q}$. This realization is particularly natural when $\widehat{\mathfrak{sl}_2(\mathbb{C})}$ is constructed from the loop algebra of $\mathfrak{sl}_2(\mathbb{C})$.

On the other hand, denoting by K_2 the Kronecker quiver (vertex set $\{1,2\}$ with two arrows from 1 to 2), Example 2.3.6 asserts that $H(\mathsf{rep}_{\mathbb{F}_q}(K_2))$ contains as a subalgebra the positive part of the standard Chevalley presentation of $U_{\sqrt{q}}(\widehat{\mathfrak{sl}_2(\mathbb{C})})$. To explain this connection, note that while the abelian categories $\mathsf{coh}(\mathbb{P}^1_{/\mathbb{F}_q})$ and $\mathsf{rep}_{\mathbb{F}_q}(K_2)$ are not equivalent—the latter is finite length, the former is not—Beilinson [**Bei78**] proved that there is an equivalence of their derived categories:

$$D^b(\mathsf{coh}(\mathbb{P}^1_{/\mathbb{F}_q})) \simeq D^b(\mathsf{rep}_{\mathbb{F}_q}(K_2)). \tag{2.8}$$

This leads to the idea that there should exist a derived generalization of the Hall algebra construction which, when applied to the category (2.8), realizes the full quantum group $U_{\sqrt{q}}(\widehat{\mathfrak{sl}_2(\mathbb{C})})$, whereas the hearts $\mathsf{rep}_{\mathbb{F}_q}(K_2)$ and $\mathsf{coh}(\mathbb{P}^1_{/\mathbb{F}_q})$ realize only (non-conjugate) positive parts thereof. See [**PX00, Cra10, Bri13**] for work in this general direction.

Hall algebras of elliptic curves have also been studied in detail; see [**BS12, Sch12c**]. Hall algebras of curves of higher genus remain largely mysterious; see [**KSV17**].

REMARK 2.3.8. In many of the examples above, the Hall algebras arise from the base change to $\mathbb{F}_q$ of a category $\mathcal{C}_\mathbb{Z}$ defined over $\mathbb{Z}$. What is sometimes called the Hall algebra is an algebra A over the Laurent polynomial ring which interpolates between the reduction:

$$H(\mathcal{C}_\mathbb{Z} \otimes_\mathbb{Z} \mathbb{F}_q) \simeq A \otimes_{\mathbb{Z}[q,q^{-1}]} \mathbb{C}.$$

One important subtlety in the subject is that the existence of such an algebra A is not guaranteed.

We turn to examples of Hall algebras arising from non-exact categories.

EXAMPLE 2.3.9. The proto-exact category $\mathsf{vect}_{\mathbb{F}_1}$ of finite-dimensional vector spaces over $\mathbb{F}_1$, as defined in Example 1.2.2, is finitary. The Hall algebra

$$H(\mathsf{vect}_{\mathbb{F}_1}) \simeq \bigoplus_{n \in \mathbb{Z}_{\geq 0}} \mathbb{C} \cdot 1_n$$

has multiplication

$$1_n \cdot 1_m = \binom{n+m}{n} 1_{n+m}.$$

This algebra is sometimes called the *divided power algebra* because of the identification with the integral subring $\mathbb{Z}\langle \frac{x^n}{n!} : n \geq 0 \rangle \subset \mathbb{C}[x]$. The structure constant $\binom{n+m}{n}$ is the Euler characteristic of the complex Grassmannian $\mathrm{Gr}(n, n+m)$. In particular, this example is a $q \to 1$ limit of Example 2.3.4 in a rather obvious sense.

EXAMPLE 2.3.10 ([**Szc12**]). Let $\mathsf{rep}_{\mathbb{F}_1}(Q)$ be the category of representations of an acyclic quiver Q in $\mathsf{vect}_{\mathbb{F}_1}$. The category $\mathsf{rep}_{\mathbb{F}_1}(Q)$ inherits a proto-exact structure from $\mathsf{vect}_{\mathbb{F}_1}$ and is finitary. The Hall algebra $H(\mathsf{rep}_{\mathbb{F}_1}(Q))$ was studied in detail by Szczesny, who proved that the assignment $E_i \mapsto 1_{S_i}$ extends to an algebra homomorphism

$$U^+(\mathfrak{g}_Q) \to H(\mathsf{rep}_{\mathbb{F}_1}(Q)). \tag{2.9}$$

Here $U^+(\mathfrak{g}_Q)$ denotes the positive part of the (classical) enveloping algebra of $\mathfrak{g}_Q$. In general, the map (2.9) is neither surjective nor injective, even for Dynkin quivers, in contrast to Example 2.3.6. To see the basic problem in a simple example, let Q be the inward orientation of the Dynkin diagram of type D_4, so that $\mathfrak{g}_Q \simeq \mathfrak{so}_8(\mathbb{C})$. When q is a prime power, the isomorphism class of the indecomposable representation

$$\begin{array}{ccccc} & & \mathbb{F}_q & & \\ & & \big\downarrow{\scriptstyle \binom{1}{1}} & & \\ \mathbb{F}_q & \xrightarrow{\binom{1}{0}} & \mathbb{F}_q^2 & \xleftarrow{\binom{0}{1}} & \mathbb{F}_q \end{array}$$

realizes a non-simple positive root of $U^+_{\sqrt{q}}(\mathfrak{so}_8(\mathbb{C}))$. However, observe that this representation is not defined over $\mathbb{F}_1$: the matrix $\binom{1}{1}$ does not correspond to a morphism in $\mathsf{vect}_{\mathbb{F}_1}$. In this example, non-surjectivity of the map (2.9) can be traced back to this observation. On the other hand, if Q is an orientation of a Dynkin quiver of type A, where all indecomposable representations can be realized using only identity and zero matrices, the map (2.9) is an isomorphism. This illustrates the difficulty in establishing a precise relationship between $H(\mathsf{rep}_{\mathbb{F}_1}(Q))$ and a $q \to 1$ limit of $H(\mathsf{rep}_{\mathbb{F}_q}(Q))$.

EXAMPLE 2.3.11 ([**Szc12**]). Let $\mathbb{F}_1[t]$ be the free unital monoid with 0, generated by t, and $\mathbb{F}_1[t, t^{-1}]$ its localization at the origin. The affine line $\mathbb{A}^1$, seen as a monoid scheme, is $\operatorname{Spec} \mathbb{F}_1[t]$. Then $\mathbb{P}^1_{/\mathbb{F}_1}$, the *projective line over* $\mathbb{F}_1$, is the pushout of the diagram obtained by applying Spec to the diagram

$$\begin{array}{ccc} & & \mathbb{F}_1[t] \\ & & \downarrow \\ \mathbb{F}_1[t] & \hookrightarrow & \mathbb{F}_1[t, t^{-1}]. \end{array}$$

The category $\mathsf{coh}(\mathbb{P}^1_{/\mathbb{F}_1})$ of (normal) coherent sheaves on $\mathbb{P}^1_{/\mathbb{F}_1}$ is proto-exact and finitary. Arguing much as in the case of $\mathbb{P}^1_{/\mathbb{F}_q}$ (see Example 2.3.7), Szczesny proved that $H(\mathsf{coh}(\mathbb{P}^1_{/\mathbb{F}_1}))$ contains as a subalgebra a non-standard Borel subalgebra of $U(\mathcal{L}\mathfrak{sl}_2(\mathbb{C}))$. In fact, there is an isomorphism

$$U(\mathcal{L}\mathfrak{gl}_2(\mathbb{C})^+ \oplus \kappa) \simeq H(\mathsf{coh}(\mathbb{P}^1_{/\mathbb{F}_1})),$$

where κ is the abelian Lie algebra on generators c_n, $n \in \mathbb{Z}_{\geq 0}$. Geometrically, c_n corresponds to the cyclic sheaf which globalizes the $\mathbb{F}_1[t]$-module $C_n = \{\{z_i\}_{i \in \mathbb{Z}/n\mathbb{Z}}, *\}$ on which t acts by $t \cdot z_i = z_{i+1}$ and 0 sends everything to $*$.

Elliptic Hall algebras over $\mathbb{F}_1$ were studied by Yanagida [**Yan17**], who proved that they recover the specialized elliptic Hall algebras of Morton–Samuelson [**MS17**] and hence skein algebras of tori. Over $\mathbb{F}_1$, the curve case stops here: curves of genus greater than one do not have monoid scheme models.

In a different direction, Hall algebras of toric monoid schemes were studied by Szczesny [**Szc18**] and Jun–Szczesny [**JS24**]. This is one approach to the largely unexplored theory of Hall algebras of higher-dimensional varieties.

EXAMPLE 2.3.12 ([**EJS20**]). Let Mat_* be the category of pointed matroids and their strong maps. Given a matroid M, denote by E_M the union of the non-zero elements of its ground set with a marked point $*$. Given a subset $S \subset E_M$, write $M_{|S}$ and M/S for the restriction and contraction of M along S, respectively. See [**Oxl92**] for an introduction to matroids.

The category Mat_* admits a finitary proto-exact structure in which inflations are strong maps $f : N \to M$ which factor as $N \xrightarrow{\sim} M_{|S} \hookrightarrow M$ for some subset S of E_M and deflations are strong maps $f : N \to M$ which factor as $N \twoheadrightarrow N/S \xrightarrow{\sim} M$ for some subset S of E_N. The Hall algebra $H(\mathsf{Mat}_*)$ is a (Hopf) algebra dual of Schmitt's matroid-minor Hopf algebra [**Sch94**], which is of central importance in algebraic combinatorics. This algebra is not obviously related to a quasi-classical limit of some other algebra, in contrast to Examples 2.3.9–2.3.11.

Relatedly, the proto-exact structure on Mat_* allows for the definition of its algebraic K-theory $K_\bullet(\mathsf{Mat}_*)$ via the isomorphism (1.3). Eppolito, Jun and Szczesny [**EJS20**] proved that $K_\bullet(\mathsf{Mat}_*)$ naturally contains as a subgroup the stable homotopy groups $\pi^s_\bullet(\mathbb{S})$ of the sphere spectrum.

We end this section with another choice of theory with transfer ℓ.

EXAMPLE 2.3.13. Let $\ell(-) = K_0([-, \mathsf{vect}_\mathbb{C}]_c)$ be the functor which assigns to an essentially small groupoid the Grothendieck group of its category of finitely supported complex representations. The construction of a theory with transfer based on ℓ is given in [**DK19**, §8.3]. The Hall algebra $H(\mathcal{S}_\bullet(\mathsf{vect}_{\mathbb{F}_q}); \ell)$ is isomorphic to

$$\bigoplus_{n=0}^{\infty} K_0(\mathsf{rep}_{\mathbb{C}}(\mathrm{GL}_n(\mathbb{F}_q)))$$

with parabolic induction as the product. This algebra was studied in detail by Green [**Gre55**] and Zelevinsky [**Zel81**], who proved that it is a polynomial algebra on the cuspidal representations of the tower of groups $\mathrm{GL}_n(\mathbb{F}_q)$, $n \geq 0$.

2.4. A non-finitary example: the cohomological Hall algebra. Consider the non-finitary abelian category $\mathsf{rep}_{\mathbb{C}}(Q)$ of finite-dimensional complex representations of a quiver Q. We consider $\mathcal{S}_\bullet(\mathsf{rep}_{\mathbb{C}}(Q))$ as a simplicial locally finite Artin stack [**DK19**, §8.5].

Let $\ell(-) = H^\bullet(-;\mathbb{Q})$ be singular cohomology. See [**Beh04**] for an introduction to the cohomology of algebraic stacks and, in particular, for the interpretation of the singular cohomology of a quotient stack $[X/G]$ as the G-equivariant cohomology of X. The Hall algebra $H(Q) := H(\mathcal{S}_\bullet(\mathsf{rep}_{\mathbb{C}}(Q));\ell)$, called the *cohomological Hall algebra* of Q, was introduced by Kontsevich and Soibelman [**KS11**] as a foundational object in Donaldson–Thomas theory, that is, the virtual counting of stable objects in 3-dimensional Calabi–Yau categories. To be concrete, consider the case of the quiver Q_m with a single node and $m \geq 0$ loops. In this case, there is an isomorphism

$$\mathcal{S}_1(\mathsf{rep}_{\mathbb{C}}(Q_m)) \simeq \bigsqcup_{d=0}^{\infty} [\mathfrak{gl}_d(\mathbb{C})^{\oplus m} / \mathrm{GL}_d(\mathbb{C})].$$

Here $[\mathfrak{gl}_d(\mathbb{C})^{\oplus m} / \mathrm{GL}_d(\mathbb{C})]$ is the global quotient Artin stack obtained from the simultaneous conjugation action of $\mathrm{GL}_d(\mathbb{C})$ on the vector space $\mathfrak{gl}_d(\mathbb{C})^{\oplus m}$ of m-tuples of square matrices. The underlying $(\mathbb{Z}\times\mathbb{Z})$-graded vector space of $H(Q_m)$ decomposes according to the dimension d of a representation:

$$H(Q_m) = \bigoplus_{d=0}^{\infty} H_d(Q_m).$$

Since $\mathfrak{gl}_d(\mathbb{C})^{\oplus m}$ is $\mathrm{GL}_d(\mathbb{C})$-equivariantly contractible, the cohomology of the stack $[\mathfrak{gl}_d(\mathbb{C})^{\oplus m} / \mathrm{GL}_d(\mathbb{C})]$ reduces to the $\mathrm{GL}_d(\mathbb{C})$-equivariant cohomology of a point:

$$H_d(Q_m) \simeq H^\bullet(B\,\mathrm{GL}_d(\mathbb{C})) \simeq \mathbb{Q}[x_1, \dots, x_d]^{\mathfrak{S}_d}, \qquad |x_i| = 2.$$

Using localization in equivariant cohomology, the (m-dependent) product of $f_1 \in H_{d'}(Q_m)$ and $f_2 \in H_{d''}(Q_m)$ is seen to be

$$f_1 \cdot f_2 = \sum_{\pi \in \mathfrak{sh}_{d',d''}} \pi\Big(f_1(x'_1, \dots, x'_{d'}) f_2(x''_1, \dots, x''_{d''}) \prod_{l=1}^{d''} \prod_{k=1}^{d'} (x''_l - x'_k)^{m-1}\Big).$$

Here $\mathfrak{sh}_{d',d''}$ is the set of 2-shuffles of size (d', d''), that is, elements $\pi \in \mathfrak{S}_{d'+d''}$ which satisfy

$$\pi(1) < \cdots < \pi(d'), \qquad \pi(d'+1) < \cdots < \pi(d'+d'').$$

By a direct study of this multiplication, Efimov [**Efi12**] proved that there exists a $(\mathbb{Z} \times \mathbb{Z})$-graded vector space of the form $V_{Q_m} = V^{\mathrm{prim}}_{Q_m} \otimes \mathbb{Q}[u]$ and a graded algebra isomorphism[2] $\mathrm{Sym}(V_{Q_m}) \simeq H(Q_m)$. Moreover, Efimov proved that each homogeneous summand $V^{\mathrm{prim}}_{Q_m,d} \subset V^{\mathrm{prim}}_{Q_m}$ is finite-dimensional. The vector space

[2]We are ignoring a dimension-dependent shift (involving the Euler form) of the cohomological grading of $H(Q_m)$ which is required for this graded algebra isomorphism.

$V^{\mathrm{prim}}_{Q_m}$ is the *cohomological Donaldson–Thomas invariant* of Q_m. Its graded finite-dimensionality was expected in view of its conjectured geometric interpretation. Indeed, it was later proved that $V^{\mathrm{prim}}_{Q_m}$ encodes the intersection cohomology groups of the closure of the moduli space of stable representations of Q_m in the moduli space of semistable representations [**FR18**,**MR19**]. In this way, structure theorems about the cohomological Hall algebra of a quiver translate to geometric properties of its moduli space of representations. These structural and geometric results generalize to symmetric quivers Q with potential W via *critical* cohomological Hall algebras [**KS11**, **Dav17**, **DM20**], where $V^{\mathrm{prim}}_{(Q,W)}$ is also given an interpretation as the long-sought Lie algebra of BPS states [**HM98**].

The linearization ℓ can be varied in a number of ways. For example, taking ℓ to be so-called motivic functions yields the *motivic Hall algebra* of Q, cf. [**Joy07**,**KS08**, **Bri12**]. This is an extremely large algebra, much larger than the cohomological Hall algebra. The motivic Hall algebra encodes the motivic Donaldson–Thomas invariants of Q (see [**JS12**]) and it has recently found connections to vertex operator algebras [**Joy18**]. See [**Lat21**] for similar connections in the cohomological setting.

3. Functoriality between Hall algebras

One advantage of using 2-Segal spaces as a framework for understanding Hall algebras is that clearer statements concerning functoriality of the Hall algebra construction can be introduced. The purpose of this section is to survey material which relates to questions of functoriality. A summary of statements pertaining to Hall (co)algebras is found in Theorem 3.3.1 below.

Although we have adapted notation to agree with the rest of this document, the technical background for the exposition in this section largely follows [**GCKT18a**]. See also the contribution of Hackney [**Hac24**] to this volume. In particular, we work with the skeleton of the simplicial category Δ and simplicial objects $X_\bullet : \Delta^{\mathrm{op}} \to \mathscr{S}$ take values in the large ∞-category of ∞-groupoids (as modeled by Kan complexes).

Recall that an map $\alpha : [k] \to [n]$ in Δ is *active* when $\alpha(0) = 0$ and $\alpha(k) = n$. Active maps are generated by the codegeneracies $s^i : [n+1] \to [n]$ and the inner coface maps $d^i : [n-1] \to [n]$, $0 < i < n$. Also recall that a map $F : Y_\bullet \to X_\bullet$ between simplicial objects is *cartesian* with respect to a map $f : [m] \to [n]$ in Δ if the following diagram is cartesian.

$$\begin{array}{ccc} Y_n & \xrightarrow{f} & Y_m \\ {\scriptstyle F_n}\downarrow & \lrcorner & \downarrow{\scriptstyle F_m} \\ X_n & \xrightarrow[f]{} & X_m. \end{array}$$

Definition 3.0.1. Let $F : Y_\bullet \to X_\bullet$ be a simplicial map. For each $n \geq 1$, let $\alpha_n : [1] \to [n]$ be the active map determined by $\alpha_n(0) := 0$ and $\alpha_n(1) := n$ and set $(X\sigma)_n := (X(\pi_1), \ldots, X(\pi_n))$ and $(Y\sigma)_n := (Y(\pi_1), \ldots, Y(\pi_n))$, where $\pi_i : [1] \to [n]$ is the map determined by $\pi_i(0) := i-1$ and $\pi_i(1) := i$.

(1) The morphism F is *CULF* if for each $n \in \mathbb{Z}_{\geq 1}$ the following diagram is cartesian:

$$\begin{array}{ccc} Y_1 & \xleftarrow{Y(\alpha_n)} & Y_n \\ {\scriptstyle F_1}\downarrow & & \downarrow{\scriptstyle F_n} \\ X_1 & \xleftarrow{X(\alpha_n)} & X_n. \end{array}$$

(2) The morphism F is *IKEO* if $F_0 : Y_0 \to X_0$ is a (weak) equivalence and for each $n \in \mathbb{Z}_{\geq 1}$ the following diagram is cartesian:

$$\begin{array}{ccc} Y_n & \xrightarrow{(Y\sigma)_n} & Y_1 \times Y_1 \times \cdots \times Y_1 \\ {\scriptstyle F_n}\downarrow & & \downarrow{\scriptstyle F_1^{\times n}} \\ X_n & \xrightarrow{(X\sigma)_n} & X_1 \times X_1 \times \cdots \times X_1. \end{array}$$

CULF maps are those which are cartesian with respect to active maps [**GCKT18a**, Lemma 4.1]. The word CULF is a composite acronym combining C and ULF, the two of which stand for Conservative and Unique Lifting Factorizations, respectively, which have their standard meaning when $X_\bullet$ is the nerve of a category. For an extended discussion of the CULF material discussed below, see [**GCKT18a**, §4] and [**Lei12**, §5]. On the other hand, the acronym IKEO combines IK and EO which are Inner Kan and Equivalence on Objects in long form. This definition of IKEO recently appeared in [**GCKT24**, Lemma 3.1.1]; compare with [**GCKT18a**, Equation (20)].

3.1. Groupoids and incidence algebras. If I is an ∞-groupoid, then there is a slice category (or overcategory) $\mathscr{S}_{/I}$ with objects given by maps of ∞-groupoids $X \to I$. A map $g : I \to J$ induces two maps between slice categories: a pushforward $g_! : \mathscr{S}_{/I} \to \mathscr{S}_{/J}$ and a pullback $g^* : \mathscr{S}_{/J} \to \mathscr{S}_{/I}$. The pushforward $g_! : \mathscr{S}_{/I} \to \mathscr{S}_{/J}$ is defined by $g_!(a) := a \circ g$ for $a : X \to I$ a groupoid map. The pullback is defined to send $a : X \to J$ to its pullback $g^*(a) : g^*(X) \to I$ over I:

$$\begin{array}{ccc} g^*(X) & \xrightarrow{i} & X \\ {\scriptstyle g^*(a)}\downarrow & \lrcorner & \downarrow{\scriptstyle a} \\ I & \xrightarrow{g} & J. \end{array} \tag{3.1}$$

In this way, any diagram of ∞-groupoids of the form

$$I \xleftarrow{f} M \xrightarrow{g} J$$

induces maps between associated slice categories,

$$g_! \circ f^* : \mathscr{S}_{/I} \to \mathscr{S}_{/J} \qquad \text{and} \qquad f_! \circ g^* : \mathscr{S}_{/J} \to \mathscr{S}_{/I}.$$

These maps are used in the definition below to introduce the structural maps of incidence (co)algebras. In addition, we are implicitly using the natural isomorphisms $\mathscr{S}_{/I \times J} \cong \mathscr{S}_{/I} \otimes \mathscr{S}_{/J}$ (see [**GCKT18c**, §2.10]).

DEFINITION 3.1.1.

(1) If $X_\bullet : \Delta^{\mathrm{op}} \to \mathscr{S}$ is a 2-Segal space, then the *incidence coalgebra* $(\mathscr{S}_{/X_1}, \Delta, \epsilon)$ is determined by the spans

$$X_1 \xleftarrow{\partial_1} X_2 \xrightarrow{(\partial_2,\partial_0)} X_1 \times X_1 \qquad \text{and} \qquad X_1 \xleftarrow{s_0} X_0 \xrightarrow{\mathrm{can}} \mathrm{pt},$$

where $\mathrm{can} : X_0 \to \mathrm{pt}$ is the canonical map to the one-point space pt. In more detail, $\Delta : \mathscr{S}_{/X_1} \to \mathscr{S}_{/X_1 \times X_1}$ is given by $\Delta(a) := (\partial_2, \partial_0)_! \circ \partial_1^*(a)$ and $\epsilon : \mathscr{S}_{/X_1} \to \mathcal{S}$ is defined to be $\epsilon(a) := \mathrm{can}_! \circ s_0^*(a)$.

(2) If $X_\bullet : \Delta^{\mathrm{op}} \to \mathscr{S}$ is a 2-Segal space, then the *incidence algebra* $(\mathscr{S}_{/X_1}, m, \iota)$ is determined by the adjoint spans: $m(a) := (\partial_1)_! \circ (\partial_2, \partial_0)^*(a)$ and $\iota(a) := (s_0)_! \circ \mathrm{can}^*$.

These definitions can be compared to those found in Section 2.1. In more detail, the span which determines the coproduct Δ and product m above is identical to equation (2.1) and the span which determines the counit ϵ and unit ι is identical to equation (2.7). The axioms of 2-Segal spaces allow one to prove that the incidence coproduct Δ and incidence product m are coassociative and associative, respectively.

Recall that if $F : Y_\bullet \to X_\bullet$ is a map between simplicial objects in ∞-groupoids, then there is a map $F_1 : Y_1 \to X_1$. The proposition below shows that the CULF and IKEO conditions on F translate to homomorphism conditions on the pushforward map $F_! := (F_1)_! : \mathscr{S}_{/Y_1} \to \mathscr{S}_{/X_1}$ between slice categories.

PROPOSITION 3.1.2.

(1) *If $F : Y_\bullet \to X_\bullet$ is a CULF map between 2-Segal spaces, then $F_! : \mathscr{S}_{/Y_1} \to \mathscr{S}_{/X_1}$ is a coalgebra homomorphism.*

(2) *If $F : Y_\bullet \to X_\bullet$ is an IKEO map between 2-Segal spaces, then $F_! : \mathscr{S}_{/Y_1} \to \mathscr{S}_{/X_1}$ is an algebra homomorphism.*

PROOF. For part (1), we follow [**GCKT18a**, Lemma 8.2]. First suppose that $F : Y_\bullet \to X_\bullet$ is a CULF map. By virtue of the definitions above there is a commutative diagram

$$\begin{array}{ccccc} Y_1 & \xleftarrow{\alpha} & Y_n & \xrightarrow{\sigma} & Y_1^{\times n} \\ {\scriptstyle F_1}\downarrow & & \downarrow{\scriptstyle F_n} & & \downarrow{\scriptstyle F_1^{\times n}} \\ X_1 & \xleftarrow{\alpha'} & X_n & \xrightarrow{\sigma'} & X_1^{\times n}. \end{array} \tag{3.2}$$

where $\alpha := Y(\alpha_n)$, $\alpha' := X(\alpha_n)$, $\sigma := (Y\sigma)_n$ and $\sigma' := (X\sigma)_n$ in Definition 3.0.1. When $n = 0$ the left-hand square is cartesian because F is assumed to be conservative and when $n \geq 1$ the left-hand square is cartesian because F is ULF; see Definition 3.0.1 above.

Applying Beck–Chevalley to the diagram (3.2) gives the relation

$$(F_n)_! \circ \alpha^* \cong \alpha'^* \circ (F_1)_!. \tag{3.3}$$

Since the n-ary coproduct $\Delta_X^{(n)}$ is given by $\sigma'_! \circ \alpha'$, composing the right-hand side with $\sigma'_!$ gives

$$\sigma'_! \circ \alpha'^* \circ (F_1)_! \cong \Delta_X^{(n)} \circ (F_1)_!. \tag{3.4}$$

On the other hand, composing the left-hand side with $\sigma'_!$ and using $\sigma'_! \circ (F_n)_! \cong (\sigma' \circ F_n)_!$ gives

$$(F_1^{\times n} \circ \sigma)_! \circ \alpha^* \cong (F_1^{\times n})_! \circ (\sigma_! \circ \alpha^*) \cong (F_1^{\times n})_! \circ \Delta_Y^{(n)}, \tag{3.5}$$

where $\Delta_Y^{(n)}$ is the n-fold composition of the coproduct from Definition 3.1.1. Combining equation (3.4) with equations (3.3) and (3.5) implies

$$\Delta_X^{(n)} \circ (F_1)_! \cong (F_1^{\times n})_! \circ \Delta_Y^{(n)}.$$

Hence, $(F_1)_!$ is a homomorphism of incidence coalgebras.

For part (2), there is an analogous argument. From Proposition 3.0.1, the IKEO condition on F implies that the right-hand square of the diagram (3.2) is cartesian. Applying Beck–Chevalley to this square gives the relation

$$(F_n)_! \circ \sigma^* \cong \sigma'^* \circ (F_1^{\times n})_!. \tag{3.6}$$

Composing the right-hand side with $\alpha'_!$ gives

$$\alpha'_! \circ (\sigma'^* \circ (F_1^{\times n})_!) \cong (\alpha'_! \circ \sigma'^*) \circ F_1^{\times n})_! \cong m_X^{(n)} \circ (F_1^{\times n})_!,$$

where $m_X^{(n)}$ is the n-fold composition of the incidence product m in Definition 3.1.1. On the other hand, the left-hand side is

$$\begin{aligned}\alpha'_! \circ ((F_n)_! \circ \sigma^*) &\cong (\alpha' \circ F_n)_! \circ \sigma^* \\ &\cong (F_1 \circ \alpha)_! \circ \sigma^* \\ &\cong (F_1)_! \circ (\alpha_! \circ \sigma^*) \\ &\cong (F_1)_! \circ m_Y^{(n)}.\end{aligned}$$

□

3.2. Linearizing ∞-groupoids. A similar argument to Theorem 2.2.3 gives the derived Hall algebra (or numerical incidence algebra) in Proposition 3.2.1 below, see also [**Toë06**, §2] and [**GCKT18b**, §7.8]. So a 2-Segal space $X_\bullet$ allows one to construct an incidence algebra and Proposition 3.2.2 below shows that, when suitable finiteness conditions are satisfied, there is a functorial transition from the setting of incidence algebras to numerical incidence algebras.

Call an ∞-groupoid B *locally finite* when the groups $\pi_i(B,b)$ are finite for all $b \in B$ and $i \in \mathbb{Z}_{\geq 1}$, and for each $b \in B$ there exists $n \in \mathbb{Z}_{\geq 1}$ such that $\pi_i(B,b) = 0$ for all $i > n$. An ∞-groupoid B is *finite* when B is locally finite and $\pi_0(B)$ is a finite set. The *homotopy cardinality* $\|B\| \in \mathbb{Q}$ of a finite ∞-groupoid B is given by

$$\|B\| := \sum_{b \in \pi_0(B)} \prod_{i=1}^{\infty} |\pi_i(B,b)|^{(-1)^i}.$$

This will allow us to extend the definition of pushforward f_* from Section 2.3.

To an ∞-groupoid B we associate the vector space $\ell(B) := \mathrm{Fun}_c(\pi_0(B), \mathbb{C})$ of finitely supported functions. If $f : X \to Y$ is a map between locally finite ∞-groupoids and f satisfies the condition that for any $y \in \pi_0(Y)$, the homotopy fibre $F_y := \{y\} \times_Y^h X$ of f at y has finitely many connected components ($|\pi_0(F_y)| < \infty$), then there is a pullback map $f^* : \ell(Y) \to \ell(X)$ given by $f^*(\varphi)(x) := \varphi(f(x))$ (see [**Toë06**, Definition 2.5]). If $f : X \to Y$ is a map between locally finite ∞-groupoids, then there is a pushforward map $f_* : \ell(X) \to \ell(Y)$ given by

$$f_*(\varphi)(y) := \sum_{z \in \pi_0(F_y)} \varphi(i(z)) \prod_{i>0} |\pi_i(F_y, z)|^{(-1)^i}$$

where the map $i : \pi_0(F_y) \to \pi_0(X)$ is induced by the inclusion of the homotopy fibre F_y. This formula makes sense for any such map $f : X \to Y$ between locally finite ∞-groupoids because the function $\varphi \in \ell(X)$ is assumed to have finite support and any map between locally finite ∞-groupoids must have locally finite fibres. When B is finite, this assignment recovers homotopy cardinality when $\pi : B \to \mathrm{pt}$, by $\pi_*(1_B) = \|B\|$ where $1_B := \sum_{b\in\pi_0(B)} \delta_b$. The pullback f^* and pushforward f_* assignments determine contravariant and covariant functors respectively which satisfy the Beck–Chevalley property.

Denote by $\mathscr{S}^{\mathrm{lf}} \subset \mathscr{S}$ the full subcategory of locally finite ∞-groupoids (spaces) and $\mathsf{Span}(\mathscr{S}^{\mathrm{lf}})^{\mathrm{f}} \subset \mathsf{Span}(\mathscr{S}^{\mathrm{lf}})$ the subcategory of spans

$$S \xleftarrow{p} M \xrightarrow{q} T$$

for which the map p has homotopy fibres with finitely many connected components (as for f^* above). The assignments in the previous paragraph determine the functor ℓ introduced by the proposition below. The proof is by analogy with material in Section 2.3. See also [**Toë06**, §2].

PROPOSITION 3.2.1 ([**GCKT18c**, Proposition 8.2]). *There is a monoidal functor $\ell : \mathsf{Span}(\mathscr{S}^{\mathrm{lf}})^{\mathrm{f}} \to \mathsf{Vect}_{\mathbb{C}}$ from the category of finite spans of locally finite ∞-groupoids to the category of complex vector spaces which is determined by the assignments $B \mapsto \ell(B)$ and*

$$(S \xleftarrow{p} M \xrightarrow{q} T) \mapsto (q_* p^* : \ell(S) \to \ell(T)).$$

The next proposition constructs a map $\tilde{\ell}$ from the category of finite slice categories $\mathscr{F}_{/I}$ and spans to $\mathsf{Vect}_{\mathbb{C}}$. When I is a locally finite ∞-groupoid, for our purposes the *finite slice category* $\mathscr{F}_{/I} \subset \mathscr{S}_{/I}$ is the full subcategory of the slice category consisting of objects $a : X \to I$ for which X is a finite ∞-groupoid. If $\mathscr{F}_{/I} \subset \mathscr{S}_{/I}$ is a finite slice category and $a : X \to I$, then set $\tilde{\ell}(a) := a_*(1_X) \in \ell(I)$ where $1_X = \sum_{i\in\pi_0(I)} \delta_i$. Recall from Section 3.1 that if $g : I \to J$ is a map of ∞-groupoids, then there are maps $g^* : \mathscr{F}_{/J} \to \mathscr{F}_{/I}$ and $g_! : \mathscr{F}_{/I} \to \mathscr{F}_{/J}$. We set $\tilde{\ell}(g^*) := g^* : \ell(J) \to \ell(I)$ and $\tilde{\ell}(g_!) := g_* : \ell(I) \to \ell(J)$.

PROPOSITION 3.2.2. *Suppose that I and J are finite ∞-groupoids. Then the assignments above respect maps in the sense that the following properties hold:*

(1) *If $g : I \to J$ is an ∞-groupoid map and $a : X \to I$, then*

$$\tilde{\ell}(g_!)(a_*(1_X)) = g_*(\tilde{\ell}(a)).$$

(2) *If $g : I \to J$ is an ∞-groupoid map and $a : X \to J$, then*

$$\tilde{\ell}(g^*(a)) = g^*(\tilde{\ell}(a)).$$

PROOF. For (1): Recall that $g_!(a) = g \circ a : X \to I \to J$. We have made the assignment $\tilde{\ell}(g_! : \mathscr{F}_{/I} \to \mathscr{F}_{/J}) = g_* : \ell(I) \to \ell(J)$, so the map $\tilde{\ell}(g_!) : \ell(I) \to \ell(J)$ takes the vector $a_*(1_X) \in \ell(I)$ to $\tilde{\ell}(g_!)(a_*(1_X)) \in \ell(J)$. Now observe that $\tilde{\ell}(g_!)(a_*(1_X)) = g_*(a_*(1_X)) = g_*(\tilde{\ell}(a)) \in \ell(J)$, where the second equation follows from the definition of $\tilde{\ell}(a)$.

For (2): Recall that $g^*(a) : g^*(X) \to I$ is the pullback in the cartesian diagram (3.1). We will also use the inclusion $i : g^*(X) \to X$ from diagram (3.1). Now we have $\tilde{\ell}(g^* : \mathscr{F}_{/J} \to \mathscr{F}_{/I}) = g^* : \ell(J) \to \ell(I)$, so $\tilde{\ell}(g^*(a)) = (g^*(a))_*(1_{g^*(X)}) =$

$g^*(a)_*(i^*(1_X)) \overset{(*)}{=} (g^*a_*)(1_X) = g^*(a_*(1_X)) = g^*(\tilde{\ell}(a))$ where the Beck–Chevalley property is used to justify $(*)$. □

For other details see [**GCKT18b, GCKT18c**].

In this way the construction of the numerical Hall algebra factors through incidence algebras when the span defining the product satisfies sufficient finiteness conditions.

3.3. From incidence algebras to Hall algebras. When $\mathcal{C}$ is a proto-exact category, the Waldhausen construction $\mathcal{S}_\bullet(\mathcal{C})$ is 2-Segal (Theorem 1.3.1). There is an incidence coalgebra and an incidence algebra,

$$(\mathscr{S}_{/S_1(\mathcal{C})}, \Delta, \epsilon) \qquad \text{and} \qquad (\mathscr{S}_{/S_1(\mathcal{C})}, m, \iota).$$

When these spans are suitably finite, the Hall (co)algebra introduced in Section 2.3 can be obtained from the incidence (co)algebra by the groupoid linearization procedure of Section 3.2 above. This allows us to translate Proposition 3.1.2 into a statement about functoriality of the Hall (co)algebra. An exact functor $F : \mathcal{A} \to \mathcal{B}$ induces a map $F_\bullet : S_\bullet(\mathcal{A}) \to S_\bullet(\mathcal{B})$ between Waldhausen constructions. We analyze what the CULF and IKEO conditions on $F_\bullet$ entail. This is the subject of the theorem below.

THEOREM 3.3.1. *Suppose that $F : \mathcal{A} \to \mathcal{B}$ is an exact functor between finitary abelian categories. The corresponding map $F_\bullet : \mathcal{S}_\bullet(\mathcal{A}) \to \mathcal{S}_\bullet(\mathcal{B})$ between 2-Segal spaces induces homomorphisms between corresponding Hall coalgebras or Hall algebras when $F_\bullet$ is CULF or IKEO, respectively. These conditions correspond to*

(1) *CULF: If $F(Y) \in Ob(\mathcal{B})$ is in the image of F, then so must be all of the subobjects and quotients of $F(Y)$. That is, for all $X' \subset F(Y)$, there exists unique X in $Ob(\mathcal{A})$ such that $F(X) = X'$ (and there exists unique Y/X such that $F(Y/X) = F(Y)/X'$).*
(2) *IKEO: If $F(Z)$ and $F(X)$ are in the image of F and Y' is an extension of $F(Z)$ by $F(X)$ in $\mathcal{B}$, then there is a unique extension Y of Z by X in $\mathcal{A}$ such that $F(Y) = Y'$.*

PROOF. Let us describe the kind of map that F induces in more detail. The functor $F : \mathcal{A} \to \mathcal{B}$ induces a map $F_\bullet : S_\bullet(\mathcal{A}) \to S_\bullet(\mathcal{B})$ for which $F_1 : S_1(\mathcal{A}) \to S_1(\mathcal{B})$ is the map $F_1([X]) := [F(X)]$ on isomorphism classes of objects $[X] \in S_1(\mathcal{A})$. There is now a map between slice categories $F_! : \mathscr{S}_{/S_1(\mathcal{A})} \to \mathscr{S}_{/S_1(\mathcal{B})}$ (we set $F_1 := (F_1)_!$) which is defined by $F_!(a) := F_1 \circ a$ for $a : X \to S_1(\mathcal{A}) \in \mathscr{S}_{/S_1(\mathcal{A})}$ as in Section 3.1 above. By Proposition 3.1.2 above, the map $F_!$ is a coalgebra homomorphism when $F_\bullet$ is CULF and an algebra homomorphism when $F_\bullet$ is IKEO. Recall that the finite slice categories $\mathscr{F}_{/I} \subset \mathscr{S}_{/I}$ consist of objects $a : X \to I$ for which X is a finite ∞-groupoid. Restricting the map $F_!$ to the finite slice category produces a map $F_!|_{\mathscr{F}_{/S_1(\mathcal{A})}} : \mathscr{F}_{/S_1(\mathcal{A})} \to \mathscr{S}_{/S_1(\mathcal{B})}$. Notice that when X is a finite ∞-groupoid, $F_!(a) : X \xrightarrow{F_1 \circ a} S_1(\mathcal{B})$ resides in the finite slice category $\mathscr{F}_{/S_1(\mathcal{B})}$. Thus F induces a map $F_! : \mathscr{F}_{/S_1(\mathcal{A})} \to \mathscr{F}_{/S_1(\mathcal{B})}$ (now setting $F_! := F_!|_{\mathscr{F}_{/S_1(\mathcal{A})}}$). Proposition 3.2.2 above demonstrates functoriality of the map $\tilde{\ell}(F_!) : \ell(S_1(\mathcal{A})) \to \ell(S_1(\mathcal{B}))$ between the linearizations used by the Hall algebra construction.

Now we examine the special $n = 2$ cases of the CULF and IKEO conditions. This suffices to determine the conditions in our premises, since the data of the Hall (co)algebra involves only the 2-truncation of $\mathcal{S}_\bullet$. Conversely, it is easy to check that these conditions suffice to determine maps between Hall (co)algebras after appropriate finiteness conditions are imposed.

When $n = 2$, the CULF condition as spelled out in Definition 3.0.1 requires that the diagram

$$\begin{array}{ccc} \mathcal{S}_1(\mathcal{A}) & \xleftarrow{\partial_1} & \mathcal{S}_2(\mathcal{A}) \\ {\scriptstyle F_1}\downarrow & & \downarrow{\scriptstyle F_2} \\ \mathcal{S}_1(\mathcal{B}) & \xleftarrow{\partial_1} & \mathcal{S}_2(\mathcal{B}) \end{array}$$

is cartesian. So there is an isomorphism

$$\gamma : \mathcal{S}_2(\mathcal{A}) \xrightarrow{\sim} \mathcal{S}_1(\mathcal{A}) \times_{\mathcal{S}_1(\mathcal{B})} \mathcal{S}_2(\mathcal{B}).$$

This amounts to the requirement that the map $F_2 : \mathcal{S}_2(\mathcal{A}) \to \mathcal{S}_2(\mathcal{B})$ given by (3.7)

$$\left(0 \to X \xrightarrow{f} Y \to Z \to 0\right) \quad \mapsto \quad \left(0 \to F(X) \xrightarrow{F(f)} F(Y) \to F(Z) \to 0\right)$$

where $Z := \operatorname{coker}(f)$, induced by the exact functor F must satisfy the following additional property: If

$$0 \to X' \to F(Y) \to Z' \to 0$$

is a short exact sequence in $\mathcal{B}$, then there exist unique X and Z in $S_1(\mathcal{A})$ such that $F_1(X) = X'$ and $F_1(Z) = Z'$. In words, the image of F is closed under subobjects and quotients.

On the other hand, the IKEO condition from Definition 3.0.1 translates to the requirement that the diagram

$$\begin{array}{ccc} \mathcal{S}_2(\mathcal{A}) & \xrightarrow{(\partial_2,\partial_0)} & \mathcal{S}_1(\mathcal{A}) \times \mathcal{S}_1(\mathcal{A}) \\ {\scriptstyle F_2}\downarrow & & \downarrow{\scriptstyle F_1 \times F_1} \\ \mathcal{S}_2(\mathcal{B}) & \xrightarrow{(\partial_2,\partial_0)} & \mathcal{S}_1(\mathcal{B}) \times \mathcal{S}_1(\mathcal{B}) \end{array}$$

is cartesian. (Note that $\mathcal{S}_0(\mathcal{A})$ and $\mathcal{S}_0(\mathcal{B})$ are both the trivial ∞-groupoid.) So there is an isomorphism

$$\gamma : \mathcal{S}_2(\mathcal{A}) \xrightarrow{\sim} \mathcal{S}_2(\mathcal{B}) \times_{\mathcal{S}_1(\mathcal{B}) \times \mathcal{S}_1(\mathcal{B})} (\mathcal{S}_1(\mathcal{A}) \times \mathcal{S}_1(\mathcal{A}))$$

which translates to the following additional requirement on the morphism from equation (3.7): If

$$0 \to X \to Y \to Z \to 0$$

is a short exact sequence in $\mathcal{A}$, then there is a unique extension of $F(Z)$ by $F(X)$ in $\mathcal{B}$ which is $F(Y)$. □

Slightly different statements can be found in the literature. See, for example, [**CS20**, Proposition 5.3].

4. Representations of Hall algebras via relative 2-Segal spaces

The focus of this section is the theory of relative 2-Segal spaces, as formulated in [**Wal16, You18**]. From the perspective of this paper, the main feature of relative 2-Segal spaces is that they lead naturally to modules over Hall algebras.

4.1. Relative 2-Segal spaces.

DEFINITION 4.1.1. Let $X_\bullet$ be a 2-Segal space. A morphism $F_\bullet : Y_\bullet \to X_\bullet$ of simplicial spaces is *relative 2-Segal* if $Y_\bullet$ is 1-Segal and, for every $n \geq 2$ and $0 \leq i < j \leq n$, the outside square of the diagram

$$\begin{array}{ccc} Y_n & \longrightarrow & Y_{\{0,\dots,i,j,\dots,n\}} \\ \downarrow & & \downarrow \\ Y_{\{i,\dots,j\}} & \longrightarrow & Y_{\{i,j\}} \\ {\scriptstyle F_{\{i,\dots,j\}}}\downarrow & & \downarrow{\scriptstyle F_{\{i,j\}}} \\ X_{\{i,\dots,j\}} & \longrightarrow & X_{\{i,j\}} \end{array} \tag{4.1}$$

is cartesian.

As for 2-Segal spaces, we have formulated the relative 2-Segal condition in a manner which makes their relevance to modules over Hall algebras most apparent. More intrinsic formulations can be found in [**Wal16, You18**]. As pointed out by Carlier [**Car20**, §2.1], given a 2-Segal space $X_\bullet$, a morphism of simplicial spaces $F_\bullet : Y_\bullet \to X_\bullet$ is relative 2-Segal if and only if $Y_\bullet$ is 1-Segal and $F_\bullet$ is CULF in the sense of Definition 3.0.1.

Let $F_\bullet : Y_\bullet \to X_\bullet$ be relative 2-Segal. Arguing as in Section 2.1, the span

$$\begin{array}{ccccc} & & Y_1 & & \\ & {\scriptstyle (F_1,\partial_0)}\swarrow & & \searrow{\scriptstyle \partial_1} & \\ X_1 \times Y_0 & & & & Y_0 \end{array}$$

gives Y_0 the structure of a left (X_1, m)-module object in $\mathsf{Span}(\mathscr{S})$. Specifically, the first non-trivial 1-Segal condition on $Y_\bullet$ and the first non-trivial instance of diagram (4.1) imply associativity of the (X_1, m)-action.

We give two classes of examples of relative 2-Segal spaces, both of which are relative variants of the Waldhausen $\mathcal{S}_\bullet$-construction of Section 1.3. Further examples can be found in [**Wal16, You18**].

EXAMPLE 4.1.2 ([**Wal16**], [**You18**]). Let $\mathcal{C}$ be a k-linear abelian category with Grothendieck group $K_0(\mathcal{C})$. Following Bridgeland [**Bri07**, §2], a *stability function* on $\mathcal{C}$ is a group homomorphism $Z : K_0(\mathcal{C}) \to \mathbb{C}$ such that

$$Z(A) \in \mathbb{H}_+ := \left\{ m e^{\sqrt{-1}\pi\phi} \mid m \in \mathbb{R}_{>0},\ \phi \in (0,1] \right\} \subset \mathbb{C}$$

for all non-zero objects $A \in \mathcal{C}$. Write $\phi(A) \in (0,1]$ for the argument of $Z(A)$. A non-zero object $A \in \mathcal{C}$ is called *Z-semistable* if $\phi(A') \leq \phi(A)$ for all non-zero subobjects $A' \subset A$. The full subcategory $\mathcal{C}_\phi^{\mathrm{ss}} \subset \mathcal{C}$ of Z-semistable objects of argument $\phi \in (0,1]$ (together with the zero object) is abelian whence defines a 2-Segal space $\mathcal{S}_\bullet(\mathcal{C}_\phi^{\mathrm{ss}})$ by Theorem 1.3.1.

Let $\Phi : \mathcal{C} \to \mathsf{vect}_k$ be a left-exact functor. A *framed object* is a pair (M, s) consisting of an object $M \in \mathcal{C}$ and a section $s \in \Phi(M)$ (see [**Soi16**, §4]). A morphism of framed objects $(M, s) \to (M', s')$ is a pair $(\pi, \lambda) \in \mathrm{Hom}_{\mathcal{C}}(M, M') \times k$ which satisfies $\Phi(\pi)(s) = \lambda s'$. A framed object (M, s) is called *stable framed* if M

is Z-semistable and $\phi(A) < \phi(M)$ for all non-zero proper subobjects $A \subset M$ for which $s \in \Phi(A) \subset \Phi(M)$.

Let $\mathcal{S}_{\bullet}^{\text{st-fr}}(\mathcal{C}_{\phi}^{\text{ss}})$ be the simplicial groupoid whose n-simplices are diagrams of the form

$$\begin{array}{ccccccccc}
0 & \rightarrowtail & A_{\{0,1\}} & \rightarrowtail & \cdots & \rightarrowtail & A_{\{0,n\}} & \rightarrowtail & (M_0, s_0) \\
 & & \downarrow & & & & \downarrow & & \downarrow \\
 & & 0 & \rightarrowtail & \cdots & \rightarrowtail & A_{\{1,n\}} & \rightarrowtail & (M_1, s_1) \\
 & & & & & & \downarrow & & \downarrow \\
 & & & & \ddots & & \vdots & & \vdots \\
 & & & & & & \downarrow & & \downarrow \\
 & & & & & & 0 & \rightarrowtail & (M_n, s_n) \\
 & & & & & & & & \downarrow \\
 & & & & & & & & 0,
\end{array}$$

where each (M_i, s_i) is a stable framed object and the underlying diagram, obtained by forgetting the framing data $s_0, \dots, s_n$, is an object of $\mathcal{S}_{n+1}(\mathcal{C}_{\phi}^{\text{ss}})$. In the above diagram a morphism $A_{\{i,n\}} \to (M_i, s_i)$ is simply a morphism $A_{\{i,n\}} \to M_i$. The simplicial morphism $F_{\bullet} : \mathcal{S}_{\bullet}^{\text{st-fr}}(\mathcal{C}_{\phi}^{\text{ss}}) \to \mathcal{S}_{\bullet}(\mathcal{C}_{\phi}^{\text{ss}})$ which forgets the rightmost column is relative 2-Segal.

EXAMPLE 4.1.3 ([**You18**]). Let $\mathcal{C}$ be a proto-exact category with duality (P, Θ), that is, a proto-exact duality functor $P : \mathcal{C}^{\text{op}} \to \mathcal{C}$ and a natural double dual isomorphism $\Theta : \text{id}_{\mathcal{C}} \Rightarrow P \circ P^{\text{op}}$ whose components satisfy $P(\Theta_U) \circ \Theta_{P(U)} = \text{id}_{P(U)}$, $U \in \mathcal{C}$. A *symmetric form* in $\mathcal{C}$ is a pair (M, ψ_M) consisting of an object $M \in \mathcal{C}$ and an isomorphism $\psi_M : M \xrightarrow{\sim} P(M)$ which satisfies $P(\psi_M) \circ \Theta_M = \psi_M$. An *isometry* of symmetric forms $\phi : (M, \psi_M) \to (N, \psi_N)$ is an isomorphism $\phi : M \to N$ which satisfies $P(\phi) \circ \psi_N \circ \phi = \psi_M$. The *hermitian groupoid* $\mathcal{C}^h$ is the groupoid of symmetric forms and their isometries.

Let (N, ψ_N) be a symmetric form in $\mathcal{C}$. The *orthogonal* $U^{\perp}$ of an inflation $i : U \rightarrowtail N$ is defined to be the pullback

$$\begin{array}{ccc}
U^{\perp} & \overset{k}{\dashrightarrow} & N \\
\downarrow & & \downarrow{\scriptstyle P(i) \circ \psi_N} \\
0 & \rightarrowtail & P(U).
\end{array}$$

The inflation i is called *isotropic* if $P(i) \circ \psi_N \circ i$ is zero and the induced morphism $U \to U^{\perp}$ is an inflation. In this case, if $\mathcal{C}$ is exact, then on the pushout

$$\begin{array}{ccc}
U & \rightarrowtail & U^{\perp} \\
\downarrow & & \downarrow{\scriptstyle \pi} \\
0 & \dashrightarrow & M
\end{array}$$

there exists a unique symmetric form ψ_M which satisfies $P(k)\psi_N k = P(\pi)\psi_M \pi$. All known proto-exact categories are known to satisfy this property, which we henceforth assume. We call (M, ψ_M) the *isotropic reduction of N by U* and denote it by $N /\!\!/ U$.

Following Shapiro and Yao [**SY96**], let $\mathcal{R}_\bullet(\mathcal{C})$ be the simplicial groupoid whose n-simplices $\mathcal{R}_n(\mathcal{C})$ consist of diagrams

(4.2)

$$\begin{array}{ccccccccccccc}
0 & \rightarrowtail & A_{\{0,1\}} & \rightarrowtail \cdots \rightarrowtail & A_{\{0,n\}} & \rightarrowtail & A^{\perp}_{\{0,n\}} & \rightarrowtail \cdots \rightarrowtail & A^{\perp}_{\{0,1\}} & \rightarrowtail & (M_0, \psi_0) \\
 & & \downarrow & & \downarrow & & \downarrow & & \downarrow & & \downarrow \\
 & & 0 & \rightarrowtail \cdots \rightarrowtail & A_{\{1,n\}} & \rightarrowtail & A^{\perp}_{\{1,n\}} & \rightarrowtail \cdots \rightarrowtail & (M_1, \psi_1) & \rightarrowtail & P(A^{\perp}_{\{0,1\}}) \\
 & & & & & & & & \downarrow & & \downarrow \\
 & & & & & & & \ddots & \vdots & & \vdots \\
 & & & & & & & & \downarrow & & \downarrow \\
 & & & & & & & & P(A^{\perp}_{\{1,n\}}) & \rightarrowtail & P(A^{\perp}_{\{0,n\}}) \\
 & & & & & & & & \downarrow & & \downarrow \\
 & & & & & & & & P(A_{\{1,n\}}) & \rightarrowtail & P(A_{\{0,n\}}) \\
 & & & & & & & & \downarrow & & \downarrow \\
 & & & & & & & & \vdots & & \vdots \\
 & & & & & & & & \downarrow & & \downarrow \\
 & & & & & & & & 0 & \rightarrowtail & P(A_{\{0,1\}}) \\
 & & & & & & & & & & \downarrow \\
 & & & & & & & & & & 0
\end{array}$$

such that

(i) each object (M_i, ψ_i) is a symmetric form,
(ii) upon forgetting ψ_i, $0 \leq i \leq n$, the diagram is an object of $\mathcal{S}_{2n+1}(\mathcal{C})$,
(iii) each inflation $A_{\{i,j\}} \rightarrowtail (M_i, \psi_i)$ is isotropic with orthogonal $A^{\perp}_{\{i,j\}} \rightarrowtail (M_i, \psi_i)$,
(iv) for every $0 \leq i \leq j \leq n$, the symmetric form (M_j, ψ_j) is isometric to the reduction $(M_i, \psi_i) /\!\!/ A_{\{i,j\}}$,
(v) the diagram below the diagonal $x = y$ is the P-image of the diagram above the diagonal.

For example, $\mathcal{R}_0(\mathcal{C}) \simeq \mathcal{C}^h$ while $\mathcal{R}_1(\mathcal{C})$ consists of diagrams

(4.3)
$$\begin{array}{ccccccc}
0 & \rightarrowtail & A_{\{0,1\}} & \rightarrowtail & A_{\{0,1'\}} & \rightarrowtail & (M_0, \psi_0) \\
 & & \downarrow & & \downarrow & & \downarrow \\
 & & 0 & \rightarrowtail & (M_1, \psi_1) & \rightarrowtail & A_{\{1,0'\}} \\
 & & & & \downarrow & & \downarrow \\
 & & & & 0 & \rightarrowtail & A_{\{1',0'\}} \\
 & & & & & & \downarrow \\
 & & & & & & 0
\end{array}$$

which present (M_1, ψ_1) as the isotropic reduction of (M_0, ψ_0) by $A_{\{0,1\}}$. In general, objects of $\mathcal{R}_n(\mathcal{C})$ are isotropic n-flags together with presentations of all subquotients and subreductions.

The forgetful map $F_\bullet : \mathcal{R}_\bullet(\mathcal{C}) \to \mathcal{S}_\bullet(\mathcal{C})$ which sends a diagram (4.2) to the subdiagram on the objects $\{A_{\{p,q\}}\}_{0 \leq p \leq q \leq n}$ is relative 2-Segal. The proof boils down to a symmetric form of the Third Isomorphism Theorem, asserting a canonical isometry

$$(M, \psi_M) /\!\!/ U \simeq ((M, \psi_M) /\!\!/ V) /\!\!/ (U/V)$$

for a chain of isotropic subobjects $V \rightarrowtail U \to (M, \psi_M)$. In the context of ∞-categories, the relative 2-Segal property of $F_\bullet$ was proven recently by Gödicke [**Göd24**].

The $\mathcal{R}_\bullet$-construction was originally introduced in the context of Grothendieck–Witt theory. Indeed, for an exact category $\mathcal{C}$, the Grothendieck–Witt group $GW_i(\mathcal{C})$ is the ith homotopy group of the homotopy fibre over $0 \in \mathcal{C}$ of the map $|F_\bullet| : |\mathcal{R}_\bullet(\mathcal{C})| \to |\mathcal{S}_\bullet(\mathcal{C})|$ (see [**Sch10**]). For studies of the (Grothendieck–)Witt theory of proto-exact categories, see [**ELY22a, ELY22b**].

4.2. Hall modules via the $\mathcal{R}_\bullet$-construction. Consider again the setting of Section 2.3, so that the proto-exact category $\mathcal{C}$ is finitary and $\ell = \mathrm{Fun}_c(\pi_0(-), \mathbb{C})$. We consider the relative 2-Segal space $\mathcal{R}_\bullet(\mathcal{C}) \to \mathcal{S}_\bullet(\mathcal{C})$ of Example 4.1.3. Then $M(\mathcal{C}) := M(\mathcal{R}_\bullet(\mathcal{C}); \ell)$ is the complex vector space with basis $\{1_W^P\}_{W \in \pi_0 \mathcal{C}^h}$ labelled by isometry classes of symmetric forms in $\mathcal{C}$. The analogue of Lemma 2.3.3 holds, so that the morphisms (F_1, ∂_0) and ∂_1 which appear in the span

$$\begin{array}{ccccc} & \mathcal{R}_1(\mathcal{C}) & & & \\ {\scriptstyle (F_1,\partial_0)} \swarrow & & \searrow {\scriptstyle \partial_1} & & \\ \mathcal{S}_1(\mathcal{C}) \times \mathcal{R}_0(\mathcal{C}) & & \mathcal{R}_0(\mathcal{C}) \end{array}, \qquad \begin{array}{ccccc} & \text{Diag. (4.3)} & \\ \swarrow & & \searrow \\ (U, (M_1, \psi_1)) & & (M_0, \psi_0) \end{array}$$

are smooth and proper with respect to ℓ, respectively. Hence, $M(\mathcal{C})$ becomes a left $H(\mathcal{C})$-module with action determined by

$$1_U \star 1_M^P = \sum_{N \in \pi_0 \mathcal{C}^h} d_{U,M}^N 1_N^P.$$

The structure constant $d_{U,M}^N$ is equal to the cardinality of the set

$$\{U' \subset N \mid U' \simeq U \text{ is isotropic in } N,\ N /\!\!/ U' \simeq M\}.$$

We consider two particular instances of this construction.

EXAMPLE 4.2.1. We consider a symplectic counterpart of Example 2.3.4. Give $\mathsf{vect}_{\mathbb{F}_q}$ the standard $\mathbb{F}_q$-linear duality functor $P = \mathrm{Hom}_{\mathbb{F}_q}(-, \mathbb{F}_q)$ and double dual isomorphism $\Theta_U = -\mathrm{ev}_U$ the negative of the canonical evaluation isomorphism. Symmetric forms are then symplectic vector spaces. The resulting Hall module is

$$M(\mathsf{vect}_{\mathbb{F}_q}) = \bigoplus_{m=0}^{\infty} \mathbb{C} \cdot 1_{2m}^P,$$

where 1_{2m}^P denotes the basis vector corresponding to the symplectic vector space of dimension $2m$. The module structure is

$$1_n \star 1_{2m}^P = q^{\frac{n(n+1)}{2}} \begin{bmatrix} n+m \\ n \end{bmatrix}_q 1_{n+m}^P,$$

where now the structure constant counts the number of $\mathbb{F}_q$-rational points of the isotropic Grassmannian $\mathrm{IGr}(n, 2(m+n))$.

Taking instead $\Theta_U = \mathrm{ev}_U$ results in symmetric forms which are orthogonal vector spaces. The resulting Hall module $M(\mathsf{vect}_{\mathbb{F}_q})$ is a direct sum of $H(\mathsf{vect}_{\mathbb{F}_q})$-modules labelled by the Witt group, that is, the parity of the dimension of orthogonal vector spaces, and has structure constants which count $\mathbb{F}_q$-rational points of orthogonal Grassmannians.

EXAMPLE 4.2.2 ([**You16**]). We consider a relative counterpart of Example 2.3.6. Let σ be an anti-involution of a quiver Q, that is, σ consists of self-bijections of the sets of vertices and arrows of Q such that if $\alpha : i \to j$ is an arrow in Q, then so too is $\sigma(\alpha) : \sigma(j) \to \sigma(i)$. Define a duality structure on $\mathsf{rep}_{\mathbb{F}_q}(Q)$ by $P = (-)^\vee \circ \sigma^*$, where $(-)^\vee$ denotes vertex-wise $\mathbb{F}_q$-linear duality and Θ is the vertex-wise evaluation isomorphism. Symmetric forms in $\mathsf{rep}_{\mathbb{F}_q}(Q)$ are orthogonal counterparts of quiver representations. See [**DW02**]. More general duality structures exist which incorporate various signs associated to vertices and arrows and give symmetric forms which are mixed orthogonal/symplectic representations of Q. By incorporating the coalgebra and comodule structures of $H(Q)$ and $M(Q)$ (the latter has been neglected in this article), one checks that the $H(Q)$-module $M(Q)$ lifts to a module over the *reduced σ-analogue* of the quantum group $U_{\sqrt{q}}(\mathfrak{g}_Q)$, as introduced by Kashiwara and Enomoto [**EK08**]. With this interpretation, $M(Q)$ decomposes into highest-weight modules labelled by the cuspidal elements of $M(Q)$. The $\mathbb{F}_1$-analogue of this example is studied in [**You21**].

The next example varies the theory with transfer ℓ.

EXAMPLE 4.2.3. Working in the context of Example 2.3.13, consider $\mathsf{vect}_{\mathbb{F}_q}$ with the standard orthogonal or symplectic duality structure, as described in Example 4.2.1. The resulting module $M(\mathcal{R}_\bullet(\mathsf{vect}_{\mathbb{F}_q}); \ell)$ is isomorphic to

$$\bigoplus_{m=0}^{\infty} K_0(\mathsf{rep}_{\mathbb{C}}(G_m(\mathbb{F}_q))),$$

where $G_m = \mathrm{O}_m$ (resp. $G_m = \mathrm{Sp}_{2m}$) if $\Theta = \mathrm{ev}$ (resp. $\Theta = -\mathrm{ev}$) with $H(\mathcal{S}_\bullet(\mathsf{vect}_{\mathbb{F}_q}); \ell)$ acting by parabolic induction. Van Leeuwen [**vL91**] proved that $M(\mathcal{R}_\bullet(\mathsf{vect}_{\mathbb{F}_q}); \ell)$ is minimally generated over $H(\mathcal{S}_\bullet(\mathsf{vect}_{\mathbb{F}_q}); \ell)$ by the set of cuspidal representations of the tower of groups $G_m(\mathbb{F}_q)$, $m \geq 0$.

Finally, we give a non-finitary example.

EXAMPLE 4.2.4 ([**You20**]). We consider a relative counterpart of Section 2.4. Give $\mathsf{rep}_{\mathbb{C}}(Q)$ a duality structure as in Example 4.2.2. The resulting *cohomological Hall module* $M(Q) = M(\mathcal{R}_\bullet(\mathsf{rep}_{\mathbb{C}}(Q)); H^\bullet(-;\mathbb{Q}))$ plays a central role in the orientifold Donaldson–Thomas theory of Q, that is, the mixed orthogonal/symplectic generalization of Donaldson–Thomas theory [**You15, You20**]. Again, we restrict attention to the m-loop quiver Q_m. Fix the duality $P = \mathrm{Hom}_{\mathbb{C}}(-, \mathbb{C})$ and double dual isomorphism $\Theta = -\mathrm{ev}$. There is an isomorphism of stacks

$$\mathcal{R}_0(\mathsf{rep}_{\mathbb{C}}(Q_m)) \simeq \bigsqcup_{e=0}^{\infty} [\mathfrak{sp}_{2e}(\mathbb{C})^{\oplus m} / \mathrm{Sp}_{2e}(\mathbb{C})],$$

with $\mathrm{Sp}_{2e}(\mathbb{C})$ acting by conjugation, so that the underlying $(\mathbb{Z}\times\mathbb{Z})$-graded vector space of $M(Q_m)$ is

$$M(Q_m) = \bigoplus_{e=0}^{\infty} M_{2e}(Q_m),$$

where

$$M_{2e}(Q_m) \simeq H^{\bullet}(B\,\mathrm{Sp}_{2e}(\mathbb{C})) \simeq \mathbb{Q}[z_1^2,\dots,z_e^2]^{\mathfrak{S}_e}, \qquad |z_i| = 2.$$

Keeping the notation of Section 2.4, the action of $f \in H_d(Q_m)$ on $g \in M_{2e}(Q_m)$ is

$$f \star g = 2^{(m-1)d} \sum_{\pi\in\mathfrak{sh}^{\sigma}_{d,e}} \pi\Big[f(x_1,\dots,x_d)g(z_1,\dots,z_e)\times \Big(\prod_{i=1}^{d}(-x_i)\prod_{1\le i<j\le d}(-x_i-x_j)\prod_{i=1}^{d}\prod_{j=1}^{e}(x_i^2-z_j^2)\Big)^{m-1}\Big],$$

where $\mathfrak{sh}^{\sigma}_{d,e} = \mathbb{Z}_2^d \times \mathfrak{sh}_{d,e}$ is the set of signed 2-shuffles of size (d,e). The signs $\mathbb{Z}_2^d$ act on the first d variables by multiplication by signs. There is an analogue of Efimov's theorem in this relative setting. Namely, define a $(2\mathbb{Z}_{\geq 0}\times\mathbb{Z})$-graded vector space by

$$W^{\mathrm{prim}}_{Q_m} = M(Q_m)/(H(Q_m)_+ \star M(Q_m)),$$

where $H(Q_m)_+$ is the augmentation ideal of $H(Q_m)$, and pick a graded splitting so as to view $W^{\mathrm{prim}}_{Q_m}$ as a subspace of $M(Q_m)$. The vector space $W^{\mathrm{prim}}_{Q_m}$ is the *cohomological orientifold Donaldson–Thomas invariant of* Q. Then, for each $e \in 2\mathbb{Z}_{\geq 0}$, there is a $(\mathbb{Z}_{\geq 0}\times\mathbb{Z})$-graded subalgebra $H(Q_m)(e) \subset H(Q_m)$ such that the action map

$$\star : \bigoplus_{e\in 2\mathbb{Z}_{\geq 0}} H(Q_m)(e) \otimes_{\mathbb{Q}} W^{\mathrm{prim}}_{Q_m,e} \to M(Q_m) \tag{4.4}$$

is an isomorphism of graded vector spaces. Moreover, $H(Q_m)(e)$ can be described with sufficient understanding of $V^{\mathrm{prim}}_{Q_m}$. This isomorphism can be used to inductively determine $W^{\mathrm{prim}}_{Q_m,e}$. While the isomorphism (4.4) remains conjectural for general symmetric quivers, it is known that W^{prim}_{Q} is isomorphic to the Chow group of the moduli space of stable symmetric representations of Q (see [**FY18**]). A motivic version of this example is the focus of the works [**You15, You21, Bu23**].

4.3. Hall modules via framing.

EXAMPLE 4.3.1. We work with the stable framed $\mathcal{S}_{\bullet}$-construction of Example 4.1.2. Subject to smoothness and properness conditions, for each argument $\phi \in [0,1)$ there results a module $M^{\text{st-fr}}_{\phi}(\mathcal{C}) := M(\mathcal{S}^{\text{st-fr}}_{\bullet}(\mathcal{C}^{\mathrm{ss}}_{\phi});\ell)$ over the Hall algebra $H^{\mathrm{ss}}_{\phi}(\mathcal{C}) := H(\mathcal{S}_{\bullet}(\mathcal{C}^{\mathrm{ss}}_{\phi});\ell)$.

For example, let $\mathcal{C} = \mathsf{rep}_{\mathbb{C}}(Q)$. Sending a representation to its dimension vector defines a surjective group homomorphism $K_0(\mathsf{rep}_{\mathbb{C}}(Q)) \to \mathbb{Z}^{Q_0}$. A tuple $\zeta = (\zeta_i)_{i\in Q_0} \in \mathbb{H}_+^{Q_0}$ defines a function $\mathbb{Z}^{Q_0} \to \mathbb{C}, d \mapsto \sum_{i\in Q_0}\zeta_i d_i$, which induces a stability function Z on $\mathsf{rep}_{\mathbb{C}}(Q)$. For any $f \in \mathbb{Z}^{Q_0}_{\geq 0}$, the functor $\Phi : U \mapsto \bigoplus_{i\in Q_0}\mathrm{Hom}_{\mathbb{C}}(\mathbb{C}^{f_i}, U_i)$ is a framing. Working in the cohomological context of Section 2.4 and Example 4.2.4, the resulting cohomological $H^{\mathrm{ss}}_{\phi}(Q)$-modules $M^{\text{st-fr}}_{\phi}(Q)$ play an important role in Donaldson–Thomas theory [**Soi16, Fra16**]. With knowledge of the structure of the algebra $H^{\mathrm{ss}}_{\phi}(Q)$, for example as described in Section 2.4, the module $M_\phi^{\text{st-fr}}(Q)$ can be used to inductively determine topological invariants of the smooth moduli space of stable framed representations of Q, also known as non-commutative Hilbert schemes. From a different perspective, $M_\phi^{\text{st-fr}}(Q)$ can be seen to approximate $H_\phi^{\text{ss}}(Q)$ when the components of the framing vector f are sufficiently large. This has the technical benefit that, for sufficiently large f, the underlying vector space of $M_\phi^{\text{st-fr}}(Q)$ is the cohomology of a smooth variety, as opposed to a stack. This idea was used to great effect to study $H_\phi^{\text{ss}}(Q)$ in [**DM20**]. Similar ideas can be applied in the context of moduli spaces of sheaves over a smooth projective curve [**MR15**] or surface [**DPS22**].

References

[Bar16] C. Barwick, *On the algebraic K-theory of higher categories*, J. Topol. **9** (2016), no. 1, 245–347, DOI 10.1112/jtopol/jtv042. MR3465850

[Beh04] K. Behrend, *Cohomology of stacks*, Intersection theory and moduli, ICTP Lect. Notes, XIX, Abdus Salam Int. Cent. Theoret. Phys., Trieste, 2004, pp. 249–294. MR2172499

[Bei78] A. A. Beĭlinson, *Coherent sheaves on* $\mathbf{P}^n$ *and problems in linear algebra* (Russian), Funktsional. Anal. i Prilozhen. **12** (1978), no. 3, 68–69. MR509388

[BG16] A. Berenstein and J. Greenstein, *Primitively generated Hall algebras*, Pacific J. Math. **281** (2016), no. 2, 287–331, DOI 10.2140/pjm.2016.281.287. MR3463039

[BK01] P. Baumann and C. Kassel, *The Hall algebra of the category of coherent sheaves on the projective line*, J. Reine Angew. Math. **533** (2001), 207–233, DOI 10.1515/crll.2001.031. MR1823869

[Bri07] T. Bridgeland, *Stability conditions on triangulated categories*, Ann. of Math. (2) **166** (2007), no. 2, 317–345, DOI 10.4007/annals.2007.166.317. MR2373143

[Bri12] T. Bridgeland, *An introduction to motivic Hall algebras*, Adv. Math. **229** (2012), no. 1, 102–138, DOI 10.1016/j.aim.2011.09.003. MR2854172

[Bri13] T. Bridgeland, *Quantum groups via Hall algebras of complexes*, Ann. of Math. (2) **177** (2013), no. 2, 739–759, DOI 10.4007/annals.2013.177.2.9. MR3010811

[BS12] I. Burban and O. Schiffmann, *On the Hall algebra of an elliptic curve, I*, Duke Math. J. **161** (2012), no. 7, 1171–1231, DOI 10.1215/00127094-1593263. MR2922373

[Bu23] C. Bu, *Enumerative invariants in self-dual categories. I. Motivic invariants*, `arXiv:2302.00038`, 2023.

[Büh10] T. Bühler, *Exact categories*, Expo. Math. **28** (2010), no. 1, 1–69, DOI 10.1016/j.exmath.2009.04.004. MR2606234

[Car20] L. Carlier, *Incidence bicomodules, Möbius inversion and a Rota formula for infinity adjunctions*, Algebr. Geom. Topol. **20** (2020), no. 1, 169–213, DOI 10.2140/agt.2020.20.169. MR4071369

[Cra10] T. Cramer, *Double Hall algebras and derived equivalences*, Adv. Math. **224** (2010), no. 3, 1097–1120, DOI 10.1016/j.aim.2009.12.021. MR2628805

[CS20] B. Cooper and P. Samuelson, *The Hall algebras of surfaces I*, J. Inst. Math. Jussieu **19** (2020), no. 3, 971–1028, DOI 10.1017/s1474748018000324. MR4094715

[Dav17] B. Davison, *The critical CoHA of a quiver with potential*, Q. J. Math. **68** (2017), no. 2, 635–703, DOI 10.1093/qmath/haw053. MR3667216

[Del48] S. Delsarte, *Fonctions de Möbius sur les groupes abeliens finis* (French), Ann. of Math. (2) **49** (1948), 600–609, DOI 10.2307/1969047. MR25463

[DK19] T. Dyckerhoff and M. Kapranov, *Higher Segal spaces*, Lecture Notes in Mathematics, vol. 2244, Springer, Cham, 2019, DOI 10.1007/978-3-030-27124-4. MR3970975

[DM20] B. Davison and S. Meinhardt, *Cohomological Donaldson-Thomas theory of a quiver with potential and quantum enveloping algebras*, Invent. Math. **221** (2020), no. 3, 777–871, DOI 10.1007/s00222-020-00961-y. MR4132957

[DPS22] D.-E. Diaconescu, M. Porta, and F. Sala, *Cohomological Hall algebras and their representations via torsion pairs*, `arXiv:2207.08926`, 2022.

[Dri87] V. G. Drinfel′d, *A new realization of Yangians and of quantum affine algebras* (Russian), Dokl. Akad. Nauk SSSR **296** (1987), no. 1, 13–17; English transl., Soviet Math. Dokl. **36** (1988), no. 2, 212–216. MR914215

[DW02] H. Derksen and J. Weyman, *Generalized quivers associated to reductive groups*, Colloq. Math. **94** (2002), no. 2, 151–173, DOI 10.4064/cm94-2-1. MR1967372

[DW17] H. Derksen and J. Weyman, *An introduction to quiver representations*, Graduate Studies in Mathematics, vol. 184, American Mathematical Society, Providence, RI, 2017, DOI 10.1090/gsm/184. MR3727119

[Dyc18] T. Dyckerhoff, *Higher categorical aspects of Hall algebras*, Building bridges between algebra and topology, Adv. Courses Math. CRM Barcelona, Birkhäuser/Springer, Cham, 2018, pp. 1–61. MR3793857

[Efi12] A. I. Efimov, *Cohomological Hall algebra of a symmetric quiver*, Compos. Math. **148** (2012), no. 4, 1133–1146, DOI 10.1112/S0010437X12000152. MR2956038

[EJS20] C. Eppolito, J. Jun, and M. Szczesny, *Proto-exact categories of matroids, Hall algebras, and K-theory*, Math. Z. **296** (2020), no. 1-2, 147–167, DOI 10.1007/s00209-019-02429-z. MR4140735

[EK08] N. Enomoto and M. Kashiwara, *Symmetric crystals for* $\mathfrak{gl}_\infty$, Publ. Res. Inst. Math. Sci. **44** (2008), no. 3, 837–891. MR2451612

[ELY22a] J. N. Eberhardt, O. Lorscheid, and M. B. Young, *Algebraic K-theory and Grothendieck-Witt theory of monoid schemes*, Math. Z. **301** (2022), no. 2, 1407–1445, DOI 10.1007/s00209-021-02919-z. MR4418325

[ELY22b] J. N. Eberhardt, O. Lorscheid, and M. B. Young, *Group completion in the K-theory and Grothendieck-Witt theory of proto-exact categories*, J. Pure Appl. Algebra **226** (2022), no. 8, Paper No. 107018, 32, DOI 10.1016/j.jpaa.2022.107018. MR4372795

[FGK+21] M. Feller, R. Garner, J. Kock, M. U. Proulx, and M. Weber, *Every 2-Segal space is unital*, Commun. Contemp. Math. **23** (2021), no. 2, Paper No. 2050055, 6, DOI 10.1142/S0219199720500558. MR4201031

[FR18] H. Franzen and M. Reineke, *Semistable Chow-Hall algebras of quivers and quantized Donaldson-Thomas invariants*, Algebra Number Theory **12** (2018), no. 5, 1001–1025, DOI 10.2140/ant.2018.12.1001. MR3840869

[Fra16] H. Franzen, *On cohomology rings of non-commutative Hilbert schemes and CoHa-modules*, Math. Res. Lett. **23** (2016), no. 3, 805–840, DOI 10.4310/MRL.2016.v23.n3.a12. MR3533197

[FY18] H. Franzen and M. B. Young, *Cohomological orientifold Donaldson-Thomas invariants as Chow groups*, Selecta Math. (N.S.) **24** (2018), no. 3, 2035–2061, DOI 10.1007/s00029-018-0415-1. MR3816498

[Gab72] P. Gabriel, *Unzerlegbare Darstellungen. I* (German, with English summary), Manuscripta Math. **6** (1972), 71–103; correction, ibid. 6 (1972), 309, DOI 10.1007/BF01298413. MR332887

[GCKT18a] I. Gálvez-Carrillo, J. Kock, and A. Tonks, *Decomposition spaces, incidence algebras and Möbius inversion I: Basic theory*, Adv. Math. **331** (2018), 952–1015, DOI 10.1016/j.aim.2018.03.016. MR3804694

[GCKT18b] I. Gálvez-Carrillo, J. Kock, and A. Tonks, *Decomposition spaces, incidence algebras and Möbius inversion II: Completeness, length filtration, and finiteness*, Adv. Math. **333** (2018), 1242–1292, DOI 10.1016/j.aim.2018.03.017. MR3818099

[GCKT18c] I. Gálvez-Carrillo, J. Kock, and A. Tonks, *Homotopy linear algebra*, Proc. Roy. Soc. Edinburgh Sect. A **148** (2018), no. 2, 293–325, DOI 10.1017/S0308210517000208. MR3777576

[GCKT24] I. Gálvez-Carrillo, J. Kock, A. Tonks, *Convex decomposition spaces and Crapo complementation formula*, `arXiv:2409.03742v1`, 2024.

[Göd24] J. Gödicke, *An ∞-Category of 2-Segal spaces*, `arXiv:2407.13357`, 2024.

[Gre55] J. A. Green, *The characters of the finite general linear groups*, Trans. Amer. Math. Soc. **80** (1955), 402–447, DOI 10.2307/1992997. MR72878

[Gre95] J. A. Green, *Hall algebras, hereditary algebras and quantum groups*, Invent. Math. **120** (1995), no. 2, 361–377, DOI 10.1007/BF01241133. MR1329046

[Hac24] P. Hackney, *The decomposition space perspective*, `arXiv:2409.19061`, 2024.

[Hal59] P. Hall, *The algebra of partitions*, Proceedings of the 4th Canadian Mathematical Congress, Banff (1959), 147–159.

[HM98] J. A. Harvey and G. Moore, *On the algebras of BPS states*, Comm. Math. Phys. **197** (1998), no. 3, 489–519, DOI 10.1007/s002200050461. MR1652775

[Joy07] D. Joyce, *Configurations in abelian categories. II. Ringel-Hall algebras*, Adv. Math. **210** (2007), no. 2, 635–706, DOI 10.1016/j.aim.2006.07.006. MR2303235

[Joy18] D. Joyce, *Ringel–Hall style vertex algebra and Lie algebra structures on the homology of moduli spaces*, Available at `https://people.maths.ox.ac.uk/joyce/hall.pdf`, 2018.

[JS12] D. Joyce and Y. Song, *A theory of generalized Donaldson-Thomas invariants*, Mem. Amer. Math. Soc. **217** (2012), no. 1020, iv+199, DOI 10.1090/S0065-9266-2011-00630-1. MR2951762

[JS24] J. Jun and M. Szczesny, *Hall Lie algebras of toric monoid schemes*, Selecta Math. (N.S.) **30** (2024), no. 2, Paper No. 22, 57, DOI 10.1007/s00029-023-00913-3. MR4700426

[Kac90] V. G. Kac, *Infinite-dimensional Lie algebras*, 3rd ed., Cambridge University Press, Cambridge, 1990, DOI 10.1017/CBO9780511626234. MR1104219

[Kap97] M. M. Kapranov, *Eisenstein series and quantum affine algebras*, J. Math. Sci. (New York) **84** (1997), no. 5, 1311–1360, DOI 10.1007/BF02399194. Algebraic geometry, 7. MR1465518

[KS06] S.-J. Kang and O. Schiffmann, *Canonical bases for quantum generalized Kac-Moody algebras*, Adv. Math. **200** (2006), no. 2, 455–478, DOI 10.1016/j.aim.2005.01.002. MR2200853

[KS08] M. Kontsevich and Y. Soibelman, *Stability structures, motivic Donaldson-Thomas invariants and cluster transformations*, `arXiv:0811.2435`, 2008.

[KS11] M. Kontsevich and Y. Soibelman, *Cohomological Hall algebra, exponential Hodge structures and motivic Donaldson-Thomas invariants*, Commun. Number Theory Phys. **5** (2011), no. 2, 231–352, DOI 10.4310/CNTP.2011.v5.n2.a1. MR2851153

[KSV17] M. Kapranov, O. Schiffmann, and E. Vasserot, *The Hall algebra of a curve*, Selecta Math. (N.S.) **23** (2017), no. 1, 117–177, DOI 10.1007/s00029-016-0239-9. MR3595890

[Lat21] A. Latyntsev, *Cohomological Hall algebras and vertex algebras*, `arXiv:2110.14356`, 2021.

[Lei12] T. Leinster, *Notions of Möbius inversion*, Bull. Belg. Math. Soc. Simon Stevin **19** (2012), no. 5, 911–935. MR3009022

[Lus10] G. Lusztig, *Introduction to quantum groups*, Modern Birkhäuser Classics, Birkhäuser/Springer, New York, 2010. Reprint of the 1994 edition, DOI 10.1007/978-0-8176-4717-9. MR2759715

[Mac95] I. G. Macdonald, *Symmetric functions and Hall polynomials*, 2nd ed., Oxford Mathematical Monographs, The Clarendon Press, Oxford University Press, New York, 1995. With contributions by A. Zelevinsky; Oxford Science Publications. MR1354144

[MR15] S. Mozgovoy and M. Reineke, *Intersection cohomology of moduli spaces of vector bundles over curves*, `arXiv:1512.04076`, 2015.

[MR19] S. Meinhardt and M. Reineke, *Donaldson-Thomas invariants versus intersection cohomology of quiver moduli*, J. Reine Angew. Math. **754** (2019), 143–178, DOI 10.1515/crelle-2017-0010. MR4000572

[MS17] H. Morton and P. Samuelson, *The HOMFLYPT skein algebra of the torus and the elliptic Hall algebra*, Duke Math. J. **166** (2017), no. 5, 801–854, DOI 10.1215/00127094-3718881. MR3626565

[Nak01] H. Nakajima, *Quiver varieties and finite-dimensional representations of quantum affine algebras*, J. Amer. Math. Soc. **14** (2001), no. 1, 145–238, DOI 10.1090/S0894-0347-00-00353-2. MR1808477

[OR24] V. Ozornova and M. Rovelli, *Waldhausen's $S_\bullet$-construction*, `arXiv:2412.17398`, 2024.

[Oxl92] J. G. Oxley, *Matroid theory*, Oxford Science Publications, The Clarendon Press, Oxford University Press, New York, 1992. MR1207587

[PX00] L. Peng and J. Xiao, *Triangulated categories and Kac-Moody algebras*, Invent. Math. **140** (2000), no. 3, 563–603, DOI 10.1007/s002220000062. MR1760751

[Qui73] D. Quillen, *Higher algebraic K-theory. I*, Algebraic K-theory, I: Higher K-theories (Proc. Conf., Battelle Memorial Inst., Seattle, Wash., 1972), Lecture Notes in Math., Vol. 341, Springer, Berlin-New York, 1973, pp. 85–147. MR338129

[Rez01] C. Rezk, *A model for the homotopy theory of homotopy theory*, Trans. Amer. Math. Soc. **353** (2001), no. 3, 973–1007, DOI 10.1090/S0002-9947-00-02653-2. MR1804411

[Rie94] C. Riedtmann, *Lie algebras generated by indecomposables*, J. Algebra **170** (1994), no. 2, 526–546, DOI 10.1006/jabr.1994.1351. MR1302854

[Rin90a] C. M. Ringel, *Hall algebras*, Topics in algebra, Part 1 (Warsaw, 1988), Banach Center Publ., vol. 26, PWN, Warsaw, 1990, pp. 433–447. MR1171248

[Rin90b] C. M. Ringel, *Hall algebras and quantum groups*, Invent. Math. **101** (1990), no. 3, 583–591, DOI 10.1007/BF01231516. MR1062796

[Sch94] W. R. Schmitt, *Incidence Hopf algebras*, J. Pure Appl. Algebra **96** (1994), no. 3, 299–330, DOI 10.1016/0022-4049(94)90105-8. MR1303288

[Sch10] M. Schlichting, *The Mayer-Vietoris principle for Grothendieck-Witt groups of schemes*, Invent. Math. **179** (2010), no. 2, 349–433, DOI 10.1007/s00222-009-0219-1. MR2570120

[Sch12a] O. Schiffmann, *Lectures on canonical and crystal bases of Hall algebras* (English, with English and French summaries), Geometric methods in representation theory. II, Sémin. Congr., vol. 24, Soc. Math. France, Paris, 2012, pp. 143–259. MR3202708

[Sch12b] O. Schiffmann, *Lectures on Hall algebras* (English, with English and French summaries), Geometric methods in representation theory. II, Sémin. Congr., vol. 24, Soc. Math. France, Paris, 2012, pp. 1–141. MR3202707

[Sch12c] O. Schiffmann, *On the Hall algebra of an elliptic curve, II*, Duke Math. J. **161** (2012), no. 9, 1711–1750, DOI 10.1215/00127094-1593362. MR2942792

[Sch18] O. G. Schiffmann, *Kac polynomials and Lie algebras associated to quivers and curves*, Proceedings of the International Congress of Mathematicians—Rio de Janeiro 2018. Vol. II. Invited lectures, World Sci. Publ., Hackensack, NJ, 2018, pp. 1393–1424. MR3966814

[Soi16] Y. Soibelman, *Remarks on cohomological Hall algebras and their representations*, Arbeitstagung Bonn 2013, Progr. Math., vol. 319, Birkhäuser/Springer, Cham, 2016, pp. 355–385, DOI 10.1007/978-3-319-43648-7_13. MR3618057

[Ste01] E. Steinitz, *Zur Theorie der Abel'schen Gruppen*, Jahresbericht der Deutschen Mathematiker-Vereinigung **9** (1901), 80–85.

[Ste21] W. H. Stern, *2-Segal objects and algebras in spans*, J. Homotopy Relat. Struct. **16** (2021), no. 2, 297–361, DOI 10.1007/s40062-021-00282-8. MR4266208

[Ste24] W. H. Stern, *Perspectives on the 2-Segal Conditions*, this volume, 2024.

[SVdB01] B. Sevenhant and M. Van Den Bergh, *A relation between a conjecture of Kac and the structure of the Hall algebra*, J. Pure Appl. Algebra **160** (2001), no. 2-3, 319–332, DOI 10.1016/S0022-4049(00)00078-5. MR1836006

[SY96] J. M. Shapiro and D. Yao, *Hermitian $\mathcal{U}$-theory of exact categories with duality functors*, J. Pure Appl. Algebra **109** (1996), no. 3, 323–330, DOI 10.1016/0022-4049(95)00093-3. MR1388702

[Szc12] M. Szczesny, *Representations of quivers over $\mathbb{F}_1$ and Hall algebras*, Int. Math. Res. Not. IMRN **10** (2012), 2377–2404, DOI 10.1093/imrn/rnr113. MR2923170

[Szc18] M. Szczesny, *The Hopf algebra of skew shapes, torsion sheaves on $\mathbb{A}^n_{/\mathbb{F}_1}$, and ideals in Hall algebras of monoid representations*, Adv. Math. **331** (2018), 209–238, DOI 10.1016/j.aim.2018.03.032. MR3804676

[Toë06] B. Toën, *Derived Hall algebras*, Duke Math. J. **135** (2006), no. 3, 587–615, DOI 10.1215/S0012-7094-06-13536-6. MR2272977

[vL91] M. A. A. van Leeuwen, *An application of Hopf-algebra techniques to representations of finite classical groups*, J. Algebra **140** (1991), no. 1, 210–246, DOI 10.1016/0021-8693(91)90154-Z. MR1114914

[Wal85] F. Waldhausen, *Algebraic K-theory of spaces*, Algebraic and geometric topology (New Brunswick, N.J., 1983), Lecture Notes in Math., vol. 1126, Springer, Berlin, 1985, pp. 318–419, DOI 10.1007/BFb0074449. MR802796

[Wal16] T. Walde, *Hall monoidal categories and categorical modules*, `arXiv:1611.08241`, 2016.

[Yan17] S. Yanagida, *Elliptic Hall algebra on $\mathbb{F}_1$*, `arXiv:1708.08881`, 2017.

[You15] M. B. Young, *Self-dual quiver moduli and orientifold Donaldson-Thomas invariants*, Commun. Number Theory Phys. **9** (2015), no. 3, 437–475, DOI 10.4310/CNTP.2015.v9.n3.a1. MR3399924

[You16] M. B. Young, *The Hall module of an exact category with duality*, J. Algebra **446** (2016), 291–322, DOI 10.1016/j.jalgebra.2015.09.013. MR3421095

[You18] M. B. Young, *Relative 2-Segal spaces*, Algebr. Geom. Topol. **18** (2018), no. 2, 975–1039, DOI 10.2140/agt.2018.18.975. MR3773745

[You20] M. B. Young, *Representations of cohomological Hall algebras and Donaldson-Thomas theory with classical structure groups*, Comm. Math. Phys. **380** (2020), no. 1, 273–322, DOI 10.1007/s00220-020-03877-z. MR4166629

[You21] M. B. Young, *Degenerate versions of Green's theorem for Hall modules*, J. Pure Appl. Algebra **225** (2021), no. 4, Paper No. 106557, 21, DOI 10.1016/j.jpaa.2020.106557. MR4148355

[Zel81] A. V. Zelevinsky, *Representations of finite classical groups*, Lecture Notes in Mathematics, vol. 869, Springer-Verlag, Berlin-New York, 1981. A Hopf algebra approach. MR643482

Department of Mathematics, University of Iowa, 14 MacLean Hall, Iowa City, Iowa 52242-1419

Email address: `ben-cooper@uiowa.edu`

Department of Mathematics and Statistics, Utah State University, Logan, Utah 84322

Email address: `matthew.young@usu.edu`

Contemporary Mathematics
Volume **838**, 2026
https://doi.org/10.1090/conm/838/16739

Cyclic polytopes, orientals, and correspondences: Some aspects of Higher Segal spaces

Tobias Dyckerhoff

ABSTRACT. We discuss the role of higher Segal spaces at the interface of cyclic polytopes, orientals, and higher correspondences. Along the way we review examples from algebraic K-theory, show how cyclic polytopes provide a geometric model for the definition of orientals, and establish a characterization of higher Segal spaces as lax monads in higher correspondence categories.

1. Introduction

These are extended notes on talks given at the Banff workshop on "Higher Segal Spaces and their Applications to Algebraic K-Theory, Hall Algebras, and Combinatorics". Many of the results and ideas presented here are in some form present in the literature while we try to tie some loose ends or discuss slightly different perspectives here and there. We give a rough overview of the contents:

§2. Discussion of the geometry and combinatorics of cyclic polytopes
§3. Definition, basic theory, and examples of higher Segal spaces
§4. Geometric construction of Street's orientals [**Str87**] via cyclic polytopes
§5. Some key results on higher Segal spaces based on results of Rambau [**Ram97**] on cyclic polytopes
§6. Description of the (∞, ω)-category of correspondences via barycentric subdivision
§7. Characterization of higher Segal spaces as certain lax monads in higher correspondence categories
§8. Alternative characterization of higher Segal spaces via higher excision due to T. Walde
§9. Some further perspectives

Acknowledgments. I thank the organizers of the workshop, namely Julie Bergner, Joachim Kock, and Maru Sarazola, for their initiative to bring together the community and for creating a fruitful atmosphere for the many insightful exchanges that happened during the workshop. I am indebted to Mikhail Kapranov for all the shared insights over the years — in particular, his proposal to use cyclic polytopes as a foundation for the theory of higher Segal spaces is a cornerstone of everything presented in this article. I am very grateful to an anonymous referee for the careful reading and the many suggestions that helped improve the text. Last but not least, I acknowledge support by the Deutsche Forschungsgemeinschaft (DFG, German

Research Foundation) under Germany's Excellence Strategy - EXC 2121 "Quantum Universe" - 390833306 and the Collaborative Research Center - SFB 1624 "Higher structures, moduli spaces and integrability" - 506632645.

2. Cyclic polytopes

We recall some basics of cyclic polytopes. See [**Ram97**], for more background and detailed proofs.

DEFINITION 2.1. Let $d \geq 0$ and consider the map

$$\nu\colon \mathbb{R} \hookrightarrow \mathbb{R}^d,\ t \mapsto (t, t^2, t^3, \ldots, t^d).$$

For a finite subset $S \subset \mathbb{R}$, we define the d-dimensional cyclic polytope

$$C(S, d) \subset \mathbb{R}^d$$

on S to be the convex hull of the set $\nu(S) \subset \mathbb{R}^d$.

For any subset $I \subset S$ of cardinality $d+1$, the convex hull of $\nu(I)$ is a d-simplex in $\mathbb{R}^d$; in other words, the points in $\nu(S)$ are in general position. This holds, since the determinant of the van-der-Monde matrix, with rows given by the affine coordinate vectors

$$(1, \nu(i)) = (1, \nu(i)_1, \nu(i)_2, \ldots, \nu(i)_d)$$

of the points $\nu(i)$, $i \in M$, is nonzero. In other words, no $d+1$ vertices of $C(S, d)$ lie on an affine hyperplane. This implies that $C(S, d)$ is a simplicial polytope, i.e., all boundary faces are simplices.

REMARK 2.2. It is a key feature that cyclic polytopes on the same vertex set in different dimensions are related by projection maps: the map

$$\pi\colon \mathbb{R}^d \to \mathbb{R}^{d-1}, (t_1, \ldots, t_d) \mapsto (t_1, \ldots, t_{d-1})$$

maps $C(S, d)$ onto $C(S, d-1)$ cf. Figure 1).

We will typically study the cyclic polytope on the set

$$S = [n] = \{0, 1, \ldots, n\} \subset \mathbb{R}$$

for varying $n \geq 0$. The cyclic polytopes $C([n], d)$ are simplicial polytopes, i.e., their boundary is a simplicial complex. We introduce some terminology and notation to speak about simplicial complexes.

DEFINITION 2.3. Given a set S, we define an *abstract simplicial complex* K on S to be a subset of the set $\mathcal{P}^*(S)$ of nonempty finite subsets of S satisfying the condition:

- If $\sigma \in K$ and $\emptyset \neq \tau \subset \sigma$, then $\tau \in K$.

For any arbitrary subset $K \subset \mathcal{P}^*(S)$ we define $\langle K \rangle \subset \mathcal{P}^*(S)$ to be the smallest abstract simplicial complex containing K, obtained by adjoining all nonempty subsets of the sets in K.

Let $I \subset [n]$ be a subset of cardinality d. We discuss a simple criterion to decide whether the $(d-1)$-simplex $|\Delta^I|$ given by the convex hull of $\nu(I)$ is a boundary simplex of $C([n], d)$. Namely, this will be the case if all points in $\nu([n] \setminus I)$ lie on the same side of the affine hyperplane $H_I \subset \mathbb{R}^d$ spanned by $\nu(I)$. We derive a beautiful criterion for testing this. First, note that the projection π determines a canonical labelling of the two sides of the hyperplane H_I as follows. Consider the fiber of π over a point $x \in \mathbb{R}^{d-1}$, i.e., a line of the form

$$L_x = \{x\} \times \mathbb{R} \subset \mathbb{R}^d.$$

Then L_x intersects H_I in precisely one point (x, t_0) so that $\mathbb{R}^d \setminus H_I$ splits into an *upper half* H_I^+ containing points (x,t) with $t > t_0$ and a *lower half* H_I^- containing points (x,t) with $t < t_0$. To determine on which side of H_I a given point $y \in \mathbb{R}^d$ lies, we form the van-der-Monde matrix V with rows

$$(1, \nu(i)), \quad i \in I,$$

ordered according to the linear order of I, along with the last row $(1, y)$. Then we have:

(a) $\det(V) > 0$ iff $y \in H_I^+$,
(b) $\det(V) = 0$ iff $y \in H_I$,
(c) $\det(V) < 0$ iff $y \in H_I^-$.

If $y \in H_I^+$ (resp. $y \in H_I^-$), we say *y lies above (resp. below) H_I*. Finally, combining this observation with the fact that the determinant changes sign when swapping two rows, we arrive at *Gale's eveness criterion* [**Gal63**]:

(1) A point $\nu(j) \in \mathbb{R}^d$ corresponding to a *gap* $j \in [n] \setminus I$ lies above (resp. below) H_I iff the cardinality of the set
$$\{i \in I \mid i > j\}$$
is even (resp. odd). In this case, we refer to the gap j itself as *even* (resp. *odd*). The simplex $|\Delta^I|$ is a boundary simplex if one of the following conditions holds:
  (a) All gaps $j \in [n] \setminus I$ are even. In this case, we have $C([n], d) \subset H_I \cup H_I^+$.
  (b) All gaps $j \in [n] \setminus I$ are odd. In this case, we have $C([n], d) \subset H_I \cup H_I^-$.

Therefore, we observe that, due to the canonical notions of *above* and *below* derived from the projection map π, the boundary of the cyclic polytope $C([n], d)$ decomposes into an upper and lower hemisphere.

DEFINITION 2.4. Let $n \geq 0$. A subset $I \subset [n]$ is called *even* (resp. *odd*) if, for every $j \in [n] \setminus I$, the cardinality of the set

$$\{i \in I \mid i > j\}$$

is even (resp. odd). We further introduce the abstract simplicial complexes on $[n]$

$$\mathcal{L}([n], d-1) := \langle J \subset [n] \mid J \text{ even of cardinality } d \rangle \subset \mathcal{P}^*([n]),$$
$$\mathcal{U}([n], d-1) := \langle J \subset [n] \mid J \text{ odd of cardinality } d \rangle \subset \mathcal{P}^*([n]),$$

called the *lower* and *upper boundary* of $C([n], d)$, where we use the notation $\langle K \rangle$ from Definition 2.3.

For an abstract simplicial complex K on $[n]$, we denote by $|K| \subset \mathbb{R}^d$ its d-dimensional geometric realization, given by the union of all convex hulls of the sets $\nu(I)$ for $I \in K$. Typically, we will only consider the geometric realization when $|K| \subset \mathbb{R}^d$ is a *geometric simplicial complex*, i.e., any given pair of geometric simplices is disjoint or intersects in another geometric simplex. Combining the above observations, we arrive at the following statements.

PROPOSITION 2.5. *Let $n, d \geq 1$. Then, the following hold:*

(1) *The geometric realizations* $|\mathcal{L}([n], d-1)| \subset \mathbb{R}^d$ *and* $|\mathcal{U}([n], d-1)| \subset \mathbb{R}^d$ *are geometric simplicial complexes.*
(2) *The cyclic polytope* $C([n], d)$ *is a simplicial polytope with boundary given by*

$$\partial C([n], d) = |\mathcal{L}([n], d-1)| \cup |\mathcal{U}([n], d-1)| \subset \mathbb{R}^d.$$

(3) *The restrictions of the projection map* $\pi\colon \mathbb{R}^d \to \mathbb{R}^{d-1}$ *provide homeomorphisms*

$$|\mathcal{U}([n], d-1)| \xrightarrow{\cong} C([n], d-1)$$

and

$$|\mathcal{L}([n], d-1)| \xrightarrow{\cong} C([n], d-1).$$

(4) *Under the homeomorphisms from* (3), *the geometric simplicial complexes get identified with triangulations of* $C([n], d-1)$, *again given by the abstract simplicial complexes* $\mathcal{L}([n], d-1)$ *and* $\mathcal{U}([n], d-1)$ *but geometrically realized in* $\mathbb{R}^{d-1}$.

REMARK 2.6. Informally speaking, the two triangulations of $C([n], d-1)$ from Proposition 2.5 arise by "looking" at the simplicial boundary of the polytope $C([n], d)$ from above (resp. below), with respect to the direction given by the last coordinate of $\mathbb{R}^d$ (cf. Figure 1). Putting this more formally: for every $x \in C([n], d-1)$, the line

$$\{x\} \times \mathbb{R} \subset \mathbb{R}^d$$

intersects the polytope $C([n], d)$ in an interval $\{x\} \times [l_x, u_x]$. Then the union of the points $\{(x, l_x)\}$ (resp. $\{(x, u_x)\}$) forms the geometric realization of the simplicial complex $\mathcal{L}([n], d-1)$ (resp. $\mathcal{U}([n], d-1)$).

EXAMPLE 2.7. Note that we have $C([n], n) = |\Delta^n|$. Thus in this case, Proposition 2.5 yields a decomposition

$$\partial|\Delta^n| = |\mathcal{L}([n], n-1)| \cup |\mathcal{U}([n], n-1)|$$

of the boundary of the n-simplex where $\mathcal{L}([n], n-1)$ and $\mathcal{U}([n], n-1)$ correspond to the union of the faces opposite to vertices of the same parity. These give rise to two different triangulations of the polytope $C([n], n-1)$ depicted for $n = 4$ in Figure 2.

Of crucial relevance for much of what follows will be a partial ordering on the collection of d-simplices contained in the cyclic polytope $C([n], d)$ which we will now discuss. For $n \geq d \geq 1$, consider the set $S([n], d) \subset \mathcal{P}([n])$ of subsets of cardinality $d+1$. For $I \in S([n], d)$ we denote

$$\Delta^I := \langle \{I\} \rangle \subset \mathcal{P}^*([n]).$$

The geometric realization $|\Delta^I| = C(I, d)$ in $\mathbb{R}^d$ can be viewed as a subsimplex of the cyclic polytope $C([n], d)$. As explained in Example 2.7, the boundary of $|\Delta^I|$ decomposes into lower and upper hemisphere

$$\partial|\Delta^I| = |(\Delta^I)^-| \cup |(\Delta^I)^+|$$

comprised of the even and odd faces of Δ^I, respectively. For $I, J \in S([n], d)$ with $|I \cup J| = d+2$, we declare $I \prec J$ if

$$I \cap J \in (\Delta^I)^+ \cap (\Delta^J)^-.$$

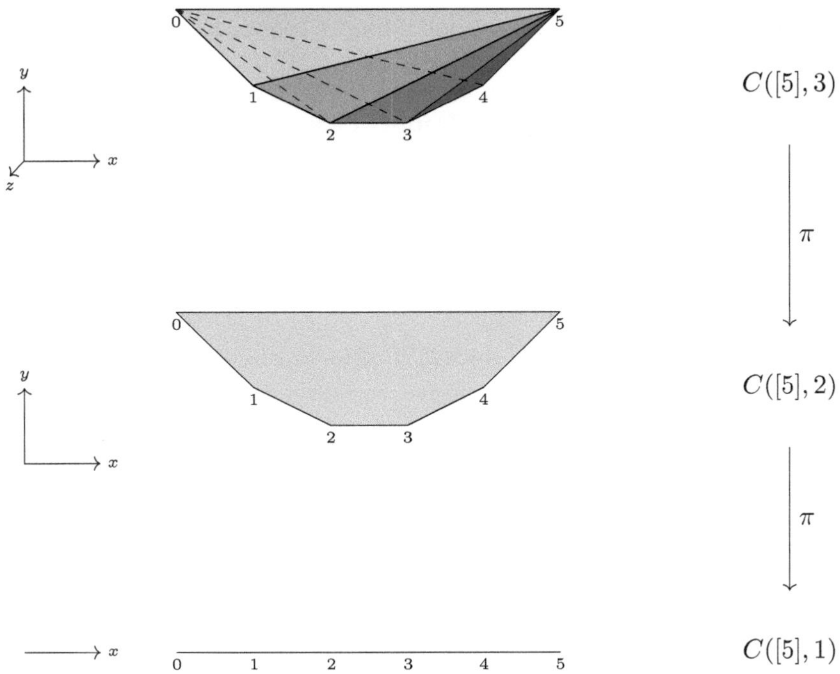

FIGURE 1. The cyclic polytopes $C([5],3)$, $C([5],2)$, and $C([5],1)$, related via the projection maps π.

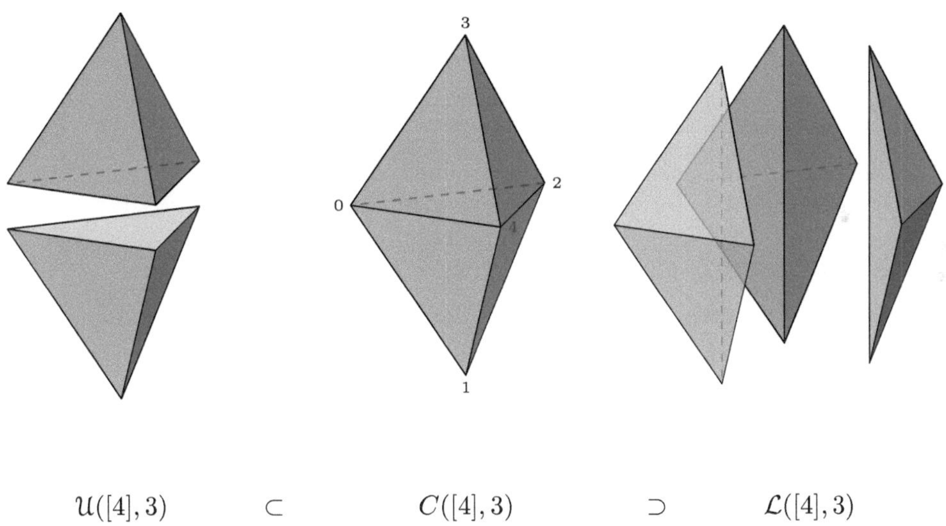

FIGURE 2. The two triangulations of $C([4],3)$ corresponding to the upper boundary $\mathcal{U}([4],3)$ and lower boundary $\mathcal{L}([4],3)$ of $|\Delta^4|$, respectively.

Note that this means that the affine hyperplane H spanned by $I \cap J$ intersects $|\Delta^I|$ and $|\Delta^J|$ in the common face $F = |\Delta^{I \cap J}|$ with $|\Delta^I| \setminus F$ lying below and $|\Delta^J| \setminus F$ lying above H. Within this context, we have the following important result which we refer to as Rambau's lemma (cf. [**Ram97**, Corollary 5.9]):

LEMMA 2.8. *The transitive closure of the relation $\preceq$ on the set $S([n], d)$ defines a partial order. In particular, there are no cycles of simplices related by $\prec$.*

PROOF. To prove the statement, we construct an order-preserving map from $S([n], d)$ into a totally ordered set (which implies the claim). Namely, define the map

$$\epsilon \colon S([n], d) \to \{o, *, e\}^{n+1}, I \mapsto (\epsilon_0, \epsilon_1, \dots, \epsilon_n)$$

where

$$\epsilon_k = \begin{cases} o & \text{if } k \text{ is an odd gap of } I, \\ * & \text{if } k \in I, \\ e & \text{if } k \text{ is an even gap of } I. \end{cases}$$

Interpreting the elements of $\{o, *, e\}^{n+1}$ as words of length $n+1$ in the alphabet $\{o, *, e\}$, we equip it with the lexicographic order induced by $o < * < e$. It is then immediate to verify that the map ϵ is order preserving. □

3. Higher Segal objects

We have now collected the necessary preliminaries to define the objects of interest in this article: the higher Segal objects. We will freely use the language of ∞-categories with [**Lur09**] as a standard reference. Let $\mathcal{C}$ be an ∞-category and let

$$X \colon \Delta^{\mathrm{op}} \to \mathcal{C}$$

be a simplicial object in $\mathcal{C}$.

REMARK 3.1. A (coherent) diagram valued in an ∞-category $\mathcal{C}$ indexed by an ordinary category I is formally defined as a functor of ∞-categories (i.e., a map of simplicial sets) $\mathrm{N}(I) \to \mathcal{C}$. To avoid clumsy notation, we will often leave the passage to the nerve implicit and simply write $I \to \mathcal{C}$ instead.

REMARK 3.2. It will be (notationally) convenient to enlarge the parametrizing category Δ slightly as follows: let $\widetilde{\Delta}$ denote the category of finite nonempty subsets of $\mathbb{N}$, equipped with the linear order induced from $\mathbb{N}$, and weakly monotone maps. There is a canonical functor given on objects by

$$p \colon \widetilde{\Delta} \to \Delta, I = \{i_0 < i_1 < \dots < i_n\} \mapsto [n]. \tag{3.3}$$

We may thus extend the domain of the simplicial object X from Δ to $\widetilde{\Delta}$ by pulling back along p. Typically, this will be left implicit and we simply write X_I to denote $X_{p(I)}$.

CONSTRUCTION 3.4. Let $X \colon \Delta^{\mathrm{op}} \to \mathcal{C}$ be a simplicial object valued in an ∞-category $\mathcal{C}$, let $n \geq 0$ and let $K \subset \mathcal{P}^*([n])$ be an abstract simplicial complex on the set $[n]$. We may consider K as a poset with respect to inclusion of subsets and form the poset $K^{\triangleright}$ by adjoining a maximal element to K. We obtain functors

$$K^{\triangleright} \xrightarrow{q} \widetilde{\Delta} \xrightarrow{p} \Delta$$

where p is the functor from (3.3) and q is defined via $I \mapsto I$ on K while the maximal element of $K^{\triangleright}$ is mapped to $[n]$. Pulling back the simplicial object X along $p \circ q$, we obtain a cone

$$(K^{\triangleright})^{\mathrm{op}} \to \mathcal{C}$$

over the base diagram $X|K^{\mathrm{op}}$ with tip X_n. Therefore, assuming that the limit

$$X_K := \lim X|K^{\mathrm{op}}$$

exists in $\mathcal{C}$, we obtain a canonical map

$$X_n \longrightarrow X_K. \tag{3.5}$$

DEFINITION 3.6. Let $\mathcal{C}$ be an ∞-category with finite limits and let $X\colon \Delta^{\mathrm{op}} \to \mathcal{C}$ be a simplicial object in $\mathcal{C}$. Let $d \geq 1$. Then X is called

(1) lower d-Segal if, for every $n > d$, the map $X_n \to X_{\mathcal{L}([n],d)}$ from (3.5) for $K = \mathcal{L}([n], d)$ is an equivalence in $\mathcal{C}$,
(2) upper d-Segal if, for every $n > d$, the map $X_n \to X_{\mathcal{U}([n],d)}$ from (3.5) for $K = \mathcal{U}([n], d)$ is an equivalence in $\mathcal{C}$.

We refer to the maps in (1) and (2) as *higher Segal maps* and to the simplicial objects satisfying them as *higher Segal objects in* $\mathcal{C}$. If X is both lower and upper d-Segal then we say X is *d-Segal*.

REMARK 3.7. The formulation of Definition 3.6 in terms of ∞-categories subsumes several previous formulations (cf. [**DK19**]):

(1) When $\mathcal{C}$ is the nerve of an ordinary category, then the limit cones simply become limit cones in the sense of ordinary category theory.
(2) When $\mathcal{C}$ is the ∞-categorical localization of a model category C, and X arises from a strict simplicial object in C, then the higher Segal conditions can be expressed in terms of homotopy limits instead of ∞-categorical limits (cf. [**Lur09**]).

EXAMPLE 3.8. Let $X\colon \Delta^{\mathrm{op}} \to \mathcal{C}$ be a simplicial object in an ∞-category $\mathcal{C}$. We give some examples of Segal conditions in low dimensions:

(1) X is lower 1-Segal if, for every $n \geq 2$, the map

$$X_n \longrightarrow X_{\{0,1\}} \times_{X_{\{1\}}} X_{\{1,2\}} \times_{X_{\{2\}}} \cdots \times_{X_{\{n-1\}}} X_{\{n-1,n\}}$$

is an equivalence in $\mathcal{C}$. Therefore, lower 1-Segal objects are the familiar *Segal objects*.

(2) X is upper 1-Segal if, for every $n \geq 2$, the map

$$p_n\colon X_n \to X_{\{0,n\}}$$

is an equivalence in $\mathcal{C}$. We claim that X is essentially constant, i.e., equivalent to the pullback of a single object in $\mathcal{C}$ along the constant functor $\Delta \to *$.

To show this, observe first that, by two-out-of-three applied to the diagram

$$\begin{array}{ccc} X_2 & \xleftarrow{\;s_0\;} & X_1 \\ {\scriptstyle d_1}\big\downarrow & \swarrow_{\mathrm{id}} & \\ X_1, & & \end{array}$$

the degeneracy map s_0 is an equivalence (since $d_1 = p_2$ is an equivalence by assumption). Further, the degeneracy map $X_0 \to X_1$ is a retract of $s_0 \colon X_1 \to X_2$ so that it is an equivalence as well. Thus, the two face maps $X_1 \to X_0$ are equivalences since they are sections of the degeneracy $X_0 \to X_1$. Finally, by composing this equivalence with the equivalence p_n, we obtain that, for every $n \geq 1$, the map

$$X_n \to X_{\{0\}} = X_0$$

corresponding to the inclusion of $\{0\}$ in $[n]$ is an equivalence. But this finally implies that the total degeneracy map

$$X_0 \to X_n$$

corresponding to the unique map $[n] \to [0]$ is an equivalence. A further application of two-out-of-three implies that X maps all morphisms in Δ to equivalences in $\mathcal{C}$. The claim now follows, since the geometric realization of Δ (i.e., the ∞-categorical localization along all edges) is contractible.

(3) The generating simplices of $\mathcal{L}([n],2)$ are $\{0,1,2\},\{0,2,3\},\ldots,\{0,n-1,n\}$ so that, by a basic cofinality argument, we observe that X is lower 2-Segal if and only if, for every $n \geq 3$, the map

$$X_n \to X_{\{0,1,2\}} \times_{X_{\{0,2\}}} X_{\{0,2,3\}} \times_{X_{\{0,3\}}} \cdots \times_{X_{\{0,n-1\}}} X_{\{0,n-1,n\}}$$

is an equivalence, i.e., if and only if the simplicial object $P^{\lhd}X$ obtained from X by pulling back along the join functor

$$\Delta \longrightarrow \Delta, [n] \mapsto [0] * [n]$$

is lower 1-Segal. Similarly, X is upper 2-Segal if and only if the pullback $P^{\rhd}X$ of X along the join functor

$$\Delta \longrightarrow \Delta, [n] \mapsto [n] * [0]$$

is lower 1-Segal. By the path space criterion of [**DK19, GCKT18**] it follows, that X is 2-Segal in the sense of [**DK19**] if and only if X is lower and upper 2-Segal in the sense of Definition 3.6.

(4) The first lower and upper 3-Segal conditions are equivalent to the maps

(3.9)
$$X_4 \to \lim \left\{ \begin{array}{ccccc} X_{\{0,1,2,3\}} & \longrightarrow & X_{\{1,2,3\}} & \longleftarrow & X_{\{1,2,3,4\}} \\ \searrow & \searrow & \downarrow & \swarrow & \swarrow \\ X_{\{0,1,3\}} & \longrightarrow & X_{\{1,3\}} & \longleftarrow & X_{\{1,3,4\}} \\ & \nwarrow & \uparrow & \nearrow & \\ & & X_{\{0,1,3,4\}} & & \end{array} \right\}$$

and

$$X_4 \to X_{\{0,2,3,4\}} \times_{X_{\{0,2,4\}}} X_{\{0,1,2,4\}}$$

being equivalences, respectively (cf. Figure 2). We may interpret the 3-Segal maps

$$X_{\mathcal{U}([4],3)} \longleftarrow X_4 \longrightarrow X_{\mathcal{L}([4],3)}$$

as comprising a $(3,2)$ "Pachner correspondence" (cf. §9.1) relating the limits over $\mathcal{U}([4],3)$ and $\mathcal{L}([4],3)$, respectively.

REMARK 3.10. In Theorem 5.4, we will show that if X is d-Segal then, for *every* triangulation $T \subset \mathcal{P}^*([n])$ of $C([n],d)$, $n > d$, the map

$$X_n \longrightarrow X_T$$

is an equivalence in $\mathcal{C}$.

Gale's eveness criterion paired with a basic cofinality statement lead to the following *path space criteria*, generalizing similar results for $d = 2$ observed independently in [**DK19**] and [**GCKT18**].

DEFINITION 3.11. Let $\mathcal{C}$ be an ∞-category and $X\colon \Delta^{\mathrm{op}} \to \mathcal{C}$ a simplicial object in $\mathcal{C}$. We denote by $\mathrm{P}^{\triangleleft}X$ the simplicial object obtained by pulling back along the functor

$$\Delta \to \Delta, [n] \mapsto [0] * [n]$$

where $*$ denotes the join. $\mathrm{P}^{\triangleleft}X$ is referred to as the *initial path space* of X (for Kan complexes, it models the space of paths with fixed starting point). Analogously, we denote by $\mathrm{P}^{\triangleright}X$ the simplicial object obtained by pulling back along the functor

$$\Delta \to \Delta, [n] \mapsto [n] * [0].$$

$\mathrm{P}^{\triangleright}X$ is referred to as the *final path space* of X (for Kan complexes, it models the space of paths with fixed endpoint).

PROPOSITION 3.12. *Let $\mathcal{C}$ be an ∞-category, $X\colon \Delta^{\mathrm{op}} \to \mathcal{C}$ a simplicial object, and let $d \geq 0$.*

(a) *Assume d is even. Then:*
  (1) *X is lower d-Segal if and only if $\mathrm{P}^{\triangleleft}X$ is lower $(d-1)$-Segal.*
  (2) *X is upper d-Segal if and only if $\mathrm{P}^{\triangleright}X$ is lower $(d-1)$-Segal.*
(b) *Assume d is odd. Then the following conitions are equivalent:*
  (i) *X is upper d-Segal.*
  (ii) *$\mathrm{P}^{\triangleleft}X$ is upper $(d-1)$-Segal.*
  (iii) *$\mathrm{P}^{\triangleright}X$ is lower $(d-1)$-Segal.*

We provide a class of examples of higher Segal spaces: let

$$\mathbb{\Delta} \subset \mathbb{C}\mathrm{at}$$

denote the full 2-subcategory of the 2-category of small categories spanned by the standard ordinals $[n]$, $n \geq 0$. For $0 \leq k \leq n$, denote by $\mathbb{\Delta}([k],[n])$ the mapping category in $\mathbb{\Delta}$, which can be identified with the poset of monotone $k+1$-tuples in $[n]$ with the componentwise order induced from $[n]$.

Let $\mathcal{A}$ be an abelian category and define $\mathcal{S}_n^{\langle k\rangle}(\mathcal{A})$ to be the full subcategory

$$\mathcal{S}_n^{\langle k\rangle}(\mathcal{A}) \subset \mathrm{Fun}(\mathbb{\Delta}([k],[n]),\mathcal{A})$$

consisting of those diagrams $A\colon \mathbb{\Delta}([k],[n]) \to \mathcal{A}$ such that

(1) for every noninjective map $\tau\colon [k] \to [n]$, we have

$$A_\tau \cong 0,$$

(2) for every injective map $\sigma\colon [k+1] \to [n]$, the sequence

$$0 \longrightarrow A_{d_k\sigma} \longrightarrow A_{d_{k-1}\sigma} \longrightarrow \dots \longrightarrow A_{d_0\sigma} \longrightarrow 0 \tag{3.13}$$

induced by the sequence

$$d_k\sigma \to d_{k-1}\sigma \to \cdots \to d_0\sigma$$

in $\mathbb{A}([k],[n])$, is exact. Note that (1) along with the commutativity of the square

$$\begin{array}{ccc} d_i\sigma & \longrightarrow & d_{i-1}\sigma \\ \downarrow & & \downarrow \\ s_{i-2}d_{i-2}d_i\sigma & \longrightarrow & d_{i-2}\sigma. \end{array}$$

in $\mathbb{A}([k],[n])$ imply that the sequence (3.13) is indeed a complex.

Fixing $k \geq 0$, the various categories $\mathcal{S}_n^{\langle k\rangle}(\mathcal{A})$, $n \geq 0$, form a strict simplicial object

$$\Delta^{\mathrm{op}} \longrightarrow \mathrm{Cat}$$

in the category of categories. Postcomposing it with the nerve $\mathrm{N}\colon \mathrm{Cat} \to \mathrm{Set}_\Delta$ and then passing to coherent nerves N_Δ, we obtain a coherent diagram

$$\mathcal{S}_\bullet^{\langle k\rangle}(\mathcal{A})\colon \mathrm{N}(\Delta^{\mathrm{op}}) \longrightarrow \mathrm{Cat}_\infty$$

which we refer to as the k-dimensional Waldhausen $\mathcal{S}_\bullet$-construction (traditionally pronounced: "S dot construction") of $\mathcal{A}$.

THEOREM 3.14. *Let $k \geq 0$. Then the simplicial object $\mathcal{S}^{\langle k\rangle}(\mathcal{A})$ is $2k$-Segal.*

- For $k = 1$, this statement was observed independently in [**DK19**] and [**GCKT18**]. From the perspective in [**DK19**], it (along with its many variants) served as the main motivation for introducing 2-Segal spaces, namely as a means to investigate various Hall algebra constructions. In this case, the result is an immediate consequence of the path space criterion: first define the thickened nerve

$$\widetilde{\mathrm{N}}(\mathcal{A})\colon \Delta^{\mathrm{op}} \longrightarrow \mathrm{Cat}, [n] \mapsto \mathrm{Fun}([n],\mathcal{A}).$$

  Then
  - $\mathrm{P}^{\lhd}\mathcal{S}^{\langle 1\rangle}(\mathcal{A})$ is equivalent to the simplicial object obtained from $\widetilde{\mathrm{N}}(\mathcal{A})$ by passing, for each $n \geq 0$, to the full subcategory of $\mathrm{Fun}([n],\mathcal{A})$ spanned by the composable chains of monomorphisms in $\mathcal{A}$.
  - $\mathrm{P}^{\rhd}\mathcal{S}^{\langle 1\rangle}(\mathcal{A})$ is equivalent to the simplicial object obtained from $\widetilde{\mathrm{N}}(\mathcal{A})$ by passing, for each $n \geq 0$, to the full subcategory of $\mathrm{Fun}([n],\mathcal{A})$ spanned by the composable chains of epimorphisms in $\mathcal{A}$.

  Both path spaces are lower 1-Segal (i.e., classical Segal) objects so that $\mathcal{S}^{\langle 1\rangle}(\mathcal{A})$ is 2-Segal by Proposition 3.12.
- For $k = 2$, we observe that

$$\mathrm{P}^{\lhd}\mathrm{P}^{\rhd}\mathcal{S}^{\langle 2\rangle}(\mathcal{A}) = \mathrm{P}^{\rhd}\mathrm{P}^{\lhd}\mathcal{S}^{\langle 2\rangle}(\mathcal{A}) \simeq \widetilde{\mathrm{N}}(\mathcal{A})$$

  which is lower 1-Segal, so that $\mathcal{S}^{\langle 2\rangle}(\mathcal{A})$ is upper 3-Segal. By Theorem 5.1, it follows that $\mathcal{S}^{\langle 2\rangle}(\mathcal{A})$ is 4-Segal.
- For $k > 2$, the proof still uses path space criteria, but is more involved [**Pog17**].

REMARK 3.15. In [**BOO**$^+$**18**, **BOO**$^+$**21**], the categorical input data for the $\mathcal{S}_\bullet$-construction is generalized to a suitable class of "stable" double categories. In this context, it turns out that the $\mathcal{S}_\bullet$-construction furnishes an equivalence between such double categories and 2-Segal objects.

4. Orientals

As indicated in §2, the analysis of the projection maps

$$\pi_d \colon C([n], d) \to C([n], d-1)$$

offers a geometric perspective on the structure of triangulations of cyclic polytopes. Here the term "perspective" can be taken quite literal — a crucial point of our discussion was that two canonical triangulations of $C([n], d-1)$ become "visible" when looking at $C([n], d)$ from above and below with respect to the last coordinate of $\mathbb{R}^d$. Remarkably, as first pointed out in [**KV91a**], it turns out that these two ways of looking at simplicial complexes form above and below have categorical relevance: they define the source and target, respectively, of a higher-dimensional composition law.

As a first hint towards this interpretation, one can note that the cyclic polytope $C([n], n) \subset \mathbb{R}^n$ can be regarded as the geometric realization of the n-simplex Δ^n which we consider as an abstract simplicial complex on the set $[n]$. As explained in Example 2.7, the boundary $S := \partial\Delta^n$ decomposes into a lower hemisphere

$$S^- := (\partial\Delta^n)^- \subset \Delta^n$$

and an upper hemisphere

$$S^+ := (\partial\Delta^n)^+ \subset \Delta^n.$$

Passing to geometric realizations, by Proposition 2.5, the projection map π identifies these simplicial hemispheres with triangulations $\mathcal{L}([n], n-1)$ and $\mathcal{U}([n], n-1)$ of the cyclic polytope $C([n], n-1)$. Under this identification, the equatorial $(n-2)$-sphere $S^- \cap S^+$ maps to the boundary of $C([n], n-1)$. We may now repeat this procedure in one dimension lower: the simplicial boundary T of $C([n], n-1)$ decomposes into a lower hemisphere T^- and an upper hemisphere T^+. Interpreting all resulting abstract simplicial complexes as subcomplexes of Δ^n, we arrive at the incidence diagram

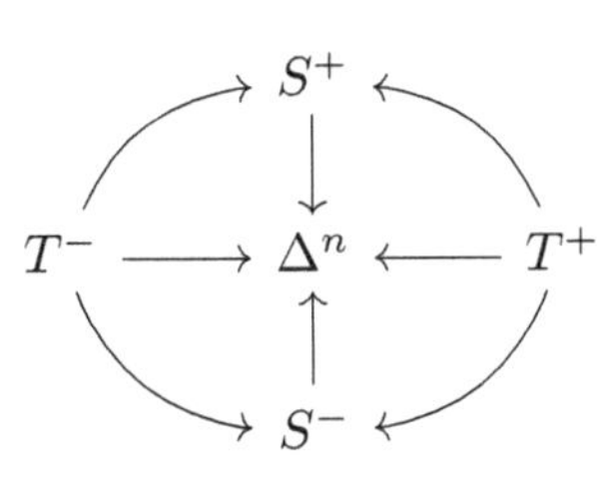

where

$$S^- \cup S^+ = \partial\Delta^n,$$
$$T^- \cup T^+ = \partial S^- = \partial S^+ = S^- \cap S^+.$$

What we have thus unpacked here looks like a globular 2-cell

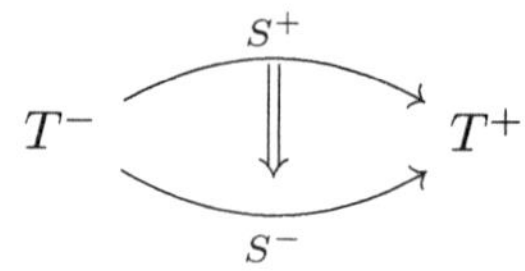

in a suitably defined combinatorial higher "cobordism category" of certain simplicial subcomplexes of Δ^n. As we will now explain, it turns out that this higher categorical structure is a geometric incarnation of Street's oriental: the free ω-category on the n-simplex.

To explain this, we first discuss some classical perspectives on strict (∞, ω)-categories. Recall that an ω-category is a set A which comes equipped with a family $\{(s_n, t_n, *_n)\}_{n \in \mathbb{N}}$ of maps comprising, for every $n \in \mathbb{N}$,

- an n-dimensional source and target map $s_n \colon A \to A$ and $t_n \colon A \to A$, respectively,
- a composition map
$$*_n \colon A \times_A A \to A$$
where the fiber product is defined as
$$A \times_A A := \{(a, a') \in A | s_n(a) = t_n(a')\},$$

satisfying globularity, unitality, associativity, and commutativity constraints (see [**Str87**]). Given an ω-category A, we may introduce, for $n \in \mathbb{N}$, the set

$$A_{\leq n} := \{a \in A | a = s_n(a)\} \subset A,$$

yielding a globular set

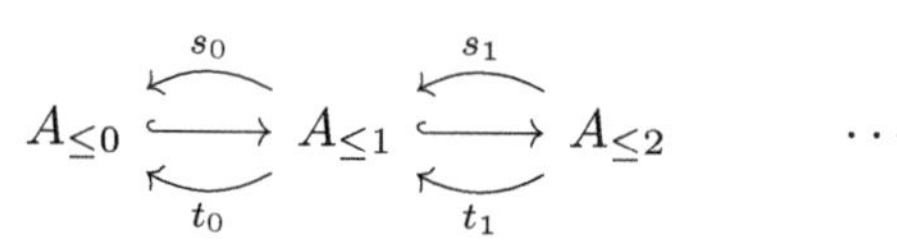

which, when equipped with the restricted composition laws $*_n$, provides an equivalent way to package the data of an ω-category.

DEFINITION 4.1. Let $n \geq d \geq 0$.

(1) By a simplicial subcomplex of $C([n], d)$, we mean an abstract simplicial complex K on $[n]$ consisting of subsets $\sigma \subset [n]$ of cardinality $\leq d+1$ such that, for every $\sigma, \tau \in K$, we have
$$|\sigma| \cap |\tau| = |\sigma \cap \tau|$$
where the geometric realization $|\sigma|$ denotes the convex hull of $\nu(\sigma) \subset \mathbb{R}^d$. For a simplicial subcomplex K of $C([n], d)$ we denote by $|K|$ its geometric realization, defined as the union of the geometric realization of its simplices.

(2) For a simplicial subcomplex K of $C([n], d)$ consider the projection map $\pi_d \colon \mathbb{R}^d \to \mathbb{R}^{d-1}$. Then K is called d-admissible if, for every $x \in \pi_d(|K|)$, we have
$$\{(x, t) \mid t \in \mathbb{R}\} \cap |K| = \{x\} \times [l_x, u_x]$$
for $l_x, u_x \in \mathbb{R}$, $l_x \leq u_x$.

Given a d-admissible subcomplex K of $C([n], d)$, it is straightforward to see that the set of points of the form (x, l_x), $x \in \pi_d(|K|)$ is the geometric realization of a subcomplex $K^- \subset K$ such that K^- defines a simplicial subcomplex of $C([n], d-1)$. Similarly, the points of the form (x, u_x) form the geometric realization of a subcomplex $K^+ \subset K$ such that K^+ defines a simplicial subcomplex of $C([n], d-1)$. Generalizing the situation of Example 2.6, the simplicial complexes K^+ and K^- can be "seen" when looking at $|K|$ from above and below, respectively.

This allows us to give the following recursive definition:

Definition 4.2. A simplicial subcomplex K of $C([n], d)$ is called *admissible*, if

(1) K is d-admissible,
(2) K^- and K^+ are admissible as subcomplexes of $C([n], d-1)$.

Example 4.3. Let K be a simplical subcomplex of $C([n], d)$ whose geometric realization defines a triangulation of $C([n], d)$, such as the complexes $\mathcal{L}([n], d)$ and $\mathcal{U}([n], d)$ from Definition 2.4. Then, as explained in Remark 2.6, K is d-admissible and we have $K^- = \mathcal{L}([n], d-1)$ and $K^+ = \mathcal{U}([n], d-1)$. In particular, by induction K is admissible.

Definition 4.4. We define an ω-category $(G_\bullet, *_\bullet)$ given by the following data.

(1) For each $n \geq 0$ and each $n \geq d \geq 0$, let G_d be the set of admissible subcomplexes of $C([n], d)$ and define $G_k := G_n$ for $k > n$.
(2) The globular set structure of $G_\bullet$ is given by the diagram

$$G_0 \underset{t_0}{\overset{s_0}{\leftleftarrows}} \hookrightarrow G_1 \underset{t_1}{\overset{s_1}{\leftleftarrows}} \hookrightarrow G_2 \qquad \cdots$$

where
- $s_d(K) = K^-$,
- $t_d(K) = K^+$,
- $u(K) = K$,

for $n \geq d \geq 0$, and $s_d = t_d = u = \mathrm{id}$, for $d > n$.
(3) The composition law $*\bullet$ on $G_\bullet$ is defined as follows: for $K, L \in G_{d+1}$ with $s_d(K) = t_d(L)$, we let

$$K *_d L = K \cup L.$$

Theorem 4.5. *The ω-category $(G_\bullet, *_\bullet)$ defined in Definition 4.4 is equivalent to Street's oriental $\mathbb{O}_n$, i.e., the free ω-category on the n-simplex.*

There are different ways to prove Theorem 4.5: one possibility is to verify that the category $\mathbb{O}_n$ is freely generated by the collection of simplices Δ^I, $I \subset [n]$ (Each such simplex gives rise to a $(|I|-1)$-dimensional morphism from $(\Delta^I)_-$ to $(\Delta^I)_+$).

The following lemma (cf. [**Ram97**, Proposition 5.11]) is the key step in this method of proof. It is of independent interest to us, as it will be used in our investigation of higher Segal conditions and their relation to correspondence categories.

Lemma 4.6. *Let $n \geq d \geq 1$. Let K be an admissible subcomplex of $C([n], d)$ and assume that $K^- \neq K^+$. Then there exists an admissible subcomplex L of $C([n], d)$ and a d-simplex Δ^I, $I \subset [n]$, such that*

1. $(\Delta^I)^- \subset L^+$,
2. $L \cup \Delta^I = K$.

In other words, K is obtained from L by stacking the simplex Δ^I along its lower hemisphere on top of L.

PROOF. Since $K^- \neq K^+$, there exists at least one d-simplex Δ^J in K. Either $(\Delta^J)^+ \subset K^+$, or we find another d-simplex $\Delta^{J'}$ in K such that

$$J \prec J'.$$

Since the transitive closure of $\preceq$ is a partial order by Lemma 2.8, this chain of "stacked" simplices must terminate at a simplex I in K such that $(\Delta^I)^+ \subset K^+$. Removing I along with the interior of its upper hemisphere from K yields the desired complex L. It is clear that L is admissible with $L^- = K^-$. □

We provide a proof sketch of Theorem 4.5. Iterated applications of Lemma 4.6 imply that an admissible subcomplex K of $C([n], d)$ can be written as a composite

$$K_1 *_{d-1} K_2 *_{d-1} \cdots *_{d-1} K_r$$

where each K_i is an admissible subcomplex of $C([n], d)$ containing exactly one d-simplex σ_i. Each K_i can in turn be written as

$$K_i = S_i *_{d-2} \sigma_i *_{d-2} S_i'$$

with S_i and S_i' admissible subcomplexes of $C([n], d-1)$ where S_i^+ is the lower equatorial hemisphere of σ_i and $(S_i')^-$ is the upper equatorial hemisphere of σ_i. This implies that, given an ω-category A, any extension of an ω-functor $G_{\leq d-1} \to A$ to G_d is uniquely determined by its values on d-simplices. Vice versa, it is straightforward to see that an assignment on the collection of d-simplices, compatible with their sources and targets, defines an extension along $G_{\leq d-1} \subset G_{\leq d}$: different choices of the factorization

$$K_1 *_{d-1} K_2 *_{d-1} \cdots *_{d-1} K_r \tag{4.7}$$

get assigned the same values, as can be directly seen from the 2-categorical interchange law (Eckmann-Hilton argument) in (the 2-category of $(d-2)$-morphisms in) A. Therefore G satisfies the universal property characterizing the oriental.

REMARK 4.8. The higher categorical relevance of cyclic polytopes was first pointed out in [**KV91b**] where triangulations of cyclic polytopes are identified with pasting schemes in the oriental. An extension to the geometric description of *all* morphisms in the oriental given by Theorem 4.5 was established in [**Gov16**].

5. The interplay of higher Segal conditions

In this section, we outline a proof of the following result from [**Pog17**]:

THEOREM 5.1. *Let $\mathcal{C}$ be an ∞-category, $X\colon \Delta^{\mathrm{op}} \to \mathcal{C}$ a simplicial object, and let $d \geq 0$. Assume that X is lower d-Segal* or *upper d-Segal. Then, for every $k > d$, X is k-Segal.*

For the proof, we first establish some preparatory results.

LEMMA 5.2. *Let $\mathcal{C}$ be an ∞-category with finite limits and let $X\colon \Delta^{\mathrm{op}} \to \mathcal{C}$ be a lower $(d-1)$-Segal object in $\mathcal{C}$. Let K, L as in Lemma 4.6. Then the inclusion $L \subset K$ induces an equivalence*

$$X_K \xrightarrow{\simeq} X_L.$$

PROOF. We first note that the inclusion

$$i\colon L \cup \{I\} \subset K$$

is cofinal: for every simplex $J \in K \setminus L \cup \{I\}$, the slice category J/i consists of the single object $J \subset I$. Second, we observe that the assumption that X be lower $(d-1)$-Segal implies that

$$X|(L \cup \{I\})^{\mathrm{op}}$$

is a right Kan extension of its restriction to L^{op}. Indeed, the relevant slice category for the pointwise formula of the value of the Kan extension at I identifies precisely with the poset $(\Delta^I)^-$ so that the lower $(d-1)$-Segal condition

$$X_I \simeq \lim X|\mathcal{L}(I, d-1)^{\mathrm{op}}$$

implies the claim. □

COROLLARY 5.3. *Let $\mathcal{C}$ be an ∞-category with finite limits and let X be a lower $(d-1)$-Segal object in $\mathcal{C}$. Then X is d-Segal.*

PROOF. Let $n > d$ and $T \subset \mathcal{P}^*([n])$ be an abstract simplicial complex corresponding to a triangulation of $C([n], d)$. Then, using the notation from Lemma 4.6, we have that T is admissible with $T^+ \neq T^-$. Applying Lemma 4.6 successively, we obtain a finite sequence $K_1, \ldots, K_m$ of admissible subcomplexes with

$$K_1 \subset K_2 \subset \cdots \subset K_m = T$$

such that

- K_{i+1} is obtained from K_i by stacking a d-simplex on top of K_i (more precisely, $K_{i+1} = K$ and $K_i = L$ in Lemma 4.6),
- $K_1 = K_1^+ = K_1^- = T^- = \mathcal{L}([n], d-1)$.

From Lemma 5.2 we then obtain a commutative diagram

$$\begin{array}{ccccccccc} X_n & & & & & \xrightarrow{\quad\simeq\quad} & & & X_{K_1} \\ \downarrow & & & & & & & & \\ X_T & \xrightarrow{\simeq} & X_{K_{m-1}} & \xrightarrow{\simeq} & X_{K_{m-2}} & \xrightarrow{\simeq} & \cdots & \xrightarrow{\simeq} & X_{K_1} \end{array}$$

in $\mathcal{C}$. Therefore, the map $X_n \to X_T$ is an equivalence so that, setting $T = \mathcal{L}([n], d)$ and $T = \mathcal{U}([n], d)$, we obtain, in particular, that X is lower and upper d-Segal. □

PROOF OF THEOREM 5.1. The Theorem with hypothesis lower d-Segal follows by induction from Corollary 5.3. For the upper d-Segal hypothesis, the proof follows precisely the same strategy, but instead of interpolating between the triangulation T and its lower boundary T^- by removing simplices from top to bottom, we remove simplices from bottom to top to connect to T^+. The arguments adapt to this procedure mutatis mutandis. □

THEOREM 5.4. *Let $\mathcal{C}$ be an ∞-category with finite limits, and let X be a d-Segal object in $\mathcal{C}$. Then, for every $n > d$, and every triangulation K of the cyclic polytope $C([n], d)$, the corresponding map*

$$X_n \to X_K$$

from (3.5) *is an equivalence in $\mathcal{C}$.*

Again, we need some basic results from the theory of cyclic polytopes. To set the stage, let T be a triangulation of $C([n],d)$ and let $I \in S([n],d+1)$ be a $d+1$-simplex such that $(\Delta^I)^- \subset T$. Then $R = T \cup \Delta^I$ is an admissible subcomplex of $C([n],d+1)$ such that $T' = R^+$ is another triangulation of $C([n],d)$, called the *bistellar flip* of T along I. We write $T \lhd T'$.

LEMMA 5.5. *Let $\mathcal{C}$ be an ∞-category with finite limits and let $X\colon \Delta^{\mathrm{op}} \to \mathcal{C}$ be a simplicial object. Let $T \lhd T'$ be triangulations of $C([n],d)$ related by a bistellar flip.*

(1) *Assume that X is lower d-Segal. Then the morphism*

$$X_R \to X_T$$

is an equivalence in $\mathcal{C}$.

(2) *Assume that X is upper d-Segal. Then the morphism*

$$X_R \to X_{T'}$$

is an equivalence in $\mathcal{C}$.

PROOF. This is a special case of Lemma 5.2. □

PROOF OF THEOREM 5.4. Suppose that T is a triangulation of $C([n],d)$. Then by [**Ram97**, Theorem 1.1(i)] there exists a sequence of bistellar flips

$$T \lhd T_1 \lhd \cdots \lhd \mathcal{U}([n],d)$$

connecting T with the upper triangulation of $C([n],d)$. Combining Lemma 5.5 with the equivalence

$$X_n \xrightarrow{\simeq} X_{\mathcal{U}([n],d)}$$

we deduce that $X_n \longrightarrow X_T$ is an equivalence as well. □

6. Higher correspondences

The goal of this section is to establish an interpretation of the higher Segal conditions in terms of higher correspondence categories. Our discussion is based on the classical simplicial combinatorics of barycentric subdivision. Consider the functor

$$\Delta \to \mathrm{Set}_\Delta,\, [n] \mapsto \mathrm{N}(\mathcal{P}^*([n])^{\mathrm{op}})$$

where $\mathcal{P}^*([n])$ denotes the poset of nonempty subsets of $[n]$. Its left Kan extension along the Yoneda embedding $\Delta \to \mathrm{Set}_\Delta$ is the left adjoint of an adjunction (cf. [**Kan57**] for a classical appearance)

$$\mathrm{sd}\colon \mathrm{Set}_\Delta \longrightarrow \mathrm{Set}_\Delta \colon \mathrm{co}_\infty\,.$$

Note that this is different from the *edgewise* subdivision; see Remark 6.7 below.

CONSTRUCTION 6.1. Let $\mathcal{C}$ be an ∞-category with limits. Then the vertices of $\mathrm{co}_\infty(\mathcal{C})$ coincide with the vertices of $\mathcal{C}$ while an edge in $\mathrm{co}_\infty(\mathcal{C})$ corresponds to a diagram

$$x_0 \longleftarrow x_{01} \longrightarrow x_1$$

in $\mathcal{C}$. A 2-simplex in $\mathrm{co}_\infty(\mathcal{C})$ corresponds to a coherent diagram

$$
\begin{array}{ccccccc}
 & & & x_1 & & & \\
 & & \nearrow & \uparrow & \nwarrow & & \\
 & x_{01} & \longleftarrow & x_{012} & \longrightarrow & x_{12} & \\
 \swarrow & & \swarrow & \downarrow & \searrow & & \searrow \\
x_0 & \longleftarrow & & x_{02} & & \longrightarrow & x_2
\end{array}
\tag{6.2}
$$

in $\mathcal{C}$. Note that the diagram (6.2) gives rise to a diagram

$$
\begin{array}{ccccc}
 & & x_{01} \times_{x_1} x_{12} & & \\
 & \swarrow & \uparrow & \searrow & \\
x_0 & \longleftarrow & x_{012} & \longrightarrow & x_2 \\
 & \nwarrow & \downarrow & \nearrow & \\
 & & x_{02} & &
\end{array}
$$

which we may interpret as a globular 2-cell in a higher category of correspondences: a correspondence (2-morphism) between the correspondences (1-morphisms)

$$x_0 \longleftarrow x_{01} \times_{x_1} x_{12} \longrightarrow x_2$$

and

$$x_0 \longleftarrow x_{02} \longrightarrow x_2.$$

In a similar fashion, every n-simplex

$$x \colon \mathcal{P}^*([n])^{\mathrm{op}} \to \mathcal{C}$$

gives rise to an n-dimensional globe whose d-dimensional equator is given by

$$
\begin{array}{ccccc}
 & & \lim x|\mathcal{L}([n], d) & & \\
 & \swarrow & & \searrow & \\
\lim x|\mathcal{L}([n], d-1) & & & & \lim x|\mathcal{U}([n], d-1) \\
 & \nwarrow & & \nearrow & \\
 & & \lim x|\mathcal{U}([n], d) & &
\end{array}
\tag{6.3}
$$

while the top n-dimensional cell is given by

$$
\begin{array}{ccccc}
 & & \lim x|\mathcal{L}([n], n-1) & & \\
 & \swarrow & \uparrow & \searrow & \\
\lim x|\mathcal{L}([n], n-2) & \longleftarrow & x_{\{0,1,\ldots,n\}} & \longrightarrow & \lim x|\mathcal{U}([n], n-2) \\
 & \nwarrow & \downarrow & \nearrow & \\
 & & \lim x|\mathcal{U}([n], n-1). & &
\end{array}
$$

The legs of the correspondence

$$\lim x|\mathcal{L}([n], n-1) \xleftarrow{\ l\ } x_{\{0,1,\ldots,n\}} \xrightarrow{\ u\ } \lim x|\mathcal{U}([n], n-1) \tag{6.4}$$

are precisely the upper, resp. lower, $(n-1)$-Segal maps induced by the higher Segal cones of Definition 3.6.

DEFINITION 6.5. Let $\mathcal{C}$ be an ∞-category with limits and let σ_x be an n-simplex of $\mathrm{co}_\infty(\mathcal{C})$ corresponding to a diagram

$$x\colon \mathcal{P}^*([n])^{\mathrm{op}} \to \mathcal{C}.$$

Then σ_x is called *lower* (resp. *upper*) *thin* if the map l (resp. u) in (6.4) is an equivalence in $\mathcal{C}$. We call σ_x *thin* if it is both lower and upper thin.

The following result is due to J. Gödicke, Q. Ho, and W. Stern, and a detailed proof will appear in a forthcoming paper by these authors. Here, we provide a sketch of an argument emphasizing the connection to orientals.

PROPOSITION 6.6. *Let $\mathcal{C}$ be an ∞-category with limits. Then the simplicial set $\mathrm{co}_\infty(\mathcal{C})$, stratified with the thin simplices from Definition 6.5, is a complicial set (cf.* [**Ver08, Rie18**]*).*

PROOF. We set $\mathcal{D} := \mathrm{co}_\infty(\mathcal{C})$ and sketch how to verify the filling conditions that need to be satisfied (cf. Definition 2.13 in [**Rie18**]). Let $n \geq 3$, $0 \leq i \leq n$, and let

$$\tau\colon \Lambda^n_i \to \mathcal{D}$$

be a horn in $\mathcal{D}$ such that all simplices in Λ^n_i containing the vertices $\{i-1, i, i+1\} \cap [n]$ are thin. We denote by $\mathcal{P}^n_i$ the poset of nondegenerate simplices of Λ^n_i. The horn $\tau\colon \Lambda^n_i \to \mathcal{D}$ then corresponds to a diagram

$$x\colon (\mathcal{P}^n_i)^{\mathrm{op}} \to \mathcal{C}.$$

Consider the full inclusions of posets

$$(\mathcal{P}^n_i)^{\mathrm{op}} \overset{\alpha}{\hookrightarrow} (\mathcal{P}^n_i)^{\mathrm{op}} \cup \{[n]\} \overset{\beta}{\hookrightarrow} \mathcal{P}^*([n])^{\mathrm{op}}.$$

We define

(1) $x'\colon (\mathcal{P}^n_i)^{\mathrm{op}} \cup \{[n]\} \to \mathcal{C}$ to be the right Kan extension of x along α, so that

$$x'([n]) \simeq \lim_{I \in \mathcal{P}^n_i} x_I,$$

(2) $x''\colon \mathcal{P}^*([n])^{\mathrm{op}} \to \mathcal{C}$ to be the left Kan extension of x' along β, so that

$$x''(\partial_i) \simeq x'([n]).$$

Then x'' corresponds to an n-simplex $\sigma\colon \Delta^n \to \mathcal{D}$ extending τ. Using the imposed thinness conditions, it can now be shown that the simplex σ is indeed thin.

The second type of filling condition a complicial set needs to satisfy is the following: for each n-simplex $\sigma\colon \Delta^n \to \mathcal{D}$ and $0 \leq i \leq n$ such that

1. σ maps all subsimplices containing $\{i-1, i, i+1\} \cap [n]$ to thin simplices, and
2. σ maps the faces $\partial_{i-1}\Delta^n$ and $\partial_{i+1}\Delta^n$ to thin simplices,

then the simplex $\sigma \circ \partial_i$ is thin.

To verify this condition suppose that the face $\partial_i\Delta^n$ belongs to the lower hemisphere $\mathcal{L}([n], n-1)$ of Δ^n and let

$$x\colon \mathcal{P}^*([n])^{\mathrm{op}} \to \mathcal{C}$$

be the diagram corresponding to σ. We consider the diagram

$$\begin{array}{ccccc} & & \lim x|\mathcal{L}([n], n-1) & & \\ & \swarrow^{s_l} & \uparrow s & \searrow^{t_l} & \\ \lim x|\mathcal{L}([n], n-2) & \longleftarrow & x_{\{0,1,\dots,n\}} & \longrightarrow & \lim x|\mathcal{U}([n], n-2) \\ & \nwarrow_{t_u} & \downarrow t & \nearrow_{s_u} & \\ & & \lim x|\mathcal{U}([n], n-1). & & \end{array}$$

where s and t are equivalences, due to the thinness of σ. By decomposing $\mathcal{U}([n], n-1)$ as in (4.7) and using the assumption that all faces in $\mathcal{U}([n], n-1)$ are thin, we deduce that t_u and s_u are equivalences so that, by two out of three, the maps s_l and t_l are equivalences as well. Now decomposing $\mathcal{L}([n], n-1)$ as in (4.7) as well, and letting K_i denote the subcomplex containing the face ∂_i, we deduce, again by two out of three, that the legs in the correpondence

$$x|(K_i)^- \xleftarrow{\simeq} x|K_i \xrightarrow{\simeq} x|(K_i)^+$$

are equivalences. Finally, the imposed thinness conditions on σ imply that the latter correspondence is equivalent to the correspondence

$$x|(\partial_i\Delta^n)^- \longleftarrow x|\partial_i\Delta^n \longrightarrow x|(\partial_i\Delta^n)^+$$

so that its legs must be equivalences as well, showing that the simplex $\sigma \circ \partial_i$ is thin. □

According to Proposition 6.6, we may interpret $\mathrm{co}_\infty(\mathcal{C})$ as a model for the (∞, ω)-category of higher correspondences in $\mathcal{C}$ (cf. [**Lou22**]). Various interesting truncations of $\mathrm{co}_\infty(\mathcal{C})$ will be relevant for us:

- Denote by $\mathrm{co}_n(\mathcal{C}) \subset \mathrm{co}_\infty(\mathcal{C})$ the simplicial subset consisting of those simplices all of whose d-subsimplices for $d > n$ are thin. Then $\mathrm{co}_n(\mathcal{C})$ models an (∞, n)-category of higher correspondences where all correspondences above globular dimension n (cf. (6.3)) are invertible.
- Denote by $\mathrm{co}_n^l(\mathcal{C}) \subset \mathrm{co}_n(\mathcal{C})$ the simplicial subset consisting of those simplices all of whose $(n-1)$-subsimplices are lower thin. Then $\mathrm{co}_n^l(\mathcal{C})$ models an (∞, n)-category of higher correspondences where both legs of all correspondences above globular dimension n (cf. (6.3)) are invertible and, in addition, the lower legs of the correspondences in globular dimension $n-1$ are invertible. In particular, by inverting the lower leg, we may interpret such a correspondence as a morphism from the lower to the upper hemisphere of dimension $n-2$.
- Similarly, denote by $\mathrm{co}_n^u(\mathcal{C}) \subset \mathrm{co}_n(\mathcal{C})$ the simplicial subset consisting of those simplices all of whose $(n-1)$-subsimplices are upper thin. In this case, by inverting the upper leg of a correspondence in globular dimension $n-1$, we may interpret such a correspondence as a morphism from the upper to the lower hemisphere of dimension $n-2$.

REMARK 6.7. In [**DK19**], another variant of higher correspondence categories is used when studying 2-Segal spaces. There, we used the *edgewise* subdivision in the form of an adjunction

$$\mathrm{tw}\colon \mathrm{Set}_\Delta \longleftrightarrow \mathrm{Set}_\Delta : \overline{\mathrm{co}}$$

instead of the *barycentric* subdivision to construct an ∞-category of correspondences (a.k.a. spans). The relation between the two constructions is that $\overline{\mathrm{co}}$ models an $(\infty, 2)$-category (all simplices in dimensions > 2 are already thin) which is equivalent to the $(\infty, 2)$-category $\mathrm{co}_2^u(\mathcal{C})$ as defined here. This already suggests that the construction $\overline{\mathrm{co}}$ is too limited when studying higher Segal spaces in dimensions > 2.

REMARK 6.8. Note that the Segal conditions are closely related to the thinness conditions in Definition 6.5: both are given in terms of limits of diagrams parameterized by the poset of facets of the upper (resp. lower) hemispheres of cyclic polytopes. A precise relation, along with its structural relevance, will be discussed in §7.

7. Monads

In this section, we explain how to identify higher Segal objects with certain lax monads in higher correspondences categories. Variations of this perspective have been established in [**Ste21, Göd24, GG19**].

In our approach, we will show how to construct from *any* simplicial object in $\mathcal{C}$ a lax monad in the (∞, ω)-category $\mathrm{co}_\infty(\mathcal{C})$ of correspondences in $\mathcal{C}$. The lower and upper n-Segal conditions are then responsible for yielding monads in the (∞, n)-categorical truncations $\mathrm{co}_n^l(\mathcal{C})$ and $\mathrm{co}_n^u(\mathcal{C})$, respectively. Finally, we then show that this construction establishes an equivalence between suitably defined ∞-categories. There doesn't seem to be a standard concept of a lax monad available within the framework of (∞, ω)-categories so we provide a (somewhat ad-hoc) definition suitable for our purposes. With more work, it will be possible to put our constructions in a model independent context, but we do not worry about this here.

To put our chosen model of a monad in context, we recall (one possible definition of) the notion of a monad valued in an $(\infty, 2)$-category $\mathcal{D}$. Informally, a monad consists of choices of

- an object x in $\mathcal{D}$,
- an endomorphism $M\colon x \to x$ in $\mathcal{D}$,
- natural transformations $\mu\colon M \circ M \Rightarrow M$ and $\epsilon\colon \mathrm{id}_x \Rightarrow M$,
- a coherent system of higher associativity and unitality constraints.

Passing to the monoidal ∞-category of endomorphisms $\mathcal{E}^\otimes = \mathcal{D}(x, x)$, a monad can be identified with an associative algebra in $\mathcal{E}^\otimes$. Thinking of $\mathcal{E}^\otimes$ as modelled by a coCartesian fibration q over Δ^{op}, one may define such algebra objects as sections that map convex maps in Δ to q-coCartesian edges (cf. [**Lur17**, 4.1]). We provide a reformulation which can be adapted to our (∞, ω)-categorical context: to this end, we replace $\mathcal{E}^\otimes \to \Delta^{\mathrm{op}}$ by a relative subdivision as follows.

Let $\kappa\colon \mathrm{tot}(\mathcal{P}^*) \to \Delta$ denote the covariant Grothendieck construction of the functor

$$\mathcal{P}^*\colon \Delta \to \mathrm{Cat},\ [n] \mapsto \mathcal{P}^*([n]) \tag{7.1}$$

where $\mathcal{P}^*([n])$ denotes the poset of nonempty subsets of $[n]$. Further define the functor

$$\lambda\colon \mathrm{tot}(\mathcal{P}^*) \to \Delta, ([n], I = \{0 \le i_0 < i_1 < \cdots < i_k \le n\}) \mapsto [k]. \tag{7.2}$$

Then define $\mathrm{rsd}(q)\colon \mathrm{rsd}(\mathcal{E}^\otimes) \to \Delta^{\mathrm{op}}$ via the adjunction

$$\mathrm{Hom}_{\Delta^{\mathrm{op}}}(K, \mathrm{rsd}(\mathcal{E}^\otimes)) \cong \mathrm{Hom}_{\Delta^{\mathrm{op}}}(K \times_{\Delta^{\mathrm{op}}} \mathrm{tot}(\mathcal{P}^*), \mathcal{E}^\otimes)'$$

where

(1) the fiber product over Δ^{op} is taken with respect to κ and equipped with the map to Δ^{op} induced by λ,

(2) the symbol $'$ indicates that we take the subset consisting of maps

$$f\colon K \times_{\Delta^{\mathrm{op}}} \mathrm{tot}(\mathcal{P}^*) \longrightarrow \mathcal{E}^{\otimes}$$

satisfying the following condition: for every vertex $x \in K_0$, the restriction of f to $\{x\} \times_{\Delta^{\mathrm{op}}} \mathrm{tot}(\mathcal{P}^*)$ maps convex inclusions $I \subset J$ to q-coCartesian edges in $\mathcal{E}^{\otimes}$.

EXAMPLE 7.3. A vertex of $\mathrm{rsd}(\mathcal{E}^{\otimes})$ over $[2] \in \Delta$ corresponds to a diagram in $\mathcal{E}^{\otimes}$

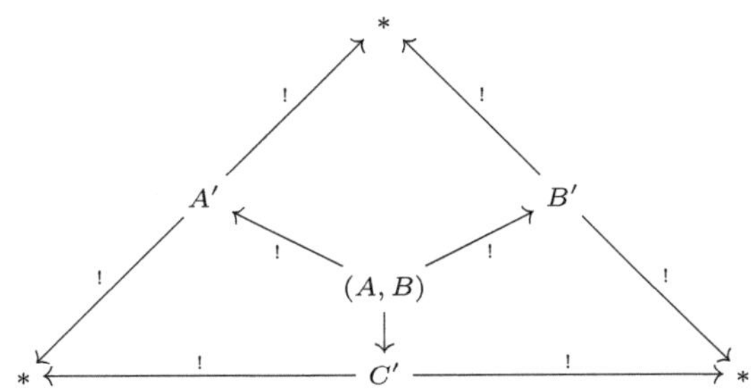

where the edges marked ! are coCartesian: for example the edge $(A, B) \to A'$ covers the convex edge $[2] \leftarrow [1]$ given by the face map ∂_2 and corresponds to an equivalence $A \to A'$. On the other hand, the edge $(A, B) \to C$, which covers the nonconvex edge ∂_1, is *not* coCartesian and corresponds to a morphism $A \otimes B \to C$ in $\mathcal{E}$.

PROPOSITION 7.4. *Let $q\colon \mathcal{E}^{\otimes} \to \Delta^{\mathrm{op}}$ be a monoidal ∞-category. Then there is an equivalence between the ∞-categories of*

(1) *sections of q mapping convex edges in Δ to q-coCartesian edges in $\mathcal{E}^{\otimes}$ (i.e., associative algebras in $\mathcal{E}^{\otimes}$),*

(2) *sections of $\mathrm{rsd}(q)$ mapping injective maps in Δ to $\mathrm{rsd}(q)$-coCartesian edges in $\mathrm{rsd}(\mathcal{E}^{\otimes})$.*

PROOF. By definition, sections of $\mathrm{rsd}(q)$ can be identified with diagrams

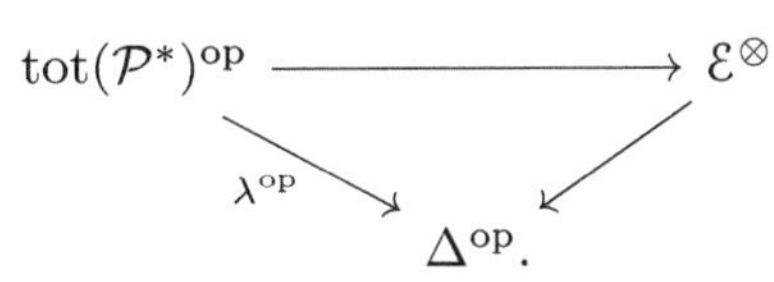

The statement then follows from Lemma 7.11. □

To obtain an analogous concept of a *lax monad* in the (∞, ω)-category $\mathrm{co}_{\infty}(\mathcal{C})$, we first replace the defining simplicial set by a simplicial ∞-category: while the set of n-simplices $\mathrm{co}_{\infty}(\mathcal{C})$ is defined by the formula

$$\mathrm{co}_{\infty}(\mathcal{C})_n = \mathrm{Hom}_{\mathrm{Set}_{\Delta}}(\mathrm{sd}(\Delta^n), \mathcal{C}),$$

we now define

$$\underline{\mathrm{co}_{\infty}}(\mathcal{C})_n = \underline{\mathrm{Hom}}_{\mathrm{Set}_{\Delta}}(\mathrm{sd}(\Delta^n), \mathcal{C}) \tag{7.5}$$

to be the simplicial set of maps which, since $\mathcal{C}$ is an ∞-category, will itself be an ∞-category. We thus obtain a strict functor

$$\underline{\mathrm{co}_\infty}(\mathcal{C})\colon \Delta^{\mathrm{op}} \longrightarrow \mathrm{Set}_\Delta$$

taking values in ∞-categories. We finally pass to the Grothendieck construction (relative nerve) to obtain a coCartesian fibration

$$\pi\colon \mathrm{tot}(\underline{\mathrm{co}_\infty}(\mathcal{C})) \to \Delta^{\mathrm{op}} \tag{7.6}$$

which will play an analogous role to the fibration $\mathrm{rsd}(q)$ above.

DEFINITION 7.7. A *lax monad* in $\underline{\mathrm{co}_\infty}(\mathcal{C})$ is a section of π which maps all injective maps in Δ to π-coCartesian edges.

REMARK 7.8. Further note that the fibration π encodes more than simply the (∞, ω)-category $\mathrm{co}_\infty(\mathcal{C})$ but rather a double categorical refinement that keeps track of morphisms in the original category $\mathcal{C}$ and leads to the right amount of "laxness" to make direct connection to higher Segal objects in $\mathcal{C}$. Hence our definition of monad should be regarded as dependent on this additional data.

THEOREM 7.9. *Let $\mathcal{C}$ be an ∞-category with limits. Then, the following hold:*

(1) *There is an equivalence of ∞-categories*

$$\mathrm{Fun}(\Delta^{\mathrm{op}}, \mathcal{C}) \xrightarrow{\simeq} \{\textit{monads in } \underline{\mathrm{co}_\infty}(\mathcal{C})\},\ X \mapsto M_X. \tag{7.10}$$

(2) *Under the equivalence* (7.10), *lower and upper d-Segal objects in $\mathcal{C}$ get identified with monads in $\underline{\mathrm{co}^l_d}(\mathcal{C})$ and $\underline{\mathrm{co}^r_d}(\mathcal{C})$, respectively.*

PROOF. As above, let $\mathrm{tot}(\mathcal{P}^*) \to \Delta$ denote the covariant Grothendieck construction of the functor (7.1). Unravelling the defining formula (7.5) of $\underline{\mathrm{co}_\infty}(\mathcal{C})$, we observe that a section of (7.6) corresponds to a functor

$$\mathrm{tot}(\mathcal{P}^*)^{\mathrm{op}} \to \mathcal{C}.$$

Pullback along the functor λ from (7.2) provides a functor of ∞-categories

$$\lambda^*\colon \mathrm{Fun}(\Delta^{\mathrm{op}}, \mathcal{C}) \to \mathrm{Fun}(\mathrm{tot}(\mathcal{P}^*)^{\mathrm{op}}, \mathcal{C}).$$

For a simplicial object $X \in \mathrm{Fun}(\Delta^{\mathrm{op}}, \mathcal{C})$, the section of π corresponding to $\lambda^*(X)$ maps injective maps in Δ to coCartesian edges: this reduces to the fact that, for $\phi\colon [m] \to [n]$ injective, and $I \in \mathcal{P}^*([m])$, the morphism $([m], I) \to ([n], \phi(I))$ in $\mathrm{tot}(\mathcal{P}^*)$ is mapped under λ to an identity morphism in Δ. The statement of (1) now follows from Lemma 7.11 below. The statement of (2) is immediate from the discussion in §6, since higher Segal conditions for a simplicial object X translate directly into thinness conditions for the monad M_X. □

LEMMA 7.11. *The functor*

$$\lambda\colon \mathrm{tot}(\mathcal{P}^*) \to \Delta$$

from the proof of Theorem 7.9 *exhibits Δ as an ∞-categorical localization of* $\mathrm{tot}(\mathcal{P}^*)$ *along the set of morphisms of the form*

$$\phi\colon ([m], I) \to ([n], J)$$

where $\phi\colon [m] \to [n]$ is a morphism in Δ such that $\phi(I) = J$.

PROOF. Let $\mathcal{C}$ be an ∞-category. We first claim that every object of $Y \in \text{Fun}(\text{tot}(\mathcal{P}^*), \mathcal{C})$ admits a right Kan extension along λ. This follows from the pointwise formula: to determine the value of the right Kan extension at $[k] \in \Delta$, we need to compute a limit over the slice category $[k]/\lambda$. But this category has an initial object given by $([k],[k])$, so that $\lambda_*(Y)_k \simeq Y([k],[k])$. We thus obtain an adjunction of ∞-categories

$$\lambda^* \colon \text{Fun}(\Delta, \mathcal{C}) \longrightarrow \text{Fun}(\text{tot}(\mathcal{P}^*), \mathcal{C}) \colon \lambda_*$$

where λ_* denotes the right Kan extension functor along λ. The fact that λ^* is fully faithful now immediately follows from the fact that the counit $\text{id} \to \lambda_*\lambda^*$ is an equivalence. Vice versa, the essential image of λ^* is spanned by the collection of objects on which the unit $\lambda^*\lambda_* \to \text{id}$ acts as an equivalence. Again, by the above computation of the slice categories, it follows that the action of the unit on $Y \in \text{Fun}(\text{tot}(\mathcal{P}^*), \mathcal{C})$ evaluated at $([n], I = \{i_0, \dots, i_k\})$ is the morphism

$$Y([k],[k]) \longrightarrow Y(([n], I)).$$

The fact that any functor Y in the image of λ^* needs to invert, more generally, morphisms of the type specified in the statement of the Lemma follows by two out of three. □

REMARK 7.12. In [**FGK$^+$21**], it is shown that every 2-Segal space automatically satisfies certain unitality constraints which were previously required explicitly in the definition of so-called *unital* 2-Segal spaces. It would be very interesting to study analogous phenomena for higher Segal conditions and the relevance for the lax monads of this section.

8. Higher excision

In this section, we explain a characterization of the higher Segal conditions from §3 in terms of higher-dimensional excision conditions as they appear, for instance, in Goodwillie calculus. These results are due to T. Walde [**Wal20**].

Let $\mathcal{C}$ be an ∞-category and let

$$X \colon \Delta^{\text{op}} \to \mathcal{C}$$

be a simplicial object in $\mathcal{C}$. Then, it is straightforward to show that the following are equivalent:

(1) X is a Segal object (i.e., a lower 1-Segal object, in our sense).
(2) For every $0 \leq k \leq n$, the square

$$\begin{array}{ccc} X_{\{0,1,\dots,n\}} & \longrightarrow & X_{\{0,1,\dots,k\}} \\ \downarrow & & \downarrow \\ X_{\{k,k+1\dots,n\}} & \longrightarrow & X_{\{k\}} \end{array}$$

is a pullback square in $\mathcal{C}$ (following the conventions of Remark 3.2).
(3) X maps biCartesian squares in Δ to Cartesian squares in $\mathcal{C}$.

The latter characterization has some particular appeal, since it reformulates the Segal conditions in term of intrinsically categorical propertiesf Δ. This raises the question whether there is an analogous characterization of the higher Segal conditions.

EXAMPLE 8.1. Unravelling the first lower 3-Segal condition, stating that the cone

$$X|(\mathcal{L}([4],3)^{\triangleright})^{\mathrm{op}}$$

is a limit cone in $\mathcal{C}$, we observe that it translates verbatim to the statement that the cube

$$
\begin{array}{ccccccc}
X_{\{0,1,2,3,4\}} & & \longrightarrow & & X_{\{1,2,3,4\}} & & \\
 & \searrow & & & & \searrow & \\
\downarrow & & X_{\{0,1,3,4\}} & \longrightarrow & \downarrow & & X_{\{1,3,4\}} \\
X_{\{0,1,2,3\}} & & \longrightarrow & & X_{\{1,2,3\}} & & \downarrow \\
 & \searrow & \downarrow & & & \searrow & \\
 & & X_{\{0,1,3\}} & \longrightarrow & & & X_{\{1,3\}}
\end{array}
\tag{8.2}
$$

is Cartesian (cf. (3.9)). This is in analogy to the first lower 1-Segal condition saying that the square

$$
\begin{array}{ccc}
X_{\{0,1,2\}} & \longrightarrow & X_{\{1,2\}} \\
\downarrow & & \downarrow \\
X_{\{0,1\}} & \longrightarrow & X_{\{1\}}
\end{array}
$$

is a pullback square. Further, we note that the cube in Δ, parametrizing the cube (8.2), is a strongly biCartesian cube (i.e., all its faces are biCartesian).

THEOREM 8.3. *Let $\mathcal{C}$ be an ∞-category with limits, let $X\colon \Delta^{\mathrm{op}} \to \mathcal{C}$ be a simplicial object in $\mathcal{C}$, and let $d \geq 1$. Then the following are equivalent:*

(1) *X is a lower $(2d-1)$-Segal object.*
(2) *X maps each strongly biCartesian $(d+1)$-cube in Δ to a Cartesian cube in $\mathcal{C}$.*

For the proof we refer to [**Wal20**].

9. Further perspectives

9.1. Coherent Pachner moves. Let $1 \leq k \leq n$, let $S \subset \partial\Delta^{n+1}$ be an abstract simplicial complex which is the union of k faces, and let S' be the union of the complementary faces. A $(k, n+1-k)$-*Pachner move* on a triangulated piecewise linear n-manifold M is the procedure of replacing a subcomplex of M, identified with S, by S'.

Now let $\mathcal{C}$ be an ∞-category with limits, and let $X\colon \Delta^{\mathrm{op}} \to \mathcal{C}$ be a d-Segal object.

We may view the datum

$$S = \mathcal{L}([d+1],d) \hookrightarrow \Delta^{d+1} \hookleftarrow \mathcal{U}([d+1],d) = S'$$

of the lower and upper hemisphere as specifying a type of $(\frac{d+1}{2}, \frac{d+1}{2})$ (resp. $(\frac{d+2}{2}, \frac{d}{2})$, depending on the parity of d) Pachner move which we refer to as a Segal-Pachner move. The lowest d-Segal maps comprise a correspondence

$$X_S \xleftarrow{\simeq} X_{d+1} \xrightarrow{\simeq} X_{S'}$$

whose legs are equivalences, which we may interpret as the statement that the object X_S is "invariant" under the Pachner move that replaces S by S'. The higher simplices of X along with the higher d-Segal maps provide higher coherence data corresponding to higher associativity constraints.

More globally, given a linearly ordered set V and an abstract n-dimensional simplicial complex T on V whose geometric realization is a manifold M, we may obtain correspondences

$$X_T \xleftarrow{\simeq} X_B \xrightarrow{\simeq} X_{T'}$$

associated to Segal-Pachner moves thus establishing the invariance of X_T (here B is obtained by attaching an $n+1$-simplex to T).

To turn this construction into a useful invariant of the manifold M, one needs to address

(1) the dependence on the linear order of V,
(2) the fact that we only have access to the Segal-Pachner moves (and not more general Pachner moves).

In dimension 2, it turns out that there is a satisfying resolution of these issues [**DK18**, **Dyc17**, **DK15**], leading to a combinatorial framework for the construction of correspondence-valued TFTs, conceptually perhaps culminating in a version of the cobordism hypothesis for these types of TFTs established in [**Ste21**].

In dimensions bigger than 2, the construction of useful analogous higher-dimensional TFT-type manifold invariants from higher Segal objects with additional symmetries is a very interesting open problem.

9.2. Additive coefficients and 2-categorical methods. In [**DJW19**], the classical Dold–Kan correspondence is used to analyze higher Segal conditions for simplicial abelian groups as follows:

THEOREM 9.1. *Let $X\colon \Delta^{\mathrm{op}} \to \mathcal{A}$ be a simplicial object valued in an abelian category $\mathcal{A}$ and $m \geq 1$. Then the following are equivalent:*

(1) *X is $2m$-Segal.*
(2) *X has unique outer horn fillers above dimension m, i.e., for every $n > m$, the horn restriction maps*

$$X_n \to X_{\Lambda^n_0} \text{ and } X_n \to X_{\Lambda^n_n}$$

are isomorphisms in $\mathcal{A}$.
(3) *The normalized chain complex $C(X)$ associated to X under the Dold–Kan correspondence is m-truncated, i.e., $C(X)_n \cong 0$, for $n > m$.*

In [**Dyc21**], a categorified variant of the Dold-Kan correspondence is established, furnishing an equivalence

$$\mathcal{C}\colon \mathcal{S}t_{\mathbb{\Delta}} \longrightarrow \mathrm{Ch}_{\geq 0}(\mathcal{S}t)$$

between 2-simplicial objects in the $(\infty,2)$-category $\mathcal{S}t$ of stable ∞-categories and connective chain complexes in $\mathcal{S}t$. An interesting feature of this correspondence is the appearance of the simplex 2-category $\mathbb{\Delta}$ defined as the full sub 2-category of the 2-category of categories spanned by the standard ordinals. A 2-simplicial object is then a 2-functor

$$\mathcal{X}\colon \mathbb{\Delta}^{\mathrm{op}} \to \mathcal{S}t$$

comprising an underlying ordinary simplicial object, along with presentations of the 2-morphisms in $\mathbb{\Delta}$ as natural transformations in $\mathcal{S}t$.

EXAMPLE 9.2. The 2-morphisms among the nondegenerate subsimplices of the 2-simplex can be depicted as follows:

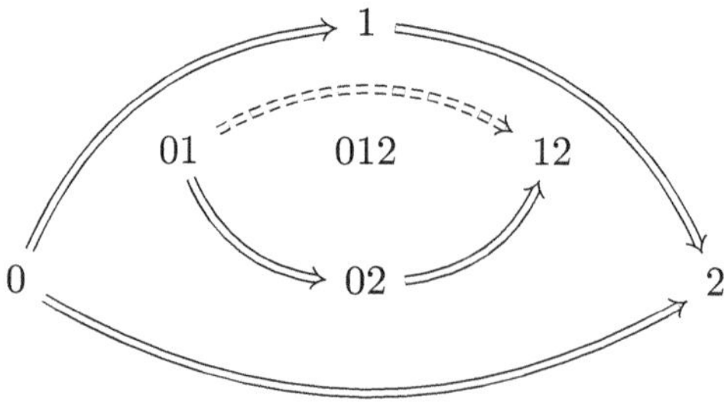

where the dashed arrow signifies that the composite factors through the degenerate edge 11.

For an ordinary simplicial object, the value of X on Λ_0^2 would be defined as the limit of the diagram

$$\begin{array}{ccc} X_{01} & & X_{02} \\ & \searrow \quad \swarrow & \\ & X_0. & \end{array}$$

The additional 2-categorical data exhibited in Example 9.2 suggests that one may define the value of a 2-simplicial object $\mathcal{X}$ on this horn as a partially lax limit of the marked diagram

$$\begin{array}{ccc} \mathcal{X}_{01} & \longrightarrow & \mathcal{X}_{02} \\ & {\scriptstyle +}\searrow \quad \swarrow{\scriptstyle +} & \\ & \mathcal{X}_0. & \end{array}$$

Here, the markings indicate the edges over which the corresponding triangle in the limit cone is supposed to commute up to natural 2-isomorphism, while over unmarked edges, we only require commutativity up to a possibly noninvertible 2-morphism.

In work in progress, we establish a suitable formalism to analyze the corresponding horn filling conditions, and establish the following result:

THEOREM 9.3. *Let $\mathcal{X} \in \mathcal{S}t_{\mathbb{\Delta}}$ be a 2-simplicial stable ∞-category.*

(1) *Then $\mathcal{X}$ is an* outer Kan complex *in the sense that for every $n \geq 1$,*
   (a) *the map $\mathcal{X}_n \longrightarrow \mathcal{X}_{\mathbb{\Lambda}_0^n}$ admits a fully faithful right adjoint (left localization),*
   (b) *the map $\mathcal{X}_n \longrightarrow \mathcal{X}_{\mathbb{\Lambda}_n^n}$ admits a fully faithful left adjoint (right localization).*
(2) *The following are equivalent:*
   (a) *for every $n > m$, the map $\mathcal{X}_n \xrightarrow{\simeq} \mathcal{X}_{\mathbb{\Lambda}_0^n}$ is an equivalence,*
   (b) *for every $n > m$, the map $\mathcal{X}_n \xrightarrow{\simeq} \mathcal{X}_{\mathbb{\Lambda}_n^n}$ is an equivalence,*
   (c) *the categorified normalized chain complex $\mathcal{C}(\mathcal{X})$ is m-truncated,*
   (d) *$\mathcal{X}|\Delta$ is $2m$-Segal.*

REMARK 9.4. We expect that the categorified Dold-Kan correspondence along with Theorem 9.3 generalize from $\mathcal{S}t$ to lax additive $(\infty, 2)$-categories which are idempotent complete in a suitable sense. It would also be very interesting to establish general relations between 2-horn filling conditions and higher Segal conditions beyond this lax additive context.

References

[BOO+18] J. E. Bergner, A. M. Osorno, V. Ozornova, M. Rovelli, and C. I. Scheimbauer, *2-Segal sets and the Waldhausen construction*, Topology Appl. **235** (2018), 445–484, DOI 10.1016/j.topol.2017.12.009. MR3760213

[BOO+21] J. E. Bergner, A. M. Osorno, V. Ozornova, M. Rovelli, and C. I. Scheimbauer, *2-Segal objects and the Waldhausen construction*, Algebr. Geom. Topol. **21** (2021), no. 3, 1267–1326, DOI 10.2140/agt.2021.21.1267. MR4299667

[DJW19] T. Dyckerhoff, G. Jasso, and T. Walde, *Simplicial structures in higher Auslander-Reiten theory*, Adv. Math. **355** (2019), 106762, 73, DOI 10.1016/j.aim.2019.106762. MR3994443

[DK15] T. Dyckerhoff and M. Kapranov, *Crossed simplicial groups and structured surfaces*, Stacks and categories in geometry, topology, and algebra, Contemp. Math., vol. 643, Amer. Math. Soc., Providence, RI, 2015, pp. 37–110, DOI 10.1090/conm/643/12896. MR3381470

[DK18] T. Dyckerhoff and M. Kapranov, *Triangulated surfaces in triangulated categories*, J. Eur. Math. Soc. (JEMS) **20** (2018), no. 6, 1473–1524, DOI 10.4171/JEMS/791. MR3801819

[DK19] T. Dyckerhoff and M. Kapranov, *Higher Segal spaces*, Lecture Notes in Mathematics, vol. 2244, Springer, Cham, 2019, DOI 10.1007/978-3-030-27124-4. MR3970975

[Dyc17] T. Dyckerhoff, *$\mathbb{A}^1$-homotopy invariants of topological Fukaya categories of surfaces*, Compos. Math. **153** (2017), no. 8, 1673–1705, DOI 10.1112/S0010437X17007205. MR3705272

[Dyc21] T. Dyckerhoff, *A categorified Dold-Kan correspondence*, Selecta Math. (N.S.) **27** (2021), no. 2, Paper No. 14, 35, DOI 10.1007/s00029-021-00618-5. MR4227866

[FGK+21] M. Feller, R. Garner, J. Kock, M. U. Proulx, and M. Weber, *Every 2-Segal space is unital*, Commun. Contemp. Math. **23** (2021), no. 2, Paper No. 2050055, 6, DOI 10.1142/S0219199720500558. MR4201031

[Gal63] D. Gale, *Neighborly and cyclic polytopes*, Proc. Sympos. Pure Math., Vol. VII, Amer. Math. Soc., Providence, RI, 1963, pp. 225–232. MR152944

[GCKT18] I. Gálvez-Carrillo, J. Kock, and A. Tonks, *Decomposition spaces, incidence algebras and Möbius inversion I: Basic theory*, Adv. Math. **331** (2018), 952–1015, DOI 10.1016/j.aim.2018.03.016. MR3804694

[GG19] A. Gal and E. Gal, *Higher Segal spaces and Lax $\mathbb{A}_\infty$-algebras*, arXiv:1905.03376, 2019.

[Göd24] J. Gödicke, *An ∞-category of 2-Segal Spaces*, arXiv:2407.13357, 2024.

[Gov16] A. Govzman, *Cyclic polytopes and higher categories*, Master thesis, University of Bonn, 2016.

[Kan57] D. M. Kan, *On c. s. s. complexes*, Amer. J. Math. **79** (1957), 449–476, DOI 10.2307/2372558. MR90047

[KV91a] M. M. Kapranov and V. A. Voevodsky, *Combinatorial-geometric aspects of polycategory theory: pasting schemes and higher Bruhat orders (list of results)* (English, with French summary), Cahiers Topologie Géom. Différentielle Catég. **32** (1991), no. 1, 11–27. International Category Theory Meeting (Bangor, 1989 and Cambridge, 1990). MR1130400

[KV91b] M. M. Kapranov and V. A. Voevodsky, *Combinatorial-geometric aspects of polycategory theory: pasting schemes and higher Bruhat orders (list of results)* (English, with French summary), Cahiers Topologie Géom. Différentielle Catég. **32** (1991), no. 1, 11–27. International Category Theory Meeting (Bangor, 1989 and Cambridge, 1990). MR1130400

[Lou22] F. Loubaton, *The complicial model of (∞, ω)-categories*, arXiv:2207.08504, 2022.

[Lur09] J. Lurie, *Higher topos theory*, Annals of Mathematics Studies, vol. 170, Princeton University Press, Princeton, NJ, 2009, DOI 10.1515/9781400830558. MR2522659

[Lur17] J. Lurie, *Higher Algebra*, Preprint, September 2017.

[Pog17] T. Poguntke, *Higher Segal structures in algebraic K-theory*, arXiv:1709.06510, 2017.

[Ram97] J. Rambau, *Triangulations of cyclic polytopes and higher Bruhat orders*, Mathematika **44** (1997), no. 1, 162–194, DOI 10.1112/S0025579300012055. MR1464385

[Rie18] E. Riehl, *Complicial sets, an overture*, 2016 MATRIX annals, MATRIX Book Ser., vol. 1, Springer, Cham, 2018, pp. 49–76. MR3792516

[Ste21] W. H. Stern, *2-Segal objects and algebras in spans*, J. Homotopy Relat. Struct. **16** (2021), no. 2, 297–361, DOI 10.1007/s40062-021-00282-8. MR4266208

[Str87] R. Street, *The algebra of oriented simplexes*, J. Pure Appl. Algebra **49** (1987), no. 3, 283–335, DOI 10.1016/0022-4049(87)90137-X. MR920944

[Ver08] D. Verity, *Complicial sets characterising the simplicial nerves of strict ω-categories*, Mem. Amer. Math. Soc. **193** (2008), no. 905, xvi+184, DOI 10.1090/memo/0905. MR2399898

[Wal20] T. Walde, *Higher Segal spaces via higher excision*, Theory Appl. Categ. **35** (2020), Paper No. 28, 1048–1086. MR4124491

Universität Hamburg, Fachbereich Mathematik, Bundesstrasse 55, 20146 Hamburg
Email address: tobias.dyckerhoff@uni-hamburg.de

Contemporary Mathematics
Volume **838**, 2026
https://doi.org/10.1090/conm/838/16740

A combinatorial construction of homology via ACGW categories

Maru Sarazola, Brandon T. Shapiro, and Inna Zakharevich

ABSTRACT. This paper introduces the reader to the key ideas and techniques for working with ACGW categories, a recently introduced framework for algebraic K-theory. To do so, we focus on how homology theory generalizes to ACGW categories, particularly in the central example of finite sets. We show how the ACGW formalism can be used to produce various classical homological algebra results such as the Snake Lemma and long exact sequences of relative pairs.

1. Introduction

The main goal of this paper is to explain how to construct homological algebra in the case of finite sets. In order to extend the notion of homology beyond algebraic contexts, it is important to consider exactly what makes homology possible. The standard definition of the homology of a chain complex is

$$H_n = \ker d_n / \operatorname{im} d_{n+1}.$$

Thus in order to define homology, we must at least have images, kernels, and quotients. In an algebraic setting this is always possible in the context of an *abelian category*. To go beyond this it is necessary to analyze exactly how the category of finite sets fails to be abelian.[1] Functions between finite sets always have well-defined images, and a function splits the codomain into the image of the function and the complement to the image (which we can consider to be a "cokernel"). Moreover, if we use *partially defined functions* (i.e. functions that are only defined on a subset of the domain, which we call the *domain of definition*) then the "kernel" is also well-defined: it is simply the complement of the domain of definition.

This intuitive definition has several downsides, all coming from the fact that the *kernel* of a surjective map does not satisfy the kinds of properties that we would like. For example, if finite sets worked in complete analogy with abelian categories, then for a surjective map $f\colon A \twoheadrightarrow B$ it would be the case that $|A| = |B| + |\ker f|$. If f is injective on its domain of definition this holds, but it does not hold otherwise.

It turns out that many of the proofs in homological algebra rely on a simple duality structure: that in an abelian category there is, for every object A, a natural bijection between injections into A and surjections out of A. When considering

Zakharevich was supported in part by a Simons Foundation Fellowship, and by NSF CAREER DMS-1846767.

[1]This is somewhat akin to asking "how is a raven like a writing desk?"

partial functions which are injective on their domain of definition, we get a similar duality in finite sets: total maps f into A correspond, up to isomorphism, with surjective maps out of A with domain $A \setminus \operatorname{im}(f)$ (see example 14). Thus any result that relies only on this observation should be generalizable into a context where this is the case. This leads us to the notion of an ACGW category.[2]

ACGW categories aim to generalize abelian categories categories by requiring only the properties that make these so particularly well-suited for algebraic K-theory. Their key insight lies in the fact that a K-theory machinery for abelian categories only truly sees the monomorphisms and epimorphisms, along with the bicartesian squares between them. In particular, it is not necessary to be able to compose monomorphisms and epimorphisms with one another, which strongly suggests that these two types of morphisms should be part of the structure of a double category. Thus, ACGW categories are double categories with additional data and properties that axiomatize the conditions required to obtain a K-theory machinery analogous to the ones for abelian categories due to Quillen [**Qui73**] and Waldhausen [**Wal85**].

The main appeal of this new framework is that it generalizes the notion of short exact sequences to key non-additive settings such as finite sets and varieties, where the notion of complements replaces that of kernels and cokernels. This makes it possible to study finite sets and varieties as if they were the objects of an abelian category, for the purposes of K-theory. Aside from setting the stage for further study of derived motivic measures and of the K-theory of varieties [**CZ22**], this new framework has also been used by Haesemeyer and Weibel [**HW21**] and by Coley and Weibel [**CW23**] to develop the K-theory of partially cancellative A-sets.

As they retain much of the intuition present in the context of abelian categories, we find that many proofs of classical results in K-theory translate almost verbatim to the setting of ACGW categories and some of their variants—CGW, ECGW categories. For instance, ACGW categories satisfy versions of the Dévissage and Localization Theorems [**CZ22**]. On the other hand, the ECGW categories of the first and second named authors [**SS**] admit an $S_\bullet$-construction that produces a K-theory spectrum, as well as versions of Additivity and the Fibration Theorem. The goal of this article is to show how ACGW categories also behave like abelian categories with another context in mind: homological algebra. Our main result can be phrased as follows.

THEOREM 1. *Let $\mathbb{A}$ be an ACGW category, and write $\mathbf{Ch}_{\mathbb{A}}$ for the category of chain complexes over $\mathbb{A}$. For all n there exists a functor $H_n\colon \mathbf{Ch}_{\mathbb{A}} \longrightarrow \mathbb{A}^\flat$ (where $\mathbb{A}^\flat$ is a certain ordinary category associated to $\mathbb{A}$); see Proposition* 65 *and Theorem* 68. *Moreover, the functors H_n satisfy standard theorems of homology theories: the Snake Lemma (Theorem* 56*), and the Long Exact Sequence of a relative pair (Theorem* 60*).*

This definition of homology also yields a natural definition of quasi-isomorphism (see Section 4.4).

[2]The acronym "ACGW" stands for "abelian CGW." The acronym "CGW" is simultaneously a tribute and a joke. The tribute is to many of the people whose work inspired the paper (Cisinski, Grayson, Waldhausen, Weibel, among others). The joke is in relation to the fact that originally we called these "inexact" categories, which is a poor name for something of which "exact category" is an example. It was then suggested that we use "close enough for government work," a phrase that used to mean "very precise" but in recent times has stronger connotations of sloppiness.

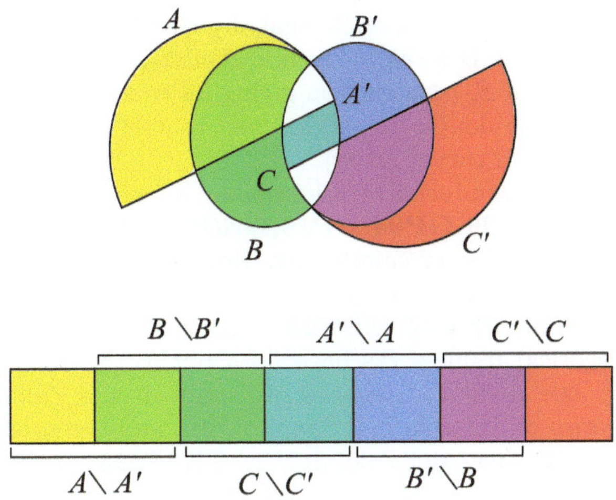

FIGURE 1.1. A picture of the Snake Lemma

When $\mathbb{A}$ is the ACGW category associated to an abelian category $\mathcal{A}$, we have that $\mathbb{A}^\flat = \mathcal{A}$, so the theorem recovers the standard homological algebra results—excellent accounts of these can be found in [**Wei94, Rot09**], and more precise references are given in Section 4 when we recall these classical results.

As we hope that this article can also be used as a friendly introduction to ACGW categories, we choose to forgo a fully abstract treatment in lieu of providing more intuition. Our proofs are given in the context of any ACGW category, but in place of fully formal arguments we rely on the key diagrams illustrating each result and intuition from the example of finite sets which exhibit the ACGW formalism in a manner that allows for illustrations of the homological notions we introduce. For instance, the Snake Lemma in the case of $\mathbb{F}$**inSet** can be visualized using the pair of pictures in Figure 1.1.

The connection to 2-Segal spaces. 2-Segal spaces, introduced by Dyckerhoff and Kapranov [**DK19**] and Gálvez-Carrillo, Kock, and Tonks [**GCKT18**], encode structures which may not have a well-defined composition, but where composition is associative whenever defined. This notion has a wealth of examples with quite different mathematical flavors, in no small part due to the fact that the definition of a 2-Segal space admits several formulations which carry their own intuition. Aside from the original references, a new reader might wish to consult the articles in this volume by Hackney [**Hac**] and Stern [**Ste**] which introduce these perspectives with minimal prerequisites.

A notable source of examples consists of proto-exact categories [**DK19**, Section 2.4], which give rise to 2-Segal spaces via their $S_\bullet$-construction. This is analogous to the classical $S_\bullet$-construction of an exact category, introduced by Waldhausen as a framework for algebraic K-theory [**Wal85**]—an introductory survey of this construction can be found in this volume in the article by Ozornova and Rovelli [**OR**]. Inspired by this fact, Bergner, Osorno, Ozornova, Rovelli and Scheimbauer

[**BOO⁺18, BOO⁺21**] studied the question: what is a natural input for an $S_\bullet$-construction that will always produce 2-Segal objects? For the case of 2-Segal sets, they found that an answer is given by certain *double-categorical structures*; more precisely, that there is an $S_\bullet$-construction giving an equivalence between the category of augmented stable double categories, and that of 2-Segal sets—an account of this is given in Rovelli's survey [**Rov**].

Double categories, introduced by Ehresmann [**Ehr63**] and greatly developed by Grandis and Paré [**GP99, GP04, GP19, Gra19**], have been the focus of much work within the category theory community. However, it was much more recently that they started to be studied in connection to homotopy theory; see for instance [**FPP08, FP10, Mos20, MSV22, MSV23**], among others. In a similar vein, different double-categorical frameworks have been put forth for the purposes of algebraic K-theory: the *ACGW categories* of Campbell and Zakharevich [**CZ22**], the *ECGW categories* of Sarazola and Shapiro [**SS**], and the *squares categories* of Campbell, Kuijper, Merling, and Zakharevich [**CKMZ23**]. In light of the previous work, it then seems natural to ask: what is the connection between these new double-categorical structures and 2-Segal objects?

The case of squares categories is studied in recent work of the first named author and Calle [**CS24**]. They define an $S_\bullet$-construction for squares categories, and show that squares categories satisfying certain "stability" conditions give rise to 2-Segal spaces. Work in progress by Bergner along with the second and third named authors shows that the $S_\bullet$-construction for ACGW and ECGW categories (by means of the more general CGW categories) always produces a 2-Segal space, and CGW categories live among the natural inputs to the $S_\bullet$-construction producing 2-Segal groupoids from [**BOO⁺21**]. Anticipating this result, we hope that this paper can serve as a gentle introduction to the key properties and techniques that are used in the ACGW setting.

Organization. This paper is organized as follows. In Section 2 we explain the working features of an ACGW category, and explain the ACGW structure associated to finite sets. In Section 3 we define chain complexes over an ACGW category and construct the homology functors H_i, and explain them in detail in the special case of finite sets. In Section 4 we prove all the components of Theorem 1.

Disclaimer about strict equality vs. equality up to isomorphism. In much of the paper we are concerned with injections into finite sets and surjections out of finite sets. These would more clearly be described as "inclusions of subsets" and "projections down to quotient sets," with suitable disclaimers about working up to isomorphism in more general cases. In the interest of brevity and intuition generation, we often omit such statements, but functionally speaking when discussing bijections between "injections" and "surjections" it is important to include "up to isomorphism" when not working with explicit models, subsets, and quotient sets.

Acknowledgements. The authors would like to thank the Banff International Research Station for hosting the workshop *Segal Spaces and their Applications to Algebraic K-Theory, Hall Algebras, and Combinatorics*. We would also like to thank Shruthi Sridhar-Shapiro for making the illustrations in this paper for us, and the two anonymous referees as well as Joachim Kock for their detailed reading and thoughtful comments which prompted us to significantly improve our introduction.

2. ACGW categories

In this section we describe ACGW categories and their properties. Their aim is to axiomatize several features common to both algebraic categories (where we have access to a notion of (co)kernels) and to combinatorial categories (where we have a notion of complements). As a consequence, their definition can be quite technical, especially for readers who are not already acquainted with the language of double categories. We do not give a complete, formal definition of ACGW categories. Instead, we present an intuitive approach that only highlights the essential features, with an emphasis on the examples of finite sets and finite pointed sets. A thorough reader may find the definition in [**CZ22**, Definition 5.6]; generalizations for the interested reader can be found in [**CZ22**, Definition 2.5] and [**SS**, Definitions 2.4, 4.1].

A double category is a collection of objects with *two* different category structures between them, as well as data about how these two different category structures relate to one another. We give an informal definition here; a reader unfamiliar with double categories might wish to consult [**Gra19**, Definition 3.1.1] for a more detailed treatment.

Definition 2. A *double category* $\mathbb{A}$ consists of objects $A, B, A', B', \ldots$, horizontal morphisms $A \rightarrowtail B$, vertical morphisms $A \circ\!\!\rightarrow A'$, and squares

$$\begin{array}{ccc} A & \rightarrowtail & B \\ \circ\!\!\downarrow & \circlearrowleft & \circ\!\!\downarrow \\ A' & \rightarrowtail & B' \end{array}$$

together with associative and unital compositions for horizontal morphisms and vertical morphisms, as well as two such composition laws for squares (giving horizontal and vertical pastings) that must satisfy an interchange law. Following the terminology of [**CZ22**], we will refer to squares as *pseudo-commutative squares.*

While in general a pseudo-commutative square is not necessarily uniquely determined by the morphisms on its boundary, in all of our examples each diagram of the above shape either *is* a square in the double category structure, or it is *not* a square in the double category structure.

Notation 3. Given objects A, B in a double category $\mathbb{A}$, we denote by $\mathrm{Hor}(A, B)$ the category whose objects are the horizontal morphisms $A \rightarrowtail B$ in $\mathbb{A}$, and whose morphisms are given by squares in $\mathbb{A}$. Similarly, we use $\mathrm{Ver}(A, B)$ for the category of vertical morphisms $A \circ\!\!\rightarrow B$ and squares in $\mathbb{A}$.

Remark 4. We will use blackboard bold for double categories and ACGW categories, and ordinary bold or calligraphic font for ordinary categories. Thus **FinSet** and $\mathcal{C}$ are ordinary categories, but $\mathbb{F}$**inSet** and $\mathbb{A}$ are double categories.

Definition 5. An *ACGW category* is a double category $\mathbb{A}$ with the following additional structure and properties:

- There is an object in $\mathbb{A}$ which is initial both for the horizontal and the vertical morphisms, and both types of morphisms are monic; we denote such an initial object by $\varnothing$.

- For each pair of objects A, B in $\mathbb{A}$, the set of isomorphisms $A \cong B$ in the vertical and horizontal categories agree in a suitable sense (see [**CZ22**, Definition 2.5],[**SS**, Definition 1.9]).
- For every object $A \in \mathbb{A}$ there is a pair of maps

$$c\colon \bigcup_{B\in\mathbb{A}} \mathrm{Hor}(B,A) \rightleftarrows \bigcup_{C\in\mathbb{A}} \mathrm{Ver}(C,A) :k.$$

  which are inverses up to isomorphisms on B and C. The function c is the "cokernel," or "complement," and the function k is the "kernel." For a horizontal morphism $B \rightarrowtail A$ we write $A/\!/B \circ\!\!\rightarrow A$ for its cokernel, and for a vertical morphism $C \circ\!\!\rightarrow A$ we write $A\backslash\!\backslash C \rightarrowtail A$ for its kernel.
- The pair above is natural in the following sense. For every vertical morphism $B \circ\!\!\rightarrow D$, the bijection c extends to a bijection

$$c\colon \left\{ \begin{array}{ccc} A & \rightarrowtail & B \\ \downarrow & \circlearrowleft & \downarrow \\ C & \rightarrowtail & D \end{array} \right\} \longrightarrow \left\{ \text{pullbacks} \quad \begin{array}{ccc} X & \circ\!\!\rightarrow & B \\ \downarrow & & \downarrow \\ Y & \circ\!\!\rightarrow & D \end{array} \right\} \tag{6}$$

  An analogous statement holds for k for any horizontal morphism $C \rightarrowtail D$.

  As a consequence of this, given any diagram

$$A \rightarrowtail C \leftarrow\!\!\circ\, B$$

  there exists a unique (up to unique isomorphism) pseudo-commutative square

$$\begin{array}{ccc} A \oslash_C B & \rightarrowtail & B \\ \downarrow & \circlearrowleft & \downarrow \\ A & \rightarrowtail & C \end{array}$$

Again, we emphasize that this definition is only an intuitive one. Formally, the assignments c and k are part of the data of an ACGW category, and we need to be slightly more careful when describing their codomain categories, something we glossed over here. These functors are also required to be equivalences of categories. Finally, there are a host of additional axioms one can ask for, that allow us to encode notions like pushouts/pullbacks, image factorizations, weak equivalences, and so on; these give rise to more specialized variants of ACGW categories that we will not discuss here, but that the reader can find in [**CZ22**, **SS**].

REMARK 7. The condition that vertical and horizontal isomorphisms in $\mathbb{A}$ agree means that in $\mathbb{A}$ there is a single notion of "isomorphism," unlike in general double categories. This allows us to state properties "up to (unique) isomorphism" without ambiguity.

As a first example, we describe how one can view an abelian category as an ACGW category; we will return to this example often as we introduce homological notions later in the paper.

EXAMPLE 8. Any abelian category $\mathcal{A}$ gives rise to an ACGW category in a natural way. The double category $\mathbb{A}$ to consider has as objects the objects of $\mathcal{A}$, as horizontal morphisms the monomorphisms, and as vertical morphisms the opposite of the epimorphisms; that is, $\mathbb{A}$ has a vertical morphism $A \circ\!\!\rightarrow B$ for each epimorphism $B \twoheadrightarrow A$ in $\mathcal{A}$. Squares are given by the commutative squares in $\mathcal{A}$; that is, the diagram in $\mathbb{A}$ as below left is a square in the double category precisely when the corresponding diagram in $\mathcal{A}$ as below right is commutative.

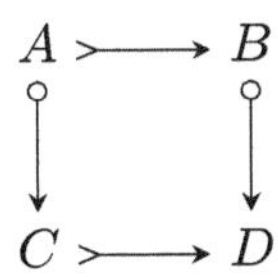

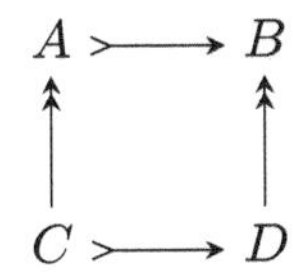

The object $\varnothing$ is the zero object, and the functors c and k consist of taking cokernels and kernels, respectively.

Since our goal is to present several classical homological algebra results using the language of ACGW categories, we introduce a combinatorial example that will serve as one source of intuition.

EXAMPLE 9. The ACGW category $\mathbb{F}$**inSet** has

objects: finite sets,
horizontal morphisms: injections,
vertical morphisms: injections, and
pseudo-commutative squares: squares which are pullbacks when considered as a diagram in **FinSet**.

The object $\varnothing$ is the empty set. Both c and k will be the same: for any injection $A \longrightarrow B$ we take the inclusion $B\backslash A \longrightarrow B$ (i.e. the inclusion of the complement). Then the duality between pseudo-commutative squares and pullback squares states that, if $B, C \subseteq D$ then

$$B \cap C = B\backslash\left((D\backslash C) \cap B\right).$$

The construction $\oslash$ is simply the pullback in the ordinary category **FinSet**.

All of the relevant squares in this double category can be visualized using pictures such as those below. For each picture on the left, we include on the right the various types of diagrams in the ACGW category of sets it could correspond to.

(10)

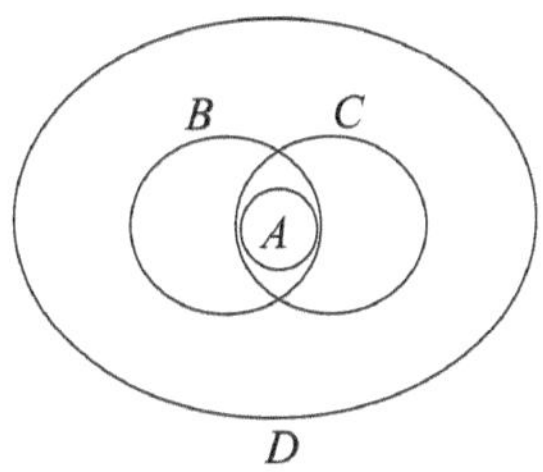

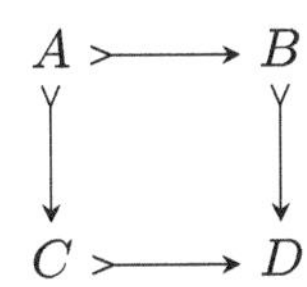

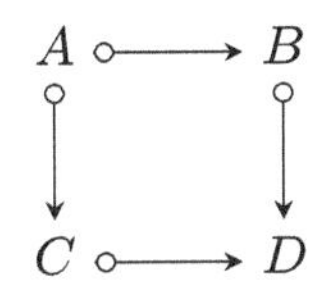

In a commuting square of inclusions (of either arrow type), the set A includes into the intersection of B and C but may not cover all of $B \cap C$.
(11)

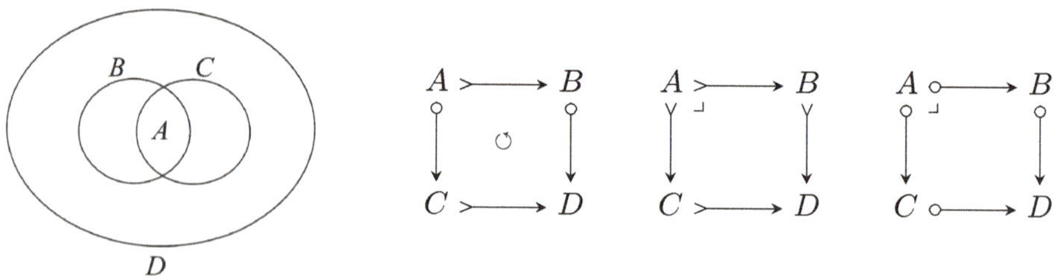

In a cartesian square (be it of horizontal morphisms, vertical morphisms, or a mix) we moreover have $A = B \cap C$.

For the pseudo-commutative square in (11), its complement square in the horizontal direction (which is a cartesian square in the underlying horizontal category) models the intersection of B and $D \backslash C$ in D.

(12)

In this picture B is shaded red, $D \backslash C$ shaded blue, and their intersection $B \backslash A$ shaded purple.

REMARK 13. The primary references [**CZ22**, **SS**] focus on the example of finite sets, as they are concerned with K-theory, which requires finiteness. By contrast, nothing in this paper requires finiteness, so all results about $\mathbb{F}\mathbf{inSet}$ are equally true in the ACGW category $\mathbb{S}\mathbf{et}$; in the interest of being faithful to our sources we stay with the finite set example.

EXAMPLE 14. The ACGW category $\mathbb{F}\mathbf{inSet}_*$ has

objects: finite pointed sets,
horizontal morphisms: pointed injections,
vertical morphisms: pointed surjections which are injective away from the preimage of the basepoint, and
pseudo-commutative squares: squares which are commutative when considered in $\mathbf{FinSet}_*$ (analogously to the abelian case).

The vertical morphisms are precisely the cokernels of inclusions, which send the image of an inclusion to the basepoint. Dually, the kernel of such a surjection is the inclusion of the preimage of the basepoint.

The ACGW category $\mathbb{F}\mathbf{inSet}_*$ is isomorphic[3] to the ACGW category $\mathbb{F}\mathbf{inSet}$ via the double functor $\mathbb{F}\mathbf{inSet} \longrightarrow \mathbb{F}\mathbf{inSet}_*$ taking a set A to A_+ (adding a disjoint

[3]Technically, to make this fully correct we would need to say that they are equivalent, in the sense of [**SS**, Definition 1.10]. However, if you (for example) model finite sets as finite subsets of positive integers, and model finite pointed sets as finite subsets of nonnegative integers with 0 as the basepoint, then this functor is an honest isomorphism of categories.

basepoint). On horizontal morphisms this simply takes a function f to f_+. On vertical morphisms it takes an injection $g\colon I \longrightarrow J$ to the surjection $\widetilde{g}\colon J_+ \longrightarrow I_+$ given by $\widetilde{g}(j) = g^{-1}(j)$ if j is in the image of g, and $g(j) = *$ otherwise.

The following gives a natural way to assign an ordinary category to any ACGW category.

DEFINITION 15. Given an ACGW category $\mathbb{A}$, there is an ordinary category which is the *flattening* of it, denoted $\mathbb{A}^\flat$. This category has

objects: the objects of $\mathbb{A}$,

morphisms from A to B: isomorphism classes of diagrams

$$A \longleftarrow\!\!\circ\ \overline{A} \rightarrowtail B,$$

composition of morphisms: $A \leftarrow\!\circ\ \overline{A} \rightarrowtail B$ and $B \leftarrow\!\circ\ \overline{B} \rightarrowtail C$ compose to

$$A \longleftarrow\!\!\circ\ \overline{A} \oslash_B \overline{B} \rightarrowtail C.$$

In other words, we take a category of spans with "legs in different directions" and use the existence of $\oslash$ to compose.

Every morphism $A \leftarrow\!\circ\ \overline{A} \rightarrowtail B$ in $\mathbb{A}^\flat$ has a *kernel* (given by the kernel of $\overline{A} \circ\!\!\rightarrow A$) and an *image*, given by $\overline{A}$. A *zero morphism*, denoted 0, is the unique morphism where $\overline{A} = \varnothing$; there is a zero morphism between any two objects, and the composite of any morphism with a zero morphism is a zero morphism.

If $\mathbb{A}$ is the ACGW category associated to an abelian category $\mathcal{A}$, then $\mathbb{A}^\flat \cong \mathcal{A}$. Indeed, the morphisms in $\mathbb{A}^\flat$ are essentially the epi-mono factorization of the morphisms in $\mathcal{A}$.

REMARK 16. There is an important difference between the ACGW category associated to an abelian category and the ACGW category $\mathbb{F}$**inSet**. Morphisms in **FinSet** have associated epi-mono factorizations, so it may be tempting to try and define an ACGW structure where the vertical morphisms correspond to opposite surjections. In the algebraic context this works perfectly: maps are "equifibered": to determine whether a morphism is injective it suffices to check the preimage of 0. But for finite sets this is no longer true. For instance, what is the kernel of the function $\{1, 2, 3, 4, 5\} \longrightarrow \{1, 2\}$ taking 1 and 2 to 1 and 3, 4, 5 to 2?

Moreover, there is no longer a bijection between isomorphism classes of surjections out of A (i.e. quotients of A) and isomorphism classes of injections into A (i.e. subobjects of A)—there are far more surjections. In finite sets there are four injective functions into $A = \{1, 2\}$, but *five* surjective partial functions out of A. Even if we restrict to only fully-defined surjective functions, these are in bijection with partitions, and injective functions are in bijection with subsets; in the case of the five-element set there are 32 subsets and 52 partitions. Thus it is impossible to construct a duality of the type we would like for homology if we do not restrict to functions that are injective on their domain of definition. (See also the discussion of representations of quivers over $\mathrm{Vect}(\mathbb{F}_1)$ in [**Szc12**].)

3. Homology in an ACGW category

Our goal is to use the duality structure present in the definition of an ACGW category in order to define chain complexes and homology.

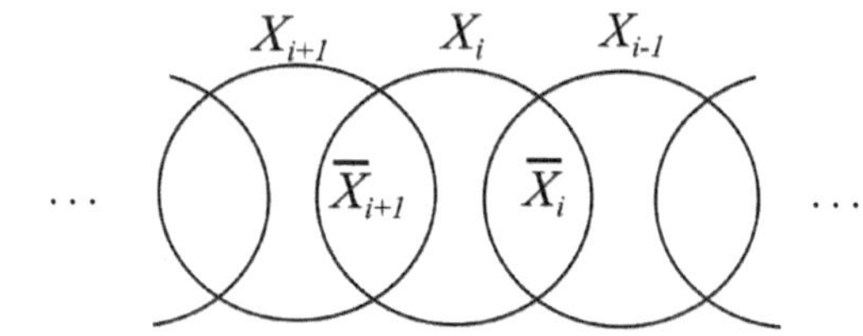

FIGURE 3.1. A chain complex

3.1. Chain complexes and homology.

DEFINITION 17. Let $\mathbb{A}$ be an ACGW category. A *chain complex in* $\mathbb{A}$ is a sequence of morphisms $d_i \colon A_i \longrightarrow A_{i-1}$ in $\mathbb{A}^\flat$ such that $d_i d_{i+1} = 0$ for all i.

Unwinding this definition, a chain complex consists of a diagram in $\mathbb{A}$ of the form

$$\cdots X_{i+1} \leftarrow\!\!\!-\!\!\circ\ \overline{X}_{i+1} \rightarrowtail X_i \leftarrow\!\!\!-\!\!\circ\ \overline{X}_i \rightarrowtail X_{i-1} \cdots$$

along with squares in $\mathbb{A}$ of the form below.

$$\begin{array}{ccc} \varnothing & \rightarrowtail & \overline{X}_i \\ \downarrow\!\circ & \circlearrowleft & \downarrow\!\circ \\ \overline{X}_{i+1} & \rightarrowtail & X_i \end{array}$$

EXAMPLE 18. Suppose that $\mathbb{A}$ is the ACGW category associated to an abelian category $\mathcal{A}$. Then $\mathbb{A}^\flat = \mathcal{A}$. Moreover, the zero morphism in $\mathbb{A}^\flat$ is the zero morphism in $\mathcal{A}$, so the definition is exactly the usual definition of a chain complex in $\mathcal{A}$.

EXAMPLE 19. A chain complex in $\mathbb{F}\mathbf{inSet}$ has the form of a sequence of spans of inclusions.

$$\cdots X_{i+1} \leftarrow\!\!\!-\!\!\circ\ \overline{X}_{i+1} \rightarrowtail X_i \leftarrow\!\!\!-\!\!\circ\ \overline{X}_i \rightarrowtail X_{i-1} \cdots$$

Consider what it means for $d_n d_{n+1} = 0$ to hold. Suppose that we have two morphisms in $\mathbb{F}\mathbf{inSet}^\flat$:

$$A \leftarrow\!\!\!-\!\!\circ\ \overline{A} \rightarrowtail B \leftarrow\!\!\!-\!\!\circ\ \overline{B} \rightarrowtail C.$$

Up to isomorphism, this is the data of three sets A, B, C with specified intersections $A \cap B = \overline{A}$ and $B \cap C = \overline{B}$. The composite of these two morphisms is then $A \leftarrow\!\!\!-\!\!\circ\ \overline{A} \cap \overline{B} \rightarrowtail C$. This will be the zero morphism exactly when $\overline{A} \cap \overline{B} = \varnothing$, or in other words, when the transition sets $\overline{A}$ and $\overline{B}$ do not intersect in B.

Then, a chain complex consists of a sequence of sets X_i equipped with intersections $\overline{X}_i = X_i \cap X_{i-1}$ such that $\overline{X}_{i+1}$ and $\overline{X}_i$ are disjoint in X_i. We can therefore visualize a chain complex as in Figure 3.1 (which coincidentally looks a bit like a real chain).

Notice that this structure gives a natural decomposition of each X_i into three subsets: $\overline{X}_{i+1}$, $\overline{X}_i$, and $X_i \backslash (\overline{X}_{i+1} \cup \overline{X}_i)$. This will be relevant in Definition 22 when we define the homology of a chain complex.

REMARK 20. We have now discussed how a chain complex of sets

$$\cdots X_{i+1} \longleftarrow\!\!\circ\ \overline{X}_{i+1} \rightarrowtail X_i \longleftarrow\!\!\circ\ \overline{X}_i \rightarrowtail X_{i-1} \cdots$$

can be interpreted combinatorially as a sequence of sets with specified intersections, or (less combinatorially, almost algebraically) as a sequence of maps between pointed sets. In fact there is also a fully algebraic interpretation of these complexes: for any ring R, chain complexes of sets describe the chain complexes of free R-modules whose differentials are composed of only projections and coproduct injections. In particular, given a chain complex of sets as above we get a chain complex of R-modules

$$\cdots \longrightarrow R^{X_{i+1}} \longrightarrow R^{X_i} \longrightarrow R^{X_{i-1}} \longrightarrow \cdots$$

where the differentials factor as the composite $R^{X_i} \longrightarrow R^{\overline{X}_i} \longrightarrow R^{X_{i-1}}$ which projects from R^{X_i} to the components in the image of $\overline{X}_i$ and inserts 0s for each element of X_{i-1} outside the image of $\overline{X}_i$. This interpretation, induced by a structure-preserving double functor $\mathbb{F}\mathbf{inSet} \longrightarrow R\text{-}\mathbb{M}\mathbf{od}$ sending A to R^A, respects homology and therefore completely determines the homological theory of finite sets. While we won't discuss it further, it encourages the interpretation of finite sets as "modules over the field with one element" suggested by Jacques Tits in [**Tit57**].

In any pseudo-commutative square as below left, its complementary square in the vertical direction has the form below right,

$$(21)\qquad \begin{array}{ccc} \varnothing & \rightarrowtail & B \\ \downarrow\!\!\circ & \circlearrowleft & \downarrow\!\!\circ \\ A & \rightarrowtail & C \end{array} \qquad\qquad \begin{array}{ccc} A & \rightarrowtail & C\backslash\!\backslash B \\ \downarrow & \lrcorner & \downarrow \\ A & \rightarrowtail & C \end{array}$$

which in particular includes a horizontal morphism from A to $C\backslash\!\backslash B = k(B \circ\!\!\rightarrow C)$. In the case of finite sets this says that if A and B are disjoint subsets of C, then there is a natural inclusion of A into $C\backslash B$. (In the case of an abelian category, this expresses the statement that if $gf = 0$ then f factors through $\ker g$.) There is similarly a vertical morphism from B to $C/\!/A = c(A \rightarrowtail C)$.

DEFINITION 22. Let $\mathbb{A}$ be an ACGW category. The *i-th homology* $H_i(X_\bullet)$ of a chain complex $X_\bullet$ in $\mathbb{A}$

$$\cdots X_{i+1} \longleftarrow\!\!\circ\ \overline{X}_{i+1} \rightarrowtail X_i \longleftarrow\!\!\circ\ \overline{X}_i \rightarrowtail X_{i-1} \cdots$$

is defined by taking complements in both directions (in either order) of the squares defining the chain condition, as in

(23)

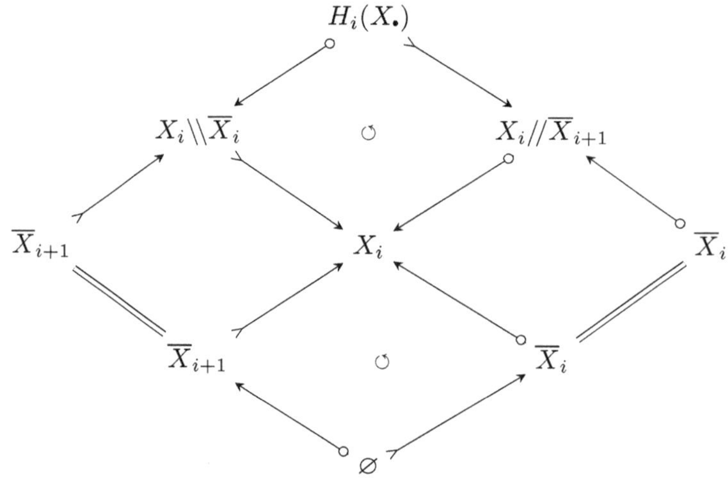

EXAMPLE 24. Let us compare this definition to the usual notion of homology for an abelian category in a bit more detail. If we consider a chain complex (with the epi-mono factorization of each differential)

(25)

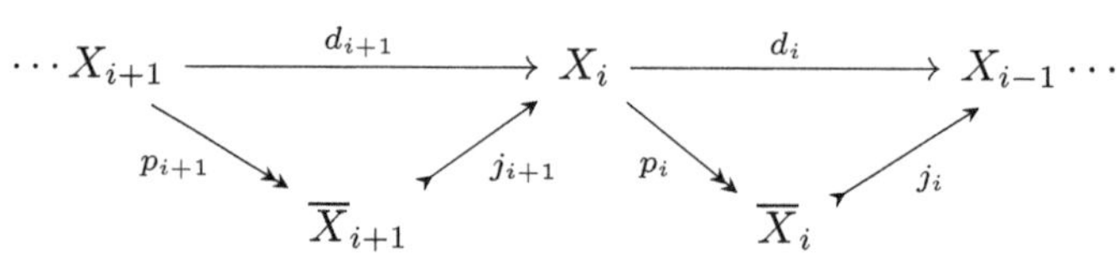

then the i-th homology is given by $H_i(X_\bullet) = \frac{\ker d_i}{\operatorname{im} d_{i+1}}$. Note that $\operatorname{im} d_{i+1} = \overline{X}_{i+1}$, and that $\ker d_i = \ker(j_i p_i) = \ker p_i$; hence we can rewrite the homology as $H_i(X_\bullet) = \frac{\ker p_i}{\overline{X}_{i+1}}$. This quotient can be captured in the diagram

(26)

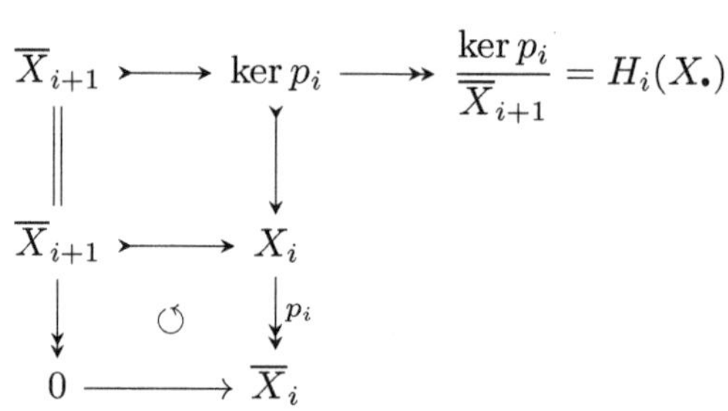

which is precisely the left half of the diagram defining homology. We start with the commutative square on the bottom given by the chain condition, take the kernels of the vertical epis, and then the cokernel of the induced inclusion $\overline{X}_{i+1} \hookrightarrow \ker p_i$. The other half of the grid diagram is redundant, but illustrates the alternative definition of homology as the kernel of the induced map $X_i/\overline{X}_{i+1} \longrightarrow \overline{X}_i$.

EXAMPLE 27. Consider now the case of **FinSet**. Translating the diagram above for the ACGW category of sets, we obtain the diagram

$$\begin{array}{ccccc} \overline{X}_{i+1} & \rightarrowtail & X_i\backslash\overline{X}_i & \leftarrow\!\!\circ & \left(X_i\backslash\overline{X}_i\right)\backslash\overline{X}_{i+1} \\ \| & & \downarrow & & \\ \overline{X}_{i+1} & \rightarrowtail & X_i & & \\ \uparrow & & \uparrow & & \\ \emptyset & \rightarrowtail & \overline{X}_i & & \end{array} \tag{28}$$

and we see that $H_i(X_\bullet) = \left(X_i\backslash\overline{X}_i\right)\backslash\overline{X}_{i+1} = X_i\backslash\left(\overline{X}_{i+1}\cup\overline{X}_i\right)$. In this case the symmetry in the definition is even clearer, as we also have $\left(X_i\backslash\overline{X}_{i+1}\right)\backslash\overline{X}_i = X_i\backslash\left(\overline{X}_{i+1}\cup\overline{X}_i\right)$.

In other words, the i-th homology measures the elements of X_i that are not reached by either differential. In pictures, we can see the homology in a chain complex of sets as in (29).

(29)

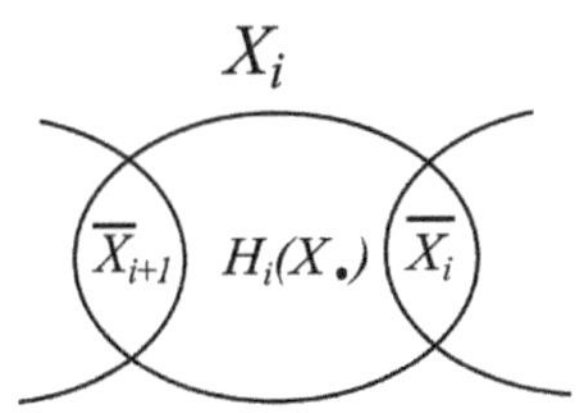

$$\cdots X_{i+1} \leftarrow\!\!\circ\ \overline{X}_{i+1} \rightarrowtail X_i \leftarrow\!\!\circ\ \overline{X}_i \rightarrowtail X_{i-1}\cdots$$

This picture provides a helpful guide for how to interpret the subsets $\overline{X}_{i+1}, \overline{X}_i$ of X_i and their complements in analogy with the algebraic setting. The inclusion $\overline{X}_{i+1} \rightarrowtail X_i$ is completely analogous to the inclusion of $\mathrm{im}(d_{i+1})$ into X_i, while the inclusion $X_i\backslash\overline{X}_i \rightarrowtail X_i$ corresponds to the inclusion of the kernel. Therefore in the picture above, $\overline{X}_{i+1}$ represents the image of a differential d_{i+1} while $\overline{X}_{i+1}\cup H_i(X_\bullet)$ represents the kernel of the next differential d_i. The set $\overline{X}_i$ in the picture can be thought of as the quotient of X_i by $\ker(d_i)$.

As usual, a definition of homology leads to a notion of exact chain complexes, which we now introduce.

DEFINITION 30. A chain complex in $\mathbb{A}$

$$\cdots X_{i+1} \longleftarrow\!\!\circ\ \overline{X}_{i+1} \rightarrowtail X_i \longleftarrow\!\!\circ\ \overline{X}_i \rightarrowtail X_{i-1}\cdots$$

is *exact* if $H_i(X_\bullet) = \varnothing$ for all i.

Unwinding this definition, $H_i(X_\bullet)$ is $\varnothing$ exactly when each subdiagram

$$\overline{X}_{i+1} \rightarrowtail X_i \leftarrow\!\!\circ\ \overline{X}_i$$

is a kernel-cokernel pair; that is, $\overline{X}_{i+1}$ is isomorphic to the kernel of $X_i \leftarrow\!\!\circ\ \overline{X}_i$ and $\overline{X}_i$ is isomorphic to the cokernel of $\overline{X}_{i+1} \rightarrowtail X_i$. In this case the horizontal

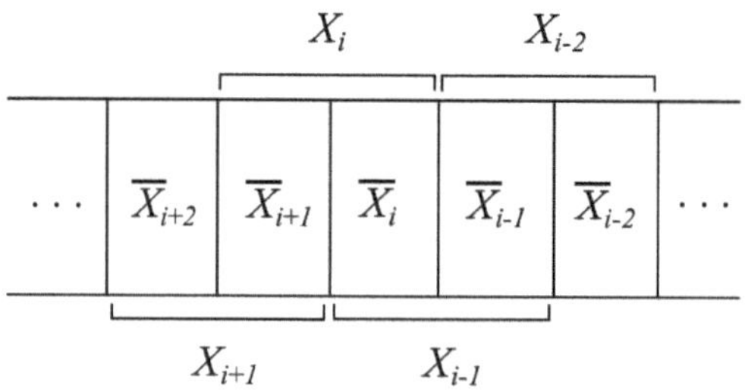

FIGURE 3.2. An exact complex

morphism $\overline{X}_{i+1} \rightarrowtail X_i \backslash\backslash \overline{X}_i$ (whose cokernel is $H_i(X_\bullet)$) is an isomorphism. This condition corresponds to the chain complex being "exact at X_i" when $\mathbb{A}$ arises from an abelian category.

EXAMPLE 31. Expressing the above condition purely in the language of sets, we see that a chain complex of sets $X_\bullet$ is exact if and only if $X_i = \overline{X}_{i+1} \sqcup \overline{X}_i$ for all i. Based on this we will draw exact complexes as in Figure 3.2, where the complex is determined by a sequence of "adjacent" but not overlapping sets $\ldots, \overline{X}_{i+1}, \overline{X}_i, \ldots$ whose successive pairs form the sets X_i.

EXAMPLE 32. Our notion of exact chain complex is compatible with the notion of "short exact sequence" that we obtain from the complement functors c and k in our ACGW framework. Recall that in an abelian category, a short exact sequence consists of a diagram

$$A \overset{i}{\rightarrowtail} B \overset{p}{\twoheadrightarrow} C$$

where i is a mono, p is an epi, $A = \ker p$, and $C = \operatorname{coker} i$. Equivalently, this is an exact chain complex concentrated in only three degrees.

In our ACGW category of finite sets, the analogue of the first description consists of a diagram

$$A \overset{i}{\rightarrowtail} B \overset{p}{\hookleftarrow} C$$

where both i and p are inclusions, $A = B \backslash C$, and $C = B \backslash A$. Then, we have that $B = A \sqcup C$, and our "short exact sequence" above is the same data as the exact complex concentrated in three degrees

$$A = A \rightarrowtail B \hookleftarrow C = C.$$

There is a standard illustration of an exact chain complex as a "zigzag of short exact sequences" which are glued together. We can draw a similar picture in an ACGW category as in (33), emphasizing how the exact complex $X_\bullet$ is built out of short exact sequences.

(33)
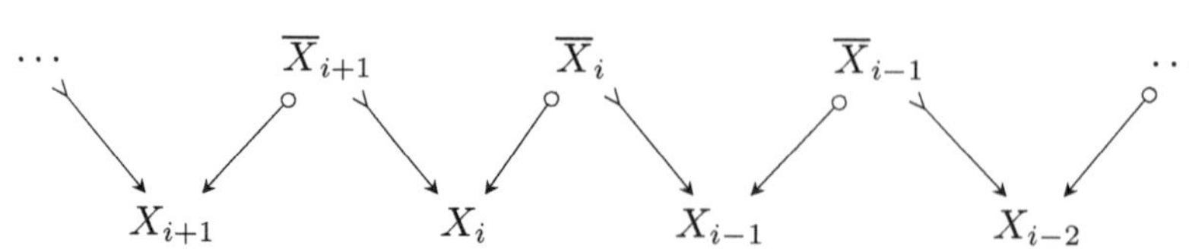

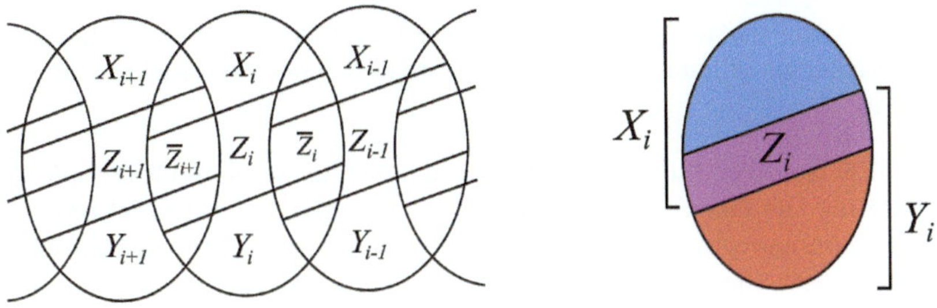

FIGURE 3.3. A morphism of chain complexes

3.2. Morphisms between chain complexes.

DEFINITION 34. A *morphism of chain complexes* in $\mathbb{A}$, written $f_\bullet\colon X_\bullet \longrightarrow Y_\bullet$, consists of morphisms $f_i\colon X_i \leftarrow\!\circ\, Z_i \rightarrowtail Y_i$ in $\mathbb{A}^\flat$ such that for each i there exists a diagram

$$\begin{array}{ccccc} X_i & \longleftarrow\!\circ & \overline{X}_i & \rightarrowtail & X_{i-1} \\ \uparrow & & \uparrow & \circlearrowleft & \uparrow \\ Z_i & \longleftarrow\!\circ & \overline{Z}_i & \rightarrowtail & Z_{i-1} \\ \downarrow & \circlearrowleft & \downarrow & & \downarrow \\ Y_i & \longleftarrow\!\circ & \overline{Y}_i & \rightarrowtail & Y_{i-1} \end{array} \tag{35}$$

which consists of two commutative squares and two pseudo-commutative squares. Note that such a diagram, if it exists, is unique: the data of the maps f_i and the differentials in the chain complex give the outside of the square, and the two pseudo-commutative squares are unique (up to unique isomorphism). This is exactly the condition that this square is commutative inside $\mathbb{A}^\flat$.

The category of chain complexes over $\mathbb{A}$ and morphisms of chain complexes is denoted $\mathbf{Ch}_{\mathbb{A}}$.

EXAMPLE 36. In the case of finite sets, this is expressing the following observation. Suppose that we are given four subsets X_i, X_{i-1}, Y_i, and Y_{i-1} of some set A. Then it must be the case that

$$(X_i \cap X_{i-1}) \cap (Y_i \cap Y_{i-1}) = (X_i \cap Y_i) \cap (X_{i-1} \cap Y_{i-1}).$$

This quadruple intersection is exactly $\overline{Z}_i$. The compatibility condition above is reverse-engineering this situation: suppose that we know that there exist sets X_i, X_{i-1}, Y_i, Y_{i-1} and we're given pairwise-intersection data $X_i \cap X_{i-1}$, $Y_i \cap Y_{i-1}$, $X_i \cap Y_i$ and $X_{i-1} \cap Y_{i-1}$ (with the additional assumption that $X_i \cap Y_{i-1}$ and $X_{i-1} \cap Y_i$ are as small as possible). Is it possible that such an A exists? The answer is yes exactly when a diagram as in (35) exists.

Pictorially, we can view a morphism of chain complexes as on the left of Figure 3.3, where each oval consists of the sets X_i and Y_i overlapping on Z_i as illustrated on the right of Figure 3.3.

While this picture is somewhat complicated, we will see that it becomes easier to work with the pictures (and the morphisms themselves) when horizontal and vertical morphisms of chain complexes are treated separately rather than combined into a span.

REMARK 37. We can also interpret our morphisms of chain complexes of sets by interpreting the morphisms in $\mathbf{FinSet}^\flat$ as injective partial functions, where the backwards monomorphism is the inclusion of the domain of definition of a partial function. Suppose that we are given a square of injective partial functions as below.

$$\begin{array}{ccc} X_i & \xrightarrow{d_i} & X_{i-1} \\ {\scriptstyle f_i}\downarrow & & \downarrow{\scriptstyle f_{i-1}} \\ Y_i & \xrightarrow[d_i']{} & Y_{i-1} \end{array}$$

The composite $f_{i-1}d_i$ is given by the map

$$x \mapsto \begin{cases} * & \text{if } x \in X_i \backslash \overline{X}_i \\ d_i x & \text{if } x \in \overline{X}_i \end{cases} \mapsto \begin{cases} * & \text{if } x \in X_i \backslash \overline{X}_i \\ * & \text{if } x \in \overline{X}_i \text{ and } d_i x \in X_{i-1} \backslash Z_{i-1} \\ f_{i-1} d_i x & \text{if } x \in \overline{X}_i \text{ and } d_i x \in Z_{i-1} \end{cases}$$

where $\overline{X}_i$ denotes the domain of definition of d_i and Z_{i-1} the domain of definition of f_{i-1}. Then, the domain of definition of the composite $f_{i-1}d_i$ is precisely the pullback

$$\begin{array}{ccc} \overline{Z}_i & \rightarrowtail & Z_{i-1} \\ \circ\!\!\downarrow & \lrcorner & \circ\!\!\downarrow \\ \overline{X}_i & \rightarrowtail & X_{i-1} \end{array}$$

The requirement that the chain map f commutes with the differentials amounts to saying that the composites $f_{i-1}d_i$ and $d_i'f_i$ have the same domain of definition and agree on that domain, which can be expressed as the diagram given in Definition 34.

We compare this definition to the usual notion for abelian categories.

EXAMPLE 38. In an abelian category $\mathcal{A}$, a morphism of chain complexes $f_\bullet \colon X_\bullet \longrightarrow Y_\bullet$ consists of a sequence of morphisms $f_i \colon X_i \longrightarrow Y_i$ such that for all i the diagram

$$\begin{array}{ccc} X_i & \xrightarrow{d_i} & X_{i-1} \\ {\scriptstyle f_i}\downarrow & & \downarrow{\scriptstyle f_{i-1}} \\ Y_i & \xrightarrow[d_i']{} & Y_{i-1} \end{array}$$

commutes. To compare it with our ACGW definition, we factor the maps as epi-mono in order to express things in ACGW language. First, we can factor the differentials to obtain a commutative diagram

(39)

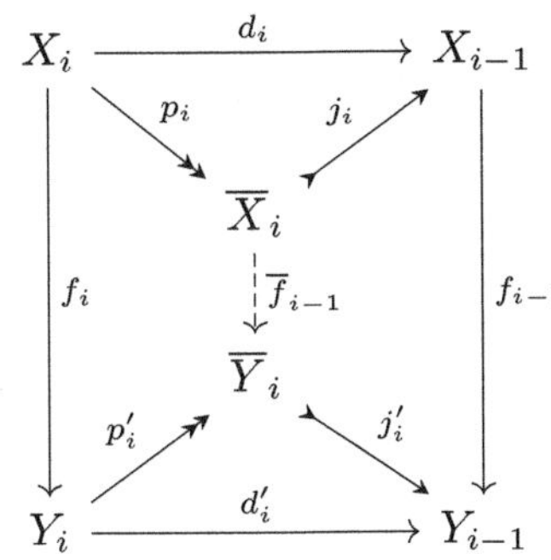

The dashed map $\overline{f}_{i-1}$ is simply the restriction of f_{i-1} to $\overline{X}_i = \operatorname{im} d_i$; the equation $f_{i-1}d_i = d'_i f_i$ ensures that the image of $\overline{f}_{i-1}$ lies inside of $\overline{Y}_i = \operatorname{im} d'_i$ as implicitly claimed in the diagram. Then, we further factor the vertical morphisms as depicted in the commutative diagram

(40)

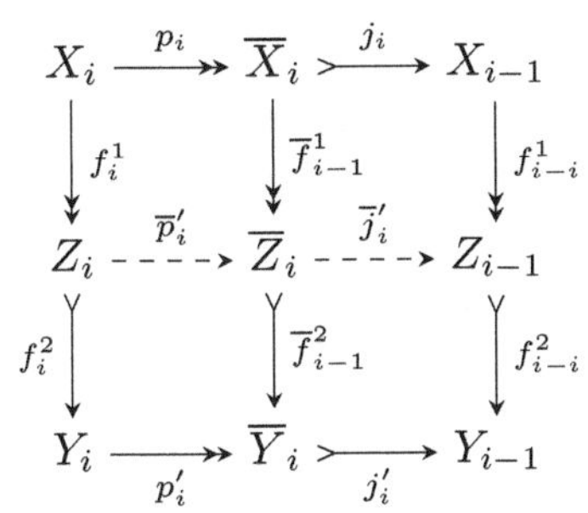

The dashed maps are the restrictions of the corresponding maps to the images, just as in the previous diagram. Finally, we have that $\overline{p}'_i$ is an epi, since $\overline{p}'_i f_i^1$ equals $\overline{f}^1_{i-1} p_i$ which is an epi. Similarly $\overline{j}'_i$ is a mono, as $f^2_{i-1}\overline{j}'_i$ equals $j'_i \overline{f}^2_{i-1}$ which is a mono. We can now see that this picture matches precisely our definition of morphism of complexes of sets.

Now that we have a definition of a morphism of chain complexes, we can use it to define an ACGW structure on $\mathbf{Ch}_{\mathbb{A}}$ as follows.

DEFINITION 41. The ACGW category $\mathbb{C}\mathbf{h}_{\mathbb{A}}$ has

objects: objects of $\mathbf{Ch}_{\mathbb{A}}$,

horizontal morphisms: a horizontal morphism $X_\bullet \rightarrowtail Y_\bullet$ is a sequence of horizontal morphisms $X_i \rightarrowtail Y_i$ for which there exists a horizontal morphism $\overline{X}_i \rightarrowtail \overline{Y}_i$ such that the resulting diagram

(42)

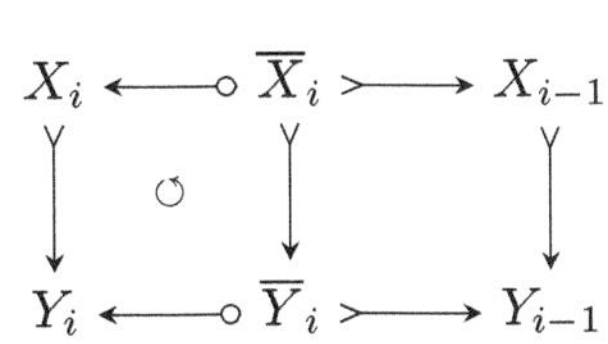

consists of a pseudo-commutative square and a commutative square.

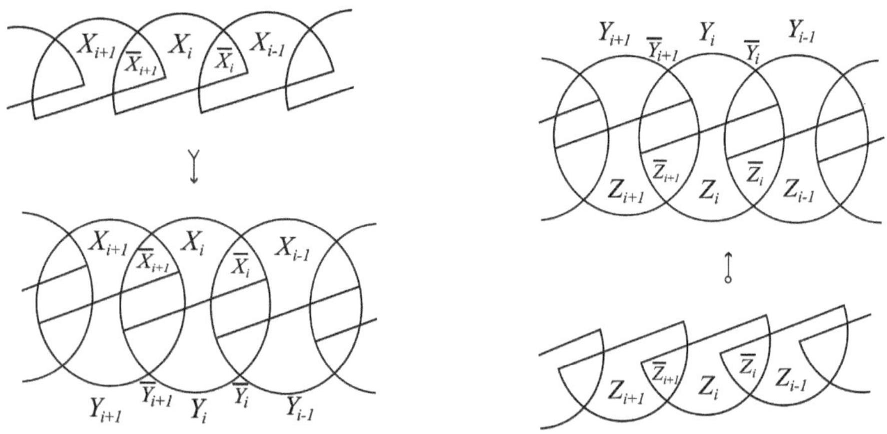

FIGURE 3.4. Examples of morphisms

vertical morphisms: a vertical morphism $Z_\bullet \circ\!\!\rightarrow Y_\bullet$ is a sequence of vertical morphisms $Z_i \circ\!\!\rightarrow Y_i$ for which there exists a vertical morphism $\overline{Z}_i \circ\!\!\rightarrow \overline{Y}_i$ such that the resulting diagram

$$
\begin{array}{ccccc}
Z_i & \leftarrow\!\!\circ & \overline{Z}_i & \rightarrowtail & Z_{i-1} \\
\downarrow{\scriptstyle\circ} & & \downarrow{\scriptstyle\circ} & \circlearrowleft & \downarrow{\scriptstyle\circ} \\
Y_i & \leftarrow\!\!\circ & \overline{Y}_i & \rightarrowtail & Y_{i-1}
\end{array}
\tag{43}
$$

consists of a commutative square and a pseudo-commutative square.

pseudo-commutative squares: squares which are pseudo-commutative at each i and each "halfway point."

We choose not to include the proof of the ACGW axioms, which is very technical and proceeds analogously to that of [**Sar21**, Theorem 5.6.16].

EXAMPLE 44. In the case of an abelian category, the horizontal and vertical morphisms of chain complexes are natural monomorphisms and (reversed) epimorphisms. In fact more generally, the horizontal and vertical morphisms of chain complexes are exactly the morphisms of chain complexes which, rather than consisting of a span at each level, contain only horizontal or vertical morphisms.

EXAMPLE 45. In the case of sets, horizontal and vertical morphisms of chain complexes can be pictured as in (46) on the left and right, respectively; a pictorial version is in Figure 3.4.

$$
\begin{array}{ccccc}
X_i & \leftarrow\!\!\circ & \overline{X}_i & \rightarrowtail & X_{i-1} \\
\downarrow{\scriptstyle\curlyvee} & \circlearrowleft & \downarrow{\scriptstyle\curlyvee} & & \downarrow{\scriptstyle\curlyvee} \\
Y_i & \leftarrow\!\!\circ & \overline{Y}_i & \rightarrowtail & Y_{i-1}
\end{array}
\qquad\qquad
\begin{array}{ccccc}
Z_i & \leftarrow\!\!\circ & \overline{Z}_i & \rightarrowtail & Z_{i-1} \\
\downarrow{\scriptstyle\circ} & & \downarrow{\scriptstyle\circ} & \circlearrowleft & \downarrow{\scriptstyle\circ} \\
Y_i & \leftarrow\!\!\circ & \overline{Y}_i & \rightarrowtail & Y_{i-1}
\end{array}
\tag{46}
$$

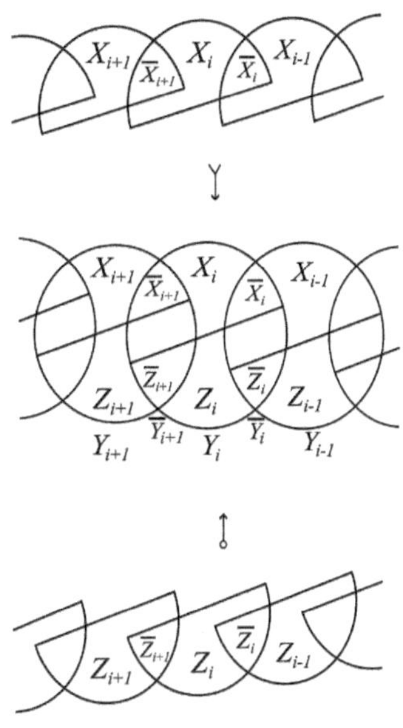

FIGURE 3.5. A short exact sequence

In both cases there is a levelwise inclusion of chain complexes which commutes with the inclusions in the differentials, but horizontal morphisms require $\overline{X}_i$ to be the entire intersection of X_i and $\overline{Y}_i$, while vertical morphisms instead require $\overline{Z}_i$ to be the intersection of Z_{i-1} and $\overline{Y}_i$. These pictures provide a convenient tool for keeping track of this, where in a chain complex pictured as overlapping circles the image of a horizontal morphism behaves as the top half of the up-right directed lines and the image of a vertical morphism behaves as the bottom half.

These pictures also illustrate how to take the complement of a horizontal or vertical morphism, since the inclusion of the levelwise complement at each level is evidently of the opposite form to the original morphism. Moreover, a chain complex $Y_\bullet$ partitioned as in Figure 3.5 can be regarded as a short exact sequence $X_\bullet \rightarrowtail Y_\bullet \leftarrow\!\circ\, Z_\bullet$ (or "extension" or "complementary pair"). These complements are not computed levelwise on the intersections, since each $\overline{Y}_i$ may include more than the union of $\overline{X}_i$ and $\overline{Z}_i$, but as the picture shows in the complement of the horizontal morphism $X_\bullet \rightarrowtail Y_\bullet$ the intersections can be computed by intersecting $\overline{Y}_i \backslash \overline{X}_i$ with $Y_{i-1} \backslash X_{i-1}$, and similarly for the complement of a vertical morphism.

We conclude this section with an alternative description of a morphism of chain complexes that factors them into horizontal and vertical morphisms. As we've explained previously, our spans of sets $X \leftarrow\!\circ\, Z \rightarrowtail Y$ encode partially defined inclusions $X \longrightarrow Y$. The same is true for chain complexes, as we show in the following result.

LEMMA 47. *A morphism of chain complexes $f_\bullet \colon X_\bullet \longrightarrow Y_\bullet$ consists of a span of inclusions of chain complexes*

$$X_\bullet \longleftarrow\!\circ\, Z_\bullet \rightarrowtail Y_\bullet.$$

More formally,

$$(\mathbb{C}\mathbf{h}_{\mathbb{A}})^{\flat} \cong \mathbf{Ch}_{\mathbb{A}}.$$

PROOF. The proof of this is straightforward, amounting only to checking that the middle row of the diagram in (48) defining a morphism of chain complexes $f_\bullet$ does in fact satisfy the chain condition.

$$\begin{array}{ccccc} X_i & \longleftarrow\!\circ & \overline{X}_i & \rightarrowtail & X_{i-1} \\ {\scriptstyle f_i^1}\uparrow & & \uparrow{\scriptstyle \overline{f}^1_{i-1}} & & \uparrow{\scriptstyle f^1_{i-i}} \\ Z_i & \longleftarrow\!\circ & \overline{Z}_i & \rightarrowtail & Z_{i-1} \\ {\scriptstyle f_i^2}\downarrow & & \downarrow{\scriptstyle \overline{f}^2_{i-1}} & & \downarrow{\scriptstyle f^2_{i-i}} \\ Y_i & \longleftarrow\!\circ & \overline{Y}_i & \rightarrowtail & Y_{i-1} \end{array} \tag{48}$$

□

4. Recovering classical results

In this section we study the notions of chain complexes, exact sequences, and homology of finite sets introduced in Section 3, to recover several classical results present in the context of abelian categories.

4.1. The Snake Lemma. Our goal in this subsection is to prove a version of the Snake Lemma for finite sets. We begin by recalling the classical statement of the Snake Lemma; a proof can be found for instance in [**Rot09**, Corollary 6.12].

THEOREM 49 (Snake Lemma). *Let $\mathcal{A}$ be an abelian category and consider a morphism of exact sequences*

$$\begin{array}{ccccccc} & & A & \longrightarrow & B & \twoheadrightarrow & C & \longrightarrow & 0 \\ & & \downarrow{\scriptstyle f} & & \downarrow{\scriptstyle g} & & \downarrow{\scriptstyle h} & & \\ 0 & \longrightarrow & A' & \rightarrowtail & B' & \longrightarrow & C' & & \end{array} \tag{50}$$

Then, there exists an exact sequence

$$\ker f \longrightarrow \ker g \longrightarrow \ker h \longrightarrow \operatorname{coker} f \longrightarrow \operatorname{coker} g \longrightarrow \operatorname{coker} h.$$

Moreover, if $A \longrightarrow B$ is a mono, then so is $\ker f \longrightarrow \ker g$, and if $B' \longrightarrow C'$ is an epi, then so is $\operatorname{coker} g \longrightarrow \operatorname{coker} h$.

In the case of finite sets, it will be easier to prove the version with a stronger assumption first, where we start with two short exact sequences. Using the description of short exact sequences from Example 32 and our definition of morphisms, we find that the statement of the Snake Lemma translates to the following.

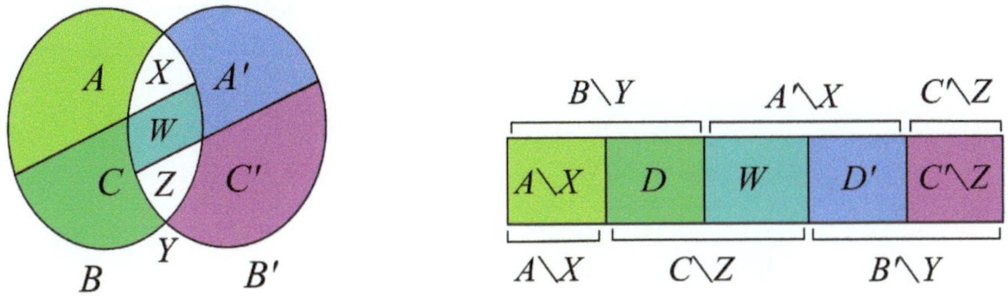

FIGURE 4.1. The Weak Snake Lemma

THEOREM 51 (Weak Snake Lemma for ACGW categories). *Given a morphism of short exact sequences in $\mathbb{A}$ as in*

(52)
$$\begin{array}{ccccc} A & \rightarrowtail & B & \leftarrow\!\!\circ & C \\ \uparrow\!\circ & \circlearrowleft & \uparrow\!\circ & & \uparrow\!\circ \\ X & \rightarrowtail & Y & \leftarrow\!\!\circ & Z \\ \downarrow & & \downarrow & \circlearrowleft & \downarrow \\ A' & \rightarrowtail & B' & \leftarrow\!\!\circ & C' \end{array}$$

there is an exact sequence of the form

(53)
$$\begin{array}{ccccccccccc} & & & D & & W & & D' & & & \\ & & \swarrow\!\circ & & \searrow\ \ \circ\!\swarrow & & \searrow\ \ \circ\!\swarrow & & \searrow & & \\ A\backslash\backslash X & \rightarrowtail & B\backslash\backslash Y & & C\backslash\backslash Z & & A'/\!/X & & B'/\!/Y & \leftarrow\!\!\circ & C'/\!/Z \end{array}$$

Before we prove this, we can summarize the essence of the proof with the picture on the left in Figure 4.1 which shows this morphism of short exact sequences in the ACGW category $\mathbb{F}$**inSet**. The exact sequence, which as in Example 31 is a sequence of subsets in the diagram whose successive unions agree with the bottom row of the zigzag diagram in the theorem, is displayed in the pictures as the sequence $A\backslash X, D = C\backslash(Y\backslash X), W = (Y\backslash X)\backslash Z, D' = A'\backslash(Y\backslash Z), C'\backslash Z$.

The goal of the proof is therefore to construct D, W, and D' in the ACGW language along with the desired maps to $B\backslash\backslash Y, C\backslash\backslash Z, A'/\!/X$, and $B'/\!/Y$.

PROOF. Taking the kernel and cokernel of the top-left square in the diagram in both directions we get a diagram of pullbacks and pseudo-commutative squares

(54)

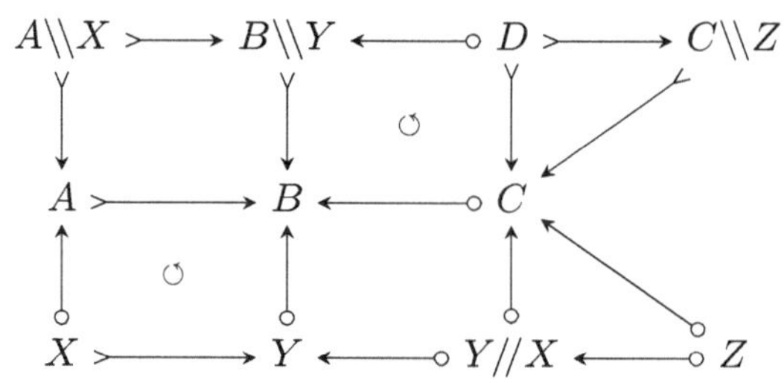

where the top row contains the beginning of our desired exact sequence. The horizontal morphism $D \rightarrowtail C\backslash\backslash Z$ is induced by the vertical morphism $Y//X \leftarrow\!\circ\, Z$ as $D \cong C\backslash\backslash(Y//X)$, based on the diagram of kernels and cokernels in (55), where we denote $W = (Y//X)\backslash\backslash Z$ (more details on how to construct this induced morphism can be found in [**CZ22**, Lemma 2.10]).

(55)

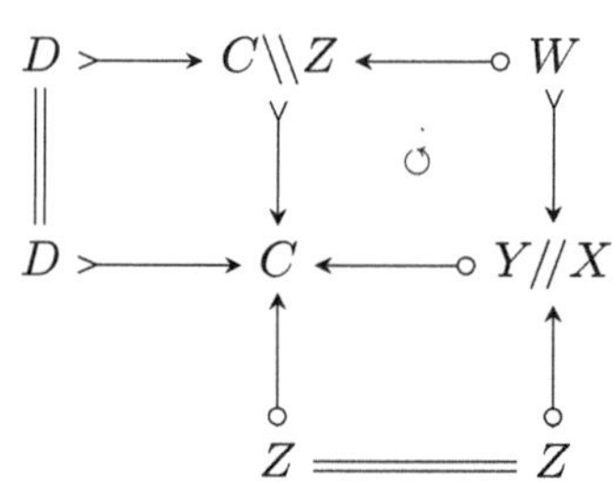

In order to show that the kernel of $W \circ\!\!\rightarrow C\backslash\backslash Z$ is naturally isomorphic to the kernel of $Y//X \circ\!\!\rightarrow C$, it suffices to check that the cokernel of $W \rightarrowtail Y//X$ is naturally isomorphic to the cokernel of $C\backslash\backslash Z \rightarrowtail C$. (This is because the upper-right square is "distinguished"; see [**CZ22**, Section 2] for a more in-depth discussion.) As this is true, we are done; note that the sequence so far is exact by construction.

The construction of D' is entirely dual to this, namely as $A'//(Y\backslash\backslash Z)$. All that remains then is to show that W can be equivalently constructed as $(Y//X)\backslash\backslash Z$ as above or dually as $(Y\backslash\backslash Z)//X$, but these are the two equivalent definitions of the homology of the sequence $X \rightarrowtail Y \leftarrow\!\circ\, Z$ from Definition 22, so the proof is complete. □

The stronger form of the Snake Lemma in an ACGW category is as follows, where the short exact sequences are replaced by sequences which are no longer exact at A and C'.

THEOREM 56 (Strong Snake Lemma for ACGW categories). *Given a morphism of exact sequences in an ACGW category* $\mathbb{A}$ *as in*

$$\begin{array}{ccccccccc} A & \leftarrow\!\!\circ & \overline{A} & \rightarrowtail & B & \leftarrow\!\!\circ & C & = & C \\ \uparrow\circ & & \uparrow\circ & \circlearrowleft & \uparrow\circ & & \uparrow\circ & \circlearrowleft & \uparrow\circ \\ X & = & X & \rightarrowtail & Y & \leftarrow\!\!\circ & Z & = & Z \\ \downarrow & \circlearrowleft & \downarrow & & \downarrow & \circlearrowleft & \downarrow & & \downarrow \\ A' & = & A' & \rightarrowtail & B' & \leftarrow\!\!\circ & \overline{C}' & \rightarrowtail & C' \end{array} \tag{57}$$

there is an exact sequence of the form

$$\begin{array}{ccccccccccc} & \overline{A}\backslash\backslash X & & D & & W & & D' & & \overline{C}'/\!/Z & \\ \swarrow\circ & & \searrow & \circ\swarrow & \searrow & \circ\swarrow & \searrow & \circ\swarrow & \searrow & \circ\swarrow & \searrow \\ A\backslash\backslash X & & B\backslash\backslash Y & & C\backslash\backslash Z & & A'/\!/X & & B'/\!/Y & & C'/\!/Z \end{array} \tag{58}$$

In this case, the picture in $\mathbb{F}$**inSet** is very similar to that for the weak version of the Snake Lemma, but allows for the exact sequence to continue beyond the union of B and B', as depicted in Figure 4.2.

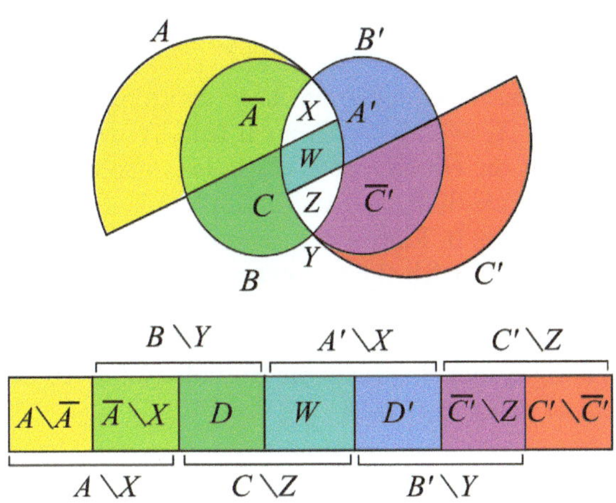

FIGURE 4.2. The Strong Snake Lemma

PROOF. Applying Theorem 51 to the middle two columns of the diagram gives all but the leftmost and rightmost morphisms in the desired exact sequence. The remaining maps are those induced on kernels and cokernels by the pullback squares in the upper left and lower right corners of the diagram in the theorem statement. $\square$

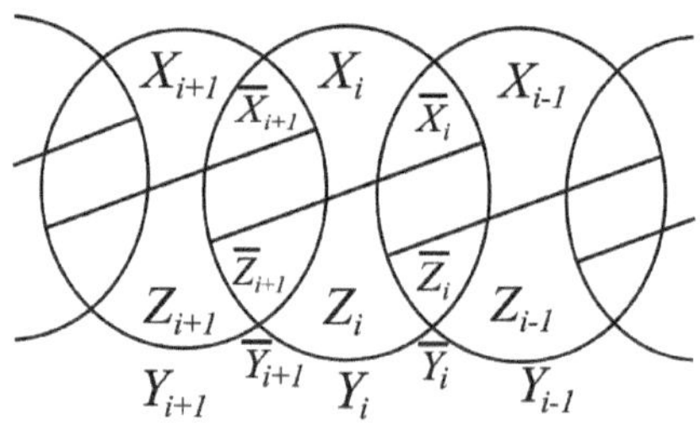

FIGURE 4.3. A chain partition

4.2. Long Exact Sequence in Homology. In this subsection, we will show how a short exact sequence of chain complexes induces a long exact sequence in homology. In order to be more precise, and to get the correct formulation in the context of finite sets, we start by recalling the classical result for abelian categories, a proof of which can be found for instance in [**Rot09**, Theorem 6.10].

THEOREM 59 (Long Exact Sequence in Homology). *Let $\mathcal{A}$ be an abelian category. Then, every short exact sequence of chain complexes in $\mathcal{A}$*

$$0 \longrightarrow X_\bullet \hookrightarrow Y_\bullet \twoheadrightarrow Z_\bullet \longrightarrow 0$$

induces a long exact sequence in homology

$$\cdots \longrightarrow H_{i+1}(Z_\bullet) \longrightarrow H_i(X_\bullet) \longrightarrow H_i(Y_\bullet) \longrightarrow H_i(Z_\bullet) \longrightarrow H_{i-1}(X_\bullet) \longrightarrow \cdots.$$

Recall that a short exact sequence of chain complexes in an ACGW category $\mathbb{A}$ is a pair of morphisms $X_\bullet \rightarrowtail Y_\bullet \leftarrow\!\circ\, Z_\bullet$ such that $X_\bullet$ is isomorphic to the kernel of the vertical morphism or, equivalently, $Z_\bullet$ is isomorphic to the cokernel of the horizontal morphism. As we discussed in Example 45, in $\mathbb{F}$**inSet** this short exact sequence can be visualized as in Figure 4.3, where each set Y_i is partitioned into X_i and Z_i in a manner consistent with the respective horizontal and vertical morphisms of chain complexes.

THEOREM 60 (Long Exact Sequence in Homology for ACGW categories). *An exact sequence of chain complexes in $\mathbb{A}$*

$$X_\bullet \rightarrowtail Y_\bullet \leftarrow\!\circ\, Y_\bullet/\!/X_\bullet$$

induces a long exact sequence (i.e. an exact chain complex) in homology

(61)
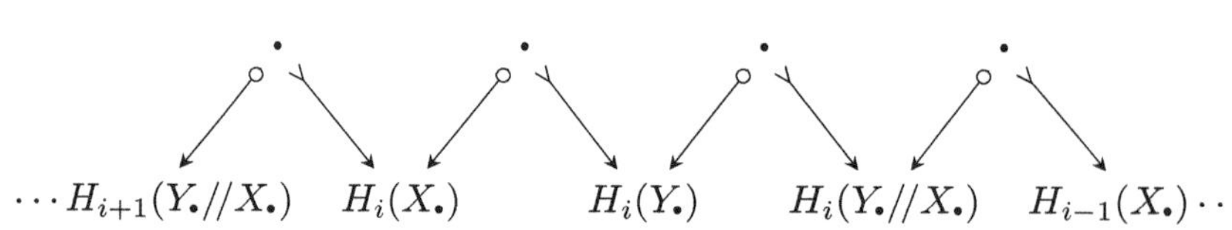

Setting $Z_\bullet := Y_\bullet/\!/X_\bullet$, this exact sequence can be visualized as in Figure 4.4.

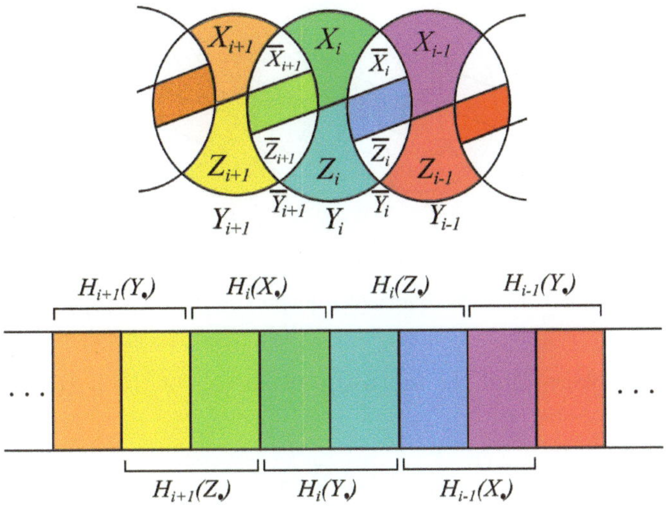

FIGURE 4.4. A long exact sequence

PROOF. Let $Z_\bullet := Y_\bullet // X_\bullet$. The data making up the exact sequence includes the diagram

$$
\begin{array}{ccccccccc}
\overline{X}_{i+1} & \rightarrowtail & \overline{Y}_{i+1}\backslash\backslash\overline{Z}_{i+1} & \rightarrowtail & \overline{Y}_{i+1} & \xleftarrow{\circ} & \overline{Z}_{i+1} & & \\
& \searrow & \downarrow & & \downarrow & \circlearrowleft & \downarrow & & \\
& & X_i & \rightarrowtail & Y_i & \xleftarrow{\circ} & Z_i & & \\
& & \uparrow & \circlearrowleft & \uparrow & & \uparrow & & \\
& & \overline{X}_i & \rightarrowtail & \overline{Y}_i & \xleftarrow{\circ} & \overline{Z}_i & & \\
& & \downarrow & & \downarrow & \circlearrowleft & \downarrow & & \\
& & X_{i-1} & \rightarrowtail & Y_{i-1} & \xleftarrow{\circ} & Z_{i-1} & & \\
& & \uparrow & \circlearrowleft & \uparrow & & \uparrow & \nwarrow & \\
& & \overline{X}_{i-1} & \rightarrowtail & \overline{Y}_{i-1} & \xleftarrow{\circ} & \overline{Y}_{i-1}//\overline{X}_{i-1} & \xleftarrow{\circ} & \overline{Z}_{i-1}
\end{array}
\tag{62}
$$

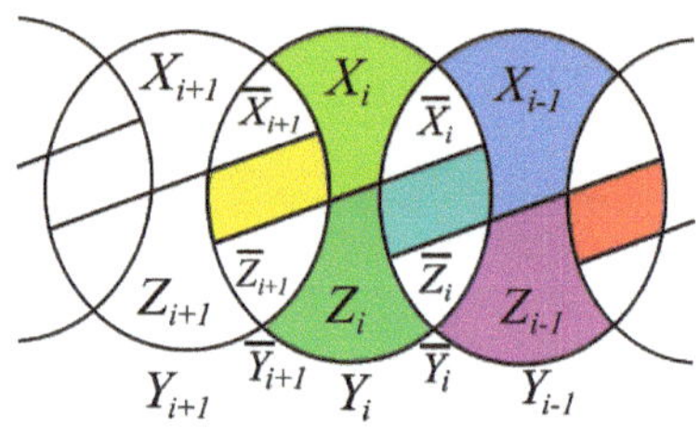

FIGURE 4.5. Picture for the diagram in (63)

which by taking cokernels of the top row and kernels of the bottom row induces a diagram

(63)

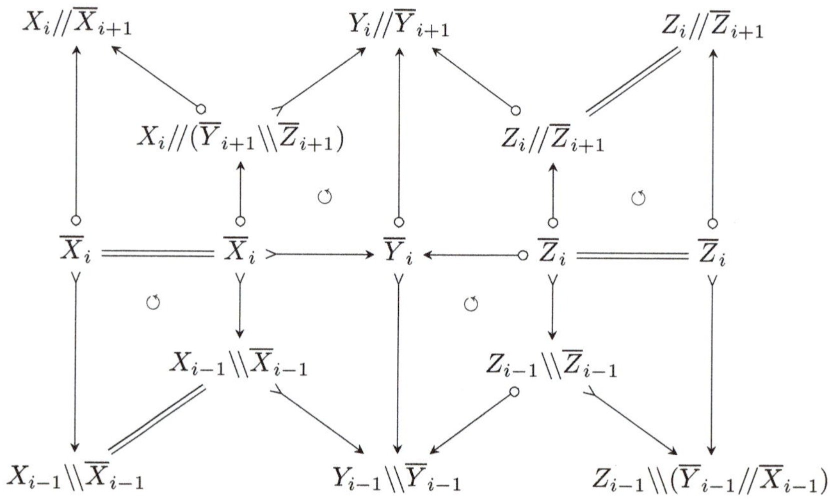

Here, the vertical map $X_i//\left(\overline{Y}_{i+1}\backslash\backslash\overline{Z}_{i+1}\right) \circ\!\!\rightarrow X_i//\overline{X}_{i+1}$ is induced by the horizontal map $\overline{X}_{i+1} \rightarrowtail \overline{Y}_{i+1}\backslash\backslash\overline{Z}_{i+1}$ in (62) by using [**CZ22**, Lemma 2.10] in a manner identical to that of the proof of Theorem 51.

This diagram is precisely of the form required in Theorem 56, and by definition of homology (Definition 22) the resulting exact sequence is between the homologies of the chain complexes as in

(64)

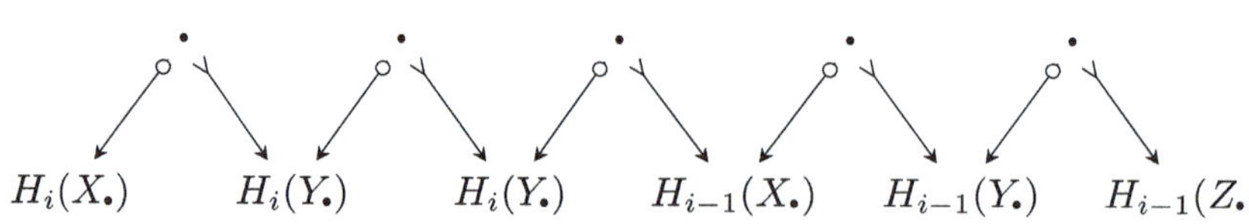

Piecing together these diagrams for all i, which can be straightforwardly checked to agree on the overlapping portions, gives the desired long exact sequence. □

To further illustrate how the strong version of the Snake Lemma applies here in the case of **FinSet**, Figure 4.5 depicts the diagram in (63), colored analogously to Figure 4.2.

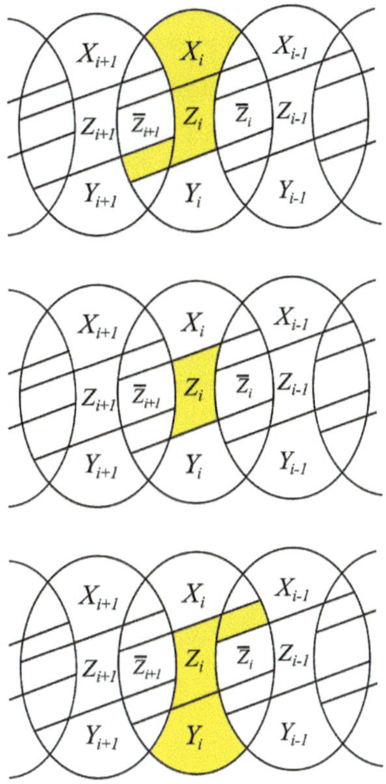

FIGURE 4.6. A morphism on homology

4.3. Homology as a functor. In Section 3, we defined homology as an assignment that takes a chain complex $X_\bullet$ in an ACGW category $\mathbb{A}$ and produces an object $H_i(X_\bullet)$ of $\mathbb{A}$ for each i. In fact, this assignment extends to maps in a functorial way, as we show in this subsection.

PROPOSITION 65. *A morphism of chain complexes $f_\bullet\colon X_\bullet \longrightarrow Y_\bullet$ in $\mathbf{Ch}_{\mathbb{A}}$ induces a morphism in $\mathbb{A}^\flat$*

$$H_i(X_\bullet) \longleftarrow\!\circ \bullet \rightarrowtail H_i(Y_\bullet)$$

Before we formally construct this span, consider Figure 3.3 illustrating a morphism of chain complexes in $\mathbb{F}$**inSet**. On the top and bottom in Figure 4.6 are $H_i(X_\bullet)$ and $H_i(Y_\bullet)$ shaded in that picture. Evidently from the pictures, these two "intersect" in the sense that $Z_i \backslash (\overline{X}_i \cup \overline{Y}_{i+1})$ includes into both and is the maximal subset of Z_i to do so (in the center of Figure 4.6).[4] The proof then merely amounts to constructing this subset in the language of ACGW categories, which we can do using Theorem 60.

[4]Due to the limitations of picture labeling, the center shaded region, while containing the label Z_i, does not make up all of Z_i which includes all of the center oval between the two diagonal lines.

PROOF. A morphism of chain complexes of the form $X_\bullet \leftarrow\!\circ\, Z_\bullet \rightarrowtail Y_\bullet$ induces two short exact sequences of chain complexes,

$$X_\bullet\backslash\backslash Z_\bullet \rightarrowtail X_\bullet \leftarrow\!\circ\, Z_\bullet \qquad \text{and} \qquad Z_\bullet \rightarrowtail Y_\bullet \leftarrow\!\circ\, Y_\bullet/\!/Z_\bullet.$$

By Theorem 60, these induce long exact sequences in homology which include, respectively, spans of the form

$$H_i(X_\bullet) \leftarrow\!\circ \bullet \rightarrowtail H_i(Z_\bullet) \qquad \text{and} \qquad H_i(Z_\bullet) \leftarrow\!\circ \bullet \rightarrowtail H_i(Y_\bullet).$$

We can then compose these two spans in the category $\mathbb{F}\mathbf{inSet}^\flat$ (that is, take the square completion of the "cospan" given by the two morphisms into $H_i(Z_\bullet)$, as given by the last item in Definition 5) to get a span $H_i(X_\bullet) \leftarrow\!\circ \bullet \rightarrowtail H_i(Y_\bullet)$. □

EXAMPLE 66. It is not possible to restrict H_i to a functor which lands in $\mathbb{A}$, even when restricting to horizontal (resp. vertical) morphisms in $\mathbb{C}\mathbf{h}_\mathbb{A}$. In other words, a horizontal chain morphism $X_\bullet \rightarrowtail Y_\bullet$ does *not* necessarily induce a horizontal morphism $H_i(X_\bullet) \rightarrowtail H_i(Y_\bullet)$. For instance, if we consider the map between chain complexes in (67) of finite sets concentrated in degrees 1,2 and 3

$$\begin{array}{ccccccccc} \emptyset & = & \emptyset & \rightarrowtail & \{a\} & \leftarrow\!\circ & \emptyset & = & \emptyset \\ \downarrow & & \downarrow & & \downarrow & & \downarrow & & \downarrow \\ \{a\} & = & \{a\} & \rightarrowtail & \{a,b\} & \leftarrow\!\circ & \{b\} & = & \{b\} \end{array} \tag{67}$$

we see that $H_2(X_\bullet) = \{a\}$ while $H_2(Y_\bullet) = \emptyset$. Thus there is no horizontal morphism $H_2(X_\bullet) \rightarrowtail H_2(Y_\bullet)$, even though there is a morphism in $\mathbf{Ch}^\flat_{\mathbb{F}\mathbf{inSet}}$.

Just as in the classical setting [**Rot09**, Proposition 6.8], taking the i-th homology is a functorial process, as we now show.

THEOREM 68. *Homology is a functor* $\mathbb{C}\mathbf{h}_\mathbb{A} \longrightarrow \mathbb{A}^\flat$.

PROOF. We have already defined homology on objects in Definition 22, and on morphisms in Proposition 65. A careful study of the proof of Proposition 65 reveals that if the morphism of chain complexes we start with is the identity, then it will be mapped to the identity span on homology (i.e. the equivalence class of the span whose legs are both identity maps). It remains to show that this assignment respects composition.

Moreover as morphisms of chain complexes $X_\bullet \longrightarrow Y_\bullet$ factor as composites of the form $X_\bullet \leftarrow\!\circ\, Z_\bullet \rightarrowtail Y_\bullet$, and the span associated to a morphism in Proposition 65 is defined as the corresponding composite of spans in $\mathbb{A}^\flat$, it suffices to check that composition is respected for composites of only horizontal morphisms. The argument for vertical morphisms will be entirely dual.

Let $f\colon X_\bullet \rightarrowtail Y_\bullet$ and $g\colon Y_\bullet \rightarrowtail Z_\bullet$ be horizontal morphisms in $\mathbb{C}\mathbf{h}_\mathbb{A}$; we will show that $H_i(g) \circ H_i(f) = H_i(g \circ f)$. Write $X_i' = X_i/\!/\overline{X}_i$, so that $H_i(X) = X_i'\backslash\backslash\overline{X}_{i+1}$ (and similarly for $H_i(Y)$ and $H_i(Z)$). To check that composition is respected, it suffices to check that it is respected when $H_i(f)$ is a single horizontal morphism, and $H_i(g)$ is a single vertical morphism (instead of a general span). The former happens when $\overline{X}_{i+1} = \overline{Y}_{i+1} \times_{Y_i'} X_i'$; the latter happens when $Y_i = Z_i$ (and thus $Y_i' = Z_i'$). We therefore focus on this case. In this case, the data of the two compositions reduces to the diagram in (69), where $H_i(g) \circ H_i(f)$ is the composite around the lower-left of the rightmost square, and $H_i(g \circ f)$ is the span around the upper-right of the rightmost square.

$$
\begin{array}{ccccccccc}
\overline{Y}_{i+1} \times_{Y_i} X'_i & \rightarrowtail & \overline{Z}_{i+1} \times_{Z'_i} X'_i & \rightarrowtail & X'_i & \leftarrow\!\circ & H_i(X) & \leftarrow\!\circ & X_i \backslash\backslash (\overline{Z}_{i+1} \times_{Z_i} X'_i) \\
\downarrow & & \downarrow \lrcorner & & \downarrow & \circlearrowleft & \downarrow & & \downarrow \\
\overline{Y}_{i+1} & \rightarrowtail & \overline{Z}_{i+1} & \rightarrowtail & Z'_i & \leftarrow\!\circ & H_i(Y) & \leftarrow\!\circ & H_i(Z)
\end{array}
\tag{69}
$$

Thus it suffices to check that the rightmost square is pseudo-commutative. The composite of the two squares on the left is a pullback square, and the composite of the two squares on the right is a pseudo-commutative square. The statement that H_i respects composition is then the statement that this implies that the rightmost square in the diagram is a pseudo-commutative square. This is part of the naturality of the duality between pullback squares as on the left and pseudo-commutative squares as on the right; on the left, the leftmost square must be a pullback square by a standard argument. □

4.4. Quasi-isomorphisms. Just like in the setting of abelian categories, we can now use our notion of homology to define what it means for a morphism of chain complexes of sets to be a quasi-isomorphism.

DEFINITION 70. A morphism $f_\bullet \colon X_\bullet \longrightarrow Y_\bullet$ of chain complexes in an ACGW category $\mathbb{A}$ is a *quasi-isomorphism* if it induces an isomorphism on homology; that is, if the associated span on homology is of the form

$$H_i(X_\bullet) \xleftarrow{\cong}\!\circ \bullet \overset{\cong}{\rightarrowtail} H_i(Y_\bullet)$$

for all i.

EXAMPLE 71. If $\mathbb{A}$ arises from an abelian category, this agrees with the usual definition that the induced morphism on homology $H_i(X_\bullet) \longrightarrow H_i(Y_\bullet)$ is an isomorphism for all i.

EXAMPLE 72. In $\mathbb{F}$**inSet**, a quasi-isomorphism can be pictured as a morphism of chain complexes in which all three of the highlighted subsets in Figure 4.6 agree. Unwinding this, the condition becomes that the sets $X_i \backslash \left(\overline{X}_i \cup \overline{X}_{i+1} \cup Z_i\right)$, $Y_i \backslash \left(\overline{Y}_i \cup \overline{Y}_{i+1} \cup Z_i\right)$, $\overline{Y}_{i+1} \cap Z_i \backslash \overline{Z}_{i+1}$, and $\overline{X}_i \cap Z_i \backslash \overline{Z}_i$ (those subsets highlighted on the top in Figure 4.7) are all empty.

A quasi-isomorphism can therefore be pictured as in Figure 4.7, where X_i is shaded blue, Y_i is shaded red, and Z_i which includes into both is shaded purple.

Given a morphism of chain complexes $f_\bullet \colon X_\bullet \longrightarrow Y_\bullet$ in an abelian category, in some cases it is possible to characterize whether $f_\bullet$ is a quasi-isomorphism without comparing the homologies of $X_\bullet$ and $Y_\bullet$. Namely, if we know that $f_\bullet$ is a monomorphism (respectively, epimorphism), then it is a quasi-isomorphism if and only if its cokernel (resp. kernel) is an exact chain complex. This characterization generalizes to any ACGW category, as we show in the following result.

PROPOSITION 73. *A horizontal morphism of chain complexes $f_\bullet \colon X_\bullet \rightarrowtail Y_\bullet$ is a quasi-isomorphism if and only if its cokernel is an exact chain complex. Likewise a vertical morphism of chain complexes is a quasi-isomorphism if and only if its kernel is exact.*

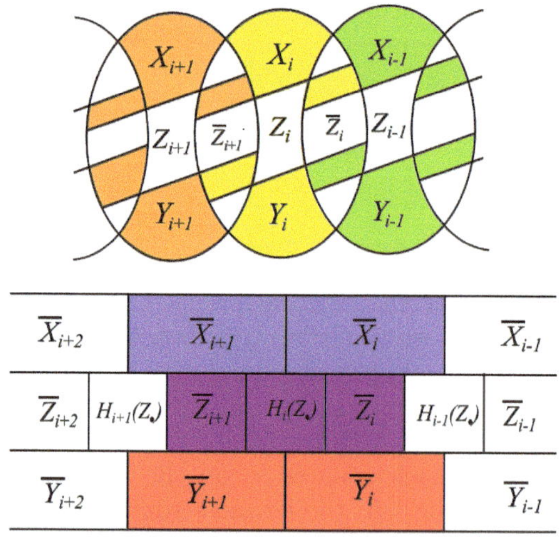

FIGURE 4.7. A quasi-isomorphism

PROOF. If we consider the short exact sequence of chain complexes

$$X_\bullet \xrightarrow{f_\bullet} Y_\bullet \longleftarrow\!\circ\; Y_\bullet//X_\bullet$$

then by Theorem 60 we get an induced long exact sequence

(74)

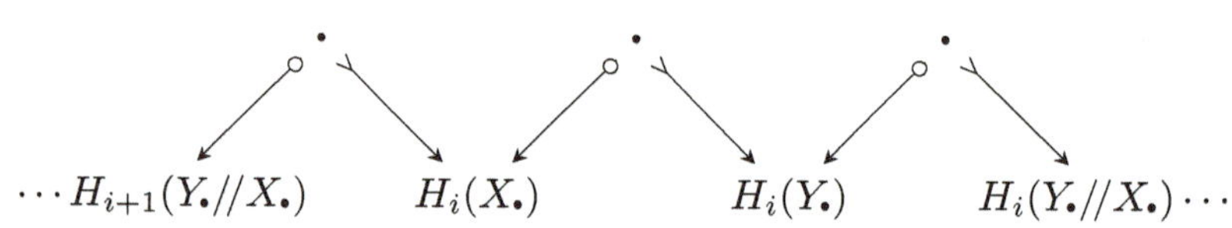

Suppose that the chain complex $Y_\bullet//X_\bullet$ is exact. Then the long exact sequence above must be of the form

(75)

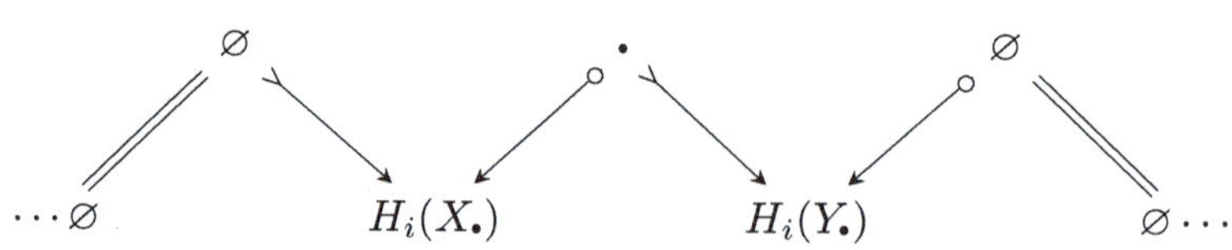

as any morphism into $\varnothing$ is an isomorphism.

As this sequence is exact, the vertical morphism $H_i(X_\bullet) \longleftarrow\!\circ\; \bullet$ must be an isomorphism as it is isomorphic to the cokernel of $\varnothing \rightarrowtail H_i(X_\bullet)$. Dually, the horizontal morphism $\bullet \rightarrowtail H_i(Y_\bullet)$ is an isomorphism as it is isomorphic to the kernel of $H_i(Y_\bullet) \longleftarrow\!\circ\; \varnothing$. This shows that $f_\bullet$ is a quasi-isomorphism.

Conversely, suppose that $f_\bullet$ is a quasi-isomorphism. Then the long exact sequence is of the form

(76)

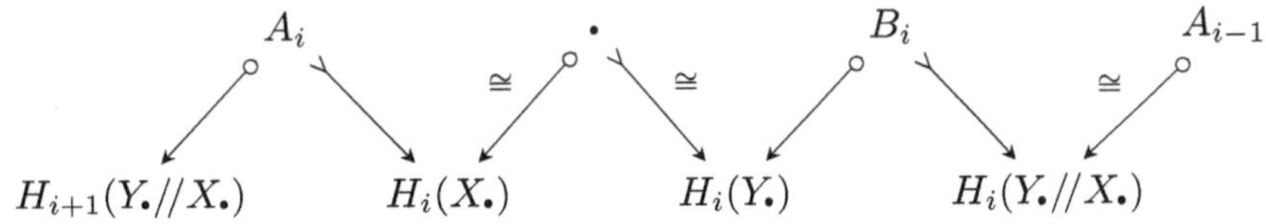

The previous argument then works in reverse, where as the kernel (resp. cokernel) of an isomorphism the object A_i (resp. B_i) must be isomorphic to $\varnothing$ for all i. We then have the short exact sequence $\varnothing \rightarrowtail H_i(Y_\bullet//X_\bullet) \leftarrow\!\circ\ \varnothing$, which by the same principle implies that both maps are isomorphisms and hence $H_i(Y_\bullet//X_\bullet) \cong \varnothing$. This shows that $Y_\bullet//X_\bullet$ is an exact chain complex.

The argument for a vertical morphism of chain complexes is entirely dual to this. ☐

REMARK 77. Let $f_\bullet \colon X_\bullet \longrightarrow Y_\bullet$ be a morphism of chain complexes, which we can express as a span

$$X_\bullet \xleftarrow{f^1_\bullet} Z_\bullet \xrightarrow{f^2_\bullet} Y_\bullet.$$

If we have that $f_\bullet$ is a quasi-isomorphism, this does *not* imply that both $f^1_\bullet$ and $f^2_\bullet$ are also quasi-isomorphisms. For instance, a counter-example is given by the span of chain complexes in (78) concentrated in degrees 1, 2 and 3.

(78)

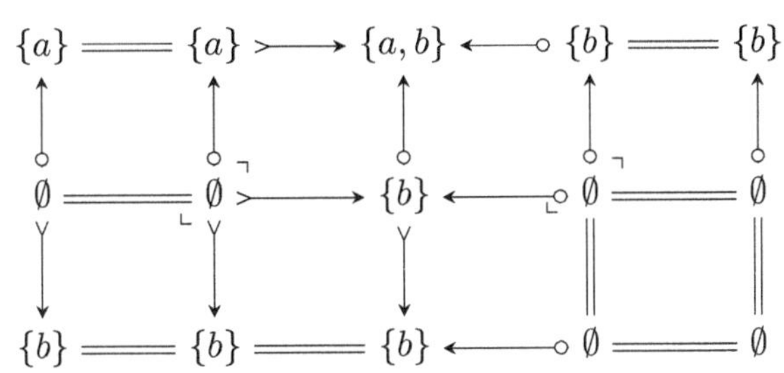

REMARK 79. In the case of finite sets, homology behaves more like chain complexes over a field, rather than chain complexes over a ring. The most striking example of this phenomenon is that, over a field, every chain complex has a quasi-isomorphism to and from a chain complex in which all differentials are zero. (This is sometimes stated as "the chain complex is quasi-isomorphic to its homology.") This can be done quite simply: for each vector space in the complex, choose a basis in such a way that it restricts to a basis of the kernel of the next differential, and to the image of the previous. Then a subset of the basis will give a basis for the homology, as well. Note that there are two possible ways of constructing a quasi-isomorphism to the homology: one by mapping the homology into the chain complex via this basis, and one quotienting out by the basis elements that do not generate the homology. (Over a ring it is generally not the case that this is possible; for a simple example consider the chain complex over $\mathbf{Z}$ which contains the single map $2 \colon \mathbf{Z}/4 \longrightarrow \mathbf{Z}/4$.)

In the case of sets this is done similarly, without the non-canonical step of choosing the basis. The homology of the chain complex of sets includes naturally into the original chain complex, giving a quasi-isomorphism. The choice of whether to make this inclusion vertical or horizontal is non-canonical, but it corresponds naturally to the choice of inclusion or quotient in the field case.

References

[BOO+18] J. E. Bergner, A. M. Osorno, V. Ozornova, M. Rovelli, and C. I. Scheimbauer, *2-Segal sets and the Waldhausen construction*, Topology Appl. **235** (2018), 445–484, DOI 10.1016/j.topol.2017.12.009. MR3760213

[BOO+21] J. E. Bergner, A. M. Osorno, V. Ozornova, M. Rovelli, and C. I. Scheimbauer, *2-Segal objects and the Waldhausen construction*, Algebr. Geom. Topol. **21** (2021), no. 3, 1267–1326, DOI 10.2140/agt.2021.21.1267. MR4299667

[CKMZ23] J. Campbell, J. Kuijper, M. Merling, and I. Zakharevich. Algebraic K-theory for squares categories, 2023. arXiv:2310.02852.

[CS24] M. Calle and M. Sarazola. Squares K-theory and 2-Segal spaces, 2024. arXiv:2409.16428.

[CW23] I. Coley and C. Weibel, *Localization, monoid sets and K-theory*, J. Algebra **614** (2023), 754–779, DOI 10.1016/j.jalgebra.2022.09.009. MR4500705

[CZ22] J. A. Campbell and I. Zakharevich, *Dévissage and localization for the Grothendieck spectrum of varieties.* part A, Adv. Math. **411** (2022), no. part A, Paper No. 108710, 80, DOI 10.1016/j.aim.2022.108710. MR4512395

[DK19] T. Dyckerhoff and M. Kapranov, *Higher Segal spaces*, Lecture Notes in Mathematics, vol. 2244, Springer, Cham, 2019, DOI 10.1007/978-3-030-27124-4. MR3970975

[Ehr63] C. Ehresmann, *Catégories structurées* (French), Ann. Sci. École Norm. Sup. (3) **80** (1963), 349–426. MR197529

[FP10] T. M. Fiore and S. Paoli, *A Thomason model structure on the category of small n-fold categories*, Algebr. Geom. Topol. **10** (2010), no. 4, 1933–2008, DOI 10.2140/agt.2010.10.1933. MR2728481

[FPP08] T. M. Fiore, S. Paoli, and D. Pronk, *Model structures on the category of small double categories*, Algebr. Geom. Topol. **8** (2008), no. 4, 1855–1959, DOI 10.2140/agt.2008.8.1855. MR2449004

[GCKT18] I. Gálvez-Carrillo, J. Kock, and A. Tonks, *Decomposition spaces, incidence algebras and Möbius inversion I: Basic theory*, Adv. Math. **331** (2018), 952–1015, DOI 10.1016/j.aim.2018.03.016. MR3804694

[GP99] M. Grandis and R. Pare, *Limits in double categories* (English, with French summary), Cahiers Topologie Géom. Différentielle Catég. **40** (1999), no. 3, 162–220. MR1716779

[GP04] M. Grandis and R. Pare, *Adjoint for double categories. Addenda to: "Limits in double categories" [Cah. Topol. Géom. Différ. Catég.* **40** *(1999), no. 3, 162–220; MR1716779]* (English, with French summary), Cah. Topol. Géom. Différ. Catég. **45** (2004), no. 3, 193–240. MR2090335

[GP19] M. Grandis and R. Paré, *Persistent double limits* (English, with English and French summaries), Cah. Topol. Géom. Différ. Catég. **60** (2019), no. 3, 255–297. MR3965316

[Gra19] M. Grandis, *Higher dimensional categories*, World Scientific Publishing Co. Pte. Ltd., Hackensack, NJ, 2020. From double to multiple categories. MR3970290

[Hac] P. Hackney. The decomposition space perspective. This volume (arXiv:2409.19061).

[HW21] C. Haesemeyer and C. A. Weibel, *The K'-theory of monoid sets*, Proc. Amer. Math. Soc. **149** (2021), no. 7, 2813–2824, DOI 10.1090/proc/15517. MR4257796

[Mos20] L. Moser. A double $(\infty, 1)$-categorical nerve for double categories. arXiv:2007.01848, 2020.

[MSV22] L. Moser, M. Sarazola, and P. Verdugo, *A 2Cat-inspired model structure for double categories* (English, with English and French summaries), Cah. Topol. Géom. Différ. Catég. **63** (2022), no. 2, 184–236. MR4552208

[MSV23] L. Moser, M. Sarazola, and P. Verdugo, *A model structure for weakly horizontally invariant double categories*, Algebr. Geom. Topol. **23** (2023), no. 4, 1725–1786, DOI 10.2140/agt.2023.23.1725. MR4602412

[OR] V. Ozornova and M. Rovelli. Waldhausen's S-construction. This volume (arXiv:2412.17398).

[Qui73] D. Quillen, *Higher algebraic K-theory. I*, Algebraic K-theory, I: Higher K-theories (Proc. Conf., Battelle Memorial Inst., Seattle, Wash., 1972), Lecture Notes in Math., Vol. 341, Springer, Berlin-New York, 1973, pp. 85–147. MR338129

[Rot09] J. J. Rotman, *An introduction to homological algebra*, 2nd ed., Universitext, Springer, New York, 2009, DOI 10.1007/b98977. MR2455920

[Rov] M. Rovelli. The $S_\bullet$-construction as an equivalence between 2-Segal spaces and stable augmented double Segal spaces. This volume (arXiv:2412.17400).

[Sar21] M. E. Sarazola Duarte, *Constructing K-Theory Spectra from Algebraic Structures with a Class of Acyclic Objects*, ProQuest LLC, Ann Arbor, MI, 2021. Thesis (Ph.D.)–Cornell University. MR4297495

[SS] M. Sarazola and B. Shapiro. Additivity and fiber sequences for combinatorial K-theory. arXiv:2107.07701.

[Ste] W. H. Stern. Perspectives on the 2-Segal conditions. This volume.

[Szc12] M. Szczesny, *Representations of quivers over $\mathbb{F}_1$ and Hall algebras*, Int. Math. Res. Not. IMRN **10** (2012), 2377–2404, DOI 10.1093/imrn/rnr113. MR2923170

[Tit57] J. Tits, *Sur les analogues algébriques des groupes semi-simples complexes* (French), Colloque d'algèbre supérieure, tenu à Bruxelles du 19 au 22 décembre 1956, Centre Belge de Recherches Mathématiques, Établissements Ceuterick, Louvain, 1957, pp. 261–289. MR108765

[Wal85] F. Waldhausen, *Algebraic K-theory of spaces*, Algebraic and geometric topology (New Brunswick, N.J., 1983), Lecture Notes in Math., vol. 1126, Springer, Berlin, 1985, pp. 318–419, DOI 10.1007/BFb0074449. MR802796

[Wei94] C. A. Weibel, *An introduction to homological algebra*, Cambridge Studies in Advanced Mathematics, vol. 38, Cambridge University Press, Cambridge, 1994, DOI 10.1017/CBO9781139644136. MR1269324

School of Mathematics, University of Minnesota, Minneapolis, Minnesota 55455
Email address: maru@umn.edu

Kerchof Hall, 141 Cabell Dr, Charlottesville, Virginia 22903
Email address: brandonshapiro@virginia.edu

Malott Hall, Cornell University, Ithaca, New York 14850
Email address: zakh@math.cornell.edu

Contemporary Mathematics
Volume **838**, 2026
https://doi.org/10.1090/conm/838/16741

Wrapped Fukaya categories and the 2-Segal conditions

Hiro Lee Tanaka

ABSTRACT. These are notes from an expository talk given at the BIRS in January 2024. After introducing Weinstein sectors, wrapped Fukaya categories, and foundational results in the computations of wrapped Fukaya categories, we sketch the emergence of the 2-Segal conditions (in the sense of Dyckerhoff and Kapranov) or, equivalently, the decomposition space axioms (in the sense of Gálvez-Carrillo, Kock, and Tonks) in the context of wrapped Fukaya categories. We end with a discussion of open questions.

1. Introduction

Simplicial objects satisfying the *2-Segal* conditions arise in varied settings. Dyckerhoff and Kapranov [**10**] articulated the 2-Segal conditions motivated by the question of defining Hall algebras for 2-periodic categories. Independently, Gálvez-Carrillo, Kock, and Tonks [**15**] articulated the *decomposition space* axioms, which are equivalent to the 2-Segal axioms,and at end motivated by understanding incidence coalgebras and their generalizations.

Though neither of the above motivations involves symplectic geometry, this introductory talk concerns a very nice connection (Theorem 1.1) to the theory of wrapped Fukaya categories in two dimensions.

The two-dimensional connection invites new results (see [**8**]) and speculation (see Section 5) about higher dimensions. Just as there are many useful generalizations of a Segal space (to model either (∞, n)-categories, or 2-Segal spaces), there are most likely many generalizations of the notion of 2-Segal space, and symplectic geometry gives a hint at one.

1.1. What are Fukaya categories? We recall that symplectic manifolds are $2n$-dimensional manifolds M equipped with a 2-form ω for which $d\omega = 0$ (closedness) such that the top exterior power ω^n defines a volume form on M (non-degeneracy). Symplectic manifolds are often motivated as a way to express the Hamiltonian formulation of classical mechanics: a function $f\colon M \to \mathbb{R}$ defines a 1-form df, and ω, because it is non-degenerate, converts df to a unique vector field X_f whose flow models the evolution of a physical system governed by the "Hamiltonian" f. We refer the reader to [**5, 30, 31**] for more on symplectic geometry.

Given this background, perhaps the least romantic characterization of Fukaya categories is as invariants of symplectic manifolds: To each reasonable symplectic manifold (M, ω), one assigns a stable ∞-category that traditionally is modeled as an

2020 *Mathematics Subject Classification.* Primary 53D37, 53D37, 18N45, 19D99.

A_∞-category linear over some unital commutative base ring. At first, and perhaps remarkably, these categories were defined and studied with no physical motivation whatsoever.

A more inspiring description of Fukaya categories emerged through *homological mirror symmetry*, a phenomenon first proposed by Kontsevich in [**21**] based on inspiration from the physics of twists of Type II string theory, for which symplectic manifolds, especially Kähler manifolds, also serve as a natural setting for targets in sigma models defining quantum field theories. The theory is incredibly rich – for example, the Geometric Langlands correspondence is expected to be an instance of homological mirror symmetry when the latter is formulated appropriately [**7, 17, 20**]. Fukaya categories are no longer for simply distinguishing symplectic manifolds; they are now a conduit for predicting and computing algebro-geometric phenomena through, or simply inspired by, seemingly completely unrelated symplectic geometry. For example, concrete, and purely algebraic, conjectures about resolutions of sheaves on toric stacks can be proven by following observations regarding Fukaya categories [**16**].) For a homotopy theorist, this is one reason to pursue Fukaya categories over the sphere spectrum: inspirations are bound to emerge for understanding spectral algebraic geometry. For background from the viewpoint of mathematical physics, see [**18, 19, 41**].

1.2. 2-Segal data. The *wrapped* Fukaya category is an invariant of symplectic manifolds that are exact, i.e., equipped with an antiderivative of the symplectic form.

It turns out that wrapped Fukaya categories are computed via a local-to-global procedure (see Remark 1.2) . In two dimensions, one can choose the local building blocks to be disks with stops – i.e., pairs $(D^2, \mathfrak{f}_n)$ where $\mathfrak{f}_n$ is a subset of $S^1 = \partial D^2$ with $n+1$ elements. (The set $\mathfrak{f}_n$ is a particular choice of a *stop* $\mathfrak{f}$ – see Lemma 3.7.) So understanding the relationships between the wrapped Fukaya categories of $(D^2, \mathfrak{f}_n)$ is, in theory, sufficient to understand the theory for all 2-dimensional objects. Such an understanding can be captured as follows.

THEOREM 1.1. *The wrapped Fukaya categories $\mathcal{W}(D^2, \mathfrak{f}_n)$ form a 2-Segal object in stable ∞-categories. The 2-Segal object coincides with the Waldhausen s-dot construction for the base ring* $\mathbf{k}$.

REMARK 1.2. The local-to-global property of wrapped Fukaya categories was called the Kontsevich cosheaf conjecture for some time, and has now been proven. As far as we know, Seidel shared the conjecture with some students at the Talbot Workshop in March 2009, and Kontsevich wrote some notes the following summer; the full proof is due to Ganatra, Pardon, and Shende [**12**]. In fact, after identifying all M with their thickenings $M \times T^*[0,1]$, the local-to-global property is simply an assertion that the wrapped Fukaya category assignment preserves finite colimits [**22**, Section 1.8]. Some discussion of the history and understanding of the conjecture can be found in [**29**].

As we will see, the above theorem breaks symmetry. It is a bit subtle to articulate the geometry of the face maps in the $S_\bullet$-construction, and in fact both the 2-dimensional geometry and the algebra compel us to consider not just the 2-Segal object above, but a canonical (para)cyclic lift. Indeed, any functorial invariant of 2-dimensional Weinstein sectors determines a (para)cyclic object, and the fact that the invariant is unchanged under deformations of Weinstein structure is precisely the 2-Segal axiom.

Theorem 1.1 appears in [**9**] and these notes are meant to give an introductory idea of why the theorem is true. Our goal is to lower the bar of entry for readers interested in any arc in this circle of ideas.

1.3. Acknowledgments. We thank the anonymous referee for very helpful suggestions. We would like to thank the organizers Julie Bergner, Joachim Kock, Mark Penney, and Maru Sarazola, of the January 2024 BIRS Workshop: Higher Segal Spaces and their Applications to Algebraic K-Theory, Hall Algebras, and Combinatorics (24w5266). The author was supported by an NSF CAREER Grant (DMS-2044557) and an Alfred P. Sloan Foundation Fellowship during the preparation of this work.

2. Defining the wrapped Fukaya category of $(D^2, \mathfrak{f})$

We are interested in perhaps the simplest, non-trivial symplectic manifold: $\mathbb{R}^2$. In this work, $\mathbb{R}^2$ as a symplectic manifold is simply the pair $(\mathbb{R}^2, \omega)$ where ω is the smooth 2-form $dydx$, i.e., the standard volume form (up to a sign). For more on symplectic geometry we refer to the sources cited in Section 1.1.

To contextualize the importance of $\mathbb{R}^2$ and D^2 in what follows, we note that every two-dimensional Weinstein sector (a symplectic manifold, possibly with boundary, equipped with a special 1-form; see Remark 2.2 and [**11**, §2]) can be glued out of D^2 with various Weinstein sector structures.

2.1. The standard Weinstein structure on D^2. Because $\mathbb{R}^2$ is contractible, any closed 2-form is exact, and in particular we may choose a 1-form θ so that $d\theta = \omega$. We choose a rotationally symmetric 1-form

$$\theta := (ydx - xdy)/2.$$

Note that along $S^1 = \partial D^2$, a vector pairs positively with θ if and only if the vector points clockwise.

Because ω is non-degenerate, θ determines a unique vector field X_θ satisfying the equation $\theta = \omega(X_\theta, -)$.[1] Explicitly, we have

$$X_\theta = (y\partial_y + x\partial_x)/2.$$

REMARK 2.1 ($\mathbb{R}^2$ is Weinstein of index 0). With respect to the standard Riemannian metric on $\mathbb{R}^2$, we may identify X_θ with the gradient vector field of the smooth function $(x^2 + y^2)/4$, a Morse function with a unique critical point of index zero.

REMARK 2.2. More generally, if X_θ admits a Lyapunov function f that is furthermore Morse, we say that the triple (M, θ, f) is a *Weinstein manifold.* (Some authors do not demand a choice of f and only abstractly demand the existence of at least one f.) Weinstein manifolds are a class of non-compact symplectic manifolds that, thanks to the Lyapunov function, allow for symplectic handlebody decompositions akin to classical Morse theory. Alan Weinstein, the namesake, first identified appropriate notions of symplectic handlebodies [**40**].

---

[1] X_θ is called the Liouville vector field.

REMARK 2.3 (The Weinstein domain). It turns out that much[2] of the symplectic geometry of $(\mathbb{R}^2, \theta)$ is interchangeable with the symplectic geometry of $(D^2, \theta|_{D^2})$. Roughly speaking, this is because $\mathbb{R}^2$ is the completion[3] of D^2 along the flow of (the symplectically well-behaved) vector field X_θ – i.e., $\mathbb{R}^2 = D^2 \bigcup \mathbb{R}_{\geq 0} S^1$.

So, we often go back and forth between the geometry of $\mathbb{R}^2$ and that of D^2. This procedure is common; when we are being careful, we often refer to D^2 as the Weinstein *domain* and $\mathbb{R}^2$ as the Weinstein *manifold.*

2.2. A_∞-operations. We fix a (unital and commutative) base ring $\mathbf{k}$ throughout.

If $\{L_i\}$ is a countable collection of connected curves in D^2 (with boundary contained in and transverse to ∂D^2, and interiors contained in the interior of D^2), assumed to be in general position, one can define cochain complexes and A_∞-operations as follows.[4]

REMARK 2.4. Not every symplectic manifold (M, ω) admits a Kähler structure [**14**, **39**], but every symplectic manifold does admit an *almost complex structure* J that plays well with ω and uniquely so up to contractible choice.[5] To ensure that the A_∞-relations hold for the operations defined below, one must ensure compactness of the space of pseudoholomorphic curves – maps $u : \Sigma \to M$ whose derivative commutes with the J on the domain Riemann surface and with J (which may vary pointwise on Σ) – and the regularity of the space of such curves (to invoke standard facts about 1-manifolds with boundary). In particular, one must devote some time ensuring that such J exist; we omit all details and refer the reader to [**11**, 3.2].

First, the *Floer cochain complex* $CF(L_i, L_j)$ is a free $\mathbf{k}$-module generated by the (necessarily finitely many) intersection points $L_i \cap L_j$. We do not articulate the grading conventions, nor the extra geometry needed to pin down the conventions; see Remark 2.5. Regardless, given an intersection point x, we define a differential $m^1 x$ as a linear combination of other intersection points y, where the coefficient in front of y is determined by a count of pseudoholomorphic strips[6] with boundary on the Lagrangians limiting to x and to y.

Note that the case $i = j$ is problematic; the intersection set is not finite, for example. In this paper, we can ignore this point as follows: we artificially place

---

[2]Much, but not all. If one is only interested in the symplectic geometry of $\mathbb{R}^2$ which, outside a compact set, is well-behaved with respect to the flow of X_θ, then one may reduce to the case of D^2 – informally, this can be done by simply flowing positively or negatively along X_θ. (Indeed, by Cartan's magic formula, $\mathcal{L}_{X_\theta}\omega = d(\iota_{X_\theta}\omega) + \iota_{X_\theta} d\omega = d(\theta) = \omega$, so the flow preserves all symplectic geometry up to some scalar dependent on the flowtime.)

[3]For any Weinstein manifold M where f has finitely many critical points, one can write M as a union of a compact codimension-zero submanifold M_0 and the positive flow of ∂M_0 along X_θ – i.e., M is always a completion of M_0.

[4]There is nothing special about D^2 and curves in D^2. For a Weinstein manifold M and a collection of exact, cylindrical Lagrangians $\{L_i\}$, one may define A_∞ operations as described here, but this would involve a more thorough discussion of almost-complex structures.

[5]An almost-complex structure is a smooth map of vector bundles $J : TM \to TM$ such that $J^2 = -1$. "Playing well" with ω has two common versions – tameness, and compatibility. These conditions are both widely used, and are different – see Remark 1.3 of [**28**] – but we do not get into the details in this exposition.

[6]In our setting of transverse curves in D^2, one may study the honest holomorphic curve equation – i.e., one seeks holomorphic maps from the strip $\mathbb{R} \times [0, 1] \subset \mathbb{C}$ to D^2 such that $\mathbb{R} \times \{0\}$ has boundary on L_i, $\mathbb{R} \times \{1\}$ has boundary on L_j, and the limits at $\{\pm\infty\} \times [0, 1]$ are constant chords $[0, 1] \to L_i \cap L_j$.

a partial order on the collection $\{L_i\}$, and declare morphisms to be non-zero only when $L_i < L_j$. (As a matter of convention, we can also choose $L_i < L_j$, but once some geometry is involved, one must take care to ensure that the convention is compatible with the geometry.)

For all $d \geq 2$, there are maps

$$m^d : CF(L_{d-1}, L_d) \otimes \ldots \otimes CF(L_0, L_1) \to CF(L_0, L_d)[2-d]$$

given by counting maps from $\{|z| \leq 1\} \setminus \{z_0, \ldots, z_d\} \subset \mathbb{C}$ to D^2 with (i) the boundary arc from z_i to z_{i+1} mapping to L_i, (ii) the punctures z_i limiting to intersection points $L_{i-1} \cap L_i$, and (iii) the map is holomorphic. The $d = 1$ case recovers the differential above.

The operations $\{m^d\}_{d\geq 1}$ render all this data into a non-unital[7] A_∞-category, where the A_∞-relations are satisfied due to standard but highly useful and technical theorems about how pseudoholomorphic disks can degenerate in a symplectic manifold. (The analytic foundations go at least as far back as Uhlenbeck's celebrated compactness theorems in gauge theory. As Taubes once remarked, almost all compactness theorems are a "riff" on Uhlenbeck's original techniques.)

However, we may formally augment this non-unital A_∞-category by formally declaring all endomorphism rings to equal the base ring $\mathbf{k}$, and declaring the generator $1 \in \mathbf{k} = \hom(L_i, L_i)$ to act as a strict unit. This algebraic trick is commonly used to avoid somewhat gnarly geometric choices one must make when the L_i are not in general position; it provides one motivation for Seidel's "directed A_∞-categories" used in, e.g., [**35**].

REMARK 2.5. Grading data for Lagrangians, which are necessary to define the gradings of generators in the hom cochain complexes, are a beautiful part of the story of setting up the Fukaya category. Indeed, homotopy theory enters the story of Fukaya categories in two ways: *algebraically*, for example to understand notions of equivalence between different A_∞-categories, and *geometrically*, in a similar way to how reduction of structure groups enter cobordism theory. It turns out that to have a $\mathbb{Z}$-graded cochain complex, one must choose certain lifts of tangential structure maps $M \to B(U/O)$ and $L \to B\mathbb{Z}$. We do not go into the details here, but for more information see [**21**, page 134] and [**34**]. Suffice it to say that lifting a map $L \to B\mathbb{Z} \simeq S^1$ to $\mathbb{R}$ involves a choice, and the algebraic shift $L[1]$ corresponds to shifting this choice by the deck transformation $+1 \in \mathbb{Z}$. We suppress the choice from the notation L.

2.3. The wrapped Fukaya category. Suppose one is given a non-empty finite subset $\mathfrak{f} \subset S^1 = \partial D^2$, called a *stop* for D^2. We now sketch the construction of an A_∞-category $\mathcal{O} = \mathcal{O}(D^2, \mathfrak{f})$. We use the auxiliary category $\mathcal{O}$ to arrive at our eventual destination of the wrapped Fukaya category $\mathcal{W} = \mathcal{W}(D^2, \mathfrak{f})$.

REMARK 2.6. The wrapped Fukaya category of $(D^2, \mathfrak{f})$ – sometimes referred to as the partially wrapped Fukaya category of D^2 stopped at $\mathfrak{f}$ – was first defined by Sylvan in his thesis [**36**]. We instead follow an approach of Abouzaid and Seidel, made explicit in work of Ganatra, Pardon, and Shende [**11**], which avoids a great deal of geometric setup and makes many arguments algebraically transparent.

[7]We have not defined endomorphisms – i.e., we have not touched the case $L_i = L_j$; and in particular we have not defined units.

There are more general definitions of stops in Weinstein domains – see the works just cited. One may contemplate the case that $\mathfrak{f} = \emptyset$, but this case is irrelevant for us.

The intuition for "wrapping" is as follows. Given a Lagrangian L, we want to push (or wrap) L forward along the boundary ∂D^2 as much as possible without crossing a stop, then compute intersections and holomorphic disks with respect to wrapped versions of our Lagrangians. This intuition was set up using quadratic wrapping in [**1**] for the wrapped Fukaya category without any stops, and the first definition of a wrapped Fukaya category with stops, due to Sylvan, also used a quadratic Hamiltonian to "wrap" a brane [**36**].

Instead of using this quadratic model, the approach of Abouzaid and Seidel, as written in [**11**], is to choose a collection of Lagrangians that have been wrapped forward linearly (not quadratically) only a finite amount, without crossing a stop; by localizing natural morphisms (the continuation maps) representing "forward wrapping," one algebraically creates a category in which Lagrangians behave as though they are being wrapped as much as possible – so long as one chooses a *cofinal* collection, i.e., a collection in which any wrapping for large enough time has been realized).

CHOICE 2.7 (Choice of positively wrapped cofinal collections). Consider the (possibly non-compact with non-empty boundary) 2-manifold $D^2 \setminus \mathfrak{f}$. There is a finite set of smooth isotopy classes of smooth, embedded, closed intervals $L \subset D^2 \setminus \mathfrak{f}$ with (i) boundary points contained in $\partial D^2 \setminus \mathfrak{f}$, (ii) interior contained in the interior of D^2, and (iii) transverse to $\partial D^2 \setminus \mathfrak{f}$. (We only consider isotopies through curves satisfying these three conditions.) Choose a representative L_0 in each isotopy class $[L]$.

Further, for each L_0, choose closed intervals $L_1, L_2, \ldots, \subset D^2$ satisfying the conditions above and smooth isotopies $\{L_k \sim L_{k+1}\}_{k \geq 0}$ satisfying the following conditions.

(1) (Positive wrapping) For all $k \geq 0$, the isotopy from L_k to L_{k+1} is *positive* in the sense that, (a) The isotopy takes place through curves meeting the conditions (i) - (iii) above, and (b) near ∂D^2, the time-derivative $\partial_t L_k$ pairs positively with the Liouville form θ. Put another way, near ∂D^2, the isotopy is strictly clockwise.
(2) (Cofinality) Given an embedded interval $L_0 \subset D^2$ with two endpoints $a_0, b_0 \in \partial D^2$, let $a_\infty, b_\infty \in \mathfrak{f}$ be the positively wrapped limits. (For example, if one moves a_0 clockwise along ∂D^2, a_∞ is defined to be the first element of $\mathfrak{f}$ one encounters.) We say that the choice of isotopies $\{L_k \sim L_{k+1}\}_{k \geq 0}$ is *cofinal* if: For all L' obtained as a (not necessarily positive, but through curves satisfying (i) - (iii)) isotopy of L_0, there exists k so that L_k is obtained as a positive isotopy of L'.[8]

Finally, let $L^{(i)}_{k_i}$ be the k_ith embedded interval appearing in a sequence as above for the representative $L^{(i)}_0$ of some interval. ($L^{(i)}_0$ and $L^{(j)}_0$ need not be in the same isotopy class if $i \neq j$.) By usual transversality arguments, we may arrange for any finite subcollection of $\{L^{(i)}_k\}_{i,k}$ to be in general position.

[8]Cofinality usually requires another condition, which is that any two morphisms $L' \to L_k$ and $L' \to L_K$ coincide upon passage to an isotopy $L' \to L_N$ for $N >> k, K$. This condition turns out to be satisfied automatically. See Lemma 3.27(iii) and Lemma 3.29 of [**11**].

CONSTRUCTION 2.8 (Unwrapped preliminary category $\mathcal{O}$). Having made Choice 2.7, we define an A_∞-category $\mathcal{O}$ as follows. Its objects are in bijection with pairs (i,k) where i is an isotopy class $[L]$, and $k \in \mathbb{Z}_{\geq 0}$. We write $L_k^{(i)}$ for the object corresponding to (i,k). We define $\mathcal{O}$ to be the directed A_∞-category where

$$\hom(L_k^{(i)}, L_{k'}^{(i')}) = \begin{cases} CF(L_k^{(i)}, L_{k'}^{(i')}) & k > k' \\ \mathbf{k} & L_k^{(i)} = L_{k'}^{(i')} \\ 0 & \text{otherwise.} \end{cases}$$

By Section 2.2, $\mathcal{O}$ is a (directed) unital A_∞-category.

REMARK 2.9 (Continuation elements C). Note that $\mathcal{O}$ uses the partial order on the collection $\{L_k^{(i)}\}$, but did not use the data of the positive isotopies. We invoke them now.

Indeed, given a positive isotopy $L \to L'$, one has a well-defined element of $H^0CF(L,L')$ called the *continuation element* or *continuation map* associated to the positive isotopy. In particular, in $\mathcal{O}$, for all $i = i'$, we have well-defined degree 0 elements in cohomology induced by the positive isotopy from $L_{k'}^{(i')}$ to $L_k^{(i)}$ for $k > k'$.

We let C denote the collection of continuation elements resulting from the positive isotopies chosen in Choice 2.7.

DEFINITION 2.10 (The wrapped Fukaya category $\mathcal{W}$). We define the *wrapped Fukaya category* of $(D^2, \mathfrak{f})$ to be the localization

$$\mathcal{W} := \mathcal{O}[C^{-1}].$$

That is, $\mathcal{W}$ is the A_∞-category obtained from $\mathcal{O}$ by universally rendering all chosen continuation maps invertible.

REMARK 2.11. The localization of an A_∞-category was constructively defined in [**11**] using work of [**26, 27**] on quotients of A_∞-categories. That this construction satisfies the ∞-categorical universal property of localization is proven in [**33, 38**].

Given $L = L_k, L' \in \operatorname{Ob}\mathcal{O} = \operatorname{Ob}\mathcal{W}$, we have the sequence of cochain complexes

$$CF^*(L_k, L') \to CF^*(L_{k+1}, L') \to CF^*(L_{k+2}, L') \to \dots$$

induced by the continuation maps associated to the positive isotopies.

LEMMA 2.12 (Lemma 3.37 of [**11**]). *The induced map*

$$(\operatorname{colim}_{k\to\infty} CF^*(L_k, L')) \to \hom_{\mathcal{W}}(L, L')$$

is a quasi-isomorphism.

Lemma 2.12 justifies the terminology of "wrapping"; for example, up to quasi-isomorphism, the generators of $\hom_{\mathcal{W}}(L,L')$ are thus an increasing union of intersections points of L_t and L' where $\{L_t\}_{t\geq 0}$ is an isotopy of L_0 "wrapping" L_0 near ∂M positively ad infinitum.[9] The lemma also tells us how to compute morphisms in $\mathcal{W}$: choose any cofinal sequence and compute the direct limit (i.e., colimit of cochain complexes).

PROPOSITION 2.13 ([**11**, Proposition 3.39]). *Up to quasi-equivalence of A_∞-categories, $\mathcal{W}$ is independent of Choice 2.7.*

[9]The reader may enjoy visualizing the example of $M = T^*S^1$ and what a cotangent fiber L of T^*S^1 looks like as L is wrapped positively.

3. Computing the wrapped Fukaya category of $(D^2, \mathfrak{f})$

In the previous section we sketched the definition of the wrapped Fukaya category of a Weinstein domain $M = D^2$ equipped with a stop $\mathfrak{f}$.

3.1. Equivalences of A_∞-categories. What does it mean to "compute" this category?

If the invariant of $(M, \mathfrak{f})$ were a chain complex, one would be happy to compute the chain complex up to quasi-isomorphism. For an A_∞-category, there are various notions of equivalence.

One is sometimes happy to compute the A_∞-category up to quasi-equivalence of A_∞-categories – i.e., up to a functor that is essentially surjective and fully faithful (meaning the map on morphism complexes is a quasi-isomorphism of cochain complexes, a generalization of quasi-equivalence of dg categories). But, this notion may be too strong for reasons both geometric and algebraic.

Here is a weaker notion of equivalence. An A_∞-category is, homotopy-theoretically, the same thing as a category enriched in chain complexes[10], so we may take the stable closure of any A_∞-category. That is, we may seek to classify $\mathcal{W}$ up to quasi-equivalence *after* completing $\mathcal{W}$ under mapping cones and shifts. In the literature, one often writes $\mathrm{Tw}\,\mathcal{W}$ for[11] the triangulated closure (i.e., stable closure) of $\mathcal{W}$, but we abuse notation and treat $\mathcal{W}$ as though it has already been completed under mapping cones and shifts. One could have in mind the collection of (dg) modules over $\mathbf{k} \oplus \mathbf{k}$ for which both factors of $\mathbf{k}$ act non-trivially; or more generally dg-modules over some base ring given by chain complexes of free modules. (Importantly, depending on the ring, this may not contain chain complexes of arbitrary projective modules).

An even weaker notion is then to take the Karoubi completion (sometimes called the idempotent completion) of $\mathcal{W}$, so that $\mathcal{W}$ not only has all shifts and mapping cones, but also for which every idempotent map splits. One can have in mind the A_∞-category of all finitely generated complexes over some fixed $\mathbf{k}$-algebra. For more, see [**24**, 4.4.5].

The weakest natural notion one often uses is that of Ind-completion $\mathrm{Ind}(\mathcal{W})$, which is not only Karoubi complete, but is in fact presentable, implying $\mathrm{Ind}(\mathcal{W})$ has all limits and colimits. In fact, $\mathrm{Ind}(\mathcal{W})$ enjoys many pleasant properties, such as amenability to adjoint functor theorems, that we do not discuss here. One can have in mind the A_∞-category of (not necessarily finitely generated) complexes over a fixed $\mathbf{k}$-algebra.

One of the most satisfying "computations" of an A_∞-category $\mathcal{A}$ is then to exhibit a quasi-equivalence between $\mathcal{A}$ and a (full sub)category $R\mathsf{Mod}$ of modules over some A_∞-algebra or dg-algebra R. Often, R is found as the endomorphism algebra of some direct sum $\oplus L_i$ of objects $L_i \in \mathrm{Ob}\,\mathcal{A}$ that generate $\mathcal{A}$ under mapping cones and shifts.[12] Depending on presentability/adjunction-related desires, one might identify $\mathcal{A}$ with the finitely generated R-modules, or lose some information and pass to an Ind-completion, identifying $\mathcal{A}$ with all R-modules.

---

[10]This is made precise in [**38**]; see especially Section 1.2 of ibid.

[11]The symbol Tw stands for "twisted," as early models of the triangulated closure used a presentation of "twisted complexes" as in work of Bondal and Kapranov [**4**].

[12]One can view this as a version of the Freyd-Mitchell embedding theorem but for dg- or A_∞-categories.

3.2. Generation.

DEFINITION 3.1. Let $\mathfrak{f} \subset D^2$ be a choice of stop, and let $a \in \mathfrak{f}$ be a point in the stop. An object $L \in \mathcal{O}$ is called a *linking disk*[13] of a if a connected arc $\gamma \subset \partial D^2$ with boundary points $\partial\gamma = \partial L$ has the property that $\gamma \bigcap \mathfrak{f} = a$.

THEOREM 3.2 ([**12**, Theorem 1.13]). *Choose a linking disk $L_a \in \operatorname{Ob}\mathcal{W}(D^2, \mathfrak{f})$ for every $a \in \mathfrak{f}$. Then the (finite) collection $\{L_a\}_{a\in\mathfrak{f}}$ generates*[14] *$\mathcal{W}$ under mapping cones and shifts.*

COROLLARY 3.3. *Let $\{L_a\}_{a\in\mathfrak{f}}$ be a collection of linking disks, and let R be the endomorphism algebra of the direct sum $\bigoplus_{a\in\mathfrak{f}} L_a$. Then we have a fully faithful embedding of A_∞-categories $\mathcal{W}(D, \mathfrak{f}) \hookrightarrow R\mathsf{Mod}$ whose image is identified with the stable subcategory generated by the modules represented by each L_a.*

Thus, up to Karoubi completion, $\mathcal{W}(D, \mathfrak{f})$ is quasi-equivalent to the A_∞-category of finitely generated R-modules.

After Ind-completion, $\operatorname{Ind}\mathcal{W}(D, \mathfrak{f})$ *is quasi-equivalent to the A_∞-category* $R\mathsf{Mod}$ *of all R-modules.*

REMARK 3.4 (Symmetries and gradings). There is a compelling symmetry to this collection of generators. For example, if one were to take the stop $\mathfrak{f} = \mathfrak{f}_n \subset \partial D^2$ to have a discrete rotational symmetry by the cyclic group C_{n+1}, we have symmetries of the triple $(D^2, \theta, \mathfrak{f}_n)$. As it turns out, this is where one needs to be careful about the notion of gradings. A full rotation of the circle does not, in fact, take an object L to itself; rather, it takes L to $L[2]$, a shift by 2.

3.3. Different generators. If one were to go ahead and compute $R = \operatorname{End}(\oplus_{a\in\mathfrak{f}} L_a)$, one would not immediately see the connection to the $S_\bullet$-construction. One can instead break some symmetry to choose a different collection of generators that makes this connection clearer.

(We are going to take a detour: by breaking symmetry, we see the $S_\bullet$-construction for the base ring $\mathbf{k}$; by studying how we broke symmetry, we realize that the $S_\bullet$-construction admits a (para)cyclic refinement.)

NOTATION 3.5 ($a \pm 1$). Given $a \in \mathfrak{f} \subset S^1$, let $a + 1$ denote the successor to a in the cyclic order of $\mathfrak{f}$. (We cyclically order by the positive – i.e., clockwise – orientation.) Likewise, we let $a - 1$ denote the predecessor to a.

It is then straightforward using Lemma 2.12 to see that the linking disks have morphisms as follows:

$$\hom_{\mathcal{W}}(L_a, L_{a+1}) \simeq \mathbf{k}, \qquad \hom_{\mathcal{W}}(L_{a+1}, L_a) \simeq 0.$$

(The quasi-isomorphism to the base ring $\mathbf{k}$, as opposed to a shift thereof, is dependent on some choices of gradings, which we suppress.) Thus, up to choice of $\mathbf{k}$-linear generator, there is a unique morphism $L_a \to L_{a+1}$ in the wrapped Fukaya

---

[13]More generally: If the Weinstein domain M has dimension $2n$, one can take $\mathfrak{f}$ to consist of Legendrian submanifolds of the contact boundary $\partial_\infty M$ of a Weinstein domain, and ask for $(D^n, \partial D^n) \cong (L, \partial L) \subset (M, \partial M)$ to be a Lagrangian disk with linking number 1 along a Legendrian component of $\mathfrak{f}$. Such Lagrangian disks are called linking disks.

[14]More generally, if M is a Weinstein manifold with stop $\mathfrak{f}$, to generate $\mathcal{W}(M, \mathfrak{f})$ one needs not only linking disks for every Legendrian component of $\mathfrak{f}$, but also every cocore to handles of M of index $(\dim M)/2$. In the case $M = D^2$ with our choice of Weinstein Morse function, there are no critical points of index 1, so no cocores are needed to generate $\mathcal{W}$.

category of $(D^2, \mathfrak{f})$. It turns out that we can compute the mapping cone of this morphism. We want to take successive mapping cones, so let us set some notation.

NOTATION 3.6 ($\mathbb{L}_{a,b}$). Let $a, b \in \mathfrak{f}$. We let $\mathbb{L}_{a,b} \in \operatorname{Ob}\mathcal{O}$ denote a compact connected curve with $\partial\mathbb{L}_{a,b}$ contained in the arcs from $a-1$ to a, and from $b-1$ to b. (For example, $\mathbb{L}_{a,a+1} = L_a$.)

The main goal of this section is to give the reader some direction for proving the following.

LEMMA 3.7. *Fix $a \in \mathfrak{f} = \mathfrak{f}_n = \{a, a+1, \dots, a+n\}$. Then (up to shifts and mapping cones) the collection of n objects*

$$\{\mathbb{L}_{a,b}\}_{b=a+1,\dots,a+n}$$

generates $\mathcal{W}(D^2, \mathfrak{f}_n)$.

The collection $\{L_a\}_{a\in\mathfrak{f}_n}$ of linking disks has cardinality one larger than the collection $\{\mathbb{L}_{a,b}\}$. It turns out the linking disks have a relation, so does not strictly speaking form a "basis" for $\mathcal{W}(D^2, \mathfrak{f})$. One also notes the symmetry-breaking in choosing the $\mathbb{L}$ generators, as a particular $a \in \mathfrak{f}$ is privileged.

We can compute morphisms between various $\mathbb{L}_{a,b}$ as follows:

PROPOSITION 3.8. *The complex $\hom_{\mathcal{W}}(\mathbb{L}_{a,b}, \mathbb{L}_{a',b'})$ has vanishing cohomology if and only if there exists an arc c in S^1 with endpoints $\{a, b\}$ for which neither endpoint of $\mathbb{L}_{a',b'}$ is in c. Otherwise, $\hom_{\mathcal{W}}(\mathbb{L}_{a,b}, \mathbb{L}_{a',b'}) \simeq \mathbf{k}$ (up to a shift depending on choice of grading).*

The following claims can be verified by testing on each generator $\{L_a\}_{a\in\mathfrak{f}}$:

PROPOSITION 3.9. *If $a = b$, then $\mathbb{L}_{a,b} \simeq 0$. That is, $\hom(\mathbb{L}_{a,b}, X)$ and $\hom_{\mathcal{W}}(X, \mathbb{L}_{a,b})$ are acyclic for any object $X \in \operatorname{Ob}\mathcal{W}$.*

PROPOSITION 3.10. *In the case $\hom_{\mathcal{W}}(\mathbb{L}_{a,b}, \mathbb{L}_{a',b'}) \simeq \mathbf{k}$, assume without generality that a' is contained in the positive arc from a to b.*[15] *Choose a $\mathbf{k}$-linear generator x of the (cohomology of the) morphism complex. Then the mapping cone of x is (up to equivalence in $\mathcal{W}(D^2, \mathfrak{f})$) the direct sum of objects*

$$\mathbb{L}_{a,b'} \oplus \mathbb{L}_{a',b}.$$

REMARK 3.11. In particular, if $a = b'$, by Proposition 3.9, the mapping cone is the object $\mathbb{L}_{a,b'}$.

Given the Propositions above, Lemma 3.7 is straightforward to prove; one can show that the collection $\{L_a\}$ is (up to equivalence in $\mathcal{W}(D^2, \mathfrak{f}_n)$) contained in the closure of $\{\mathbb{L}_{a,b}\}_{b=a+1,\dots,a+n}$ under mapping cones and shifts.

3.4. The A_n quivers.

PROPOSITION 3.12. *Let $\mathbf{k}A_n$ denote the path algebra of the A_n quiver (the graph associated to a linear poset with n elements.) Then*

$$\operatorname{End}(\oplus_{b=a+1,\dots,a+n}\mathbb{L}_{a,b}) \simeq \mathbf{k}A_n.$$

By convention, we declare $\mathbf{k}A_0$ to be the zero algebra.

[15] That is, we assume a', b' are chosen so that when moving clockwise from a to b, we encounter a' but not b'.

REMARK 3.13. $\mathbf{k}A_n$ is isomorphic to a rank n triangular matrix algebra whose diagonal entries are the base ring $\mathbf{k}$, and where for each $i < j$, the (i, j)th entry is $\hom(\mathbb{L}_{a,a+j}, \mathbb{L}_{a,a+i}) \simeq \mathbf{k}$.

The only difficulty in proving the above claim is in verifying the higher A_∞-terms in the wrapped Fukaya category. The most "hands-on" proof is to show that one can choose a cofinal wrapping sequence for each $\mathbb{L}_{a,a+i}$ such that the A_∞-terms $\{m^d\}_{d\geq 3}$ in the localization $\mathcal{O}[C^{-1}]$ all vanish, and that the m^2 terms agree with the multiplication in $\mathbf{k}A_n$. The reader can consult [**8**, §1.3.2] for another discussion.

Now let $A_n\mathsf{Mod}$ denote the category of representations of the A_n quiver – that is, the ∞-category of functors from the linear poset with n elements to the ∞-category of $\mathbf{k}$-modules.[16] Then we let $A_n\mathsf{Mod}^{\mathrm{fin}}$ be the full subcategory of perfect objects (so the functor lands in finitely presentable complexes of $\mathbf{k}$-modules). Note that $A_0\mathsf{Mod} \simeq A_0\mathsf{Mod}^{\mathrm{fin}}$ is the trivial ∞-category where all objects are zero objects.

COROLLARY 3.14. *There exists a quasi-equivalence of A_∞-categories $\mathcal{W}(D^2, \mathfrak{f}_n)$ $\simeq A_n\mathsf{Mod}^{\mathrm{fin}}$.*

4. The $S_\bullet$-construction

The $S_\bullet$-construction for the ∞-category $\mathbf{k}\mathsf{Mod}$ (or $\mathbf{k}\mathsf{Mod}^{\mathrm{fin}}$) is a simplicial object in ∞-categories

$$s_\bullet \mathbf{k}\mathsf{Mod}^{\mathrm{fin}} : \Delta^{\mathrm{op}} \to \mathcal{C}\mathrm{at}_\infty$$

where $s_n\mathbf{k}\mathsf{Mod}^{\mathrm{fin}} \simeq A_n\mathsf{Mod}^{\mathrm{fin}}$ and the face maps are localizations $d_i : A_n\mathsf{Mod}^{\mathrm{fin}} \to A_{n-1}\mathsf{Mod}^{\mathrm{fin}}$ described on objects as follows: (i) If $i \geq 1$, d_i forgets the ith chain complex X_i in the A_n quiver representation $0 = X_0 \to X_1 \to \ldots \to X_{n-1} \to X_n$. (ii) If $i = 0$, d_i passes to the A_{n-1} quiver representation obtained by quotienting by X_1: $0 \to X_2/X_1 \to \ldots \to X_n/X_1$. More generally, for any stable ∞-category $\mathcal{C}$, one can assign the object $[n] \in \Delta^{\mathrm{op}}$ to the ∞-category of functors from the A_n quiver to $\mathcal{C}$; this describes $S_\bullet$-construction of $\mathcal{C}$ (at the level of objects).

It turns out that, for any stable ∞-category $\mathcal{C}$, the domain Δ^{op} of the $S_\bullet$-construction can be upgraded. More specifically, there are well-known (not fully faithful) functors

$$\Delta \to \Lambda_\infty \to \Lambda$$

where Δ is the category of simplices, Λ is the cyclic category of Connes [**6**] and Λ_∞ is the paracyclic category of Getzler-Jones [**13**]. Though these categories all have purely combinatorial models, let us describe them (informally and) geometrically. The category Λ has objects that consist of a finite non-empty subset of S^1, and whose morphisms are modeled by isotopy classes of collisions and anti-collisions of points, along with isotopy classes of orientation-preserving diffeomorphisms of the circle respecting the subsets. The category Λ_∞ has as objects $\mathbb{Z}$-equivariant discrete subsets of $\mathbb{R}$, with morphisms modeled by isotopy classes of collisions and anti-collisions, along with isotopy classes of ($\mathbb{Z}$-equivariant) orientation-preserving diffeomorphisms of $\mathbb{R}$ respecting the subsets. Δ is then the category obtained by fixing a unit interval in $\mathbb{R}$. Informally, one may think of the above two functors as

[16] In this note, by $\mathbf{k}$-modules we mean cochain complexes of $\mathbf{k}$ modules. So an object of $A_n\mathsf{Mod}$ is a choice of cochain complexes $X_1, \ldots, X_n$, chain maps $X_i \to X_j$ for all $i < j$, homotopies between the composition $X_i \to X_j \to X_k$ and the map $X_i \to X_k$, and higher homotopies cohering these.

induced by the inclusion of a unit interval into $\mathbb{R}$, and by the usual covering map from $\mathbb{R}$ to S^1.

In this language, to extend a functor from Δ^{op} to $\Lambda_\infty^{\mathrm{op}}$ is the process of equipping a natural $\mathbb{Z}$-automorphism on each object of a simplicial object. This is one feature of a paracyclic object, and to factor a functor from Λ_∞ through Λ is to equip a $\mathbb{Z}$-equivariant structure that turns the $\mathbb{Z}$-automorphism of the nth object into a $\mathbb{Z}/n\mathbb{Z}$ symmetry. This is one feature of a cyclic object. We do not describe here the compatibility of all this with the face and degeneracy maps, which informally are modeled by the (anti)collisions of points.

By the (now no longer) "hidden" symmetry of the $S_\bullet$-construction below, we mean that indeed the $S_\bullet$-construction lifts to a functor out of Λ_∞ for kMod (and in fact, for any stable ∞-category). It lifts to a functor out of Λ if the stable ∞-category has the structure of being 2-periodic (i.e., is equipped with an automorphism from the identity functor to the shift-by-2 functor). These results can be found in [**9**, **25**, **32**]. As far as we know, the first work of Dyckerhoff-Karpranov on higher Segal spaces [**10**] was the first to predict such a hidden symmetry.

The lift to Λ in fact makes the face maps from the first paragraph of this section even more transparent; see the introduction to [**37**]. Perhaps one of the deepest theorems stemming from this hidden symmetry is that this $\mathbb{Z}$ arises as the fundamental group of S^1, and S^1 acts naturally on the collection of stable ∞-categories by the (complex) J-homomorphism of Whitehead[17]: A theorem of Lurie [**25**] states that the resulting $\mathbb{Z}$-action on the K-theory[18] of a stable ∞-category is precisely the shift by 2 functor, which is also precisely the $\mathbb{Z}$-action one finds in the paracyclic lifts of the $S_\bullet$-construction.

4.1. Discovering a hidden symmetry in the s-dot construction. Working purely algebraically, tracing through the equivalence in Proposition 3.12, d_i is also identified with a functor $\mathcal{W}(D^2, \mathfrak{f}_n) \to \mathcal{W}(D^2, \mathfrak{f}_{n-1})$ that identifies the codomain with the localization rendering the morphism

$$\mathbb{L}_{a,i} \to \mathbb{L}_{a,i+1}$$

an equivalence. Equivalently, we see that $\mathcal{W}(D^2, \mathfrak{f}_{n-1})$ is obtained from $\mathcal{W}(D^2, \mathfrak{f}_n)$ by rendering L_{a+i+1} a zero object.[19] In fact, for each i, there is a *geometrically* constructed functor $\mathcal{W}(D^2, \mathfrak{f}_n) \to \mathcal{W}(D^2, \mathfrak{f}_{n-1})$ equivalent to d_i. (The geometry is roughly inspired by the very naive observation that $\mathfrak{f}_{n-1}$ is obtained from $\mathfrak{f}_n$ by removing an element of $\mathfrak{f}_n$.[20])

To summarize, we are seeing the $S_\bullet$-construction emerge from very simple geometry: Given a collection $\mathfrak{f}_n$ of $n+1$ marked points on S^1, removing one marked point results in a functor $\mathcal{W}(D^2, \mathfrak{f}_n) \to \mathcal{W}(D^2, \mathfrak{f}_{n-1})$ modeling the face maps of the $S_\bullet$-construction.

---

[17]Confusingly, Whitehead used the letter H for what we now call the J-homomorphism.

[18]It is well-known that the s-dot construction models the K theory of a stable ∞-category: The based loop space of the geometric realization of the groupoids associated to $s_\bullet \mathcal{C}$ is one common definition of the K-theory space of $\mathcal{C}$.

[19]This is part of a more general theorem called "stop removal is localization" (Theorem 1.20 of [**12**]), a result anticipated by Sylvan [**36**].

[20]The stop removal theorem is a theorem about a geometrically defined functor from $\mathcal{W}(M, \mathfrak{f})$ to $\mathcal{W}(M, \mathfrak{f}')$ when $\mathfrak{f}' \setminus \mathfrak{f}$ is Legendrian; more details are in [**12**].

And, if all we knew prior to this discussion was the $S_\bullet$-construction, the geometry suggests a surprising fact: There is a further symmetry underlying the $S_\bullet$-construction: namely, the action of automorphisms of the Liouville sector $(D^2, \mathfrak{f}_n)$. Up to isotopy, there is a discrete such family translating the elements of $\mathfrak{f}_n$. (If $\mathfrak{f}_n$ is evenly spaced along ∂D^2, this automorphism can be modeled by the natural $2\pi/n$ rotation when restricted to the boundary.) Importantly, we also have the rotations by multiples of 2π. One can compute that a full rotation of the disk results in a shift of 2 in the grading for every Lagrangian. To put this into uniform notation, we have gleaned that the categories $\mathcal{W}(D^2, \mathfrak{f}_n)$ have natural symmetries by the cyclic groups $\langle 1/n \rangle \subset \mathbb{Q}$, with 1 acting by the shift-by-two functor. This of course respects the process of forgetting points in $\mathfrak{f}_n$; thus, the $S_\bullet$-construction should also have these symmetries.

To make all this precise without passing to some combinatorial model for the wrapped Fukaya categories is rather delicate. (While the geometry suggests the structure rather intuitively, it does not allow us to write down precise statements so easily.) The work [**37**] takes some first steps to phrase all this using only geometry.

4.2. The 2-Segal condition in symplectic geometry. The previous section merely shows that the $S_\bullet$-construction has a hidden (para)cyclic symmetry, inspired by the geometry of D^2. As we know, the s-dot construction has a further property: The 2-Segal condition. It turns out this manifests symplectically. Let us state the slogan at the outset: The (geometric) deformation invariance of the wrapped Fukaya category[21] is implied by the (algebraic) 2-Segal property of the $S_\bullet$-construction.

In two dimensions, there is a ribbon graph associated to any compact domain with a stop. Let us explain this procedure for the pair $(D^2, \mathfrak{f}_n)$.

First, we complete D^2 along X_θ, and create a non-compact boundary component for each element of $\mathfrak{f}_n$.[22] Let us call the result M_n. Then M_n is abstractly diffeomorphic to $D^2 \setminus (\gamma_1 \bigcup \cdots \bigcup \gamma_{n+1})$ where each γ_i is a closed arc inside ∂D^2. So ∂M_n is a union of $n+1$ open arcs, and we think of each point of $\mathfrak{f}_n$ as corresponding to exactly one open arc. Importantly, M_n can be equipped with a 1-form θ for which $d\theta$ is again symplectic, but, for various reasons, we only care about this 1-form up to d of a compactly supported function. Put another way, the dynamics of the dual vector field X_θ are only determined outside a compact region of M_n. We can read off a graph from X_θ by considering the locus of points in M_n that do not escape to infinity under X_θ. This embedded graph yields the ribbon graph alluded to in the previous paragraph. The combinatorial isomorphism class of Γ depends on the particular choice of θ,

For example, M_3 admits choices of θ with two distinct combinatorial isomorphism classes of Γ. One is the corolla with four leaves (i.e., a graph with a unique interval vertex, of valence 4), while the other is a trivalent graph with two internal vertices and exactly one internal edge (but with four external edges); see Figure 1. This latter graph exhibits M_3 as obtained by gluing together two copies of M_2 along an M_1.

---

[21] In dimension two, as with all examples in this note.

[22] There is a general procedure that takes a "stopped domain" such as $(D^2, \mathfrak{f}_n)$ and produces a sector – see Section 2.2 of [**23**] for a concrete construction and further references.

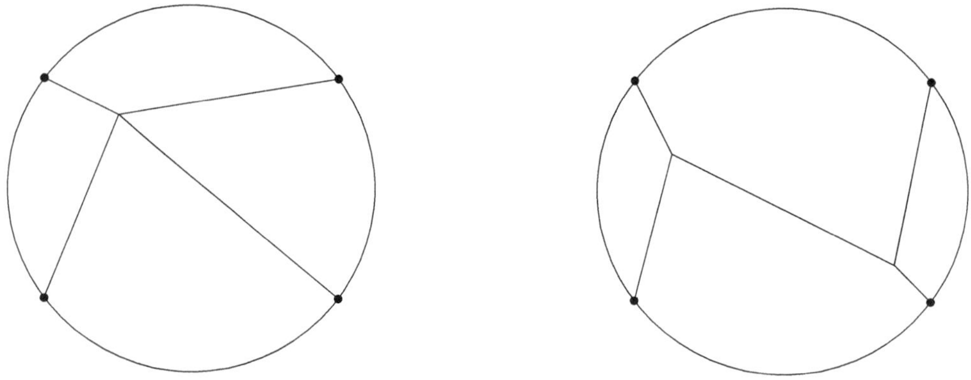

FIGURE 1. M_3 wth two different choices of θ – while we do not draw X_θ, the skeleton Γ is drawn for each choice of θ. Generic choices result in a trivalent graph as on the right; non-generic choices result in a graph having a vertex of valence 4, as of on the left.

On the other hand, by Kontsevich's cosheaf conjecture, the wrapped Fukaya category of M_3 is thus obtained as a colimit

$$\mathcal{W}(D^2, \mathfrak{f}_2) \bigcup_{\mathcal{W}(D^2,\mathfrak{f}_1)} \mathcal{W}(D^2, \mathfrak{f}_2) \simeq A_2\mathsf{Mod}^{\text{fin}} \bigcup_{A_1\mathsf{Mod}^{\text{fin}}} A_2\mathsf{Mod}^{\text{fin}}.$$

Passing to Ind-completions, using that $Pr^R \simeq Pr^L$ by the adjoint functor theorem, and checking preservation of compact objects, one can formally arrive at the equivalence

$$A_3\mathsf{Mod}^{\text{fin}} \simeq A_2\mathsf{Mod}^{\text{fin}} \times_{A_1\mathsf{Mod}^{\text{fin}}} A_2\mathsf{Mod}^{\text{fin}}$$

which is part of the 2-Segal condition of $A_\bullet\mathsf{Mod}^{\text{fin}}$.

One can, more generally, decompose the ribbon graph associated to $(D^2, \mathfrak{f}_n)$ to obtain an equivalence

$$\mathcal{W}(D^2, \mathfrak{f}_n) \simeq \operatorname{colim} \mathcal{W}(D^2, \mathfrak{f}_{n_i})$$

for any decomposition of the graph, and that the above are equivalences is logically equivalent, after passing to adjoints and checking preservation of compact objects as before, to the 2-Segal conditions

$$A_n\mathsf{Mod}^{\text{fin}} \simeq \lim A_{n_i}\mathsf{Mod}^{\text{fin}}.$$

5. Higher dimensions

The previous section, which is essentially a recap of parts of [**9**], gives a "generators and relation" approach to invariants of 2-dimensional Weinstein sectors: such 2-manifolds are generated by disks with stops, along with certain maps between, and the only relations are the 2-Segal conditions. Of course, there was the business of passing to adjoints, so one may readily argue that the most geometric relation is actually a "coSegal" condition; this was also observed in ibid.

The need to pass to adjoints was from the geometry: embeddings of ribbon graphs are dual to stop-removal. (Graph embeddings and stop removals are geometric maps whose arrows point in opposite directions.) The functors of wrapped Fukaya categories induced by these geometric operations are adjoint, and this is what allowed us to reverse the arrows when performing algebraic computations to be content with the 2-Segal conditions as articulating the deformation invariance of wrapped Fukaya categories.

There is a slightly less geometric, but fun, way to capture the directions of stop removal functors, and this is to realize that every ribbon graph has a canonical stratification (by its vertices and edges) and that this graph has a constructible sheaf of Weinstein sectors called the germ of the pair $(D^2, \mathfrak{f})$ associated to a point on the graph. (To a point interior to an edge, one associates $(D^2, \mathfrak{f}_1)$.) There is a stack $\mathsf{Broken}_{\mathrm{cyc}}$ classifying pairs $(D^2, \mathfrak{f}_n)$, as articulated in [**37**].

For a higher-dimensional Weinstein sector M, one can articulate what the appropriate generalization of a "ribbon graph" is – the so-called "core" or "skeleton" of M – and what its local singularities look like [**2**]. Moreover, one can classify the local automorphisms of these singularities [**3**]. This is the fruit of Nadler's arborealization program. In particular, one anticipates the core of M to be equipped with a map to a stack of arboreal singularities, and that this stack carries a canonical sheaf of stable ∞-categories.

(Indeed, for the 2-dimensional case, it is proven in [**37**] that a sheaf on $\mathsf{Broken}_{\mathrm{cyc}}$ is precisely a cyclic object, e.g., the $S_\bullet$-construction of some 2-periodic stable ∞-category. There is also a paracyclic version where one can remove the 2-periodicity assumption.)

Then, from this symplectic viewpoint, the higher-dimensional analogue of the 2-Segal condition is whatever condition guarantees that local deformations of an arboreal singularity leaves the wrapped Fukaya category unchanged. It is anticipated that this is articulable combinatorially using spans of trees, and a class of "Reidemeister moves" is promised in [**3**].

It is natural of course to try to compute the analogue of a "Segal object" (a simplicial object for the 2-d case, and whatever a sheaf is on the stack of arboreal singularities in general) and "Segal condition" (the 2-Segal condition in the 2-d case, and deformation invariance in general) for each dimension 4, 6, 8, ... of a symplectic manifold.

But, perhaps better, one can also ponder what combinatorial gadget arises when one articulates the sense in which any 2-dimensional symplectic manifold stabilizes (by taking direct product with $T^*[0,1]$) and understands the stable moves by which one can deform cores of Weinstein sectors. Indeed, there are natural equivalences $\mathcal{W}(M) \simeq \mathcal{W}(M \times T^*[0,1])$, so removing the dimension constraint is a natural procedure. (Considering the ∞-category of such "stabilized" sectors has already proven fruitful [**22, 23**].) Presumably the analogue of a 2-Segal object in this stabilized setting has an object for every stabilized singularity type (these are indexed by certain trees) and maps between objects satisfy certain relations reflecting deformation invariance (presumably a sequence of relations indexed by moves between trees).

References

[1] M. Abouzaid, *A geometric criterion for generating the Fukaya category*, Publ. Math. Inst. Hautes Études Sci. **112** (2010), 191–240, DOI 10.1007/s10240-010-0028-5. MR2737980

[2] D. Álvarez-Gavela, Y. Eliashberg, and D. Nadler, *Positive arborealization of polarized weinstein manifolds*, 2022, arXiv:2011.08962.

[3] D. Álvarez-Gavela, Y. Eliashberg, and D. Nadler, *Arboreal models and their stability*, J. Symplectic Geom. **21** (2023), no. 2, 331–381, DOI 10.4310/jsg.2023.v21.n2.a3. MR4653157

[4] A. I. Bondal and M. M. Kapranov, *Framed triangulated categories* (Russian), Mat. Sb. **181** (1990), no. 5, 669–683, DOI 10.1070/SM1991v070n01ABEH001253; English transl., Math. USSR-Sb. **70** (1991), no. 1, 93–107. MR1055981

[5] A. Cannas da Silva, *Lectures on symplectic geometry*, Lecture Notes in Mathematics, vol. 1764, Springer-Verlag, Berlin, 2001, DOI 10.1007/978-3-540-45330-7. MR1853077

[6] A. Connes, *Cohomologie cyclique et foncteurs* Ext^n (French, with English summary), C. R. Acad. Sci. Paris Sér. I Math. **296** (1983), no. 23, 953–958. MR777584

[7] R. Donagi and T. Pantev, *Langlands duality for Hitchin systems*, Invent. Math. **189** (2012), no. 3, 653–735, DOI 10.1007/s00222-012-0373-8. MR2957305

[8] T. Dyckerhoff, G. Jasso, and Y. Lekili, *The symplectic geometry of higher Auslander algebras: symmetric products of disks*, Forum Math. Sigma **9** (2021), Paper No. e10, 49, DOI 10.1017/fms.2021.2. MR4211943

[9] T. Dyckerhoff and M. Kapranov, *Triangulated surfaces in triangulated categories*, J. Eur. Math. Soc. (JEMS) **20** (2018), no. 6, 1473–1524, DOI 10.4171/JEMS/791. MR3801819

[10] T. Dyckerhoff and M. Kapranov, *Higher Segal spaces*, Lecture Notes in Mathematics, vol. 2244, Springer, Cham, 2019, DOI 10.1007/978-3-030-27124-4. MR3970975

[11] S. Ganatra, J. Pardon, and V. Shende, *Covariantly functorial wrapped Floer theory on Liouville sectors*, Publ. Math. Inst. Hautes Études Sci. **131** (2020), 73–200, DOI 10.1007/s10240-019-00112-x. MR4106794

[12] S. Ganatra, J. Pardon, and V. Shende, *Sectorial descent for wrapped Fukaya categories*, J. Amer. Math. Soc. **37** (2024), no. 2, 499–635, DOI 10.1090/jams/1035. MR4695507

[13] E. Getzler and J. D. S. Jones, *The cyclic homology of crossed product algebras*, J. Reine Angew. Math. **445** (1993), 161–174, DOI 10.1515/crll.1993.445.161. MR1244971

[14] R. E. Gompf, *A new construction of symplectic manifolds*, Ann. of Math. (2) **142** (1995), no. 3, 527–595, DOI 10.2307/2118554. MR1356781

[15] I. Gálvez-Carrillo, J. Kock, and A. Tonks, *Decomposition spaces, incidence algebras and Möbius inversion I: Basic theory*, Adv. Math. **331** (2018), 952–1015, DOI 10.1016/j.aim.2018.03.016. MR3804694

[16] A. Hanlon, J. Hicks, and O. Lazarev, *Resolutions of toric subvarieties by line bundles and applications*, Forum Math. Pi **12** (2024), Paper No. e24, 58, DOI 10.1017/fmp.2024.21. MR4831691

[17] T. Hausel and M. Thaddeus, *Mirror symmetry, Langlands duality, and the Hitchin system*, Invent. Math. **153** (2003), no. 1, 197–229, DOI 10.1007/s00222-003-0286-7. MR1990670

[18] K. Hori, S. Katz, A. Klemm, R. Pandharipande, R. Thomas, C. Vafa, R. Vakil, and E. Zaslow, *Mirror symmetry*, Clay Mathematics Monographs, vol. 1, American Mathematical Society, Providence, RI; Clay Mathematics Institute, Cambridge, MA, 2003. With a preface by Vafa. MR2003030

[19] C. M. Hull, *Lectures on nonlinear sigma-models and strings*, Super field theories (Vancouver, BC, 1986), NATO Adv. Sci. Inst. Ser. B: Phys., vol. 160, Plenum, New York, 1987, pp. 77–168. MR1102925

[20] A. Kapustin and E. Witten, *Electric-magnetic duality and the geometric Langlands program*, Commun. Number Theory Phys. **1** (2007), no. 1, 1–236, DOI 10.4310/CNTP.2007.v1.n1.a1. MR2306566

[21] M. Kontsevich, *Homological algebra of mirror symmetry*, Proceedings of the International Congress of Mathematicians, Vol. 1, 2 (Zürich, 1994), Birkhäuser, Basel, 1995, pp. 120–139. MR1403918

[22] O. Lazarev, Z. Sylvan, and H. L. Tanaka, *The infinity-category of stabilized liouville sectors*, 2022, `arXiv:2110.11754`.

[23] O. Lazarev, Z. Sylvan, and H. L. Tanaka, *Localization and flexibilization in symplectic geometry*, 2022, `arXiv:2109.06069`.

[24] J. Lurie, *Higher topos theory*, Annals of Mathematics Studies, vol. 170, Princeton University Press, Princeton, NJ, 2009, DOI 10.1515/9781400830558. MR2522659

[25] J. Lurie, *Rotation invariance in algebraic K-theory*, 2015, `https://www.math.ias.edu/~lurie/papers/Waldhaus.pdf`.

[26] V. Lyubashenko and O. Manzyuk, *Quotients of unital A_∞-categories*, Theory Appl. Categ. **20** (2008), No. 13, 405–496. MR2425553

[27] V. Lyubashenko and S. Ovsienko, *A construction of quotient A_∞-categories*, Homology Homotopy Appl. **8** (2006), no. 2, 157–203, DOI 10.4310/hha.2006.v8.n2.a9. MR2259271

[28] P. Massot, K. Niederkrüger, and C. Wendl, *Weak and strong fillability of higher dimensional contact manifolds*, Invent. Math. **192** (2013), no. 2, 287–373, DOI 10.1007/s00222-012-0412-5. MR3044125

[29] MathOverflow, *Has anything precise been written about the fukaya category and lagrangian skeletons?*, 2010, `https://mathoverflow.net/questions/11273/has-anything-precise-been-written-about-the-fukaya-category-and-lagrangian-skele`.

[30] D. McDuff and D. Salamon, *J-holomorphic curves and symplectic topology*, 2nd ed., American Mathematical Society Colloquium Publications, vol. 52, American Mathematical Society, Providence, RI, 2012. MR2954391

[31] D. McDuff and D. Salamon, *Introduction to symplectic topology*, 3rd ed., Oxford Graduate Texts in Mathematics, Oxford University Press, Oxford, 2017, DOI 10.1093/oso/9780198794899.001.0001. MR3674984

[32] D. Nadler, *Cyclic symmetries of A_n-quiver representations*, Adv. Math. **269** (2015), 346–363, DOI 10.1016/j.aim.2014.10.006. MR3281139

[33] Y.-G. Oh and H. L. Tanaka, *∞-categorical universal properties of quotients and localizations of A_∞-categories*, Homology Homotopy Appl. **27** (2025), no. 2, 181–207, DOI 10.4310/hha.2025.v27.n2.a9. MR4972898

[34] P. Seidel, *Graded Lagrangian submanifolds* (English, with English and French summaries), Bull. Soc. Math. France **128** (2000), no. 1, 103–149. MR1765826

[35] P. Seidel, *Fukaya categories and Picard-Lefschetz theory*, Zur. Lect. Adv. Math., Zürich: European Mathematical Society (EMS), 2008 (English).

[36] Z. Sylvan, *On partially wrapped Fukaya categories*, J. Topol. **12** (2019), no. 2, 372–441, DOI 10.1112/topo.12088. MR3911570

[37] H. L. Tanaka, *Cyclic structures and broken cycles*, 2019, `arXiv:1907.03301`.

[38] H. L. Tanaka, *Unitalities and mapping spaces in A_∞-categories*, 2024, `arXiv:2407.05532`.

[39] A. Tralle and J. Oprea, *Symplectic manifolds with no Kähler structure*, Lecture Notes in Mathematics, vol. 1661, Springer-Verlag, Berlin, 1997, DOI 10.1007/BFb0092608. MR1465676

[40] A. Weinstein, *Contact surgery and symplectic handlebodies*, Hokkaido Math. J. **20** (1991), no. 2, 241–251, DOI 10.14492/hokmj/1381413841. MR1114405

[41] E. Witten, *Mirror manifolds and topological field theory*, Mirror symmetry I, Providence, RI: American Mathematical Society, 1998, pp. 121–160 (English).

MATH DEPARTMENT, TEXAS STATE UNIVERSITY, SAN MARCOS, TEXAS

Contemporary Mathematics
Volume **838**, 2026
https://doi.org/10.1090/conm/838/16745

Cyclic Segal spaces

Julia E. Bergner and Walker H. Stern

ABSTRACT. In this survey article, we review some conceptual approaches to the cyclic category Λ, as well as its description as a crossed simplicial group. We then give a new proof of the model structure on cyclic sets, work through the details of the generalized Reedy structure on cyclic spaces, and introduce model structures for cyclic Segal spaces and cyclic 2-Segal spaces.

1. Introduction

The cyclic category Λ, originally defined by Connes [**5**], has been a useful framework for describing many different mathematical structures. It can be described as a variant of the simplex category Δ with extra morphisms, or can be described in the framework of crossed simplicial groups, which was independently introduced in [**9**] and [**14**].

More recently, cyclic structures have been studied within the context of 2-Segal spaces, as defined by Dyckerhoff and Kapranov [**8**]. While a (1-)Segal space roughly describes the structure of an up-to-homotopy topological category, in which both the set of objects and the set or morphisms are equipped with the structure of a topological space, a 2-Segal space encodes a weaker structure, in which composition need not be defined, or be unique (even up to homotopy) when it does exist. Nonetheless, this composition is still associative in an appropriate sense, and the structure of a 2-Segal space can be described via explicit combinatorial data.

Since Segal spaces and 2-Segal spaces are given by simplicial spaces, or functors from Δ^{op} to the category of spaces, we can likewise consider cyclic Segal and 2-Segal spaces as functors out of Λ^{op}. Cyclic 2-Segal spaces are of significant interest; for instance, they provide an invariant of oriented marked surfaces that associates to each such surface a topological space given by a kind of categorified state sum, as constructed in [**7**, §V.2], as well as Calabi-Yau algebras in span categories, as shown in [**21**].

2020 *Mathematics Subject Classification.* Primary 18N50; Secondary 18N40, 18N55, 18N60, 55U10, 55U35, 55U40.

The first-named author was partially supported by NSF grant DMS-1906281 and would like to thank the Isaac Newton Institute for Mathematical Sciences, Cambridge, for support and hospitality during the programme "Equivariant Homotopy Theory in Context" where work on this paper was undertaken. This work was supported by EPSRC grant no EP/R014604/1. The second-named author was supported by the NSF Research Training Group at the University of Virginia under grant DMS-1839968.

This work is preliminary to providing an answer to the following question: which cyclic 2-Segal spaces arise as the output of an $S_\bullet$-construction? For 2-Segal spaces, there is a Quillen equivalence of model categories between the model structure for 2-Segal spaces and a model structure for augmented stable double Segal spaces [**3**]. A natural follow-up is then to identify which augmented stable double Segal spaces correspond to cyclic 2-Segal spaces. We anticipate that they are described by an up-to-homotopy version of the symmetric double categories that we have shown correspond to cyclic 2-Segal sets in [**4**]. Our next goal is to establish a model structure for them, and then show that the same functors used for 2-Segal spaces, and in particular the one given by a generalization of Waldhausen's $S_\bullet$-construction, give a Quillen equivalence between them.

One of the important features of cyclic spaces is that they correspond to spaces equipped with a circle action. Thus, one motivation, aside from understanding the correspondence just described, is to give new insight into S^1-equivariant K-theory, given the central role that the $S_\bullet$-construction plays in algebraic K-theory.

The goal of this mostly expository paper is twofold. First, we give an introduction to some of the different approaches to defining and understanding the category Λ, often in parallel with the analogous perspective applied to the simplicial category Δ. Second, we describe model structures on the categories of cyclic sets and cyclic spaces, culminating in appropriate model structures for cyclic Segal spaces and cyclic 2-Segal spaces.

In the remainder of this introduction, we give a brief introduction to these ideas.

1.1. The idea of the cyclic category. Our aim in this survey is to give an introductory treatment of cyclic structures, culminating in a theory of cyclic Segal spaces. To reach the first part of this goal, we want to give multiple approaches to understanding Connes' cyclic category Λ, which was originally introduced in [**5**] to study cyclic cohomology, and the first few sections of this paper are devoted to a thorough understanding of this category. Although Connes' original approach to the cyclic category was topological, which is the same perspective taken here, in practice the cyclic category often appears in the literature in terms of a more compact description via generators and relations.

While we do, eventually, arrive at this generators-and-relations presentation, we do so after a leisurely tour through topology, first defining Λ in terms of oriented circles and homotopy classes of maps between them, and then passing to a more combinatorial description. In this portion of the paper, we draw upon the approaches of Connes [**5**] and Dyckerhoff and Kapranov [**7**, §II], and in particular we aim to make more transparent the topological intuitions that underlie the category Λ.

To arrive at the topological description of the cyclic category, it is helpful first to understand how to reformulate the simplex category Δ in an analogous topological fashion. Fundamentally, the category Δ is the category of total orders, and it is this notion that we would like to encode topologically. To do so, we can equip a finite set S with a linear order by embedding it in the oriented unit interval, as in Figure 1. If we are given two ordered sets S and T, considered as subsets of the oriented unit interval, we can consider a self-homeomorphism of the unit interval $f\colon I \to I$ that sends S into T. If we require that f is also orientation-preserving, or equivalently, strictly increasing, then f induces an injective monotone map $S \to T$.

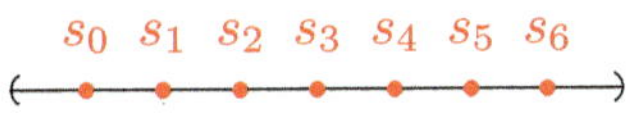

FIGURE 1. A linearly ordered set embedded in the open interval.

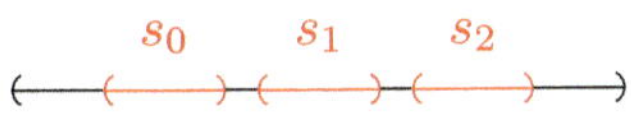

FIGURE 2. Subintervals embedded in the unit interval.

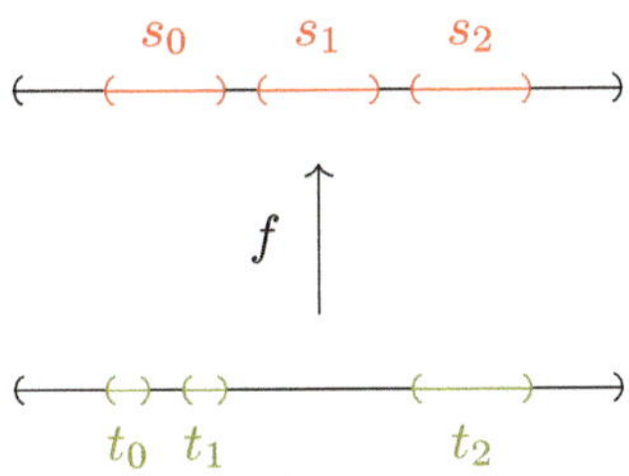

FIGURE 3. A depiction of a morphism in $\Delta_{\mathcal{T}op}$. For ease of visualization, the underlying homeomorphism of intervals is the identity.

We can thus write an injective morphism in Δ in terms of homeomorphisms of intervals.

Unfortunately, we cannot obtain the non-injective maps of Δ in this framework. We can fix this problem in two ways: either relax the requirement that f be increasing and instead ask that f be monotone, i.e., non-decreasing; or replace the points above with intervals in I. While we eventually discuss both of these potential solutions, we focus on the second here. We can think of a totally ordered set as a collection of sub-intervals with disjoint closures in the oriented unit interval I, as pictured in Figure 2. This additional data complicates the situation somewhat, since the ordered set we extract from such a collection of intervals $S \subseteq I$ is not simply the set S, but rather the set $\pi_0(S)$ of path components of S. However, in exchange for this slight complication, we gain the ability to represent any monotone map in terms of homeomorphisms of the interval.

If we have two such collections S and T of sub-intervals, any orientation-preserving homeomorphism $f\colon I \to I$ that sends S into T induces a monotone map $\pi_0(f)\colon \pi_0(S) \to \pi_0(T)$. Moreover, one can check that every monotone map $\pi_0(S) \to \pi_0(T)$ arises from such an orientation-preserving homeomorphism. For example, the map depicted in Figure 3 induces the monotone map that sends t_0 and t_1 to s_0, and sends t_2 to s_2. Formalizing this construction allows us to define a topological category $\Delta_{\mathcal{T}op}$ whose objects are pairs (I, S) and whose mapping spaces are spaces of self-homeomorphisms of intervals that preserve the chosen subsets. This topological category turns out to have contractible mapping spaces, and its homotopy category is equivalent to the simplex category Δ.

Now that we have a topological model for the simplex category, we can try to construct the cyclic category by first constructing the corresponding topological category and then taking its homotopy category. The natural question to ask, then, is “with what should we replace the oriented interval?”. Since we want the

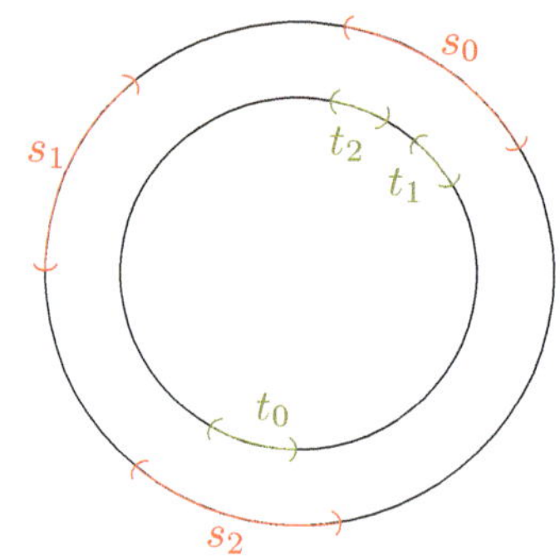

FIGURE 4. A morphism in the cyclic category, pictured in terms of the radial projection from the interior circle to the exterior circle.

cyclic category to encode cyclic group actions, which we can visualize as rotational symmetries, there is a natural choice: the circle S^1.

Following our work with the simplex category, we can naïvely extend to intervals on the circle. We can define a topological category $\Lambda_{\mathcal{T}op}$ whose objects are pairs (S^1, J) of spaces in which $J \subseteq S^1$ is a finite collection of open intervals with disjoint closures, and whose morphisms $f\colon (S^1, J) \to (S^1, K)$ are orientation-preserving homeomorphisms $f\colon S^1 \to S^1$ that map J into K.

Similarly to before, we can picture such morphisms, now viewing f as mapping points radially outward, as in Figure 4. The cyclic category Λ is then defined to be the corresponding homotopy category.

In the following table, we summarize the correspondence between the combinatorial indexing categories and their topological counterparts.

Combinatorial	**Topological**
linear order	oriented line
cyclic order	oriented circle
ordered set	collection of subintervals
order-preserving map	homotopy class of orientation-preserving homeomorphisms

We can use the more general framework of G-structured circles as described by Dyckerhoff and Kapranov in [**7**, §II] to apply the same basic intuition to access a more general class of orders, corresponding to planar Lie groups, but the perspective described above is sufficient for our purposes here.

1.2. Cyclic objects and model structures. Let us now turn to simplicial sets, given by functors $\Delta^{\mathrm{op}} \to \mathcal{S}et$, and cyclic sets $\Lambda^{\mathrm{op}} \to \mathcal{S}et$. Every cyclic set has an underlying simplicial set, induced by the inclusion functor $\Delta \to \Lambda$, and we can define a geometric realization for cyclic sets by taking the realization of the underlying simplicial set.

However, the mere existence of a cyclic structure has important ramifications for the geometric realization. Recall that the realization of a simplicial set can be obtained by gluing together copies of geometric n-simplices Δ^n, which are the geometric realizations of the representable simplicial sets $\Delta[n]$. Analogously, the realization of a cyclic set can be obtained by gluing together copies of the realizations of the cyclic representables $\Lambda[n]$. However, the realizations of $\Lambda[n]$ and $\Delta[n]$ are very different. For instance, the geometric realization of $\Delta[0]$ is a single point, but the realization of its counterpart $\Lambda[0]$ is homotopy equivalent to a circle. More generally, the geometric realization of the simplicial n-simplex $\Delta[n]$ is the topological n-simplex Δ^n, while the geometric realization of the cyclic representable $\Lambda[n]$ is $\Delta^n \times S^1$. In particular, the former is a contractible space, while the latter is not. Such subtleties carry over to the study of cyclic spaces.

In this paper, we give a new proof of a theorem of Dwyer, Hopkins, and Kan [**6**] that there is a model structure on the category of cyclic sets, and we recall its relationship to the category of topological spaces with an S^1-action. We then give several model structures on the category of cyclic spaces. First, we consider model structures in which the weak equivalences are levelwise weak equivalences of simplicial sets, given by standard approaches to model structures on categories of diagrams. In the projective model structure, fibrations are taken to be levelwise fibrations of simplicial sets, while in the injective model structure the cofibrations are given levelwise. Because Λ^{op} has the structure of a generalized Reedy structure, in the sense of Berger and Moerdijk [**2**], the category of cyclic spaces also has a generalized Reedy structure. Although the existence of this model structure is a consequence of their general theorems, here we give an explicit description for cyclic spaces and specify generating sets of cofibrations and acyclic cofibrations.

We then give model structures on the category of cyclic spaces in which the fibrant objects are cyclic Segal spaces, each given by a localization of one of the above model structures with respect to a cyclic analogue of the spine inclusions used by Rezk to describe his model structure for Segal spaces [**19**]. Finally, in a similar way we produce model structures whose fibrant objects are cyclic 2-Segal spaces, analogously to the 2-Segal space model structure of Dyckerhoff and Kapranov [**8**].

1.3. Connections with A_∞-structures. Although it is not our primary perspective, we comment briefly on the relationship between some of the structures we consider here and A_∞-structures.

Recall that one model of the topological A_∞-operad, which governs associative structures on spaces, is given by the "little intervals" operad. The reader familiar with A_∞-algebras may recognize a similar phenomenon occurring in our presentation of Δ in terms of marked intervals. There is a close relationship between the two, although we do not discuss it further here.

From another angle, composition in a Segal space or 2-Segal space is not required to be strictly associative, but rather coherently associative in a manner very much like that given by an A_∞-structure. Modeling such a coherent associativity was a key goal in the original definition of Segal spaces, but can also be understood as a motivation for the definition of 2-Segal spaces, as one aim of both [**8**] and [**10**] is to study higher categorifications of algebraic structures, such as Hall algebras in the former and incidence coalgebras in the latter.

1.4. Outline of the paper. In Section 2 we describe our topological version of the simplex category and show that its homotopy category recovers the usual simplex category Δ, and then in Section 3 we use similar methods to give a topological version of the cyclic category Λ. We take an alternate approach to these constructions and discuss duality in Sections 4 and 5. In Section 6 we investigate the category Λ from the perspective of crossed simplicial groups. Turning then to model structures, we give a new proof of the model structure on cyclic sets in Section 7. In Section 8 we consider several model structures on cyclic spaces with levelwise weak equivalences that we localize in Section 9 and 10 to obtain model structures for cyclic Segal and 2-Segal spaces.

2. A finite subset model for the simplex category

Taking the topological perspective that we began to describe in the introduction, our first goal is to give a definition of the simplex category in which the objects are discrete subsets of open intervals. Before we begin, let us fix some notation that we use throughout.

DEFINITION 2.1. An *interval space* is one of the spaces $I = [0,1], I^\circ = (0,1)$, $[0,1)$, and $(0,1]$.

We equip these interval spaces with the orientation induced by the canonical orientation on $\mathbb{R}$, and we similarly regard the unit circle S^1 with its usual orientation. For topological spaces X and Y, we denote by $C^0(X,Y)$ the space of continuous maps $X \to Y$, equipped with the compact-open topology.

For our desired topological description of the simplex category, we do not want to consider arbitrary finite subsets of I°, but instead designate a fixed subset $\underline{n}$ of cardinality $n+1$ for every $n \geq 0$. The benefit of this approach is that its homotopy category is then isomorphic to the usual simplex category.

DEFINITION 2.2. For $n \geq 0$, the *standard* $(n+1)$*-pointed interval* is the pair of topological spaces $(I^\circ, \underline{n})$, where $\underline{n} \subseteq I^\circ$ is the discrete subspace

$$\underline{n} := \left\{ \frac{1}{n+2}, \ldots, \frac{n+1}{n+2} \right\}.$$

We denote the point $\frac{k+1}{n+2} \in \underline{n}$ by $k \in I^\circ$.

With these pairs as objects, we need to specify the morphisms. As a consequence of our choice to work with discrete subsets, we need to consider maps that are less restrictive than self-homeomorphisms of the interval; we use monotone maps of intervals, which are necessarily homotopy equivalences.

DEFINITION 2.3. For interval spaces X and Y, a continuous map $f\colon X \to Y$ is *monotone* if $f(x) \leq f(z)$ in Y whenever $x \leq z$ in X. We denote by

$$\mathrm{Equiv}^+(X,Y) \subseteq C^0(X,Y)$$

the space of monotone homotopy equivalences. We denote by

$$\mathrm{Equiv}^+((I^\circ, \underline{m}), (I^\circ, \underline{n})) \subseteq \mathrm{Equiv}^+(I^\circ, I^\circ)$$

the subspace consisting of maps of pairs.

REMARK 2.4. There are several ways to draw a morphism

$$f \in \mathrm{Equiv}^+((I^\circ, \underline{m}), (I^\circ, \underline{n}))$$

schematically; we describe two of them here. On the one hand, we can draw the graph of the morphism in $I^\circ \times I^\circ$, and on the other, we can draw the mapping cylinder, together with lines indicating the image of the marked points in $I^\circ \times I$. For example, the two pictures in Figure 5 can both be used to represent the same

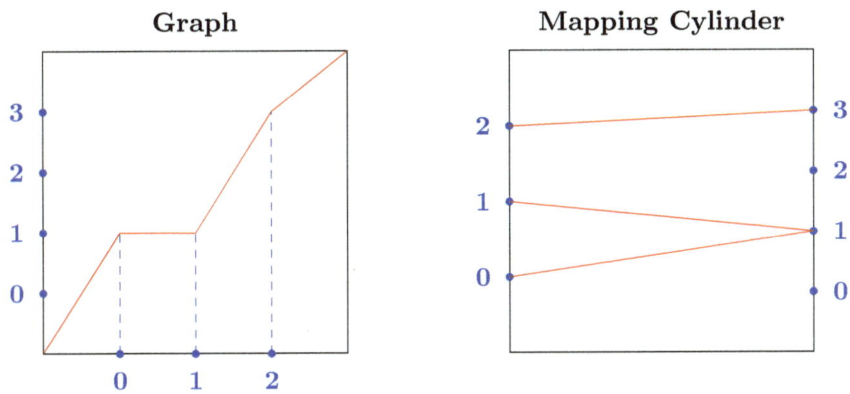

FIGURE 5. Graph an mapping cylinder depictions of the same map of marked intervals.

piecewise linear topological map $f\colon (I^\circ, \underline{2}) \to (I^\circ, \underline{3})$. While the former has the benefit of possibly being more familiar, the latter better resembles the pictures one might draw to represent maps in the simplex category. Note that these methods are merely for the purpose of visualization, and only the homotopy class of a morphism is recoverable from the mapping cylinder.

We now are able to give our chosen definition of the simplex category.

DEFINITION 2.5. The *topological simplex category* is the topological category $\Delta_{\mathcal{T}op}$ with objects $(I^\circ, \underline{n})$ for each $n \geq 0$ and morphism spaces

$$\mathrm{Map}_{\Delta_{\mathcal{T}op}}((I^\circ, \underline{n}), (I^\circ, \underline{m})) = \mathrm{Equiv}^+((I^\circ, \underline{n}), (I^\circ, \underline{m})).$$

The *simplex category* Δ is the homotopy category of $\Delta_{\mathcal{T}op}$, defined by taking path components of mapping spaces.

We now want to demonstrate that this definition is a sensible one for the simplex category, for which we recall its usual combinatorial definition.

DEFINITION 2.6. For any $n \geq 0$, the *standard ordinal* $[n]$ is the totally ordered set

$$[n] = \{0 < 1 < \cdots < n\}.$$

The simplex category Δ is the category whose objects are the standard ordinals and whose morphisms are the monotone maps between them. We denote the set of monotone maps from $[n]$ to $[m]$ by $\mathrm{Mon}([n], [m])$.

We want to establish an isomorphism between the category given by this definition and the one given in Definition 2.5. The first step is the following lemma.

LEMMA 2.7. *For any interval space X, the space* $\mathrm{Equiv}^+(X, X)$ *is contractible.*

PROOF. Straight-line homotopies provide an explicit contraction of the space $\mathrm{Equiv}^+(X, X)$. □

The following proposition follows directly.

PROPOSITION 2.8. *For any $n, m \geq 0$, there is a homotopy equivalence*

$$\mathrm{Equiv}^+((I^\circ, \underline{n}), (I^\circ, \underline{m})) \simeq \mathrm{Mon}([n], [m]),$$

where $\mathrm{Mon}([n], [m])$ *is viewed as a discrete space.*

Thus we have retrieved the more conventional definition of the simplex category from our topological definition. In what follows, we use the conventional notation in Δ, for example denoting the object corresponding to $(I^\circ, \underline{n})$ by $[n]$ and freely using the identification of $\mathrm{Hom}_\Delta([n], [m])$ with $\mathrm{Mon}([n], [m])$ provided by Proposition 2.8.

REMARK 2.9. A conventional way to present the simplex category is via a presentation by generators and relations that can be derived from the combinatorial description above. The simplex category Δ is generated by the *coface morphisms*

$$\begin{aligned} d^i \colon [n-1] &\longrightarrow [n] \\ j &\longmapsto \begin{cases} j & j < i \\ j+1 & j \geq i, \end{cases} \end{aligned}$$

that skip i in the image and the *codegeneracy morphisms*

$$\begin{aligned} s^i \colon [n+1] &\longrightarrow [n] \\ j &\longmapsto \begin{cases} j & j \leq i \\ j-1 & j > i \end{cases} \end{aligned}$$

that send two consecutive points to i. These morphisms are subject to the *cosimplicial identities*:

$$\begin{aligned} d^j \circ d_i &= d^i \circ d^{j-1} & i < j \\ s^j \circ s^i &= s^i \circ s^{j+1} & i \leq j \\ s^j \circ d^i &= \begin{cases} d^i \circ s^{j-1} & i < j \\ \mathrm{id}_{[n]} & i = j, j+1 \\ d^{i-1} \circ s^j & i > j+1. \end{cases} \end{aligned}$$

The duals of these identities are typically known as the *simplicial identities.*

We conclude this section with a discussion of the approach that we took in this section, as we repeat it frequently throughout this paper. We defined a topological category $\Delta_{\mathcal{T}op}$ with objects given by the natural numbers and morphism spaces given by spaces of monotone homotopy equivalences. The conventional description of Δ, however, is as an ordinary category, with morphism sets given by the monotone maps between finite ordered sets. We thus have a general means to compare an ordinary category, with discrete morphism spaces, to a topological category with possibly more interesting spaces of morphisms. One way to rephrase Lemma 2.7 is to say that the mapping spaces in $\Delta_{\mathcal{T}op}$ are *homotopy discrete*, in that they are homotopy equivalent to discrete spaces.

To tease further meaning from this statement, we describe some constructions for topological categories.

DEFINITION 2.10. If $\mathcal{C}$ is a topological category, its *homotopy category* is the category $\mathrm{Ho}(\mathcal{C})$ with the same objects as $\mathcal{C}$ and hom-sets given by

$$\mathrm{Hom}_{\mathrm{Ho}}(\mathcal{C})(x,y) := \pi_0 \operatorname{Map}_{\mathcal{C}}(x,y).$$

Since the functor π_0 preserves products, we can define the composition in $\mathrm{Ho}(\mathcal{C})$ to be the image under π_0 of the composition in $\mathcal{C}$. The identities are then the path components of the identity.

Since π_0 is functorial and preserves products, the construction of the homotopy category is functorial. For our purposes, the important fact is that, given a functor of topological categories $F\colon \mathcal{C} \to \mathcal{D}$, we obtain an induced functor on homotopy categories

$$\mathrm{Ho}(F)\colon \mathrm{Ho}(\mathcal{C}) \longrightarrow \mathrm{Ho}(\mathcal{D}).$$

DEFINITION 2.11. Let $\mathcal{C}$ and $\mathcal{D}$ be topological categories and $F\colon \mathcal{C} \to \mathcal{D}$ a topological functor, so that for any objects X and Y of $\mathcal{C}$, there is a continuous map

$$\operatorname{Map}_{\mathcal{C}}(X,Y) \to \operatorname{Map}_{\mathcal{D}}(FX,FY).$$

Such a functor F is a *Dwyer-Kan equivalence* if:

(1) the above map is a weak homotopy equivalence of spaces for all objects X and Y; and
(2) the induced functor $\mathrm{Ho}(F)$ on homotopy categories is essentially surjective, so that any object of $\mathrm{Ho}(\mathcal{D})$ is isomorphic to one in the image of $\mathrm{Ho}(F)$.

In the example of the simplex category, the second condition was satisfied immediately, since the two categories $\Delta_{\mathcal{T}op}$ and Δ have exactly the same set of objects. The arguments in this section can be interpreted as establishing the following result.

PROPOSITION 2.12. *The map $\Delta_{\mathcal{T}op} \to \Delta$ is a Dwyer-Kan equivalence.*

It is important to note that F being a Dwyer-Kan equivalence is strictly stronger than inducing an equivalence of homotopy categories, precisely because it captures information about the homotopy types of all of the mapping spaces.

3. The cyclic category in terms of finite subsets

We now turn to the cyclic category Λ. As in our exploration of the simplex category, we want to define specific subsets of S^1 as the objects in our topological version of the category Λ. As a first step, we define particular covering maps $\mathbb{R} \to S^1$ where, here and throughout, we identify S^1 with the unit circle in $\mathbb{C}$.

DEFINITION 3.1. For any $n \geq 0$, the $(n+1)$*-stretched universal cover* of the circle is the map

$$\begin{aligned} p_{n+1}\colon \mathbb{R} &\longrightarrow S^1 \\ x &\longmapsto \exp\left(\tfrac{2\pi i x}{n+1}\right). \end{aligned}$$

We denote by $\underline{n}$ the image of $\mathbb{Z} \subseteq \mathbb{R}$ under p_{n+1}, For simplicity, for any $0 \leq i \leq n$ we denote the image $p_{n+1}(i)$ by $i \in \underline{n} \subseteq S^1$.

As a consequence of this definition, we have the following convenient notion of monotone maps.

DEFINITION 3.2. Denote by $[0,1] \subseteq \mathbb{R}$ the closed interval. A continuous map $f : S^1 \to S^1$ is *monotone* if the lift

$$\begin{array}{ccccc} & & & & \mathbb{R} \\ & & & \overset{\tilde{f}}{\nearrow} & \big\downarrow p_1 \\ [0,1] & \xrightarrow[p_1]{} & S^1 & \xrightarrow[f]{} & S^1 \end{array}$$

is a monotone map. We denote by

$$\mathrm{Equiv}^+(S^1, S^1) \subseteq C^0(S^1, S^1)$$

the subspace consisting of monotone homotopy equivalences.

REMARK 3.3. Note that $f : S^1 \to S^1$ being monotone is equivalent to the lift

$$\begin{array}{ccccc} & & & & \mathbb{R} \\ & & & \overset{\tilde{f}}{\nearrow} & \big\downarrow p_{k+1} \\ [0,n+1] & \xrightarrow[p_{n+1}]{} & S^1 & \xrightarrow[f]{} & S^1 \end{array}$$

being monotone for *any* $k, n \geq 0$.

On an intuitive level, the idea here is that a map $f \in \mathrm{Equiv}^+(S^1, S^1)$ is a loop that traces precisely once around the target circle, possibly stopping for some time, but never backtracking. In particular, any such f has degree 1. As with the simplex category, we can restrict to maps of pairs to define the topological cyclic category, as follows.

For $n \geq 0$, denote by $\langle n \rangle$ the pair of spaces $(S^1, \underline{n})$. For $n, m > 0$, we denote by

$$\mathrm{Equiv}^+(\langle n \rangle, \langle m \rangle) = \mathrm{Equiv}^+((S^1, \underline{n}), (S^1, \underline{m})) \subseteq \mathrm{Equiv}^+(S^1, S^1)$$

the subspace consisting of maps of pairs, i.e., maps f such that $f(\langle n \rangle) \subseteq \langle m \rangle$.

DEFINITION 3.4. The *topological cyclic category* $\Lambda_{\mathcal{T}op}$ is the topological category with objects $\langle n \rangle$ for $n \geq 0$, and hom-spaces

$$\mathrm{Map}_{\Lambda_{\mathcal{T}op}}(\langle n \rangle, \langle m \rangle) = \mathrm{Equiv}^+(\langle n \rangle, \langle m \rangle).$$

Finally, for any $f \colon \langle n \rangle \to \langle m \rangle$ in $\Lambda_{\mathcal{T}op}$, we denote by $\underline{f} \colon \underline{n} \to \underline{m}$ the underlying map of sets. We often refer to $\underline{m}$ as the *underlying set* of $\langle m \rangle$.

Precisely as we did with the simplex category, we now take path components to obtain the cyclic category.

DEFINITION 3.5. The *cyclic category* Λ is the homotopy category of $\Lambda_{\mathcal{T}op}$. That is, Λ has objects $\langle n \rangle$ for $n \geq 0$ and morphisms

$$\mathrm{Hom}_\Lambda(\langle n \rangle, \langle m \rangle) = \pi_0 \, \mathrm{Equiv}^+(\langle n \rangle, \langle m \rangle).$$

Just as in the case of the simplex category, the following proposition demonstrates that passing to Λ from $\Lambda_{\mathcal{T}op}$ does not lose homotopical information.

PROPOSITION 3.6. *The mapping spaces in $\Lambda_{\mathcal{T}op}$ are homotopy discrete. In particular, the canonical functor $\Lambda_{\mathcal{T}op} \to \Lambda$ is a Dwyer-Kan equivalence.*

PROOF. The proof amounts to an examination of straight-line homotopies of induced maps on universal covers. □

3.1. Drawing morphisms. To represent a morphism in Λ visually, it suffices to draw a single map of marked circles representing that morphism. As in the case of the category Δ, there are a number of ways to represent such a map pictorally, of which we focus on two: graphs of lifts to universal covers and mapping cylinders.

Our first method takes a monotone homotopy equivalence $(S^1, \underline{n}) \to (S^1, \underline{m})$ of pairs and represents it by a morphism of intervals. Given $i \in \mathbb{Z}$ and a morphism $\widetilde{f}\colon [0, n+1] \to [i, i+m+1]$ that preserves endpoints and integer points, $\widetilde{f}$ descends to a unique map $f\colon (S^1, \underline{n}) \to (S^1, \underline{m})$ on quotients, by which we mean that the diagram

$$\begin{array}{ccc} [0, n+1] & \xrightarrow{\widetilde{f}} & [i, i+m+1] \\ {\scriptstyle p_{n+1}}\big\downarrow & & \big\downarrow{\scriptstyle p_{m+1}} \\ S^1 & \xrightarrow[f]{} & S^1 \end{array}$$

commutes.

While the lifting property of the universal cover ensures that every morphism $f \in \mathrm{Equiv}^+(\langle n\rangle, \langle m\rangle)$ can be be realized as such a quotient, the choice of $\widetilde{f}$ given f is not unique. The non-uniqueness comes precisely from the choice of $i \in \mathbb{Z}$.

LEMMA 3.7. *Given any monotone homotopy equivalence $f\colon (S^1, \underline{n}) \to (S^1, \underline{m})$ of pairs, there is a unique pair $(i, \widetilde{f})$ where $0 \le i \le m$ is an integer, and*

$$\widetilde{f}\colon [0, n+1] \longrightarrow [i, i+m+1]$$

is a monotone map that preserves endpoints and integer points, such that $\widetilde{f}$ descends to f on quotients.

PROOF. Since $p_{n+1}\colon [0, n+1] \to S^1$ is a quotient map, f is uniquely determined by the composite $\overline{f} = f \circ p_{n+1}$. The unique lifting property of the universal cover tells us that for any point $x \in p_{m+1}^{-1}(\overline{f}(0))$, there is a unique lift

$$\begin{array}{ccc} & & \mathbb{R} \\ & {\scriptstyle \widetilde{f}}\nearrow & \big\downarrow{\scriptstyle p_{m+1}} \\ [0, n+1] & \xrightarrow[\overline{f}]{} & S^1 \end{array}$$

such that $\widetilde{f}(0) = x$. Since p_{m+1} induces a bijection $[0, m+1) \to S^1$, there is a unique choice of such an x with $0 \le x < m+1$. Since f is a morphism of pairs, this unique x must be an integer, completing the proof. □

The second method of picturing a morphism $f\colon (S^1, \underline{n}) \to (S^1, \underline{m})$ in Λ is simply to draw the mapping cylinder M_f. More precisely, we consider the mapping cylinder of f as a pair (M_f, A_f), given by the pushout of pairs

$$\begin{array}{ccc} (S^1, \underline{n}) & \xrightarrow{\times\{1\}} & (S^1 \times I, \underline{n} \times I) \\ {\scriptstyle f}\big\downarrow & & \big\downarrow \\ (S^1, \underline{m}) & \longrightarrow & (M_f, A_f) \end{array}$$

The mapping cylinder (M_f, A_f) together with the canonical inclusions from $(S^1, \underline{n})$ and $(S^1, \underline{m})$ determines the homotopy class of f as a map of pairs, and thus as a morphism of Λ. Pictorially, the inclusions are indicated by the labeling of the marked points, and we draw the subspace $A_f \subseteq M_f$ in red; see Figure 7 for an example.

We conclude the section with several illustrative examples of these two ways of picturing morphisms.

EXAMPLE 3.8. As a specific example, consider a morphism $f\colon \langle 1\rangle \to \langle 2\rangle$ whose underlying map of sets is given by $0 \mapsto 2$ and $1 \mapsto 1$. Let us explain why this description uniquely determines the homotopy class f, using the more general argument above.

Since f is monotone of degree 1, the lift

$$\begin{array}{ccc} & & \mathbb{R} \\ & \overset{\widetilde{f}}{\nearrow} & \big\downarrow p_3 \\ [0,2] & \xrightarrow[f]{} & S^1 \end{array}$$

to the universal cover is non-decreasing. Since f is degree 1 and monotone, f is uniquely determined by the map

$$\widetilde{f}\colon [0,2] \longrightarrow [\widetilde{f}(0), \widetilde{f}(0)+3].$$

Since $f(0) = 2$ and $f(1) = 1$, by assumption, we can use the monotonicity of $\widetilde{f}$ to conclude that $\widetilde{f}(1) = \widetilde{f}(0)+2$. Similarly, $\widetilde{f}(2) = \widetilde{f}(0)+3$. In each of these cases, there is a unique homotopy class of endpoint-preserving monotone maps of intervals

$$[0,1] \longrightarrow [\widetilde{f}(0), \widetilde{f}(0)+2]$$

and

$$[1,2] \longrightarrow [\widetilde{f}(0)+2, \widetilde{f}(0)+3],$$

so the homotopy class of f is uniquely determined.

Let us now give our two visual depictions of the map f. Using the first method of sketching morphisms, we draw the graph of the map

$$\widetilde{f}\colon [0,2] \longrightarrow [2,5]$$

with marked points indicated on the two axes in Figure 6.

Our second method effectively amounts to drawing curves on a cylinder. In the example of the morphism $f\colon \langle 1\rangle \to \langle 2\rangle$ specified above, the resulting image is given in Figure 7.

EXAMPLE 3.9. Our second example is a morphism $g\colon \langle 2\rangle \to \langle 0\rangle$, whose underlying map of sets must send every marked point to 0. We want to argue that, in this case, the underlying map of sets does not uniquely determine the morphism g.

In Figure 8a we draw the mapping cylinder depiction of such a morphism. We have highlighted in blue the line segment in $(S^1, \underline{2})$ that gets collapsed to the point 0. Observe that we can recover g if we know two pieces of information:

(1) that g sends every marked point to 0; and
(2) that the oriented interval in $(S^1, \underline{2})$ going from 1 to 0 via the point 2 is collapsed by g. We could equivalently remember that the oriented interval in $(S^1, \underline{2})$ that goes from 0 to 1 is not collapsed.

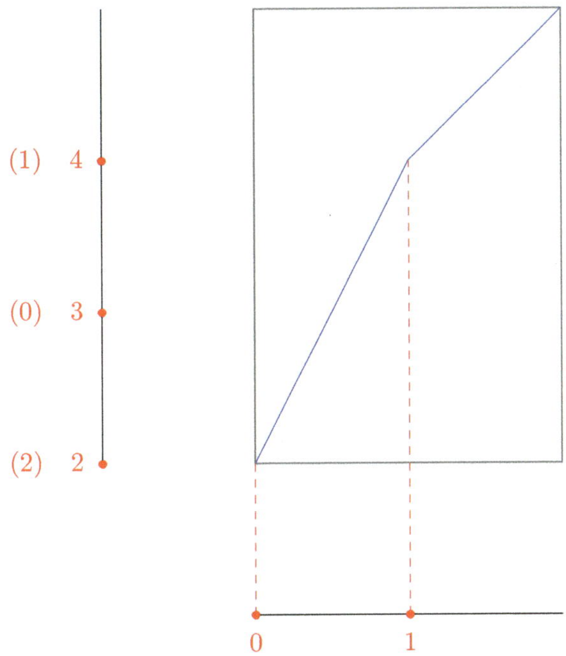

FIGURE 6. A graph depiction of the morphism $f\colon \langle 1\rangle \to \langle 2\rangle$ of Example 3.8.

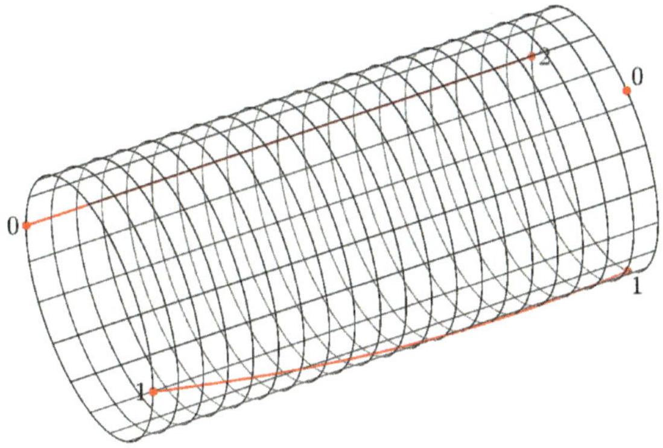

FIGURE 7. A mapping cylinder depiction of a morphism $f\colon \langle 1\rangle \to \langle 2\rangle$.

It is not hard to construct other morphisms in Λ that have the same underlying map of sets, but do not collapse the same interval. The mapping cylinder of one such is picture in Figure 8b.

For completeness, we include the graph of a lift of g in Figure 10.

EXAMPLE 3.10. As one final example, let us try to classify the possible morphisms $\langle 1\rangle \to \langle 0\rangle$. There is only one possible underlying map of sets, and so the

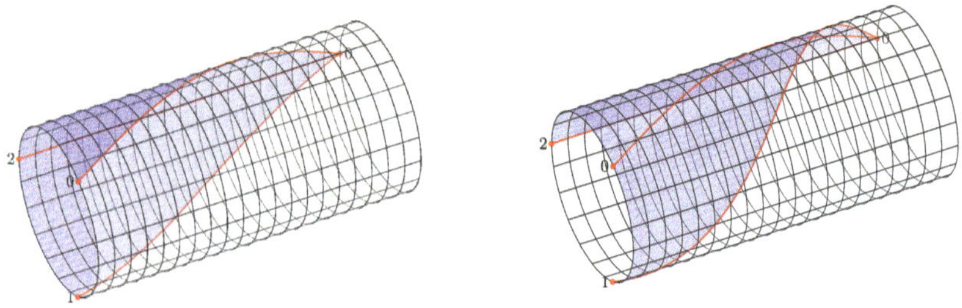

(A) The morphism g.

(B) A morphism with the same underlying map of sets as g.

FIGURE 8. Mapping cylinder depictions of two distinct morphisms $\langle 2\rangle \to \langle 0\rangle$ with the same underlying map of sets.

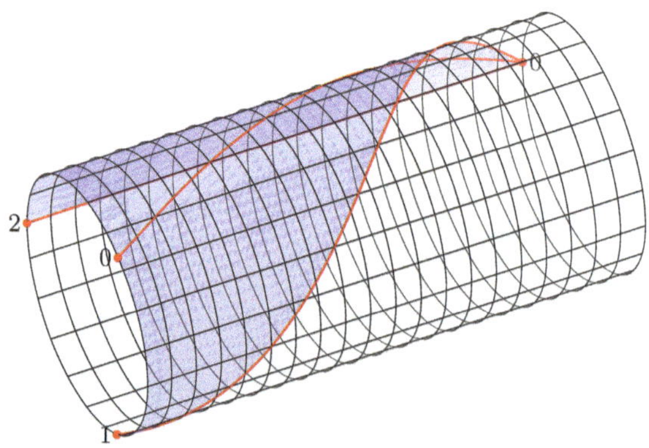

FIGURE 9. A mapping cylinder depiction of a morphism $\langle 2\rangle \to \langle 0\rangle$ with the same underlying map of sets as g, but that collapses a different interval.

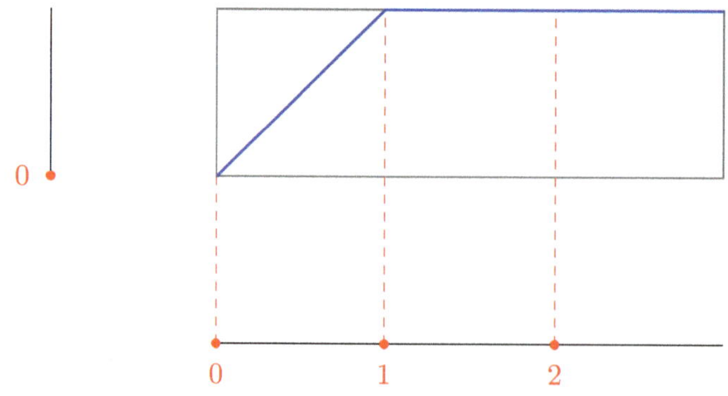

FIGURE 10. A graph depiction of a morphism $\langle r\rangle \to \langle 1\rangle$.

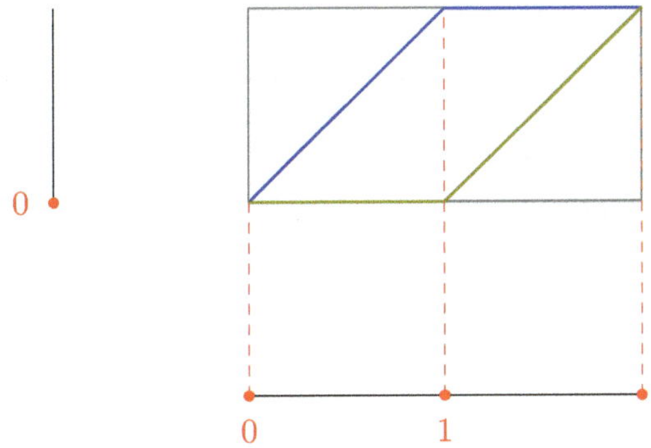

FIGURE 11. A graph depiction of lifts of the map $\langle 1\rangle \to \langle 0\rangle$.

question is which possible lifts could induce this map. Such lifts are monotone continuous maps

$$\widetilde{f}\colon [0,2] \longrightarrow [0,1]$$

such that $0 \mapsto 0$, $2 \mapsto 1$, 1 is sent to either 0 or 1. It is not hard to see that there are only two such lifts, up to homotopy; we draw their graphs in blue and green in Figure 11. Notice that, in the corresponding map of circles $f\colon (S^1, \underline{1}) \to (S^1, \underline{0})$, choosing which of these two morphisms we are representing amounts to choosing which segment of $(S^1, \underline{1})$ is not collapsed to a point by f.

3.2. Alternative approaches. We conclude with two further ways to think about the morphisms in Λ, the first of which gives a particularly nice combinatorial description of the morphisms, and the second of which can be described in terms of underlying sets.

There is, in fact, another kind of map lurking in the background of our previous discussions, one that provides one of the nicest combinatorial characterizations of Λ. Deck transformations along p_{n+1} and p_{m+1} define two $\mathbb{Z}$-actions on $\mathbb{Z}$, where $1 \in \mathbb{Z}$ acts by addition by $n+1$ and $m+1$, respectively. We denote the former $\mathbb{Z}$-set by ${}^{n\circlearrowright}\mathbb{Z}$ and the latter by ${}^{m\circlearrowright}\mathbb{Z}$. Lifting $f\colon \langle n\rangle \to \langle m\rangle$ to a morphism $\widetilde{f}$ of universal covers then yields a $\mathbb{Z}$-equivariant map

$$\widetilde{f}\colon {}^{n\circlearrowright}\mathbb{Z} \longrightarrow {}^{m\circlearrowright}\mathbb{Z}.$$

This map completely determines the path component of f. However, since multiple lifts determine the same map, multiple $\mathbb{Z}$-equivariant maps determine the same path component. To obtain a first combinatorial characterization of the morphisms in Λ, we make the following definition.

DEFINITION 3.11. We denote the set of monotone $\mathbb{Z}$-equivariant maps from ${}^{n\circlearrowright}\mathbb{Z}$ to ${}^{m\circlearrowright}\mathbb{Z}$ by $\mathbb{Z}\mathrm{Mon}({}^{n\circlearrowright}\mathbb{Z}, {}^{m\circlearrowright}\mathbb{Z})$. We define a $\mathbb{Z}$-action on $\mathbb{Z}\mathrm{Mon}({}^{n\circlearrowright}\mathbb{Z}, {}^{m\circlearrowright}\mathbb{Z})$ by

$$(1+f)(k) = f(k+n+1) = f(k) + m + 1.$$

An examination of which maps of universal covers induce the same map then yields the following result.

PROPOSITION 3.12. *Let $n, m \geq 0$. There is a canonical identification*

$$\mathrm{Hom}_\Lambda(\langle n\rangle, \langle m\rangle) \cong \mathbb{Z}\mathrm{Mon}({}^{n\circlearrowright}\mathbb{Z}, {}^{m\circlearrowright}\mathbb{Z})/\mathbb{Z}.$$

Moreover, this identification is compatible with composition.

REMARK 3.13. We can define a variant of the cyclic category whose objects are the same, but whose hom-sets are $\mathbb{Z}\mathrm{Mon}({}^{n\circlearrowleft}\mathbb{Z}, {}^{m\circlearrowleft}\mathbb{Z})$. This category is known in the literature as the *paracyclic category*, and it often denoted by Λ_∞ or $\Delta\mathbb{Z}$. See [**15**, Proposition 6.3.4(c)], [**7**, Example I.23], or [**20**, Appendix B] for more details.

Now let us consider the other combinatorial way to encode morphisms in Λ, which is in terms of morphisms of underlying sets. There is a forgetful functor

$$U\colon \Lambda \longrightarrow \mathcal{S}\mathrm{et}$$

that sends $\langle n\rangle = (S^1, \underline{n})$ to the set $\underline{n}$, canonically identified with $\{0, 1 \ldots, n\}$, and sends a morphism $f\colon \langle n\rangle \to \langle m\rangle$ to the induced map $\underline{f}\colon \underline{n} \to \underline{m}$ on marked points.

We begin by ascertaining which maps of sets are the underlying maps of morphisms in Λ.

LEMMA 3.14. *A map $\varphi\colon \underline{n} \to \underline{m}$ is the underlying map of a morphism in Λ if and only if there exist $i \in \underline{n}$ and $j \in \underline{m}$ such that φ is weakly monotone when viewed as a map*

$$\{i+1, \ldots, n, 0, \ldots, i\} \longrightarrow \{j+1, \ldots, n, 0, \ldots, j\}.$$

PROOF. The condition is necessary, since we can cut out circles at an unmarked point to obtain a map of marked intervals.

On the other hand, given such φ, i, and j, we can represent φ by a monotone map of marked intervals

$$f\colon (I^\circ, \{i+1, \ldots, i\}) \longrightarrow (I^\circ, \{j+1, \ldots, j\}).$$

The induced map between one-point compactifications can be identified with a map $\langle n\rangle \to \langle m\rangle$, and the underlying map is still φ. $\square$

NOTATION 3.15. We denote the set of maps satisfying the condition of Lemma 3.14 by $\mathrm{Cyc}(\underline{n}, \underline{m})$.

The final step in our second combinatorial description is the following.

PROPOSITION 3.16. *Let $\langle n\rangle$ and $\langle m\rangle$ be objects of Λ and let*

$$U\colon \mathrm{Hom}_\Lambda(\langle n\rangle, \langle m\rangle) \longrightarrow \mathrm{Cyc}(\underline{n}, \underline{m})$$

send $f\colon \langle m\rangle \to \langle n\rangle$ to its underlying map of sets. For $\varphi \in \mathrm{Cyc}(\underline{n}, \underline{m})$, the fiber $U^{-1}(\varphi)$ is

(1) *in bijection with $\mathbb{Z}/(n+1)$ if φ is constant; and*
(2) *a singleton if φ is non-constant.*

PROOF. First note that, as in Lemma 3.7, there is a bijection between $\mathrm{Hom}_\Lambda(\langle n\rangle, \langle m\rangle)$ and the subset of $\mathbb{Z}\mathrm{Mon}({}^{n\circlearrowleft}\mathbb{Z}, {}^{m\circlearrowleft}\mathbb{Z})$ consisting of those functions that send $0 \in {}^{n\circlearrowleft}\mathbb{Z}$ to a value between 0 and m. We identify $\mathrm{Hom}_\Lambda(\langle n\rangle, \langle m\rangle)$ with this subset and denote it by $\mathbb{Z}\mathrm{Mon}^0({}^{n\circlearrowleft}\mathbb{Z}, {}^{m\circlearrowleft}\mathbb{Z})$.

Given $\varphi \in \mathrm{Cyc}(\underline{n}, \underline{m})$, let $\psi \in \mathbb{Z}\mathrm{Mon}^0({}^{n\circlearrowleft}\mathbb{Z}, {}^{m\circlearrowleft}\mathbb{Z})$ map to φ under U. Since $0 \le \psi(0) \le m$, we must have $\psi(0) = \varphi(0)$. Since ψ is monotone and $\mathbb{Z}$-equivariant, it is uniquely determined by its restriction to a monotone map

$$\{0, \ldots, n\} \longrightarrow \{\varphi(0), \varphi(0)+1, \ldots, \varphi(0)+m+1\}.$$

Now suppose φ is nonconstant, and let $i \in \underline{n}$ be the largest value such that $\varphi(0) < \varphi(i)$. Then $\psi(i) = \varphi(i)$, since otherwise $\psi(i)$ could not both lie between

$\varphi(0)$ and $\varphi(0)+m+1$ and map to i under the quotient by the $\mathbb{Z}$-action. For any $0<k<i$, we then also have $\psi(k)=\varphi(k)$. Since the map

$$\{\varphi(i)+1,\dots,\varphi(0)+m+1\} \longrightarrow \underline{m}$$

is injective, ψ is also uniquely determined on the elements $\{i+1,\dots,n\}$, completing the proof of part (2).

On the other hand, suppose φ is constant with value $j\in\underline{m}$. Then the monotone map

$$\psi\colon \{0,\dots,n\} \longrightarrow \{\varphi(0),\varphi(0)+1,\dots,\varphi(0)+m+1\}.$$

factors through $\{\varphi(0),\varphi(0)+m+1\}$. Every monotone map

$$\{0,\dots,n\} \longrightarrow \{\varphi(0),\varphi(0)+m+1\}$$

that sends 0 to $\varphi(0)$ gives rise to an element of $\mathbb{Z}\mathrm{Mon}^0({}^{n\circlearrowleft}\mathbb{Z},{}^{m\circlearrowleft}\mathbb{Z})$ that maps to φ under U, and there are $n+1$ such maps, completing the proof of (1). □

A consequence of this proposition is another way to encode the data of a morphism in Λ, using slightly redundant information.

COROLLARY 3.17. *A morphism $\varphi\colon\langle n\rangle\to\langle m\rangle$ in Λ is equivalently given by the data of a map $\underline{\varphi}\in\mathrm{Cyc}(\underline{n},\underline{m})$ together with a choice of linear order on each fiber of $\underline{\varphi}$ that is compatible with the cyclic order on $\langle n\rangle$.*

4. An interval model and duality for the simplex category

We now turn to models based on marked subintervals and their relation to duality, starting with the simplex category in this section and then the cyclic category in the next section. We mostly forgo proofs in these sections, as they are often analogous to those for the finite subset models.

This approach gives another way to understand these categories geometrically, and it additionally provides a convenient way to understand the relationship between Δ and a subcategory ∇ that is often called the *interval category.*

4.1. The banded simplex category. To start, we look at another geometric definition.

DEFINITION 4.1. A *banded interval* is a pair of spaces (I°,J) with $J\subseteq I^\circ$, satisfying the following conditions.

- The subspace J is homeomorphic to a (nonempty) disjoint union of interval spaces,
$$J\cong\coprod_{i=0}^{K}J_i.$$
- Each of the components $J_i\subseteq J\subseteq I^\circ$ extends to an inclusion of the closed interval I into I°. That is, the closure $\overline{J_i}$ of J_i in I is homeomorphic to a closed interval.
- The closures of the J_i are disjoint in I°.

A *morphism of banded intervals* is a monotone homeomorphism $f\colon(I^\circ,J)\to(I^\circ,K)$ that is also a map of pairs. We denote the space of morphisms of banded circles by $\mathrm{Homeo}^+((I^\circ,J),(I^\circ,K))$.

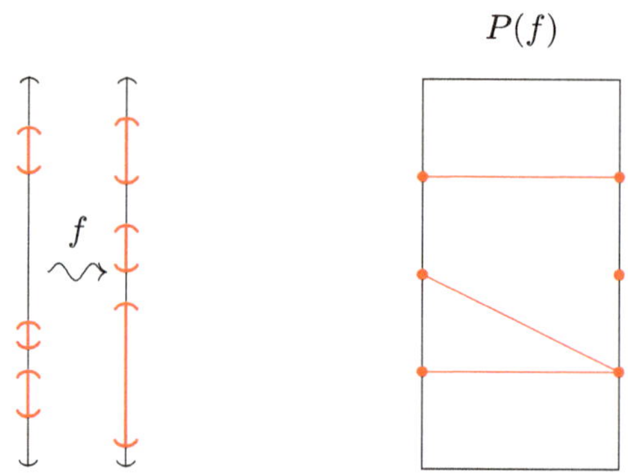

FIGURE 12. A depiction of a morphism in Δ_{Band} and its image under P.

REMARK 4.2. The latter two technical-seeming conditions guarantee that there is an isomorphism of pairs

$$(I^\circ \amalg_{\overline{J}} \pi_0(J), \pi_0(J)) \cong (I^\circ, \underline{n}),$$

where J has $n+1$ path components. It is this isomorphism that we use to define the relation between our two versions of Δ.

DEFINITION 4.3. The *banded simplex category* is the topological category Δ_{Band} whose objects are banded intervals and whose mapping spaces are

$$\operatorname{Hom}_{\Delta_{\text{Band}}}((I^\circ, J), (I^\circ, K)) = \operatorname{Homeo}^+((I^\circ, J), (I^\circ, K)).$$

If we choose, for every banded interval (I°, J), an identification

$$(I^\circ \amalg_{\overline{J}} \pi_0(J), \pi_0(J)) \cong (I^\circ, \underline{n})$$

for $n+1 = |\pi_0(J)|$ as in Remark 4.2, then every monotone homeomorphism of banded intervals $f\colon (I^\circ, J) \to (I^\circ, K)$ induces a monotone homotopy equivalence of marked intervals

$$(I^\circ, \underline{n}) \longrightarrow (I^\circ, \underline{m}).$$

This construction defines a functor

$$P\colon \Delta_{\text{Band}} \longrightarrow \Delta.$$

Pictorially, the action of the functor P on morphisms can be visualized by Figure 12.

We can summarize this discussion with the following analogue to Proposition 2.12.

PROPOSITION 4.4. *The functor $P\colon \Delta_{\text{Band}} \to \Delta$ is a Dwyer-Kan equivalence.*

4.2. The interval category and duality. Our next objective is to define a subcategory of Δ that is described in a way similar to the banded simplex category. As before, we begin with this topological description, and then give a combinatorial characterization.

DEFINITION 4.5. An *extremally banded interval* is a pair of spaces (I°, J) with $J \subseteq I^\circ$, satisfying the following conditions.

FIGURE 13. An extremally banded interval, with the bands pictured in red.

- The subspace J is homeomorphic to a (nonempty) disjoint union of interval spaces,
$$J \cong \coprod_{i=0}^{k} J_i.$$
- Each of the components $J_i \subseteq J \subseteq I^\circ \subseteq I$ extends to an inclusion of the closed interval I into $I = \overline{I^\circ}$. That is, the closure $\overline{J_i}$ of J_i in I is homeomorphic to a closed interval in $[0,1]$.
- The closures of the J_i are disjoint in I°.
- There are distinct components J_i and J_j of J such that 0 is a limit point of J_i and 1 is a limit point of J_j.

A *morphism of extremally banded intervals* consists of a monotone homeomorphism $f\colon (I^\circ, J) \to (I^\circ, K)$ that is also a map of pairs. We denote the space of morphisms of extremally banded circles by $\mathrm{Homeo}^+((I^\circ, J), (I^\circ, K))$.

Pictorially, an extremally banded interval looks like a banded interval, but with a subinterval that starts at the left, and a subinterval that ends at the right, as shown in Figure 13.

DEFINITION 4.6. The *topological interval category* ∇_{Band} has extremally banded intervals as objects and morphisms given by
$$\mathrm{Hom}_{\nabla_{\mathrm{Band}}}((I^\circ, J), (I^\circ, K)) \cong \mathrm{Homeo}^+((I^\circ, J), (I^\circ, K)).$$
The *combinatorial interval category* is the subcategory $\nabla \subseteq \Delta$ with objects $[n]$ for $n \geq 1$ and morphisms $f\colon [n] \to [m]$ that *preserve extremal elements*, in the sense that $f(0) = 0$ and $f(n) = m$.

We can construct a functor
$$Q\colon \nabla_{\mathrm{Band}} \longrightarrow \nabla$$
as follows. Given an extremally banded interval (I°, J), the orientation of I gives a unique order-preserving identification $\{0, \ldots, n\} \cong \pi_0(J)$ where $n \geq 1$ by the last part of the previous definition. We then send (I°, J) to $[n]$. We send a morphism $f\colon (I^\circ, J) \to (I^\circ, K)$ to the composite
$$[n] \xrightarrow{\cong} \pi_0(J) \xrightarrow{\pi_0(f)} \pi_0(K) \xrightarrow{\cong} [m].$$

As with our topological model for Δ, we do not lose any homotopical information by passing from the topological interval category ∇_{Band} to its combinatorial counterpart ∇, as the following proposition makes precise.

PROPOSITION 4.7. *The functor Q is a Dwyer-Kan equivalence.*

We now describe the duality between Δ and ∇, using the tools of banded intervals. Note that if (I°, J) is a banded interval, then $(I^\circ, I^\circ \setminus J)$ is an extremally banded interval. Performing this construction twice returns the original banded interval.

Additionally, if $f\colon I^\circ \to I^\circ$ is a monotone homeomorphism, then $f^{-1}\colon I^\circ \to I^\circ$ is as well. Moreover, for any homeomorphism $f \in \mathrm{Homeo}^+((I^\circ, J), (I^\circ, K))$, we have $f^{-1} \in \mathrm{Homeo}^+((I^\circ, I^\circ \backslash J), (I^\circ, I^\circ \backslash K))$. Since taking inverses of isomorphisms is functorial, we thus obtain functors

$$D_\Delta\colon \Delta_{\mathrm{Band}} \longrightarrow \nabla_{\mathrm{Band}}$$

and

$$\nabla_{\mathrm{Band}} \longrightarrow \Delta_{\mathrm{Band}}$$

given by taking complements of subsets of I° and inverses of homeomorphisms. The following first duality result is immediate, since these processes are inverse to one another.

PROPOSITION 4.8. *The functor*

$$D_\Delta\colon \Delta_{\mathrm{Band}} \longrightarrow \nabla_{\mathrm{Band}}$$

is a Dwyer-Kan equivalence.

We thus obtain the following corollary, which can be also be proved via direct combinatorial construction.

COROLLARY 4.9. *There is an equivalence of categories*

$$\Delta \longrightarrow \nabla.$$

5. An interval model and duality for the cyclic category

In this section, we translate the constructions from the previous section to the cyclic setting. In this case, the dual category ∇ is replaced by the opposite of the category Λ, which is simply Λ itself, showing that the cyclic category is actually self-dual. We use this approach to arrive at a more traditional combinatorial description of Λ and discuss its canonical factorization.

5.1. The banded cyclic category and duality. Following our banded interval model of Δ, we now construct an analogous topological model for Λ.

DEFINITION 5.1. A *banded circle* is a pair (S^1, J) of topological spaces such that $J \subseteq S^1$ is homeomorphic to a disjoint union of finitely many interval spaces whose closures in S^1 are disjoint.

A *monotone homeomorphism of banded circles* $f\colon (S^1, J) \to (S^1, K)$ is a monotone homeomorphism $f\colon S^1 \to S^1$ such that $f(J) \subseteq K$. We denote by

$$\mathrm{Homeo}^+((S^1, J), (S^1, K)) \subseteq C^0(S^1, S^1)$$

the subspace of monotone homeomorphisms of banded circles.

A schematic image of a banded circle is provided in Figure 14.

DEFINITION 5.2. The *banded cyclic category* Λ_{Band} has objects banded circles and spaces of morphisms given by $\mathrm{Homeo}^+((S^1, J), (S^1, K))$.

As with banded intervals, we can identify

$$\left(S^1 \coprod_J \pi_0(J), \pi_0(J)\right) \cong (S^1, \underline{n}).$$

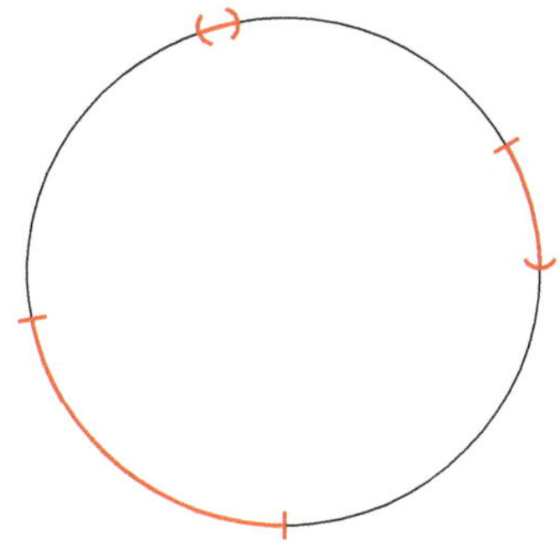

FIGURE 14. A banded circle with $J \subseteq S^1$ marked in red. Note that the intervals which make up J can be open, closed, or half-open.

This construction yields a functor

$$\Lambda_{\mathrm{Band}} \longrightarrow \Lambda$$

as in the simplicial case. Once again, the process of passing from the topological definition to the combinatorial one does not lose homotopical information.

PROPOSITION 5.3. *The functor*

$$\Lambda_{\mathrm{Band}} \longrightarrow \Lambda$$

is a Dwyer-Kan equivalence.

Now, we show that the exact same construction that yields the duality between ∇ and Δ extends to the cyclic setting. Given a banded circle (S^1, J), we can define a new banded circle (S^1, J^c), where $J^c := S^1 \setminus J$ is the complement of J in S^1. Since the inverse of a monotone homeomorphism of banded circles $f\colon (S^1, J) \to (S^1, K)$ is a monotone homeomorphism $f^{-1}\colon (S^1, K^c) \to (S^1, J^c)$ of banded circles, this construction yields a functor

$$D_\Lambda\colon \Lambda^{\mathrm{op}}_{\mathrm{Band}} \longrightarrow \Lambda_{\mathrm{Band}}.$$

It follows from construction that $D^{\mathrm{op}}_\Lambda \circ D_\Lambda$ is the identity functor, and so we obtain the following result.

PROPOSITION 5.4. *The functor*

$$D_\Lambda\colon \Lambda^{\mathrm{op}}_{\mathrm{Band}} \longrightarrow \Lambda_{\mathrm{Band}}$$

is an isomorphism of topological categories.

As in the simplicial case, we obtain duality for the usual category Λ, but the statement is more striking in this case, since we obtain that Λ is self-dual.

COROLLARY 5.5. *There is an isomorphism of categories*

$$\Lambda^{\mathrm{op}} \longrightarrow \Lambda.$$

5.2. Compactification. The topological approach from intervals also provides a way to view Δ as a subcategory of Λ. Let us denote the one-point compactification of I° by $(I^\circ)^*$ and choose a homeomorphism $\varphi\colon (I^\circ)^* \to S^1$ such that the composite

$$I^\circ \hookrightarrow (I^\circ)^* \xrightarrow[\cong]{\varphi} S^1$$

is monotone. Given a banded interval (I°, J), the pair $(S^1, \varphi(J))$ is a banded circle. Similarly, given a morphism $f\colon (I^\circ, J) \to (I^\circ, K)$ of banded intervals, the induced map on one-point compactifications

$$\iota(f)\colon\ (S^1, \varphi(J)) \longrightarrow (S^1, \varphi(K))$$

is a morphism of banded circles. We thus obtain a functor

$$\iota\colon\ \Delta_{\mathrm{Band}} \longrightarrow \Lambda_{\mathrm{Band}}.$$

PROPOSITION 5.6. *The induced functor*

$$\iota\colon\ \Delta \longrightarrow \Lambda$$

is faithful and bijective on objects.

In much the same way as with Δ, we can construct a generators-and-relations presentation of the cyclic category Λ. The proof is very similar to the one for the analogous presentation of Δ, with the additional complication that one must first show that morphisms in Λ factor uniquely as an automorphism followed by a morphism in the image of $\iota\colon \Delta \to \Lambda$.

DEFINITION 5.7. The cyclic category Λ is generated by the face maps $d^i\colon [n-1] \to [n]$ of Δ, the degeneracy maps $s^i\colon [n+1] \to [n]$ in Δ, and the cyclic shift-by-one automorphisms $\tau^n\colon [n] \to [n]$, subject to:

(1) the *cosimplicial identities*

$$d^j \circ d^i = d^i \circ d^{j-1} \qquad i < j$$

$$s^j \circ s^i = s^i \circ s^{j+1} \qquad i \leq j$$

$$s^j \circ d^i = \begin{cases} d^i \circ s^{j-1} & i < j \\ \mathrm{id}_{[n]} & i = j, j+1 \\ d^{i-1} \circ s^j & i > j+1; \end{cases}$$

and

(2) the *cyclic identities*

$$\tau \circ d^i = \begin{cases} d^{i-1} \circ \tau^{n-1} & 1 \leq i \leq n \\ d^n & i = 0 \end{cases}$$

$$\tau_n \circ s_i = \begin{cases} s^{i-1} \circ \tau^{n+1} & 1 \leq i \leq n \\ s^n \circ (\tau_{n+1})^2 & i = 0 \end{cases}$$

$$(\tau_n)^{n+1} = \mathrm{id}_{[n]}.$$

5.3. The canonical factorization. We now turn to a final property of the cyclic category that clarifies the way in which Λ is a well-behaved extension of Δ: the canonical factorization of morphisms in Λ into automorphisms and morphisms in Δ. To this end, we first describe the one-point compactification functor ι in combinatorial terms.

CONSTRUCTION 5.8. By Corollary 3.17, morphisms $\langle n\rangle \to \langle 0\rangle$ in Λ are in bijective correspondence with linear orders on the set $\underline{n}$. Such maps are the objects of the category $\Lambda_{/}\langle 0\rangle$, whose morphisms are thereby identified with monotone maps. As such, there is an equivalence of categories

$$\Delta \longrightarrow \Lambda/\langle 0\rangle$$

that sends $[n]$ to the standard linear order on $\underline{n}$. The *cyclic closure functor* is the composite

$$\Delta \longrightarrow \Lambda/\langle 0\rangle \longrightarrow \Lambda.$$

More explicitly, the cyclic closure functor sends $[n]$ to $\langle n\rangle$, and on morphisms is determined by the underlying maps of sets, together with the linear order on fibers inherited from the linear orders on the objects of Δ.

LEMMA 5.9. *For any $i, j \in \langle n\rangle$, there is a unique automorphism*

$$\varphi\colon \langle n\rangle \longrightarrow \langle n\rangle$$

such that $\varphi(i) = j$.

PROOF. Automorphisms of $\langle n\rangle$ in Λ are in bijective correspondence with isomorphisms $\psi \in \mathbb{Z}\mathrm{Mon}({}^{n\circlearrowleft}\mathbb{Z}, {}^{n\circlearrowleft}\mathbb{Z})$ such that $0 \leq \psi(i) \leq n$, by the same reasoning as in Lemma 3.7. These isomorphisms in turn can be identified with $\mathbb{Z}_{/(n+1)}$-equivariant maps $\underline{n}$ to $\underline{n}$. □

Using the inclusions constructed above, we can now prove one final key property of Λ.

PROPOSITION 5.10. *The composition map*

$$\begin{aligned} \mathrm{Hom}_\Delta([n],[m]) \times \mathrm{Aut}_\Lambda(\langle n\rangle) &\longrightarrow \mathrm{Hom}_\Lambda(\langle n\rangle, \langle m\rangle) \\ (f,\psi) &\longmapsto \iota(f)\circ\psi \end{aligned}$$

is a bijection. Thus, every morphism in Λ factors uniquely into a morphism in Δ and an automorphism of its source in Λ.

PROOF. We begin by fixing a linear order on the underlying set $\underline{m}$ of $\langle m\rangle$. Given a morphism $f : \langle n\rangle \to \langle m\rangle$ in Λ, this choice, together with the linear orders on the fibers of f guaranteed by Corollary 3.17, provides a unique linear order on the underlying set $\underline{n}$ of $\langle n\rangle$ compatible with the cyclic order and such that f is monotone. The proposition may then be proved by noting that there is a unique automorphism of $\langle n\rangle$ in the cyclic category that identifies this linear order on $\underline{n}$ with the canonical order. □

DEFINITION 5.11. The unique factorization of a morphism φ in Λ into $f_\varphi \circ \psi_\varphi$, where f_φ is a morphism in Δ and ψ_φ is an automorphism in Λ, is called the *canonical factorization*.

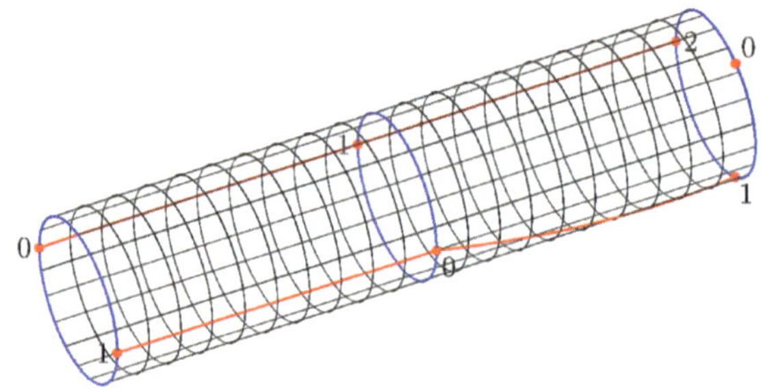

FIGURE 15. The gluing of two mapping cylinders.

REMARK 5.12. While the topological approach does not simplify the proof of the canonical factorization, it does provide good intuition for it. If we consider the morphism $\varphi\colon \langle 1\rangle \to \langle 2\rangle$ from Example 3.8, we can draw the canonical factorization as the gluing of two cylinders as in Figure 15. Notice that the labeling of the points on the central circle by 0 and 1 is the only way to label them so that the right-hand morphism is the image of a map in Δ.

6. Crossed simplicial groups and realizations

Now that we have set up several different approaches to thinking about the category Λ and its properties, we turn our attention to an axiomatization of some of these features, specifically the relationship with the simplex category Δ and the canonical factorization.

The more general notion we consider is that of *crossed simplicial groups*, introduced by Krasauskas [**14**] under the name *skew-simplicial groups*, and independently by Fiedorowicz and Loday [**9**]. While we follow the terminology from the latter, our notation instead more closely follows that of [**7**].

DEFINITION 6.1. A *crossed simplicial group* is a category $\Delta\mathfrak{G}$ satisfying the following conditions.

(1) The simplex category Δ is a subcategory of $\Delta\mathfrak{G}$ that contains all the objects of $\Delta\mathfrak{G}$.
(2) Every morphism $\varphi\colon [n] \to [m]$ in $\Delta\mathfrak{G}$ factors uniquely as $\psi\circ g$, where ψ is a morphism in Δ and $g \in \mathrm{Aut}_{\Delta\mathfrak{G}}([n])$. We call it the *canonical factorization* of φ.

We denote the automorphism groups $\mathrm{Aut}_{\Delta\mathfrak{G}}([n])$ of $\Delta\mathfrak{G}$ by $\mathfrak{G}_n$.

REMARK 6.2. Notice that this definition is suppressing a subtlety: in most cases we technically consider categories $\Delta\mathfrak{G}$ in which the subcategory from condition (1) is isomorphic, but not equal, to Δ. In practice, this distinction does not alter the way in which we work with crossed simplicial groups.

REMARK 6.3. There are two ways of defining crossed simplicial groups in the literature, either as categories $\Delta\mathfrak{G}$ as in [**9**, Definition 1.1], or as simplicial sets $\mathfrak{G}_*$ as in Construction 6.7 below, following [**14**, Definition 1.3]. These approaches are equivalent, as shown in [**14**, Theorem 1.4]. In this paper, we have chosen to define crossed simplicial groups as categories, since our main objects of study are functors out of the category Λ.

EXAMPLE 6.4. The simplest example of a crossed simplicial group is the simplex category Δ itself. In this case, the automorphism groups of the objects are trivial, so the canonical factorization does not give any new information.

EXAMPLE 6.5. There is a category $\Delta\mathfrak{S}$, called the *symmetric crossed simplicial group*, whose objects are the sets $[n]$ for $n \geq 0$, and whose morphisms are maps of sets $f\colon [m] \to [n]$ together with a choice of linear orders on the fibers $f^{-1}(i)$ for $i \in [n]$. The corresponding automorphism groups are precisely the symmetric groups. This crossed simplicial group first appeared, independently, in [**14**, Prop. 1.5] and [**9**, Prop. 3.4]. The explicit description given here is from [**7**, §I.2].

EXAMPLE 6.6. The *paracyclic category* Λ_∞ is a category whose objects are $\langle n \rangle$ for $n \geq 0$. A morphism $f\colon \langle n \rangle \to \langle m \rangle$ in Λ is a non-decreasing map of sets $f\colon \mathbb{Z} \to \mathbb{Z}$ such that

$$f(i+n+1) = f(i) + m + 1$$

for all $i \in \mathbb{Z}$. The paracyclic category is a crossed simplicial group, all of whose automorphism groups are isomorphic to $\mathbb{Z}$. This crossed simplicial group first appears in [**9**, Example 3]. The term *paracyclic*, as well as the explict description given here, comes from [**11**].

CONSTRUCTION 6.7. Let $\Delta\mathfrak{G}$ be a crossed simplicial group. Given a morphism $\varphi\colon [n] \to [m]$ in Δ, we can define a map of sets $\varphi^*\colon \mathfrak{G}_m \to \mathfrak{G}_n$ as follows. Given $g \in \mathfrak{G}_m$, the canonical factorization allows us to complete the diagram

$$\begin{array}{ccc} & & [m] \\ & & \big\downarrow{\scriptstyle \varphi} \\ [n] & \xrightarrow[g]{} & [n] \end{array}$$

uniquely to a commutative diagram

$$\begin{array}{ccc} [m] & \xrightarrow{\varphi^*(g)} & [m] \\ {\scriptstyle \psi}\big\downarrow & & \big\downarrow{\scriptstyle \varphi} \\ [n] & \xrightarrow[g]{} & [n], \end{array}$$

where ψ is a morphism in Δ. The uniqueness of the canonical factorization implies that the maps φ^* assemble into a simplicial set

$$\begin{aligned} \mathfrak{G}_*\colon \Delta^{\mathrm{op}} &\longrightarrow \mathcal{S}\mathrm{et} \\ [n] &\longmapsto \mathfrak{G}_n. \end{aligned}$$

REMARK 6.8. The simplicial set $\mathfrak{G}_*$, together with the group structures on $\mathfrak{G}_n$ and a few algebraic identities, suffices to determine the category $\Delta\mathfrak{G}$ completely. While this result is not important to our exposition here, proofs of it can be found in [**9**, Prop. 1.6] and [**14**, Theorem 1.4].

REMARK 6.9. The notation for a crossed simplicial group originally used in [**9**] was ΔG, which is also the notation used in [**15**, Ch. 6]. In the latter, the cyclic category Λ is denoted by ΔC. We have chosen to instead use the notation

$\Delta\mathfrak{G}$ from [7], since it helps distinguish the simplicial sets associated to crossed simplicial groups from other notation in use here.

NOTATION 6.10. Because we use the conventional notation Λ for the cyclic category, rather than the crossed simplicial group notation $\Delta\mathfrak{G}$, we must fix a notation for the simplicial set associated to Λ. To adhere as closely to our convention as possible, we write Λ_* for this simplicial set. We then have

$$\begin{aligned}\Lambda_0 &= \{\mathrm{id}_{[0]}\}\\ \Lambda_1 &= \{\mathrm{id}_{[1]}, \tau_1\}\\ \Lambda_2 &= \{\mathrm{id}_{[2]}, \tau_2, \tau_2^2\}\\ \vdots\ &\quad \ddots\end{aligned}$$

As we show in Lemma 6.12 below, the only non-degenerate simplices of Λ_* are $\mathrm{id}_{[0]}$ and τ_1.

There is a profound relationship between the cyclic category, cyclic objects, and circle actions, which was an initial motivation for the definition of the cyclic category by Connes [5]. This relationship extends to other crossed simplicial groups as well, and may be summarized in the following theorem. The first and third statements are proved in [9, Thm 5.3], and the second statement follows from [9, Thm 5.12].

THEOREM 6.11. *Let $\Delta\mathfrak{G}$ be a crossed simplicial group, and let $\mathfrak{G}_*$ be the associated simplicial set.*

1. *The realization $|\mathfrak{G}_*|$ is canonically equipped with the structure of a topological group.*
2. *There is a homotopy equivalence $B|\mathfrak{G}_*| \simeq |N(\Delta\mathfrak{G})|$, where $B|\mathfrak{G}_*|$ denotes the classifying space of the topological group $|\mathfrak{G}_*|$.*
3. *Given a functor $X\colon \Delta\mathfrak{G}^{\mathrm{op}} \to \mathcal{S}et$, the realization $|X|$ of the underlying simplicial object comes equipped with a continuous action of $|\mathfrak{G}_*|$, providing a functor*
$$\mathcal{S}\mathrm{et}_{\Delta\mathfrak{G}} \longrightarrow |\mathfrak{G}_*|\text{-}\mathcal{T}op$$
to spaces equipped with a continuous $|\mathfrak{G}_*|$-action.

In the case of the cyclic category Λ, the corresponding topological group is $S^1 \cong U(1) \cong SO(2)$, a fact we want to demonstrate on the level of topological spaces. First, we prove the following lemma that we mentioned above.

LEMMA 6.12. *For $n \geq 2$, every n-simplex of Λ_* is degenerate.*

PROOF. Let $s_i\colon [n+1] \to [n]$ be the ith degeneracy map in Δ. A short computation shows that $s_n^*(\tau_n^k) = \tau_{n+1}^k$ for $0 \leq k \leq n$. Moreover, $s_{n-1}^*(\tau_n^n) = \tau_{n+1}^{n+1}$, completing the proof. □

PROPOSITION 6.13. *There is a homeomorphism*
$$|\Lambda_*| \cong S^1.$$

PROOF. Expressing $|\Lambda_*|$ in terms of non-degenerate simplices, we see that it is obtained by gluing both endpoints of a single 1-simplex to a single 0-simplex. □

We do not construct the group structure on $|\Lambda_*|$ here, instead directing the interested reader to [9, §5] or [15, Ch. 7].

7. Cyclic sets and their model structure

We now turn to considering the categories $s\mathcal{S}\mathrm{et}$ of simplicial sets and $c\mathcal{S}et$ of cyclic sets. The category $s\mathcal{S}et$ of simplicial sets has as objects the functors $\Delta^{\mathrm{op}} \to \mathcal{S}et$ and as morphisms the natural transformations between them. This category has a model structure in which the weak equivalences are the weak homotopy equivalences, or maps that induce isomorphisms on all homotopy groups, the fibrations are the Serre fibrations, and the cofibrations are the retracts of cell inclusions [**17**]. In what follows, we simply denote this model structure by $s\mathcal{S}\mathrm{et}$.

Similarly, a cyclic set is given by a functor $X\colon \Lambda^{\mathrm{op}} \to \mathcal{S}et$ and is thus a simplicial set together with bijections $t_n\colon X_n \to X_n$ for every n satisfying the duals of the identities given in Definition 5.7. The category $c\mathcal{S}et$ of cyclic sets has a model structure that was originally established by Dwyer, Hopkins, and Kan [**6**]. In this section, we give a full exposition of this model structure.

We first observe that the categories $s\mathcal{S}\mathrm{et}$ and $c\mathcal{S}et$ are connected by the forgetful functor $j^*\colon c\mathcal{S}et \to s\mathcal{S}\mathrm{et}$ induced by the inclusion functor $j\colon \Delta \to \Lambda$. As a first step, let us consider the application of the functor j^* to representables.

In the category of simplicial sets, for for every integer $n \geq 0$, there is a representable simplicial set $\Delta[n]$ given by

$$\mathrm{Hom}_{\Delta}(-,[n])\colon \Delta^{\mathrm{op}} \to \mathcal{S}et.$$

Similarly, for every integer $n \geq 0$, there is a representable cyclic set $\Lambda[n]$, given by

$$\mathrm{Hom}_{\Lambda}(-,\langle n\rangle)\colon \Lambda^{\mathrm{op}} \to \mathcal{S}et.$$

By the Yoneda Lemma, these cyclic sets have the universal property that the elements $x \in X_n$ correspond uniquely to maps $c_x\colon \Lambda[n] \to X$ such that $c_x(\mathrm{id}_{\langle n\rangle}) = x$. This universal property can be encoded as a natural isomorphism

$$\mathrm{Hom}_{c\mathcal{S}\mathrm{et}}(\Lambda[n], X) \cong X_n.$$

Since the simplicial representables $\Delta[n]$ have an analogous universal property in $s\mathcal{S}\mathrm{et}$, there is a composite isomorphism

$$\mathrm{Hom}_{c\mathcal{S}et}(\Lambda[n], X) \cong X_n \cong \mathrm{Hom}_{s\mathcal{S}\mathrm{et}}(\Delta[n], j^*X).$$

Because the functor j^* preserves colimits, composing it with geometric realization provides us with a colimit-preserving functor $|j^*(-)|$ producing a topological space from a cyclic set. The key to understanding the realization of cyclic sets lies in the realization of the representables $|\Lambda[n]|$.

PROPOSITION 7.1. *The simplicial object*

$$\Delta^{\mathrm{op}} \xrightarrow{j^*\Lambda[-]} s\mathcal{S}\mathrm{et} \xrightarrow{|-|} \mathcal{T}op$$

is naturally isomorphic to $S^1 \times |\Delta[-]|$.

PROOF. We construct a map between the two, following [**9**, Prop. 5.1], and leave it to the reader to verify it is natural and a homeomorphism. A consequence of the canonical factorization from Proposition 5.10 is that we can write each k-simplex of $\Lambda[n]$ uniquely as a pair (φ, g), where $\varphi\colon [k] \to [n]$ is a morphism in Δ and g is an automorphism of $\langle k\rangle$ in Λ.

Recall from Notation 6.10 that Λ_* denotes the simplicial set associated to Λ, and from Proposition 6.13 that the realization of this simplicial set is homeomorphic to the circle. Explicitly checking the action of morphisms in Δ shows that the

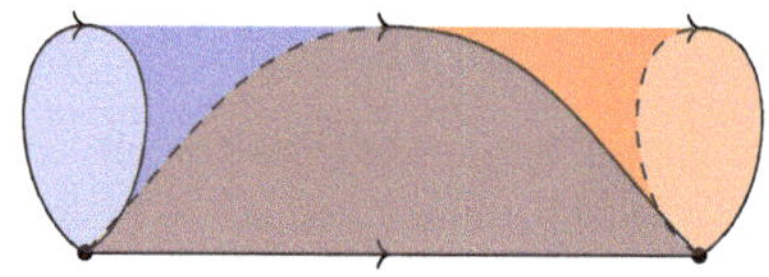

FIGURE 16. A depiction of the simplicial set $j^*\Lambda[1]$ with the two non-degenerate 2-simplices pictured in red and blue.

projection $\Lambda[n] \to \Lambda_*$ given by $(\varphi, g) \mapsto g$ is a simplicial map. On the other hand, we can define a continuous map

$$\begin{aligned} |\Lambda[n]| &\longrightarrow |\Delta[n]| \\ [(\varphi, g), x] &\longmapsto [\varphi, g \cdot x] \end{aligned}$$

where the automorphisms of $\langle k \rangle$ act on $|\Delta[k]|$ by permuting the vertices. Together, these two maps define the desired continuous maps $|\Lambda[n]| \to |\Lambda_*| \times |\Delta[n]| \cong S^1 \times |\Delta[n]|$. See Figure 16 for a depiction of $j^*\Lambda[1]$. □

Since every functor from a small category to $\mathcal{S}$et is a colimit of representables, and realization functors preserve colimits, we obtain two formulas for the realization of a cyclic set X. The first is the more common formula in terms of the underlying simplicial set:

$$|j^*X| \cong \operatorname*{colim}_{\Delta[-] \to j^*X} |\Delta[-]|,$$

where the colimit is taken over the overcategory Δ/X. The second is similar in form:

$$|j^*X| \cong \operatorname*{colim}_{\Lambda[-] \to X} |j^*\Lambda[-]|,$$

but the colimit is taken instead over the overcategory Λ/X.

These two ways of computing $|j^*X|$ suggest two different ways of thinking about the elements of X_n. On the one hand, we can think of $x \in X_n$ as an n-simplex of the underlying simplicial set. On the other, we can think of $x \in X_n$ as something we might call an "n-cyclex": a map $\Lambda[n] \to X$, or a specified copy of $S^1 \times \Delta[n]$ in the realization. One of the key distinctions between these two perspectives is that there can be multiple "cyclices" that are isomorphic, but not equal. This feature is a reflection of the fact that Λ admits non-trivial automorphisms, while Δ does not.

REMARK 7.2. The colimit formula

$$|j^*X| \cong \operatorname*{colim}_{\Lambda[-] \to X} |j^*\Lambda[-]|$$

for the realization of a cyclic set X means that we can view the realization as the left Kan extension along the Yoneda embedding of the cocyclic object that sends $\langle n \rangle$ to $|j^*\Lambda[n]|$.

There is another cocyclic object that one might use to construct an alternative to the realization, which simply sends $\langle n \rangle$ to $\Delta[n]$, and sends a morphism f in Λ to the convex map induced by the underlying map of sets of f. However, this alternative cocyclic object does not carry the same non-trivial circle actions which motivate much of the interest in cyclic sets, and so we do not discuss it further here.

Returning to our discussion of realizations of cyclic sets, Proposition 7.1 has an immediate corollary.

COROLLARY 7.3. *The induced functor* $j^*\Lambda[-]\colon \Lambda \to s\mathcal{S}\mathrm{et}$ *sends every morphism in* Λ *to a weak equivalence.*

With this corollary in place, we can now explore the homotopy theory of cyclic sets, as first described by Dwyer, Hopkins, and Kan in [**6**, 3.1]. We give a new proof here, using a recognition theorem establishing that the model structure for cyclic sets is cofibrantly generated, a fact that is implicit but not emphasized in the original proof.

Recall that, just as for simplicial sets, we have boundaries $\partial\Lambda[n] \subseteq \Lambda[n]$ for every $n > 0$, given by the union of the images of the face maps of $\Lambda[n]$, and, for every $n \geq 1$ and $0 \leq k \leq n$, the k-horn $\Lambda[n,k] \subseteq \partial\Lambda[n]$ obtained by removing the k-th face; see [**6**, §2] for more details.

THEOREM 7.4. *The category* $c\mathcal{S}et$ *of cyclic sets admits a cofibrantly generated model structure in which a map* $X \to Y$ *is:*

- *a weak equivalence or a fibration if the induced map* $j^*X \to j^*Y$ *is a weak equivalence or fibration, respectively, in* $s\mathcal{S}\mathrm{et}$*; and*
- *a cofibration if it is a retract of a (possibly transfinite) composite of cobase extensions along the inclusions* $\partial\Lambda[n] \to \Lambda[n]$ *for* $n \geq 0$.

PROOF. We claim that we can take as generating cofibrations the set

$$I = \{\partial\Lambda[n] \to \Lambda[n] \mid n \geq 0\}$$

and as generating acyclic cofibrations the set

$$J = \{\Lambda[n,k] \to \Lambda[n] \mid n \geq 1, 0 \leq k \leq n\}.$$

Using these sets, we want to verify the conditions of [**13**, 11.3.1].

First, observe that the category $c\mathcal{S}et$ has all limits and colimits, and that the weak equivalences are closed under retracts and satisfy the two-out-of-three property. It is also not hard to check that the sets I and J satisfy the small object argument, establishing condition (1) of [**13**, 11.3.1].

One can check from the definitions of fibration and acyclic fibration, or see [**6**, 3.2, 3.3], that a map in $c\mathcal{S}et$ is a fibration if and only if it has the right lifting property with respect to the maps in J, and that a map is an acyclic fibration if and only if it has the right lifting property with respect to the maps in I. It follows immediately that a map is an I-injective if and only if it is a J-injective and a weak equivalence, establishing conditions (2) and (4)(b) of [**13**, 11.3.1].

It remains to show that every J-cofibration is an I-cofibration and a weak equivalence. If a map $A \to B$ is a J-cofibration, then it has the left lifting property with respect to the fibrations, using the results from the previous paragraph. In particular, it has the left lifting property with respect to the acyclic fibrations, from which it follows, again by the results of the previous paragraph, that it is an I-cofibration. It remains to show that $A \to B$ is a weak equivalence.

Denote the left and right adjoints of j^* by $j_!$ and j_* respectively, and note that $j_!\Delta[n] \cong \Lambda[n]$. Consider a horn inclusion $\Lambda^k[n] \to \Delta[n]$. Writing each of these two simplicial sets as a colimit over its category of non-degenerate simplices, and applying $j^* \circ j_!$, we see that the morphism

$$j^*(j_!\Lambda^k[n]) \longrightarrow j^*(j_!\Delta[n])$$

is isomorphic to the morphism

$$\operatorname{colim}_{(\Delta\downarrow\Lambda^k[n])_{ND}} j^*\Lambda[r] \longrightarrow \operatorname{colim}_{(\Delta\downarrow\Delta[n])_{ND}} j^*\Lambda[r]$$

induced by the inclusion of categories of simplices. Applying the geometric realization and using Proposition 7.1 yields the morphism

$$\operatorname{colim}_{(\Delta\downarrow\Lambda^k[n])_{ND}} S^1 \times \Delta^r \longrightarrow \operatorname{colim}_{(\Delta\downarrow\Delta[n])_{ND}} S^1 \times \Delta^r.$$

However, this morphism is isomorphic to $S^1 \times |\Lambda^k[n]| \to S^1 \times \Delta^n$, which is a homotopy equivalence. As such we see that $j^*(j_!\Lambda^k[n]) \to j^*(j_!\Delta[n])$ is a weak equivalence. It is not hard to check that this morphism is also a monomorphism, and thus is an acyclic cofibration in $s\mathcal{S}$et. By adjointness, it follows that if $\pi\colon X \to Y$ is a fibration in $s\mathcal{S}$et, then the induced map $j^*(j_*X) \to j^*(j_*Y)$ is also a fibration. Now suppose that $f\colon A \to B$ is a J-cofibration, and let $\pi\colon X \to Y$ be a fibration in $s\mathcal{S}$et. A lifting problem

$$\begin{array}{ccc} j^*A & \longrightarrow & X \\ {\scriptstyle f}\downarrow & \nearrow & \downarrow{\scriptstyle \pi} \\ j^*B & \longrightarrow & Y \end{array}$$

is equivalent to a lifting problem

$$\begin{array}{ccc} A & \longrightarrow & j_*X \\ {\scriptstyle f}\downarrow & \nearrow & \downarrow{\scriptstyle j_*(\pi)} \\ B & \longrightarrow & j_*Y. \end{array}$$

However, by the preceding paragraph, the morphism $j_*(\pi)$ is a fibration, and so this lifting problem has a solution. Thus, $j^*A \to j^*B$ is an acyclic cofibration of simplicial sets, so that $A \to B$ is a weak equivalence as desired. □

Some of the motivation for this model structure is its comparison with the following model structure on S^1-spaces, or topological spaces equipped with an S^1-action. We denote by $\mathcal{T}op^{S^1}$ the category of S^1 spaces with equivariant continuous maps.

THEOREM 7.5 ([**6**, 4.1]). *The category $\mathcal{T}op^{S^1}$ of S^1-spaces admits a model structure in which a map $X \to Y$ is:*

- *a weak equivalence or a fibration whenever the underlying map of topological spaces $X \to Y$ is a weak homotopy equivalence or a Serre fibration, respectively; and*
- *a cofibration if it is a retract of a (possibly transfinite) composite of cobase extensions along the inclusions $S^1 \times |\partial\Delta[n]| \to S^1 \times |\Delta[n]|$ for $n \geq 0$.*

To compare these two model structures, it is necessary to promote the realization of cyclic sets from a functor

$$|j^*(-)|\colon c\mathcal{S}\text{et} \longrightarrow \mathcal{T}op$$

to a functor

$$L^c\colon c\mathcal{S}\text{et} \longrightarrow \mathcal{T}op^{S^1}.$$

The construction of L^c turns out to be purely formal. Since colimits in $\mathcal{T}op^{S^1}$ are preserved by the functor which sends each S^1-space to its underlying topological space, we can assign

$$L^c(X) := \operatorname*{colim}_{\Lambda[-]\to X} |j^*\Lambda[-]|$$

where the colimit is taken in $\mathcal{T}op^{S^1}$ and the realizations $|j^*\Lambda[n]| \cong S^1 \times |\Delta[n]|$ are equipped with the canonical S^1-action on the first factor. We thus obtain the following result.

PROPOSITION 7.6 ([**6**, 2.8]). *There is a functor* $L^c\colon c\mathcal{S}et \to \mathcal{T}op^{S^1}$ *such that the diagram*

$$\begin{array}{ccc} c\mathcal{S}et & \xrightarrow{L^c} & \mathcal{T}op^{S^1} \\ & \searrow^{|j^*|} & \downarrow \\ & & \mathcal{T}op \end{array}$$

commutes up to natural equivalence.

With the functor L^c in place, we can complete our discussion of the homotopy theory of cyclic sets by providing the comparison between the two model structures.

THEOREM 7.7 (). [**6**, 4.2] *The functor* $L^c\colon c\mathcal{S}et \to \mathcal{T}op^{S^1}$ *has as right adjoint the functor* $R^c = \operatorname{Hom}(L^c\Lambda[-], -)\colon \mathcal{T}op^{S^1} \to c\mathcal{S}et$. *Moreover, this pair of adjoint functors is a Quillen equivalence of model categories.*

8. Levelwise model structures on cyclic spaces

In this section, we consider functors $X\colon \Lambda^{\mathrm{op}} \to s\mathcal{S}\mathrm{et}$, and we use the standard model structure on simplicial sets [**17**] to produce model structures on the category $s\mathcal{S}\mathrm{et}^{\Lambda^{\mathrm{op}}}$ of all such functors. As is usual for categories of functors from a small category, we have:

- the *projective model structure*, in which weak equivalences and fibrations are given levelwise [**13**, 11.6.1]; and
- the *injective model structure*, in which weak equivalences and cofibrations are given levelwise [**12**], [**16**, A.2.8].

However, there is another model structure we can also consider.

One of the nice features of the category Δ^{op} is that it has the structure of a Reedy category [**18**], and as a consequence the category $s\mathcal{S}\mathrm{et}^{\Delta^{\mathrm{op}}}$ of simplicial spaces can be given the Reedy model structure. Indeed, this situation is particularly nice in that the Reedy model structure agrees with the injective model structure [**13**, 15.8.7].

The category Λ is not a Reedy category, since its objects have nontrivial automorphisms, but it does have the structure of a generalized Reedy category in the sense of Berger and Moerdijk [**2**], which we now define. Recall that a subcategory $\mathcal{D}$ of a category $\mathcal{C}$ is *wide* if it contains all the objects of $\mathcal{C}$. As an example that plays a role in this definition, we denote by $\operatorname{Iso}(\mathcal{C})$ the maximal subgroupoid of $\mathcal{C}$.

DEFINITION 8.1 ([**2**, 1.1]). Let $\mathcal{C}$ be a small category. A *generalized Reedy structure* on $\mathcal{C}$ consists of wide subcategories $\mathcal{C}^+$ and $\mathcal{C}^-$ and a degree function $d\colon \mathrm{ob}(\mathcal{C}) \to \mathbb{N}$ such that:

(1) noninvertible morphisms in $\mathcal{C}^+$ raise the degree, while those in $\mathcal{C}^-$ lower degree, and isomorphisms in $\mathcal{C}$ preserve the degree;
(2) $\mathcal{C}^+ \cap \mathcal{C}^- = \mathrm{Iso}(C)$;
(3) every morphism f of $\mathcal{C}$ can be factored as $f = gh$ with g in $\mathcal{C}^+$ and h in $\mathcal{C}^-$, and this factorization is unique up to isomorphism; and
(4) if f is a morphism of $\mathcal{C}^-$ and $f\theta = f$ for some isomorphism θ in $\mathcal{C}$, then θ is an identity map.

In the general context of crossed simplicial groups, Berger and Moerdijk show in [**2**, 2.7] that Λ has the structure of a generalized Reedy category. We give a more direct description here. We take the degree function $d\colon \mathrm{ob}(\Lambda) \to \mathbb{N}$ to be given by $\langle n \rangle \mapsto n$; then we can take Λ^+ to be the wide subcategory whose morphisms preserve or increase degree, and likewise Λ^- the wide subcategory whose morphisms preserve or lower degree.

REMARK 8.2. In [**2**, §2], Berger and Moerdijk introduce a notion of *crossed groups* over a Reedy category that generalize the crossed simplicial groups as we have described above. What we call a crossed simplicial group $\Delta\mathfrak{G}$ is, in their nomenclature, the *total category* associated to the *crossed Δ-group* $\mathfrak{G}_*$. As in Remark 6.3, this distinction reflects two different but equivalent ways of defining crossed simplicial groups.

Let $\Lambda^+(n)$ be the category whose objects are the non-invertible morphisms in Λ^+ with codomain $\langle n \rangle$ and in which a morphism from $u\colon s \to r$ to $u'\colon s' \to r$ is given by a morphism $w\colon s \to s'$ such that $u = u'w$. The automorphism group $\mathrm{Aut}(\langle n \rangle) = C_{n+1}$ acts on $\Lambda^+(n)$ by composition. For each functor $X\colon \Lambda^{\mathrm{op}} \to s\mathcal{S}\mathrm{et}$ and each $n \geq 0$, the *n-th latching object* $L_n(X)$ of X is defined to be

$$L_n(X) = \operatorname*{colim}_{\langle m \rangle \to \langle r \rangle} X_s$$

with colimit taken over $\Lambda^+(n)$. Observe that $\mathrm{Aut}(\langle n \rangle) = C_{n+1}$ acts on $L_n(X)$.

Dually, we can define the category $\Lambda^-(n)$ with objects the non-invertible morphisms in Λ^- with domain $\langle n \rangle$ and morphisms defined analogously. Again, the automorphism group $\mathrm{Aut}(\langle n \rangle) = C_{n+1}$ acts on $\Lambda^-(n)$ by precomposition. Then, given a functor $X\colon \Lambda^{\mathrm{op}} \to s\mathcal{S}\mathrm{et}$ the *n-th matching object* $M_n(X)$ of X is defined to be

$$M_n(X) = \lim_{\langle n \rangle \to \langle m \rangle} X_s$$

with limit taken over the category $\Lambda^-(n)$, and observe that it has an action of C_{n+1}.

Just as for ordinary categories, the natural maps $L_n(X) \to X_n \to M_n(X)$ induce *relative latching maps*

$$X_n \cup_{L_n(X)} L_n(Y) \to Y_n$$

and *relative matching maps*

$$X_n \to M_n(X) \times_{M_n(Y)} Y_n$$

for all $n \geq 0$.

Now, consider the category of C_{n+1}-simplicial sets, which we regard as the category of functors $C_{n+1} \to s\mathcal{S}\mathrm{et}$, where C_{n+1} is treated as a category with a single object. We can equip this category with the projective model structure, in which weak equivalences and fibrations are given levelwise. Using the discussion above, we can think of the relative latching maps and relative matching maps as morphisms in this category.

The following theorem is the specialization of [**2**, 1.6] to the case of cyclic spaces; we refer to the model structure described here as the *generalized Reedy model structure*. Observe that, in analogy with simplicial spaces, we want to take functors out of Λ^{op}; we simplify the notation in light of the isomorphism between Λ and Λ^{op} here and in what follows.

THEOREM 8.3. *There is a model structure on the category* $s\mathcal{S}\text{et}^{\Lambda}$ *in which a map* $f\colon X \to Y$ *is:*

(1) *a weak equivalence if for each* $n \geq 0$*, the induced map* $f_n\colon X_n \to Y_n$ *is a weak equivalence in* $s\mathcal{S}\text{et}^{C_{n+1}}$*;*
(2) *a cofibration if for each* $n \geq 0$*, the relative latching map*

$$X_n \cup_{L_n X} L_n Y \to Y_n$$

is a cofibration in $s\mathcal{S}\text{et}^{C_{n+1}}$*; and*
(3) *a fibration if for each* $n \geq 0$*, the relative matching map*

$$X_n \to M_n X \times_{M_n Y} Y_n$$

is a fibration in $s\mathcal{S}\text{et}^{C_{n+1}}$*.*

Note that, following Berger and Moerdijk's result, we can replace the category $s\mathcal{S}\text{et}$ with any *$\mathcal{R}$-projective model category*, or model category $\mathcal{E}$ such that for each $n \geq 0$, the category $\mathcal{E}^{C_n}$ admits the projective model structure.

Further observe that, unlike for simplicial spaces, the generalized Reedy structure on $s\mathcal{S}\text{et}^{\Lambda}$ is not the same as the injective model structure. The cofibrations described above are injective cofibrations, but not conversely. Thus, we have three distinct model structures on $s\mathcal{S}\text{et}^{\Lambda}$ with levelwise fibrations: injective, generalized Reedy, and projective.

Let us now examine the generalized Reedy structure in more detail. In the spirit of similar results for simplicial spaces, we want to describe the matching objects more explicitly, enabling us to give a description of the generating cofibrations and generating acyclic cofibrations in this model structure.

Let X be a cyclic space and $n \geq 0$. By definition, we have

$$M_n X = \lim_{\Lambda[n] \to \Lambda[m]} X_m,$$

where the limit is taken over the category $\Lambda^-(n)$. Using the representability of $\Lambda[m]$ and the duality of the category Λ, we obtain isomorphisms

$$\begin{aligned} M_n X &= \lim_{\Lambda[n] \to \Lambda[m]} \text{Map}(\Lambda[m], X) \\ &= \text{Map}\left(\operatorname*{colim}_{\Lambda[m] \to \Lambda[n]} \Lambda[m], X \right). \end{aligned}$$

EXAMPLE 8.4. When $n = 1$, we get

$$\begin{aligned} M_1(X) &= \text{Map}\left(\operatorname*{colim}_{\Lambda[0] \to \Lambda[1]} \Lambda[0], X \right) \\ &= \text{Map}\left(\Lambda[0] \amalg \Lambda[0], X \right). \end{aligned}$$

We use the notation

$$\partial\Lambda[m] := \operatorname*{colim}_{\Lambda[m]\to\Lambda[n]} \Lambda[m],$$

where the colimit is taken over the opposite of the category $\Lambda^{-}n$, which is by duality isomorphic to $\Lambda^{+}(n)$.

Now, let us consider acyclic fibrations in the generalized Reedy model structure, which are those maps of cyclic spaces $X \to Y$ such that the map $X_n \to M_nX \times_{M_nY} Y_n$ is a fibration in $s\mathcal{S}\mathrm{et}^{C_{n+1}}$. For simplicity, let $P_n := M_nX \times_{M_nY} Y_n$, and consider the pullback square defining it:

$$\begin{array}{ccc} P_n & \longrightarrow & M_n(X) \\ \downarrow & & \downarrow \\ Y_n & \longrightarrow & M_n(Y). \end{array}$$

Using the description above, we can write this diagram instead as

$$\begin{array}{ccc} P_n & \longrightarrow & \mathrm{Map}(\partial\Lambda[n], X) \\ \downarrow & & \downarrow \\ \mathrm{Map}(\Lambda[n], Y) & \longrightarrow & \mathrm{Map}(\partial\Lambda[n], Y). \end{array}$$

Now, the map $X_n \to P_n$ is an acyclic fibration in $s\mathcal{S}\mathrm{et}^{C_{n+1}}$ if and only if it has the right lifting property with respect to the maps $\partial\Delta[m] \times C_{n+1} \to \Delta[m] \times C_{n+1}$ for any $m \geq 0$. Here, we are regarding a C_{n+1}-space as a functor $Z \colon C_{n+1} \to s\mathcal{S}\mathrm{et}$, where C_{n+1} is treated as a category with one object. Having acyclic fibrations defined levelwise means that the evaluation at the single object of C_{n+1} gives an acyclic fibration of simplicial sets.

Thus, a map of cyclic spaces $X \to Y$ is an acyclic fibration if and only if a lift exists in any diagram of the form

$$\begin{array}{ccccc} \partial\Delta[m] \times C_{n+1} & \longrightarrow & \mathrm{Map}(\Lambda[n], X) & & \\ \downarrow & \nearrow & \downarrow & & \\ \Delta[m] \times C_{n+1} & \longrightarrow & P_n & \longrightarrow & \mathrm{Map}(\partial\Lambda[n], X) \\ & & \downarrow & & \downarrow \\ & & \mathrm{Map}(\Lambda[n], Y) & \longrightarrow & \mathrm{Map}(\partial\Lambda[n], Y). \end{array}$$

Applying the adjunction between mapping spaces and products, such a lift exists if and only if a lift exists in any diagram of the form

$$\begin{array}{ccc} (\partial\Delta[m] \times C_{n+1} \times \Lambda[n]) \cup (\Delta[m] \times C_{n+1} \times \partial\Lambda[n]) & \longrightarrow & X \\ \downarrow & \nearrow & \downarrow \\ \Delta[m] \times C_{n+1} \times \Lambda[n] & \longrightarrow & Y \end{array}$$

in the category of cyclic spaces. The vertical maps on the left-hand side thus provide a set of generating cofibrations for the generalized Reedy structure, where we have $m, n \geq 0$.

Using an analogous argument, we can take the set of maps

$$(V[m,k] \times C_{n+1} \times \Lambda[n]) \cup (\Delta[m] \times C_{n+1} \times \Lambda[n]) \to \Delta[m] \times C_{n+1} \times \Lambda[n]$$

as a set of generating acyclic cofibrations, where $V[m,k]$ denotes the simplicial k-horn, $m \geq 0$, $0 \leq k \leq m$, and $n \geq 0$.

The sets of generating cofibrations and generating acyclic cofibrations we have described both take the form of pushout-products

$$(A \times D) \cup (B \times C) \longrightarrow B \times D,$$

where $A \to B$ is a morphism of simplicial sets, and $C \to D$ is a morphism in $s\mathcal{S}\mathrm{et}^{\Lambda}$. Standard arguments for pushout products, such as can be found in [**13**, 9.3.4] can be used to show that, for any cofibration $A \to B$ of simplicial sets and any Reedy cofibration $C \to D$ in $s\mathcal{S}\mathrm{et}^{\Lambda}$, the corresponding pushout-product is a cofibration in $s\mathcal{S}\mathrm{et}^{\Lambda}$ and moreover is an acyclic cofibration if either $A \to B$ or $C \to D$ is a acyclic cofibration. In particular, $s\mathcal{S}\mathrm{et}^{\Lambda}$ has the structure of a simplicial model category.

We summarize the preceding discussion in the following proposition.

PROPOSITION 8.5. *The generalized Reedy model structure on* $s\mathcal{S}\mathrm{et}^{\Lambda}$ *is a cofibrantly generated simplicial model category with set of generating cofibrations*

$$\{(\partial\Delta[m] \times C_{n+1} \times \Lambda[n]) \cup) \Delta[m] \times C_{n+1} \times \partial\Lambda[n]) \to (\Delta[m] \times C_{n+1}\Lambda[n])\}$$

and set of generating acyclic cofibrations

$$\{(V[m,k] \times C_{n+1} \times \Lambda[n]) \cup (\Delta[m] \times C_{n+1} \times \Lambda[n]) \to \Delta[m] \times C_{n+1} \times \Lambda[n]\}.$$

We conclude our discussion of the generalized Reedy model structure In analogy with the case of usual Reedy categories, the generalized Reedy model structure on $s\mathcal{S}\mathrm{et}^{\Lambda}$ sits between the projective and injective model structures.

PROPOSITION 8.6. *Let* $f \colon X \to Y$ *be a Reedy fibration between cyclic simplicial sets. Then* $M_n X \to M_n Y$ *is a fibration for every* n.

Note that, since $s\mathcal{S}\mathrm{et}^{C_{n+1}}$ is equipped with the projective model structure, it does not matter whether we take this matching map to be a fibration in $s\mathcal{S}\mathrm{et}$ or a fibration in $s\mathcal{S}\mathrm{et}^{C_{n+1}}$.

PROOF. Following [**13**, Lem. 15.3.9], we define a filtration of the category $\Lambda^{-}(n)$ by declaring $F^k\Lambda^{-}(n)$ to consist of only those $\langle n \rangle \to \langle \ell \rangle$ such that $\ell \leq k$. Taking limits over this filtration gives rise to a sequence of C_{n+1}-spaces

$$\begin{array}{ccccccc}
M_n X & \longrightarrow & (M_n X)^{\leq n-2} & \longrightarrow & \cdots & \longrightarrow & (M_n X)^{\leq 0} \\
\downarrow & & \downarrow & & & & \downarrow \\
M_n Y & \longrightarrow & (M_n Y)^{\leq n-2} & \longrightarrow & \cdots & \longrightarrow & (M_n Y)^{\leq 0},
\end{array}$$

where the C_{n+1}-action is given by composing with the cyclic shifts of $\langle n \rangle$.

Given a C_{n+1}-cofibration $\iota \colon A \to B$ and a lifting problem

$$\begin{array}{ccc}
A & \longrightarrow & M_n X \\
{\scriptstyle \iota}\downarrow & \nearrow & \downarrow{\scriptstyle M_n f} \\
B & \longrightarrow & M_n Y,
\end{array}$$

we want to construct a lift inductively. First note that the category $F^0\Lambda^-(n)$ is precisely the discrete category whose objects are $\langle n\rangle \to \langle 0\rangle$. Consequently, the corresponding map $(M_nX)^{\leq 0} \to (M_nY)^{\leq 0}$ is simply

$$X_0^{n+1} \to Y_0^{n+1}$$

with C_{n+1}-action given by permuting the factors. Since $M_0X \cong * \cong M_0Y$ is the terminal object, the requirement that f be a Reedy fibration implies that $X_0 \to Y_0$ is a fibration. Thus, $(M_nX)^{\leq 0} \to (M_nY)^{\leq 0}$ is a fibration, and so we obtain a C_{n+1}-map $g_0 : B \to (M_nX)^{\leq 0}$ lifting the corresponding square.

Suppose now that we have defined $g_k \colon B \to (M_nX)^{\leq k}$ lifting the corresponding square. This map corresponds to a cone over the restricted diagram

$$X\colon\ F^k\Lambda^-(n) \longrightarrow s\mathcal{S}\mathrm{et},$$

and our aim is now to extend this cone to $F^{k+1}\Lambda^-(n)$. We denote the component of the cone at $\varphi\colon \langle n\rangle \to \langle \ell\rangle$ by μ_φ.

Let $\psi\colon \langle n\rangle \to \langle k+1\rangle$ be an object of $F^{k+1}\Lambda^-(n)$ on which μ is not defined. We wish to define a corresponding component $\mu_\psi\colon B \to X_{k+1}$. Composition with ψ defines a functor

$$I_\psi\colon\ \Lambda^-(k+1) \longrightarrow F^k\Lambda^-(n)$$

and so induces a commutative diagram

$$\begin{array}{ccc} (M_nX)^{\leq k} & \longrightarrow & M_{k+1}X \\ \downarrow & & \downarrow \\ (M_nY)^{\leq k} & \longrightarrow & M_{k+1}Y. \end{array}$$

A morphism $\mu_\psi\colon B \to X_{k+1}$ compatible with the existing cone is equivalently a lift of the induced diagram

$$\begin{array}{ccc} A & \longrightarrow & X_{k+1} \\ {\scriptstyle\iota}\downarrow & \overset{\mu_\psi}{\nearrow} & \downarrow \\ B & \longrightarrow & M_{k+1}X \times_{M_{k+1}Y} Y_{k+1} \end{array}$$

which exists since f is a Reedy fibration.

To extend the cone to all of $F^{k+1}\Lambda^-(n)$, we define $\mu_{\psi\circ\tau^\ell}$, where τ is the cyclic shift automorphism of $\langle n\rangle$, to be the composite of μ_ψ with the corresponding cyclic shift of B. We perform this procedure for each orbit of morphisms $\langle n\rangle \to \langle k+1\rangle$ under the C_{n+1}-action, thereby obtaining an extension of the cone to $F^{k+1}\Lambda^-(n)$.

By induction, we thus obtain a lift of the original diagram

$$\begin{array}{ccc} A & \longrightarrow & M_nX \\ {\scriptstyle\iota}\downarrow & \overset{\mu}{\nearrow} & \downarrow{\scriptstyle M_nf} \\ B & \longrightarrow & M_nY \end{array}$$

and so $M_nX \to M_nY$ is a fibration. □

Proposition 8.7. *The identity functors*

$$(s\mathcal{S}\mathrm{et}^\Lambda)^{\mathrm{proj}} \xrightarrow{\mathrm{id}} (s\mathcal{S}\mathrm{et}^\Lambda)^{\mathrm{Reedy}} \xrightarrow{\mathrm{id}} (s\mathcal{S}\mathrm{et}^\Lambda)^{\mathrm{inj}}$$

define left Quillen equivalences.

PROOF. Since the classes of weak equivalences in all three model structures agree, it suffices to show that the identity functors above preserve cofibrations, or that their inverse identity functors preserve fibrations.

First, consider a generating Reedy cofibration

$$\partial\Delta[m] \times C_{n+1} \times \Lambda[n] \cup \Delta[m] \times C_{n+1} \times \partial\Lambda[n] \longrightarrow \Delta[m] \times C_{n+1} \times \Lambda[n]$$

and notice that the component of this transformation at an object $\langle k \rangle$ of Λ is a pushout-product of cofibrations of simplicial sets. Consequently, the generating Reedy cofibrations are, in particular, injective cofibrations, and thus the right-hand adjunction is a Quillen functor.

For the left-hand adjunction, suppose that $f\colon X \to Y$ is a Reedy fibration. We can factor f_n as

$$X_n \longrightarrow M_n X \times_{M_n Y} Y \longrightarrow Y_n.$$

Since f is a Reedy fibration, the first of these maps is a fibration, and by Proposition 8.6, the second is also. Thus, the map $f_n\colon X_n \to Y_n$ is always a fibration of simplicial sets, and so the identity functor

$$(s\mathcal{S}\mathrm{et}^\Lambda)^{\mathrm{Reedy}} \xrightarrow{\ \mathrm{id}\ } (s\mathcal{S}\mathrm{et}^\Lambda)^{\mathrm{proj}}$$

is right Quillen, completing the proof. □

As an immediate corollary, we obtain one final property of the generalized Reedy model structure.

COROLLARY 8.8. *The projective and generalized Reedy model structures on* $s\mathcal{S}\mathrm{et}^\Lambda$ *are left proper.*

PROOF. The model structure on $s\mathcal{S}\mathrm{et}$ is left proper, and so the injective model structure is as well. Since the injective model structure is left proper, and the classes of weak equivalences in all three model structures in Proposition 8.7 are the same, it follows from the fact that the cofibrations in the projective and generalized Reedy model structures are injective cofibrations that these model structures are left proper. □

9. Model structures for cyclic Segal spaces

We now wish to localize the model structures in the previous section with respect to an analogue of the maps used to obtain the Segal space model structure on simplicial spaces. We begin by reviewing the techniques of localization of model categories, then proceed to use them to develop our desired model structure.

9.1. Local objects and localization. We understand a *localization* of a model category at a collection $\mathcal{S}$ of morphisms to be a model category with the same underlying category, but with weak equivalences obtained by formally adding $\mathcal{S}$ to the weak equivalences of the original model structure. To have better control of this procedure, we will use the framework of *left Bousfield localization* to produce our desired model structures.

Loosely speaking, the left Bousfield localization arises by formally keeping the same cofibrations as the original model structure when we add $\mathcal{S}$ to the weak equivalences. There are many ways to formally characterize left Bousfield localizations, but for our purposes the most convenient is via local objects and local equivalences.

Since the model structures we wish to consider are simplicially enriched, we will discuss the localization procedure in this setting.

In any simplicial model structure, given a weak equivalence $A \to B$ and a fibrant object X, the induced map on derived mapping spaces

$$R\mathrm{Map}(B, X) \longrightarrow R\mathrm{Map}(A, X)$$

is a weak equivalence in the model structure on simplicial sets. We can define the left Bousfield localization by reversing this observation to obtain candidate classes for the fibrant objects and weak equivalences of our localized model structure. In what follows, let $\mathcal{M}$ be a simplicial model category and S the set of morphisms at which we want to localize.

DEFINITION 9.1. A fibrant object X of $\mathcal{M}$ is *S-local* if, for every morphism $A \to B$ in S, the induced map on derived mapping spaces

$$R\mathrm{Map}(B, X) \longrightarrow R\mathrm{Map}(A, X)$$

is a weak equivalence.

A morphism $f\colon C \to D$ in $\mathcal{M}$ is an *S-local equivalence* if, for every S-local object Y, the induced map on derived mapping spaces

$$R\mathrm{Map}(C, Y) \longrightarrow R\mathrm{Map}(D, Y)$$

is a weak equivalence.

The idea is that, if the localized model structure exists, then the fibrant objects in the localized model structure must be S-local objects. Assuming that the fibrant objects are precisely the S-local objects, then the weak equivalences must be the S-local equivalences. We thus arrive at the following definition.

DEFINITION 9.2. The *left Bousfield localization* of $\mathcal{M}$ at the set S, if it exists, is the model structure uniquely determined by the requirements that

- the cofibrations are precisely the cofibrations in the original model structure $\mathcal{M}$; and
- the weak equivalences are the S-local equivalences.

The key result that we need for Bousfield localizations is the following result from [**1**, Theorem 4.46].

THEOREM 9.3. *Let $\mathcal{M}$ be a left proper, combinatorial, simplicial model category such that the sources of the generating cofibrations and acyclic cofibrations are cofibrant. Let S be a set of morphisms in $\mathcal{M}$. Then the left Bousfield localization $L_S\mathcal{M}$ of $\mathcal{M}$ at S exists and is a left proper, combinatorial, simplicial model category.*

9.2. The Segal localization. Let us first recall the localization used to obtain a model structure for Segal spaces, so that we can modify it appropriately to get a cyclic analogue.

Let $G(n)$ denote the simplicial set given by the colimit

$$\Delta[1] \overset{d_0}{\to} \Delta[0] \overset{d_1}{\leftarrow} \Delta[1] \overset{d_0}{\to} \cdots \overset{d_1}{\leftarrow} \Delta[1],$$

where there are n copies of $\Delta[1]$. There is a natural inclusion $G(n) \to \Delta[n]$; the image is sometimes called the *spine* of $\Delta[n]$. Regarding this inclusion as a map of discrete simplicial spaces, if we map into a fixed simplicial space X, then we get maps

$$X_n = \mathrm{Map}(\Delta[n], X) \to \mathrm{Map}(G(n), x) = \underbrace{X_1 \times_{X_0} \cdots \times_{X_0} X_1}_{n},$$

called *Segal maps.*

A model structure for Segal spaces is obtained by localizing either the injective or the projective model structure on simplicial spaces with respect to the maps $G(n) \to \Delta[n]$ for $n \geq 2$. As a result, the fibrant objects are precisely the simplicial spaces X that are fibrant in the underlying model structure and for which the Segal maps are weak equivalences of simplicial sets for $n \geq 2$.

We want to give a similar localization to obtain a model structure on the category of cyclic spaces in which the fibrant objects are *cyclic Segal spaces*, or cyclic spaces whose underlying simplicial spaces are Segal spaces.

First, let us make the appropriate definitions. Let $\Gamma(n)$ denote the cyclic set that is the colimit of the diagram

$$\Lambda[1] \overset{d_0}{\to} \Lambda[0] \overset{d_1}{\leftarrow} \Lambda[1] \overset{d_0}{\to} \cdots \overset{d_1}{\leftarrow} \Lambda[1],$$

which has a natural inclusion into the representable cyclic set $\Lambda[n]$. As before, we treat this map $\Gamma[n] \to \Lambda[n]$ as a map of discrete cyclic spaces.

We can now localize any one of our model structures on the category of cyclic spaces with respect to the maps $\Gamma(n) \to \Lambda[n]$ for all $n \geq 2$. It remains to show that this model structure has the desired fibrant objects, namely, the cyclic Segal spaces that are fibrant in the appropriate original model structure.

We denote the set of morphisms at which we localize by

$$\mathrm{Seg}_\Lambda := \{\Gamma(n) \to \Lambda[n]\}_{n \geq 2}.$$

To identify the fibrant objects in the localized model structure, it suffices to identify the Seg_Λ-local fibrant objects. Since the class of fibrant objects in the injective model structure contains the fibrant objects in the other two, we can thus identify the Seg_Λ-local fibrant objects in the other two model structures.

PROPOSITION 9.4. *An injectively fibrant object X of $s\mathcal{S}\mathrm{et}^\Lambda$ is Seg_Λ-local if and only if it is a cyclic Segal space.*

PROOF. Since the object X is injectively fibrant, and all objects are injectively cofibrant, the derived mapping spaces $R\,\mathrm{Map}(\Gamma(n), X)$ and $R\,\mathrm{Map}(\Lambda[n], X)$ are equivalent to the corresponding underived mapping spaces. Consequently, X is Seg_Λ-local if and only if, for every $n \geq 2$, the maps of simplicial sets

$$\mathrm{Map}(\Lambda[n], X) \longrightarrow \mathrm{Map}(\Gamma(n), X)$$

are weak equivalences. Applying the representability of $\Lambda[n]$ and the fact that Map sends colimits in the first variable to limits, we obtain a commutative diagram

$$\begin{array}{ccc} \mathrm{Map}(\Lambda[n], X) & \longrightarrow & \mathrm{Map}(\Gamma(n), X) \\ \cong\downarrow & & \downarrow\cong \\ X_n & \longrightarrow & X_1 \times_{X_0} X_1 \times_{X_0} \cdots \times_{X_0} X_1. \end{array}$$

The bottom horizontal map is precisely the Segal map for X, and thus, we see that X is Seg_Λ-local if and only if X is a cyclic Segal space. □

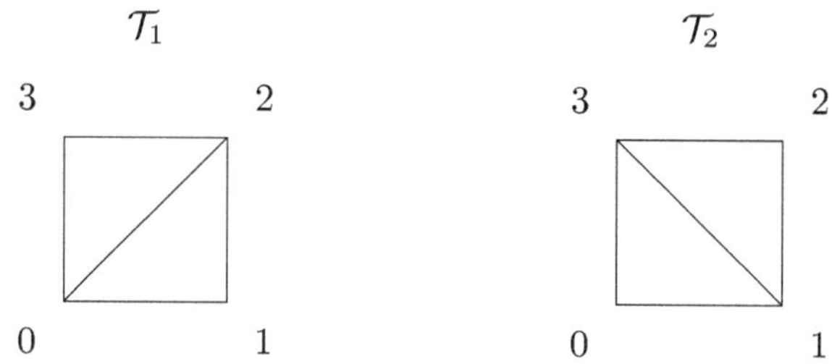

FIGURE 17. The two triangulations of the square.

Applying Theorem 9.3, we obtain our desired localization.

THEOREM 9.5. *Localizing the injective, generalized Reedy, or projective model structure on* $s\mathcal{S}\mathrm{et}^{\Lambda}$ *at the set* Seg_{Λ} *produces a left proper, combinatorial, simplicial model category whose fibrant objects are the cyclic Segal spaces that are fibrant in the original model structure.*

10. Cyclic 2-Segal spaces

We can analogously study models for cyclic 2-Segal spaces, i.e., cyclic spaces whose underlying simplicial space has the 2-Segal property property defined by Dyckerhoff and Kapranov in [**8**, §2.3].

Let us start with the definition of 2-Segal spaces, which were independently introduced under the name *decomposition spaces* by Gálvez-Carrillo, Kock, and Tonks in [**10**, §1.3]. The 2-Segal conditions generalize the Segal conditions, and there are multiple ways of seeing this generalization.

Recall from the previous section that the Segal maps are induced by the spine inclusion $G(n) \to \Delta[n]$. We can view $G(n)$ as a triangulation of the interval with precisely $n+1$ vertices, and moving up one dimension we can instead triangulate polygons that also include into $\Delta[n]$. More specifically, consider triangulations of an $(n+1)$-gon that have no additional vertices. In the case of the square, there are precisely two such triangulations, as pictured in Figure 17.

We can consider the triangles of these triangulations as 2-simplices in $\Delta[n]$, and thus consider $\mathcal{T}_1$ and $\mathcal{T}_2$ as simplicial subsets of $\Delta[n]$, consisting of those 2-simplices in the triangulation, together with their faces and degeneracies. As simplicial sets, the triangulations of the square are

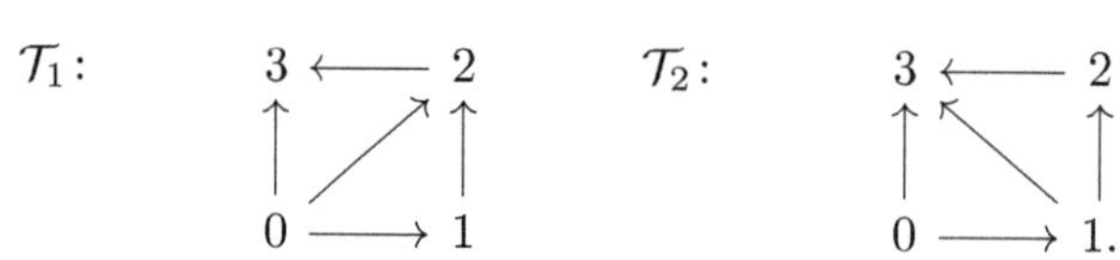

Treating these simplicial sets as discrete simplicial spaces, we can map into a fixed simplicial space X to get maps

$$\mathrm{Map}_{s\mathcal{S}\mathrm{et}^{\Delta^{\mathrm{op}}}}(\mathcal{T}_1, K) \longleftarrow \mathrm{Map}_{s\mathcal{S}\mathrm{et}^{\Delta^{\mathrm{op}}}}(\Delta[3], K) \longrightarrow \mathrm{Map}_{s\mathcal{S}\mathrm{et}^{\Delta^{\mathrm{op}}}}(\mathcal{T}_2, K)$$

precisely as we did with $G(n) \to \Delta[n]$. These induced maps can be rewritten as

$$K_2 \times_{K_1} K_2 \longleftarrow K_3 \longrightarrow K_2 \times_{K_1} K_2.$$

Note that the two pullbacks here are different, since the simplicial subsets of $\Delta[3]$ used to define them are different, as indicated by the triangulations.

More generally, let us call a triangulation of an $(n+1)$-gon which has the same vertices as that $(n+1)$-gon a *triangulation by vertices*. Then we can define a 2-Segal simplicial space as follows.

DEFINITION 10.1. A *2-Segal space* is a simplicial space K such that for every $n \geq 3$ and every triangulation of a regular $(n+1)$-gon by vertices, the induced map

$$K_n \to \underbrace{K_2 \times_{K_1} \cdots \times_{K_1} K_2}_{n-1}$$

is a weak equivalence of simplicial sets.

Roughly speaking, the condition that K be 2-Segal tells us that we can obtain the space of n-simplices of K by gluing together 2-dimensional data according to a triangulation of a polygon.

In parallel to this geometric characterization of the 2-Segal conditions, there is another, more algebraic interpretation. Compared with the up-to-homotopy categories provided by Segal spaces, we get a similar, but weaker structure from a 2-Segal space K. The simplicial space still provides spaces of "objects" K_0 and "morphisms" K_1, but if we consider the span

$$K_1 \times_{K_0} K_1 \xleftarrow{(d_2,d_0)} K_2 \xrightarrow{d_1} K_1,$$

which in a Segal space would define the composition, we see that the map (d_2, d_0) no longer need be a weak equivalence. However, this "composition" is still associative, at least up to homotopy, as can be seen from looking at the two inclusions of the triangulations of the square into $\Delta[3]$ above.

As the preceding discussion suggests, the 2-Segal conditions do indeed generalize the Segal conditions, as the following proposition makes precise.

PROPOSITION 10.2 ([**8**, Prop. 2.3.3]). *Every Segal space K is a 2-Segal space.*

While we do not replicate the proof here, we can comment on the key point. Given a triangulation by vertices $\mathcal{T}$ of an $(n+1)$-gon in the plane, and assuming the vertices of the $(n+1)$-gon are labeled counterclockwise by $0, \dots, n$, we have inclusions of simplicial sets

$$G(n) \longrightarrow \Delta[\mathcal{T}] \longrightarrow \Delta[n]$$

that, in turn, induce maps

$$\mathrm{Map}(G(n), K) \longleftarrow \mathrm{Map}(\Delta[\mathcal{T}], K) \longleftarrow \mathrm{Map}(\Delta[n], K),$$

where $\Delta[\mathcal{T}]$ is the simplicial set corresponding to the triangulation $\mathcal{T}$, using the labeling of the vertices. The right-hand map is the one we must show is an equivalence to show that K is 2-Segal, and the composite is one of the Segal maps. The proof then proceeds by using the Segal conditions to show that the left-hand map is an equivalence.

With these preliminaries in mind, we can return to cyclic spaces.

DEFINITION 10.3. A cyclic space $X\colon \Lambda^{\mathrm{op}} \to s\mathcal{S}\mathrm{et}$ is *2-Segal* if its underlying simplicial space j^*X is a 2-Segal space.

In [**8**, §5], Dyckerhoff and Kapranov describe a model structure for 2-Segal spaces, given by localizing the injective model structure on simplicial spaces at the set of maps

$$\Delta[\mathcal{T}] \longrightarrow \Delta[n]$$

for $n \geq 3$ and $\mathcal{T}$ a triangulation of the $(n+1)$-gon by vertices. The corresponding local objects are then precisely the injectively fibrant 2-Segal spaces.

We can follow a similar procedure to describe cyclic 2-Segal spaces. Let $\Lambda[\mathcal{T}]$ be the union of copies of $\Lambda[2]$ as prescribed by a triangulation of an $(n+1)$-gon for $n \geq 3$, which has an induced map to $\Lambda[n]$. We localize the levelwise model structures with respect to all such maps to obtain model structures for cyclic 2-Segal objects.

More formally, let $2\,\mathrm{Seg}_\Lambda$ denote the set of all morphisms of the form

$$\Lambda[\mathcal{T}] \to \Lambda[n]$$

for $n \geq 3$ and $\mathcal{T}$ a triangulation of the $(n+1)$-gon by vertices. Applying Theorem 9.3 as we did for Segal spaces then yields the desired model structures.

THEOREM 10.4. *Localizing the injective, generalized Reedy, or projective model structure on the category of cyclic spaces with respect to the set* $2\,\mathrm{Seg}_\Lambda$ *yields a left-proper, combinatorial, simplicial model category whose fibrant objects are the 2-Segal objects that are fibrant in the respective original model category.*

As with cyclic Segal spaces, we again can compare the resulting model structures.

PROPOSITION 10.5. *The identity functors define left Quillen equivalences*

$$L_{2\,\mathrm{Seg}_\Lambda}\left(s\mathcal{S}\mathrm{et}^\Lambda\right)^{\mathrm{proj}} \xrightarrow{\mathrm{id}} L_{2\,\mathrm{Seg}_\Lambda}\left(s\mathcal{S}\mathrm{et}^\Lambda\right)^{\mathrm{Reedy}} \xrightarrow{\mathrm{id}} L_{2\,\mathrm{Seg}_\Lambda}\left(s\mathcal{S}\mathrm{et}^\Lambda\right)^{\mathrm{inj}}$$

between localized model structures.

Moreover, Proposition 10.2 allows us to compare the 1-Segal and 2-Segal model structures.

PROPOSITION 10.6. *Fixing one of the levelwise model structures on cyclic spaces, the identity functor defines left Quillen functors on the corresponding localizations*

$$s\mathcal{S}\mathrm{et}^\Lambda \longrightarrow L_{2\,\mathrm{Seg}_\Lambda}\left(s\mathcal{S}\mathrm{et}^\Lambda\right) \longrightarrow L_{\mathrm{Seg}_\Lambda}\left(s\mathcal{S}\mathrm{et}^\Lambda\right).$$

References

[1] C. Barwick, *On left and right model categories and left and right Bousfield localizations*, Homology Homotopy Appl. **12** (2010), no. 2, 245–320, DOI 10.4310/hha.2010.v12.n2.a9. MR2771591

[2] C. Berger and I. Moerdijk, *On an extension of the notion of Reedy category*, Math. Z. **269** (2011), no. 3-4, 977–1004, DOI 10.1007/s00209-010-0770-x. MR2860274

[3] J. E. Bergner, A. M. Osorno, V. Ozornova, M. Rovelli, and C. I. Scheimbauer, *2-Segal objects and the Waldhausen construction*, Algebr. Geom. Topol. **21** (2021), no. 3, 1267–1326, DOI 10.2140/agt.2021.21.1267. MR4299667

[4] J. E. Bergner and W. H. Stern, Cyclic 2-Segal sets and the Waldhausen construction, in preparation.

[5] A. Connes, *Cohomologie cyclique et foncteurs* Ext^n (French, with English summary), C. R. Acad. Sci. Paris Sér. I Math. **296** (1983), no. 23, 953–958. MR777584

[6] W. G. Dwyer, M. J. Hopkins, and D. M. Kan, *The homotopy theory of cyclic sets*, Trans. Amer. Math. Soc. **291** (1985), no. 1, 281–289, DOI 10.2307/1999909. MR797060

[7] T. Dyckerhoff and M. Kapranov, *Crossed simplicial groups and structured surfaces*, Stacks and categories in geometry, topology, and algebra, Contemp. Math., vol. 643, Amer. Math. Soc., Providence, RI, 2015, pp. 37–110, DOI 10.1090/conm/643/12896. MR3381470

[8] T. Dyckerhoff and M. Kapranov, *Higher Segal spaces*, Lecture Notes in Mathematics, vol. 2244, Springer, Cham, 2019, DOI 10.1007/978-3-030-27124-4. MR3970975

[9] Z. Fiedorowicz and J.-L. Loday, *Crossed simplicial groups and their associated homology*, Trans. Amer. Math. Soc. **326** (1991), no. 1, 57–87, DOI 10.2307/2001855. MR998125

[10] I. Gálvez-Carrillo, J. Kock, and A. Tonks, *Decomposition spaces, incidence algebras and Möbius inversion I: Basic theory*, Adv. Math. **331** (2018), 952–1015, DOI 10.1016/j.aim.2018.03.016. MR3804694

[11] E. Getzler and J. D. S. Jones, *The cyclic homology of crossed product algebras*, J. Reine Angew. Math. **445** (1993), 161–174, DOI 10.1515/crll.1993.445.161. MR1244971

[12] A. Heller, *Homotopy theories*, Mem. Amer. Math. Soc. **71** (1988), no. 383, vi+78, DOI 10.1090/memo/0383. MR920963

[13] P. S. Hirschhorn, *Model categories and their localizations*, Mathematical Surveys and Monographs, vol. 99, American Mathematical Society, Providence, RI, 2003, DOI 10.1090/surv/099. MR1944041

[14] R. Krasauskas, *Skew-simplicial groups* (Russian, with English and Lithuanian summaries), Litovsk. Mat. Sb. **27** (1987), no. 1, 89–99. MR895962

[15] J.-L. Loday, *Cyclic homology*, Grundlehren der mathematischen Wissenschaften [Fundamental Principles of Mathematical Sciences], vol. 301, Springer-Verlag, Berlin, 1992. Appendix E by María O. Ronco, DOI 10.1007/978-3-662-21739-9. MR1217970

[16] J. Lurie, *Higher topos theory*, Annals of Mathematics Studies, vol. 170, Princeton University Press, Princeton, NJ, 2009, DOI 10.1515/9781400830558. MR2522659

[17] D. G. Quillen, *Homotopical algebra*, Lecture Notes in Mathematics, No. 43, Springer-Verlag, Berlin-New York, 1967. MR223432

[18] C. L. Reedy, Homotopy theory of model categories, unpublished manuscript, available at `http://www-math.mit.edu/~psh`.

[19] C. Rezk, *A model for the homotopy theory of homotopy theory*, Trans. Amer. Math. Soc. **353** (2001), no. 3, 973–1007, DOI 10.1090/S0002-9947-00-02653-2. MR1804411

[20] T. Nikolaus and P. Scholze, *On topological cyclic homology*, Acta Math. **221** (2018), no. 2, 203–409, DOI 10.4310/ACTA.2018.v221.n2.a1. MR3904731

[21] W. H. Stern, *2-Segal objects and algebras in spans*, J. Homotopy Relat. Struct. **16** (2021), no. 2, 297–361, DOI 10.1007/s40062-021-00282-8. MR4266208

Department of Mathematics, University of Virginia, Charlottesville, Virginia 22904
Email address: `jeb2md@virginia.edu`

Technische Universität München, CIT Department of Mathematics, Garching bei München, Germany
Email address: `walker.stern@tum.de`

Contemporary Mathematics
Volume **838**, 2026
https://doi.org/10.1090/conm/838/16742

Decomposition spaces in combinatorics

Imma Gálvez-Carrillo, Joachim Kock, and Andrew Tonks

ABSTRACT. A decomposition space (also called 2-Segal space) is a simplicial object satisfying an exactness condition weaker than the Segal condition: just as the Segal condition expresses composition, the new condition expresses decomposition. It is a general framework for incidence (co)algebras. In the present contribution, after establishing a formula for the section coefficients, we survey a large supply of examples, emphasising the notion's firm roots in classical combinatorics. The first batch of examples, similar to binomial posets, serves to illustrate two key points: (1) the incidence algebra in question is realised directly from a decomposition space, without a reduction step, and reductions are often given by CULF functors; (2) at the objective level, the convolution algebra is a monoidal structure of species. Specifically, we encounter the usual Cauchy product of species, the shuffle product of $\mathbb{L}$-species, the Dirichlet product of arithmetic species, the Joyal–Street external product of q-species and the Morrison 'Cauchy' product of q-species, and in each case a power series representation results from taking cardinality. The external product of q-species exemplifies the fact that Waldhausen's $S_\bullet$-construction on an abelian category is a decomposition space, yielding Hall algebras. The next class of examples includes Schmitt's chromatic Hopf algebra, the Faà di Bruno bialgebra, the Butcher–Connes–Kreimer Hopf algebra of trees, and several variations from operad theory. Similar structures on posets and directed graphs exemplify a general construction of decomposition spaces from directed restriction species. A short appetiser on decomposition spaces of symmetric functions is included, featuring the base change from elementary symmetric functions to monomial symmetric functions, modelled as a span of decomposition spaces. We finish by computing the Möbius function in a few cases, exhibiting a few techniques, and commenting on certain cancellations that occur in the process of taking cardinality, substantiating that these cancellations are not always possible at the objective level.

2020 *Mathematics Subject Classification.* Primary 05A19, 16T10, 06A07, 18B40, 18M80, 18N50; Secondary 18-XX, 55Pxx.

This work has received funding from grant 10.46540/3103-00099B from the Independent Research Fund Denmark and from a Spanish university requalification and mobility grant (UP2021-034, UNI/551/2021) with NextGenerationEU funds. It was supported also by research grants PID2019-103849GB-I00, PID2020-116481GB-I00, PID2020-117971GB-C22, and PID2023-149804NB-I00 (AEI/FEDER, UE) of Spain, grants 2021-SGR-0603 and 2021-SGR-1015 of Catalonia, through the Severo Ochoa and María de Maeztu Program for Centers and Units of Excellence in R&D grant number CEX2020-001084-M, and the Danish National Research Foundation through the Copenhagen Centre for Geometry and Topology (DNRF151).

0. Introduction

Decomposition spaces. The notion of decomposition space was introduced by the authors [**45–47**] as a general setting for incidence algebras and Möbius inversion, and independently under the name 2-Segal space by Dyckerhoff and Kapranov [**34**], who were motivated by homological algebra, representation theory and geometry. The inherent simplicial nature and the broad scope of applications of the notion prompted a rather abstract categorical and homotopical treatment, with the possible side effect of obscuring its firm roots in combinatorics and its attractive elementary aspects.

The purpose of the present paper is to rectify this possible shortcoming by explaining the combinatorial aspects of the basic theory through many illustrative and natural examples from classical combinatorics. From a theoretical viewpoint, the natural setting for the theory of decomposition spaces is that of simplicial ∞-groupoids, but in fact the notion of decomposition space is interesting even for simplicial sets: there are plenty of natural 'decomposition sets' that are not categories (or posets); some examples can be found in [**34**], [**9**], [**71**], [**57**]. However, it is our contention that the natural level of generality for decomposition spaces in combinatorics is that of simplicial *groupoids*, simply because many combinatorial objects have symmetries, and these are taken care of elegantly by the groupoid formalism.

From locally finite posets to Möbius categories. To motivate the notion of decomposition space, let us start with incidence coalgebras. Since the work of Joni and Rota [**59**] we know well that coalgebras in combinatorics arise from the ability to decompose structures. Very often that ability comes from something fancier, namely the ability to actually *compose* structures. A paradigmatic notion of composition is composition of arrows in a category, such as in particular a poset or a monoid. From any locally finite poset, form the free vector space on its intervals, and endow it with a coalgebra structure by defining the comultiplication as

$$\Delta([x,y]) = \sum_{x\leq m\leq y} [x,m]\otimes[m,y].$$

The same construction works for elements in a monoid with the finite decomposition property [**20**]. In an appendix to [**20**], Foata explains how any (reduced) incidence coalgebra of a poset can also be realised as the incidence coalgebra of a monoid, and conversely. However, it seems to be more fruitful to observe as Leroux [**75**] did, that both are examples of incidence coalgebras of categories. Recall that a poset can be regarded as a category in which there is at most one arrow between any two given objects. To have an interval $[x,y]$ thus means simply that $x\leq y$, and in categorical terms this means that there is an arrow from x to y. The role of elements in the interval $[x,y]$ is played by the possible two-step factorisations of the arrow $x\to y$. Recall also that a monoid is a category with only one object. Leroux showed that the notions of incidence coalgebras of posets and monoids have a common generalisation, namely to locally finite categories, meaning categories in which any given arrow admits only finitely many two-step factorisations: the incidence coalgebra of such a category is the free vector space on its arrows, with comultiplication given by

$$\Delta(f) = \sum_{b \circ a = f} a \otimes b. \tag{1}$$

The coassociativity is a consequence of the associativity of composition of arrows.

Functoriality. One important point, made by Content, Lemay, and Leroux [**23**], is that certain functors induce coalgebra homomorphisms. In modern language, these are the *CULF* functors, which stands for *conservative* and *unique lifting of factorisations*. A functor $F\colon \mathscr{C} \to \mathscr{D}$ is *conservative* if, whenever $F(a)$ is an identity, the arrow a was already an identity; see 1.5 below for more precision and discussion. Unique lifting of factorisations means that F induces, for each arrow a, a one-to-one correspondence between the factorisations of a in $\mathscr{C}$ and the factorisations of $F(a)$ in $\mathscr{D}$.

In the classical theory of posets, often it is not the raw incidence coalgebra that is most interesting, but rather a *reduced* incidence coalgebra, where two intervals are identified if they are equivalent in some specific sense (e.g. isomorphic as abstract posets). As observed in [**23**], these reductions can quite often be realised by CULF functors. For example, the functor from the poset $(\mathbb{N}, \leq)$ to the monoid $(\mathbb{N}, +)$, sending an 'arrow' $x \leq y$ to the monoid element $y - x$, is CULF and realises a classical reduction: the reduced incidence coalgebra of the poset $(\mathbb{N}, \leq)$ is precisely the raw incidence coalgebra of the monoid $(\mathbb{N}, +)$.

In the general setting of decomposition spaces, virtually *all* reduction procedures become instances of CULF functors, and furthermore, many of them are revealed to be instances of decalage (see 1.5.3 below), a general construction in simplicial homotopy theory.

Möbius inversion. Möbius inversion amounts to establishing the convolution invertibility of the zeta function; the inverse is then defined to be the Möbius function [**90**]. Leroux [**75**] established a Möbius inversion formula for any *Möbius category*. A category is Möbius when it is locally finite and when for each arrow there are only finitely many ways to write it as a composite of a chain of non-identity arrows. This notion covers both locally finite posets and monoids with the finite-decomposition property. The formula for the Möbius function is

$$\mu = \Phi_{\text{even}} - \Phi_{\text{odd}}.$$

Here $\Phi_{\text{even}} = \sum_{k \text{ even}} \Phi_k$, where $\Phi_k(f)$ is the number of decompositions of f into a chain of k composable non-identity arrows, and similarly for k odd.

Simplicial viewpoints. The importance of sequences of composable arrows suggests a simplicial viewpoint (see the glossary in Appendix B), which is fundamental to the theory of decomposition spaces (and one of the reasons the theory tends to drift into homotopy theory). Recall (see B.1.7) that the nerve of a category $\mathscr{C}$ is the simplicial set

$$N\mathscr{C}\colon \mathbb{\Delta}^{\text{op}} \to \textbf{\textit{Set}}$$

whose set of n-simplices is the set of sequences of n composable arrows in $\mathscr{C}$ (allowing identity arrows). The face maps are given by composing arrows (for the inner face maps) and by discarding arrows at the beginning or the end of the sequence (outer face maps). The degeneracy maps are given by inserting an identity map in the sequence.

Leroux's theory can be formulated in terms of simplicial sets, as already exploited by Dür [**29**], and many of the arguments then rely on certain simple pullback conditions, the first being the Segal condition, which characterises categories among simplicial sets; see B.2.3. Most importantly in our exploitation of this simplicial viewpoint, the comultiplication (1) can be written in terms of the nerve $N\mathscr{C}$ as a push-pull formula, $\Delta = (d_2, d_0)_! \circ {d_1}^*$, to be explained below.

Objective method. Möbius inversion is a versatile algebraic counting device. The fact that the formula is always given by an alternating sum illustrates one of the great features of algebra over bijective combinatorics: the existence of additive inverses. On the other hand, it is well appreciated that bijective proofs in general represent deeper insight than purely algebraic proofs.

There is a rather general method for lifting algebraic identities to bijections of sets, which one may try to apply whenever the identity takes place in the vector space with basis the set of isomorphism classes of objects. This so-called objective method, pioneered in the context of Möbius inversion by Lawvere and Menni [**73**], aspires to work directly with the combinatorial objects rather than their numbers, using linear algebra with coefficients in ***Set*** rather than a ring or field.

To illustrate this method, observe that a vector in the free vector space on a set S is just a collection of scalars indexed by (a finite subset of) S. The objective counterpart is a family of sets indexed by S, i.e. an object in the slice category $\boldsymbol{Set}_{/S}$. The notion of cardinality has a natural extension to families of finite sets: the cardinality of a family of finite sets indexed by some set B is a B-indexed family of natural numbers, and is in particular an element in the vector space with basis B. Finiteness issues enter the picture and should be taken proper care of; see below.

Linear maps at this level are given by spans $S \leftarrow M \rightarrow T$, which are, in more abstract terms, the *linear functors*, i.e. functors between slices preserving sums and certain other colimits. Indeed, the pullback formula for composition of spans turns out to correspond precisely to matrix multiplication. Spans have cardinalities, which are linear maps.

The Möbius inversion principle states an equality between certain linear maps (elements in the incidence algebra). At the objective level, such an equality can be expressed as a levelwise bijection of the spans of sets that represents those linear functors. In this way, the algebraic identity is revealed to be the cardinality of a bijection of sets, which carries much more structural information.

Lawvere and Menni [**73**] established an objective version of the Möbius inversion principle for Möbius categories in the sense of Leroux [**75**]. A trick is needed to account for the signs: where the algebraic identity states that ζ is convolution invertible with inverse $\mu = \Phi_{\text{even}} - \Phi_{\text{odd}}$:

$$\zeta * (\Phi_{\text{even}} - \Phi_{\text{odd}}) = \varepsilon,$$

to avoid the minus sign, that term has to be moved to the other side of the equation, and the equivalent statement

$$\zeta * \Phi_{\text{even}} \quad = \quad \varepsilon \quad + \quad \zeta * \Phi_{\text{odd}}$$

can be realised as an explicit bijection of sets [**73**].

From sets to groupoids. It is useful now to generalise from sets to groupoids, in order to get a better treatment of symmetries. A prominent example illustrating this is the Faà di Bruno coalgebra (treated in detail in 2.4): it ought to be the incidence coalgebra of (a skeleton of) the category of finite sets and surjections, but since finite sets have symmetries, there are too many factorisations, even of identity arrows. This problem is solved by passing to *fat nerves* (see B.2.2). The fat nerve of a category is the simplicial groupoid

$$\mathbf{N}\mathscr{C}\colon \mathbb{\Delta}^{\mathrm{op}} \to \boldsymbol{Grpd}$$

whose groupoid of n-simplices is the groupoid whose objects are sequences of n composable arrows, and whose arrows are isomorphisms at each level:

$$\begin{array}{ccccccccc} \cdot & \longrightarrow & \cdot & \longrightarrow & \cdot & \longrightarrow & \cdots & \longrightarrow & \cdot \\ \downarrow\wr & & \downarrow\wr & & \downarrow\wr & & & & \downarrow\wr \\ \cdot & \longrightarrow & \cdot & \longrightarrow & \cdot & \longrightarrow & \cdots & \longrightarrow & \cdot . \end{array}$$

The slice categories now have to be groupoid slices $\boldsymbol{Grpd}_{/X}$ instead of set slices. Linear algebra works well at this level of generality too (see Appendix A), and there is a notion of homotopy cardinality, which is invariant under homotopy equivalence. This approach was initiated by Baez and Dolan [**3**] and further developed by Baez, Hoffnung, and Walker [**5**]. A cleaner homotopy version of their formalism was introduced in [**44**], where in particular the notion of homotopy sum is exploited. The upgrade from sets to groupoids is essentially straightforward, as long as the notions involved are taken in a correct homotopy sense, as recalled in Appendix A: bijections of sets are replaced by equivalences of groupoids; the slices playing the role of vector spaces are homotopy slices, the pullbacks and fibres involved in the functors are homotopy pullbacks and homotopy fibres, and the sums are homotopy sums (i.e. colimits indexed by groupoids, just as ordinary sums are colimits indexed by sets).

Decomposition spaces and their incidence (co)algebras. The final abstraction step, which became the starting point for our work [**45–47**], and which is where the present paper starts, is to notice that coassociative coalgebras and a Möbius inversion principle can be obtained from simplicial groupoids more general than those satisfying the Segal condition. We call them *decomposition spaces*; Dyckerhoff and Kapranov [**34**] call them 2-Segal spaces.[1] Whereas the Segal condition is the expression of the ability to compose morphisms, the new condition is about the ability to decompose, which in general is easier to achieve than composability—indeed every Segal space is a decomposition space (Proposition 1.1.4).

The decomposition-space axiom on a simplicial groupoid $X\colon \mathbb{\Delta}^{\mathrm{op}} \to \boldsymbol{Grpd}$ is expressly the condition needed for a canonical coalgebra structure to be induced on the slice category $\boldsymbol{Grpd}_{/X_1}$. The comultiplication is the linear functor

$$\Delta\colon \boldsymbol{Grpd}_{/X_1} \to \boldsymbol{Grpd}_{/X_1} \otimes \boldsymbol{Grpd}_{/X_1}$$

given by the span

$$X_1 \xleftarrow{d_1} X_2 \xrightarrow{(d_2,d_0)} X_1 \times X_1$$

[1] Originally they were called *unital* 2-Segal spaces, but it was later shown [**37**] that the unitality condition is automatic. See Hackney's contribution [**56**] to this volume for lucid exposition of that issue. The term 2-Segal space may well be the most practical in the broader picture, in particular in view of the generalisations to k-Segal spaces for $k > 2$ (see Dyckerhoff's contribution [**31**] in this volume), but from the viewpoint of combinatorics we feel the decomposition space terminology has its merits.

(with reference to general simplicial notation, reviewed in Appendix B). This span can be read as saying that the comultiplication of an edge $f \in X_1$ returns the sum of all pairs of edges (a, b) that are the short edges of a triangle with long edge f. In the case that X is the fat nerve of a category, this is the homotopy sum of all pairs (a, b) of arrows with composite $b \circ a = f$, just as in (1).

Incidence coalgebras, without the need of reduction. It is likely that all incidence (co)algebras can be realised directly (without imposing a reduction) as incidence (co)algebras of decomposition spaces. The decomposition space is found by analysing the reduction step. For example, Dür [**29**] realises the q-binomial coalgebra as the reduced incidence coalgebra of the category $\boldsymbol{vect}^{\text{inj}}$ of finite dimensional vector spaces over a finite field and linear injections, by imposing the equivalence relation identifying two linear injections if their quotients are isomorphic. Trying to realise the reduced incidence coalgebra directly as a decomposition space immediately leads to Waldhausen's $S_\bullet$-construction, a basic construction in K-theory: the q-binomial coalgebra is directly the incidence coalgebra of $S_\bullet(\boldsymbol{vect})$.

Hall algebras. The q-binomial coalgebra fits into a general class of examples: for any abelian category (or even stable ∞-category [**45**]), the Waldhausen $S_\bullet$-construction is a decomposition space (which is not generally Segal). Under the appropriate finiteness conditions, the resulting incidence algebras include the Hall algebras, as well as the derived Hall algebras first constructed by Toën [**102**]. This class of examples plays a key role in the work of Dyckerhoff and Kapranov [**30**, **32–34**]; we refer to their work for the remarkable richness of the Hall algebra aspects of the theory. See also Bergner et. al [**9**], Walde [**103**], Young [**109**], Poguntke [**87**], and the contribution of Cooper and Young [**24**] for further pointers in this direction.

Organisation of the paper. In Section 1 we start out with a short, self-contained summary of the basic notions and results of the theory of decomposition spaces, emphasising combinatorial aspects: the definition in Subsection 1.1, their incidence coalgebras in 1.2, and the convolution product in 1.3. In 1.4 we introduce techniques for computing section coefficients, under suitable finiteness conditions, with a closed formula for the case of Segal spaces. In 1.5 we briefly review the notion of CULF functor, relevant because they induce coalgebra homomorphisms. We exploit decalage (a key example of CULF functor) to establish a criterion for local discreteness, essentially the situation in which the section coefficients are integral. We introduce monoidal decomposition spaces as CULF monoidal structures. These induce bialgebras instead of just coalgebras. A running example in this section is Schmitt's Hopf algebra of graphs [**95**] (called the chromatic Hopf algebra by Aguiar, Bergeron and Sottile [**1**]), an archetypical example of a coalgebra that cannot be the (raw) incidence coalgebra of a category, but is readily obtained as the incidence coalgebra of a decomposition space. It illustrates well the combinatorial meaning of the decomposition space axiom (Example 1.1.5), the mechanism by which the coalgebra structure arises (1.2.4), and the CULF monoidal structure that makes it a bialgebra (1.5.11).

In Section 2, we first go through some very basic examples, which correspond closely to power series representations of the binomial posets of Doubilet, Rota, and Stanley [**28**], and show how the objective version of these classical incidence algebras amounts to monoidal structures on various kinds of species. We emphasise decalage as a general principle behind classical reduction procedures. The case of the Joyal–Street external product of q-species leads to the general treatment of the Waldhausen $S_\bullet$-construction as a decomposition space in 2.3. In 2.4 we revisit the Faà di Bruno bialgebra. Classically it is the reduced incidence bialgebra of the poset of set partitions (reduction modulo type equivalence), but can also be obtained directly from the category of surjections. This suggests that again the reduction step is a decalage, but the relationship turns out to be more subtle: it is a CULF functor but not directly a decalage. In 2.5 we treat examples related to trees and graphs, starting with the Butcher–Connes–Kreimer Hopf algebra of trees [**22**], another example of an incidence coalgebra that cannot be the (raw) incidence coalgebra of a category. We proceed to treat operadic variations, including incidence bialgebras of general operads, as well as related constructions with directed graphs; see Manchon [**79**] and Manin [**81**]. We briefly explain how most of the examples treated in this subsection are subsumed in the notion of decomposition spaces from restriction species of Schmitt [**94**] as well as directed restriction species, treated in detail elsewhere [**49**], and comment also briefly on hereditary species (also from [**94**]) and directed hereditary species from the viewpoint of decomposition spaces. To finish Section 2, we give a brief introduction to symmetric functions from the viewpoint of decomposition spaces. The highlight in this short account is the base change from elementary symmetric functions to monomial symmetric functions, modelled at the objective level by means of an IKEO-CULF span of decomposition spaces.

In Section 3 we come to Möbius inversion, and need first to recall a few notions from [**46**]: complete decomposition spaces and nondegeneracy in 3.1, and the notion of locally finite length and the general Möbius inversion formula in 3.2. In 3.3 we compute the Möbius function in a few easy cases, and comment on certain cancellations that occur in the process of taking cardinality, substantiating that these cancellations are not possible at the objective level. These issues are related to the distinction between bijections and natural bijections.

In Appendix A we provide background on groupoids necessary to understand groupoid slices as the objective analogue of vector spaces, and linear functors and spans as the objective analogue of linear maps. We also explain how to recover the vector space level via taking homotopy cardinality.

In Appendix B we briefly recall the simplicial machinery, which is an essential tool in our undertakings, with special emphasis on the relationship with simplicial complexes. In particular we explain the nerve and the fat nerve of a small category, whereby the simplicial setting covers the cases of categories, and in particular posets and monoids.

Note. Most of the material in this paper was originally Section 5 of the large single manuscript *Decomposition spaces, incidence algebras and Möbius inversion* [**43**]. That manuscript was split into five papers that were published [**44–47, 49**] plus the present paper, which was posted to the arXiv in 2016. It was not submitted for publication at the time, since it was felt that it ought to be expanded with an account of symmetric functions from the decomposition-space viewpoint. The development of that theory suffered substantial delays, but now, on the occasion of the Banff workshop proceedings, we have finally included a very brief account (just an appetiser) of the decomposition-space viewpoint on symmetric functions as Subsection 2.6. We have also taken the opportunity to update the exposition with some remarks and pointers to some other developments that have taken place on decomposition spaces in combinatorics since 2016.

Acknowledgments. We are grateful to André Joyal, Kurusch Ebrahimi-Fard, Mark Weber, Louis Carlier, Alex Cebrian, Wilson Forero, Philip Hackney, Darij Grinberg, and Julie Bergner, for very useful feedback and help.

1. Decomposition spaces and incidence coalgebras

1.1. Segal spaces and decomposition spaces. Segal spaces and decomposition spaces are simplicial groupoids $X\colon \mathbb{\Delta}^{\mathrm{op}} \to \boldsymbol{Grpd}$ satisfying certain exactness properties. We refer to Appendix B for a glossary on simplicial groupoids.

A great many arguments and calculations throughout are performed with pullbacks. We stress again that when we say 'pullback' we mean homotopy pullback (see Appendix A), and similarly 'fibre' means homotopy fibre.

1.1.1. *Segal spaces (Segal groupoids).* A simplicial groupoid X is called a *Segal space*, or a *Segal groupoid*, when all squares of the form

$$\begin{array}{ccc} X_{n+1} & \xrightarrow{d_0} & X_n \\ {\scriptstyle d_{n+1}}\downarrow & & \downarrow{\scriptstyle d_n} \\ X_n & \xrightarrow[d_0]{} & X_{n-1} \end{array}$$

are (homotopy) pullbacks; see Appendix A.1.5.

The most important such square is

$$\begin{array}{ccc} X_2 & \xrightarrow{d_0} & X_1 \\ {\scriptstyle d_2}\downarrow & \lrcorner & \downarrow{\scriptstyle d_1} \\ X_1 & \xrightarrow[d_0]{} & X_0 \end{array} \tag{2}$$

which says that X_2 can be identified with the groupoid $X_1 \times_{X_0} X_1$ of composable pairs of 'arrows'. This condition is satisfied by the nerve or the fat nerve of a small category.

For a Segal space X, the vector space with basis $\pi_0 X_1$ has a coalgebra structure analogous to (1).

It turns out [**45**] that simplicial groupoids other than Segal spaces induce coalgebras. These are the decomposition spaces, which are characterised by a weaker exactness condition than the Segal condition. To give the explicit definitions we need first some simplicial terminology. We refer to Appendix B for notation (which is standard).

1.1.2. *Face and degeneracy maps, active and inert maps.* The simplex category $\mathbb{\Delta}$ (see Appendix B) has an active-inert factorisation system (an example of the general categorical notion of generic-free factorisation system, important in monad theory [**106**, **107**]). An arrow $a\colon [m] \to [n]$ in $\mathbb{\Delta}$ is *active* (also called *generic*) when it preserves end-points, $a(0) = 0$ and $a(m) = n$; and it is *inert* (also called *free*) if it is distance preserving, $a(i+1) = a(i)+1$ for $0 \leq i \leq m-1$. The active maps are generated by the codegeneracy maps $s^i\colon [n+1] \to [n]$ and by the *inner* coface maps $d^i\colon [n-1] \to [n]$, $0 < i < n$, while the inert maps are generated by the *outer* coface maps $d^\bot := d^0$ and $d^\top := d^n$. Every morphism in $\mathbb{\Delta}$ factors uniquely as an active map followed by an inert map. Furthermore, it is a basic fact [**45**] that active and inert maps in $\mathbb{\Delta}$ admit pushouts along each other, and the resulting maps are again active and inert. For a simplicial groupoid $X\colon \mathbb{\Delta}^{\mathrm{op}} \to \boldsymbol{Grpd}$, the images of active and inert maps in $\mathbb{\Delta}$ are again called active and inert.

1.1.3. *Decomposition spaces* [**45**]. A simplicial groupoid $X\colon \mathbb{\Delta}^{\mathrm{op}} \to \boldsymbol{Grpd}$ is called a *decomposition space* when it takes active-inert pushouts to pullbacks.

One can break this down to checking that the following simplicial-identity squares are pullbacks. In the diagrams, the indices are $n \geq 0$ and $0 \leq k \leq n$, so that all horizontal arrows are active maps (and the vertical arrows are inert maps):

$$(3)\qquad \begin{array}{ccccc} X_{n+1} & \xrightarrow{s_{k+1}} & X_{n+2} & \xleftarrow{d_{k+2}} & X_{n+3} \\ {\scriptstyle d_\bot}\downarrow \lrcorner & & \downarrow{\scriptstyle d_\bot} & & \llcorner \downarrow{\scriptstyle d_\bot} \\ X_n & \xrightarrow[s_k]{} & X_{n+1} & \xleftarrow[d_{k+1}]{} & X_{n+2} \end{array} \qquad \begin{array}{ccccc} X_{n+1} & \xrightarrow{s_k} & X_{n+2} & \xleftarrow{d_{k+1}} & X_{n+3} \\ {\scriptstyle d_\top}\downarrow \lrcorner & & \downarrow{\scriptstyle d_\top} & & \llcorner \downarrow{\scriptstyle d_\top} \\ X_n & \xrightarrow[s_k]{} & X_{n+1} & \xleftarrow[d_{k+1}]{} & X_{n+2}. \end{array}$$

The most important cases are the four squares that involve $d_1\colon X_2 \to X_1$ (corresponding to composition of arrows in a category) and $s_0\colon X_0 \to X_1$ (corresponding to the identity arrows in a category):

$$(4)\qquad \begin{array}{ccc} X_2 & \xleftarrow{d_2} & X_3 \\ {\scriptstyle d_\bot}\downarrow & \llcorner & \downarrow{\scriptstyle d_\bot} \\ X_1 & \xleftarrow[d_1]{} & X_2 \end{array} \quad \begin{array}{ccc} X_2 & \xleftarrow{d_1} & X_3 \\ {\scriptstyle d_\top}\downarrow & \llcorner & \downarrow{\scriptstyle d_\top} \\ X_1 & \xleftarrow[d_1]{} & X_2 \end{array} \quad \begin{array}{ccc} X_1 & \xrightarrow{s_1} & X_2 \\ {\scriptstyle d_\bot}\downarrow & \lrcorner & \downarrow{\scriptstyle d_\bot} \\ X_0 & \xrightarrow[s_0]{} & X_1 \end{array} \quad \begin{array}{ccc} X_1 & \xrightarrow{s_0} & X_2 \\ {\scriptstyle d_\top}\downarrow & \lrcorner & \downarrow{\scriptstyle d_\top} \\ X_0 & \xrightarrow[s_0]{} & X_1. \end{array}$$

We shall see shortly that the first two pullback squares are essential ingredients in getting coassociativity of the *incidence coalgebra* of X. The last two pullback squares express counitality, but it has turned out, by a theorem of Feller et al. [**37**] that they are automatically pullbacks if just the two first squares are pullbacks.

Although the Segal axiom squares are quite different from the decomposition space axioms, it is not difficult to prove the following, which shows that the new setting of decomposition spaces does cover the cases of nerves and fat nerves of categories.

PROPOSITION 1.1.4 ([**45**, Proposition 3.5], [**34**, Proposition 5.2.6]). *Every Segal space is a decomposition space.*

1.1.5. *Example (Schmitt's Hopf algebra of graphs).* We give an example of a decomposition space that is not a Segal space, to illustrate the combinatorial meaning of the pullback condition: it is about structures that can be decomposed but not always composed. We shall continue this example in 1.2.4, and see that it corresponds to the Hopf algebra of graphs of Schmitt [**95**].

We define a simplicial groupoid X by taking X_1 to be the groupoid of graphs, and more generally letting X_k be the groupoid of graphs with an ordered partition of the vertex set into k (possibly empty) parts. In particular we have $X_0 = 1$, the contractible groupoid, consisting only of the empty graph.

These groupoids form a simplicial object: the outer face maps delete the first or last part of the graph, and the inner face maps join adjacent parts. The degeneracy maps insert an empty part. The simplicial identities are readily checked.

Note that X is not a Segal space: for the Segal square (2)

$$\begin{array}{ccc} X_2 & \xrightarrow{d_0} & X_1 \\ {\scriptstyle d_2}\downarrow & & \downarrow{\scriptstyle d_1} \\ X_1 & \xrightarrow[d_0]{} & X_0 \end{array}$$

to be a pullback would mean that a graph with a two-part partition could be reconstructed uniquely from knowing the two parts individually. But this is not true, because the two parts individually contain no information about the edges going between them.

One can check that it *is* a decomposition space: that the square

$$\begin{array}{ccc} X_2 & \xleftarrow{d_2} & X_3 \\ {\scriptstyle d_0}\downarrow & \lrcorner & \downarrow{\scriptstyle d_0} \\ X_1 & \xleftarrow[d_1]{} & X_2 \end{array}$$

is a pullback is to say that a graph with a three-part partition ($\in X_3$) can be reconstructed uniquely from a pair of elements in X_2 with common image in X_1 (under the indicated face maps). The following picture represents elements corresponding to each other in the four groupoids.

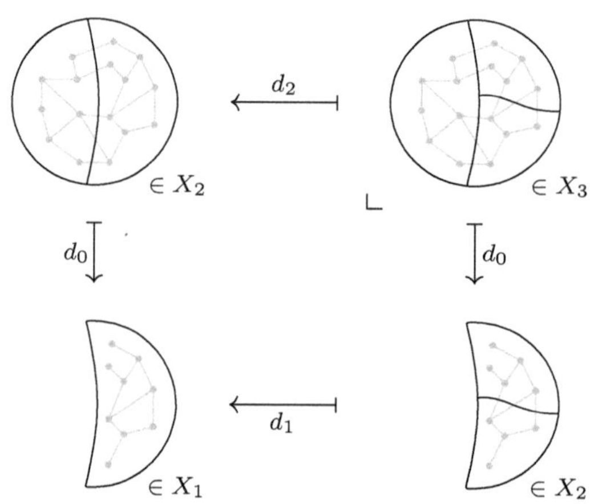

The horizontal maps join the last two parts of the partition. The vertical maps forget the first part. Note first that the diagram commutes. To reconstruct the graph with a three-part partition (upper right-hand corner), most of the information is already available in the upper left-hand corner, namely the underlying graph and all the subdivisions except the one between part 2 and part 3. But this information is precisely available in the lower right-hand corner, and their common image in X_1 says precisely how this missing piece of information is to be implanted.

1.2. Incidence coalgebras of decomposition spaces. We now turn to the incidence coalgebra (with groupoid coefficients) associated to any decomposition space, explaining the origin of the decomposition space axioms.

The incidence coalgebra associated to a decomposition space X is a comonoid object in the symmetric monoidal 2-category ***LIN*** (whose objects are groupoid slices and whose morphisms are linear functors; see A.3), and the underlying object is $\boldsymbol{Grpd}_{/X_1}$. Since $\boldsymbol{Grpd}_{/X_1} \otimes \boldsymbol{Grpd}_{/X_1} = \boldsymbol{Grpd}_{/X_1 \times X_1}$, and since linear functors are given by spans, to define a comultiplication functor is to give a span

$$X_1 \leftarrow M \rightarrow X_1 \times X_1.$$

1.2.1. *Comultiplication and counit.* For X a decomposition space, we can consider the following structure maps on $\boldsymbol{Grpd}_{/X_1}$. The span

$$X_1 \xleftarrow{d_1} X_2 \xrightarrow{(d_2,d_0)} X_1 \times X_1 \tag{5}$$

defines a linear functor, the *comultiplication*

$$\begin{aligned} \Delta \colon \boldsymbol{Grpd}_{/X_1} &\longrightarrow \boldsymbol{Grpd}_{/(X_1 \times X_1)} \\ (T \xrightarrow{t} X_1) &\longmapsto (d_2, d_0)_! \circ {d_1}^*(t). \end{aligned}$$

Likewise, the span

$$X_1 \xleftarrow{s_0} X_0 \xrightarrow{z} 1 \tag{6}$$

defines a linear functor, the *counit*

$$\begin{aligned} \varepsilon \colon \boldsymbol{Grpd}_{/X_1} &\longrightarrow \boldsymbol{Grpd} \\ (T \xrightarrow{t} X_1) &\longmapsto z_! \circ {s_0}^*(t). \end{aligned}$$

We proceed to explain that coassociativity follows from the decomposition space axiom. The coalgebra $(\boldsymbol{Grpd}_{/X_1}, \Delta, \varepsilon)$ is called the *incidence coalgebra* of the decomposition space X. (Note that in the classical incidence-algebra literature (e.g. [**90**], [**75**]), the counit is often denoted δ.)

1.2.2. *Coassociativity.* The comultiplication and counit maps on $\boldsymbol{Grpd}_{/X_1}$, defined in 1.2.1 for any simplicial groupoid X, become coassociative and counital when the decomposition space axioms hold for X. The desired coassociativity diagram (which should commute up to equivalence)

$$\begin{array}{ccc} \boldsymbol{Grpd}_{/X_1} & \xrightarrow{\quad\Delta\quad} & \boldsymbol{Grpd}_{/X_1 \times X_1} \\ {\scriptstyle\Delta}\downarrow & & \downarrow{\scriptstyle\Delta \otimes \mathrm{id}} \\ \boldsymbol{Grpd}_{/X_1 \times X_1} & \xrightarrow[\mathrm{id} \otimes \Delta]{} & \boldsymbol{Grpd}_{/X_1 \times X_1 \times X_1} \end{array}$$

is induced by the solid spans in the diagram

$$\begin{array}{ccccc} X_1 & \xleftarrow{\quad d_1 \quad} & X_2 & \xrightarrow{\quad (d_2,d_0) \quad} & X_1 \times X_1 \\ {\scriptstyle d_1}\uparrow & & \uparrow{\scriptstyle d_1} & & \uparrow{\scriptstyle d_1 \times \mathrm{id}} \\ X_2 & \xleftarrow{\quad d_2 \quad} & X_3 & \xrightarrow[(d_3, d_0 d_0)]{} & X_2 \times X_1 \\ {\scriptstyle (d_2,d_0)}\downarrow & & \downarrow{\scriptstyle (d_2 d_2, d_0)} & & \downarrow{\scriptstyle (d_2,d_0)\times \mathrm{id}} \\ X_1 \times X_1 & \xleftarrow[\mathrm{id} \times d_1]{} & X_1 \times X_2 & \xrightarrow[\mathrm{id} \times (d_2,d_0)]{} & X_1 \times X_1 \times X_1. \end{array}$$

Coassociativity will follow from the Beck–Chevalley Lemma A.3.2 if the dashed part of the diagram can be established with pullbacks as indicated. Consider the upper right-hand square: it is a pullback if and only if its composite with the first projection is a pullback:

$$\begin{array}{ccccc} X_2 & \xrightarrow{(d_\top, d_0)} & X_1 \times X_1 & \xrightarrow{\mathrm{pr}_1} & X_1 \\ {\scriptstyle d_1}\uparrow & \lrcorner & \uparrow{\scriptstyle d_1 \times \mathrm{id}} & \lrcorner & \uparrow{\scriptstyle d_1} \\ X_3 & \xrightarrow[(d_\top, d_0 d_0)]{} & X_2 \times X_1 & \xrightarrow[\mathrm{pr}_1]{} & X_2. \end{array}$$

Saying that this composite outer square $d_\top d_1 = d_1 d_\top$ is a pullback is precisely one of the first decomposition space axioms (4).

If one is just interested in coassociativity at the level of π_0, this pullback and its twin, $d_\bot d_2 = d_1 d_\bot$, are all that are needed, as was the case in the work of Toën [**102**] who dealt with the case where X is the Waldhausen $S_\bullet$-construction of a dg category. On the other hand, it is interesting to analyse when the coassociativity is actually homotopy coherent at the level of groupoid slices. It is proved in [**45**, Theorem 7.3] that we get this coherent coassociativity when all the decomposition space axioms hold:

THEOREM 1.2.3. *If X is a decomposition space then $\boldsymbol{Grpd}_{/X_1}$ has the structure of strong homotopy comonoid in the symmetric monoidal 2-category **LIN**, with the comultiplication and counit defined by the spans* (5) *and* (6).

1.2.4. *Example: Schmitt's Hopf algebra of graphs, continued.* The following coalgebra is due to Schmitt [**95**]. For a graph G, with vertex set V, and a subset $U \subset V$, define $G|U$ to be the graph whose vertex set is U, and whose graph structure is induced by restriction (that is, the edges of $G|U$ are those edges of G both of whose incident vertices belong to U). On the vector space with basis the set of isomorphism classes of graphs, define a comultiplication by the rule

$$\Delta(G) = \sum_{A+B=V} G|A \otimes G|B.$$

This coalgebra is obtained from the decomposition space in Example 1.1.5. Indeed, we have to take X_1 the groupoid of graphs, because the coalgebra has linear basis the set of isomorphism classes of graphs. Since the comultiplication sums over all ways to partition the vertex set into two parts (possibly empty), we must take X_2 to be the groupoid of graphs with a two-part partition of the vertex set. More generally, X_k is the groupoid of graphs with an ordered partition of the vertex set into k (possibly empty) parts.

Taking pullback along $d_1 \colon X_2 \to X_1$ is to consider all possible two-part partitions of a given graph, and taking lowershriek along $(d_2, d_0) \colon X_2 \to X_1 \times X_1$ is to return the graphs induced by the two parts. In conclusion, this is precisely Schmitt's comultiplication.

1.2.5. *Comultiplication of basis elements.* We proceed to spell out the effect of the comultiplication on basis elements. The slice $\boldsymbol{Grpd}_{/X_1}$ has a canonical basis $\{\ulcorner f \urcorner \colon 1 \to X_1\}_{f \in \pi_0 X_1}$. Here $\ulcorner f \urcorner \colon 1 \to X_1$ denotes the map that singles out the element $f \in X_1$, in category theory called the *name* of f. The notion of basis for slices means that every object $T \to X_1$ can be written uniquely as a homotopy sum of names; see Lemma A.2.7. Giving $\ulcorner f \urcorner$ as input to the comultiplication, and expanding the result into a homotopy sum of names, we get:

$$\begin{aligned}\Delta(\ulcorner f \urcorner) &:= \big((X_2)_f \xrightarrow{d_1^*\ulcorner f\urcorner} X_2 \xrightarrow{(d_2,d_0)} X_1 \times X_1\big) \\ &= \int^{\sigma \in (X_2)_f} \ulcorner d_2\sigma \urcorner \otimes \ulcorner d_0 \sigma \urcorner \qquad (7)\\ &= \int^{(a,b)\in X_1\times X_1} (X_2)_{f,a,b}\, \ulcorner a \urcorner \otimes \ulcorner b \urcorner \quad \in \boldsymbol{Grpd}_{/X_1} \otimes \boldsymbol{Grpd}_{/X_1}.\end{aligned}$$

Here $(X_2)_f$ is the fibre of $d_1\colon X_2 \to X_1$ over f, and similarly $(X_2)_{f,a,b}$ is the fibre of $(d_1, d_2, d_0)\colon X_2 \to X_1 \times X_1 \times X_1$ over (f, a, b). Here and throughout, 'fibre' means 'homotopy fibre'; see A.1.6.

If X is the strict nerve of a category then X_2 is the set of all composable pairs of arrows and $(X_2)_f$ is the subset of those pairs with composite f. In particular, $(X_2)_{f,a,b}$ is then either empty or a singleton, and the comultiplication reduces to the formula (1) from the introduction,

$$\Delta(f) = \sum_{ab=f} a \otimes b.$$

If X is the fat nerve of a category, or more generally if X is a Segal space (that is, $X_2 \simeq X_1 \times_{X_0} X_1$), then as in the case of the ordinary nerve we see that if f is not the composite up to isomorphism of a and b then $X_{f,a,b}$ is empty. In the case of a fat nerve, it is non-empty if and only if one can write

$$f \;=\; (\; x \xrightarrow{\cong} x' \xrightarrow{a} y \xrightarrow[\cong]{\varphi} y' \xrightarrow{b} z' \xrightarrow{\cong} z \;) \;\in\; X_1.$$

In the more general case of a Segal space, it is non-empty if and only if there exists $\sigma \in X_2$ and $\varphi \in \mathrm{Map}_{X_0}(d_0 a, d_1 b)$ such that $f \cong d_1\sigma$ and $\sigma \in X_2$ corresponds to $(a, b, \varphi) \in X_1 \times_{X_0} X_1$ in the notation of A.1.5.

LEMMA 1.2.6. *If X is a Segal space and $a, b \in X_1$ then the groupoid $(X_2)_{a,b}$ is discrete, naturally equivalent to the set of isomorphisms $\mathrm{Map}_{X_0}(d_0a, d_1b)$.*

PROOF. Since $X_2 \simeq X_1 \times_{X_0} X_1$ we can compute $(X_2)_{a,b}$ as the pullback

$$\begin{array}{ccccc} (X_1 \times_{X_0} X_1)_{a,b} & \longrightarrow & X_1 \times_{X_0} X_1 & \longrightarrow & X_0 \\ \downarrow \lrcorner & & \downarrow \lrcorner & & \downarrow \Delta \\ 1 & \xrightarrow{\ulcorner a,b \urcorner} & X_1 \times X_1 & \xrightarrow{d_0 \times d_1} & X_0 \times X_0. \end{array}$$

But the homotopy fibres of the diagonal $\Delta\colon X_0 \to X_0 \times X_0$ are (naturally equivalent to) the mapping spaces. □

PROPOSITION 1.2.7. *If X is a Segal space and $f \in X_1$ then*

$$\Delta(\ulcorner f \urcorner) = \int^{(a,b)\in X_1 \times X_1} \sum_{\varphi \in \mathrm{Map}_{X_0}(d_0 a, d_1 b)} \mathrm{Map}_{X_1}(f, \ell_{a,b}(\varphi))\, \ulcorner a \urcorner \otimes \ulcorner b \urcorner \;\in\; \boldsymbol{Grpd}_{/X_1} \otimes \boldsymbol{Grpd}_{/X_1}.$$

where $\ell_{a,b}\colon \mathrm{Map}_{X_0}(d_0a, d_1b) \simeq (X_2)_{a,b} \xrightarrow{d_1} X_1$.

Observe that each set $\mathrm{Map}_{X_1}(f, \ell_{a,b}(\varphi))$ in the sum is either empty (if $f \not\simeq \ell_{a,b}(\varphi)$) or non-canonically in bijection with the set $\mathrm{Aut}(f) = \mathrm{Map}_{X_1}(f, f)$.

PROOF. By the previous lemma, $(X_2)_{f,a,b}$ is the homotopy fibre of $\ell_{a,b}$ over f, and as the domain of $\ell_{a,b}$ is discrete this fibre is the sum of the fibres $(X_2)_{f,a,b,\varphi}$,

$$\begin{array}{ccccc} (X_2)_{f,a,b,\varphi} & \longrightarrow & (X_2)_{f,a,b} & \longrightarrow & 1 \\ \downarrow & & \downarrow & & \downarrow \ulcorner f \urcorner \\ 1 & \xrightarrow{\ulcorner \varphi \urcorner} & \mathrm{Map}_{X_0}(d_0 a, d_1 b) & \xrightarrow{\ell_{a,b}} & X_1, \end{array}$$

in which each $(X_2)_{f,a,b,\varphi} \simeq \mathrm{Map}_{X_1}(f, \ell_{a,b}(\varphi))$. □

1.2.8. *Local finiteness.* As long as we work at the objective level, where all results and proofs are naturally bijective, it is not necessary to impose any finiteness conditions. But in order to be able to take cardinality to recover numerical results (i.e. at the vector-space level), suitable finiteness conditions must be imposed. Intuitively, mimicking the local finiteness for categories, we should require that for each $n \in \mathbb{N}$ the active map $X_n \to X_1$ is finite. In the category case this requirement means that, for each arrow $f \in X_1$ and $n \in \mathbb{N}$, there are only finitely many decompositions of f into a sequence of n arrows. Technically, the appropriate definition is the following from [**46**].

A decomposition space $X\colon \mathbb{\Delta}^{\mathrm{op}} \to \boldsymbol{Grpd}$ is termed *locally finite* if X_1 is locally finite (in the sense of groupoids A.1.4) and both $s_0\colon X_0 \to X_1$ and $d_1\colon X_2 \to X_1$ are finite maps. Then the comultiplication and counit defined above are finite linear functors, and hence (by Proposition A.4.3) descend to slices of finite groupoids

$$\Delta\colon \boldsymbol{grpd}_{/X_1} \to \boldsymbol{grpd}_{/X_1} \otimes \boldsymbol{grpd}_{/X_1}, \qquad \varepsilon\colon \boldsymbol{grpd}_{/X_1} \to \boldsymbol{grpd}.$$

Here $\boldsymbol{grpd}$ is the full subcategory of finite groupoids, see A.1.4.

We can then take cardinality to obtain comultiplication and counit maps of vector spaces

$$|\Delta| : \mathbb{Q}_{\pi_0 X_1} \to \mathbb{Q}_{\pi_0 X_1} \otimes \mathbb{Q}_{\pi_0 X_1}, \qquad |\varepsilon| : \mathbb{Q}_{\pi_0 X_1} \to \mathbb{Q},$$

which are coassociative and counital, and

$$\mathcal{I}_X := (\mathbb{Q}_{\pi_0 X_1}, |\Delta|, |\varepsilon|)$$

is termed the *numerical incidence coalgebra* of X; see [**46**, 7.8].

REMARK 1.2.9. If X is the nerve of a poset P, then it is locally finite in the above sense if and only if all intervals $[x, y]$ are finite, which is the usual definition for posets [**97**]. The points in this interval parametrise precisely the two-stage factorisations of the unique arrow $x \to y$, so this condition amounts to $X_2 \to X_1$ having finite fibre over $x \to y$. (In the poset case, the conditions on X_1 and on $s_0\colon X_0 \to X_1$ are automatically satisfied, since everything is discrete.)

Examples of infinite categories that are locally finite are given by free monoids or the free category on a directed graph.

1.3. Convolution algebras.

1.3.1. *Linear dual.* If X is a decomposition space, we have seen there is a natural coassociative comultiplication on $\boldsymbol{Grpd}_{/X_1}$, the incidence coalgebra of X, which we see as an 'objectification' of the vector space $\mathbb{Q}_{\pi_0 X_1}$ underlying the classical incidence coalgebra. One may also consider the incidence (or convolution) *algebra* $\boldsymbol{Grpd}^{X_1}$, which can be obtained from the incidence coalgebra by taking the linear dual (A.3.4). Since $\boldsymbol{Grpd}_{/X_1}$ is the free homotopy-sum completion of

X_1 (just as $\mathbb{Q}_{\pi_0 X_1}$ is the 'linear-combination completion' of the set $\pi_0 X_1$), objects in $\boldsymbol{Grpd}^{X_1}$ can be regarded either as presheaves $X_1 \to \boldsymbol{Grpd}$ or as linear functors $\boldsymbol{Grpd}_{/X_1} \to \boldsymbol{Grpd}$; see A.3.4. The category $\boldsymbol{Grpd}^{X_1}$ is interpreted as an 'objectification' of the incidence algebra, denoted $\mathcal{I}^X$, which has underlying profinite-dimensional vector space $\mathbb{Q}^{\pi_0 X_1}$.

1.3.2. *Convolution.* The multiplication in the incidence algebra is the convolution product, given as the dual of the comultiplication. Consider two linear functors

$$F, G\colon \boldsymbol{Grpd}_{/X_1} \longrightarrow \boldsymbol{Grpd}$$

given by spans $X_1 \leftarrow M \to 1$ and $X_1 \leftarrow N \to 1$. Their tensor product $F \otimes G$ is then given by the span

$$X_1 \times X_1 \leftarrow M \times N \to 1.$$

Their *convolution* $F * G$ is the composite of $F \otimes G$ with the comultiplication

$$F * G\colon \boldsymbol{Grpd}_{/X_1} \xrightarrow{\Delta} \boldsymbol{Grpd}_{/X_1} \otimes \boldsymbol{Grpd}_{/X_1} \xrightarrow{F \otimes G} \boldsymbol{Grpd} \otimes \boldsymbol{Grpd} \simeq \boldsymbol{Grpd},$$

altogether given by the composite span

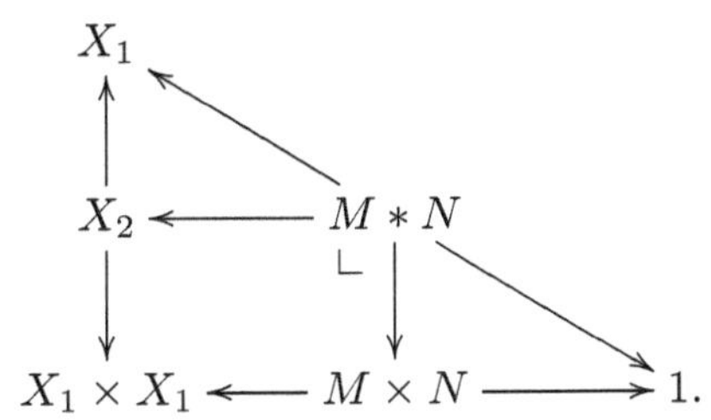

The neutral functor for the convolution product is ε.

1.3.3. *The zeta functor.* The *zeta functor*

$$\zeta\colon \boldsymbol{Grpd}_{/X_1} \to \boldsymbol{Grpd}$$

is the linear functor defined by the span

$$X_1 \xleftarrow{=} X_1 \to 1\,,$$

which, as an element of $\boldsymbol{Grpd}^{X_1}$, is the terminal presheaf.

Assuming X_1 locally finite then ζ is a finite linear functor and descends to

$$\zeta\colon \boldsymbol{grpd}_{/X_1} \to \boldsymbol{grpd}.$$

Its cardinality $\mathbb{Q}_{\pi_0 X_1} \to \mathbb{Q}$, which can be regarded as an element in the profinite-dimensional vector space $\mathbb{Q}^{\pi_0 X_1}$, is then the usual zeta function $\pi_0 X_1 \to \mathbb{Q}$ with value 1 on each 1-simplex of X.

1.4. Section coefficients.

1.4.1. *Section coefficients.* If X is a locally finite decomposition space then the homotopy cardinality of the comultiplication at the objective level

$$\begin{aligned} \boldsymbol{grpd}_{/X_1} &\longrightarrow \boldsymbol{grpd}_{/X_1 \times X_1} \\ \ulcorner f \urcorner &\longmapsto \big((X_2)_f \to X_2 \to X_1 \times X_1\big) \end{aligned}$$

yields a comultiplication in the category of vector spaces

$$\begin{aligned} \mathbb{Q}_{\pi_0 X_1} &\longrightarrow \mathbb{Q}_{\pi_0 X_1} \otimes \mathbb{Q}_{\pi_0 X_1} \\ \delta_f &\longmapsto \int^{(a,b)\in X_1\times X_1} |(X_2)_{f,a,b}|\ \delta_a \otimes \delta_b \ = \ \sum_{a,b} c^f_{a,b}\ \delta_a \otimes \delta_b \end{aligned}$$

that defines the numerical incidence coalgebra $\mathcal{I}_X$. It is just the cardinality of (7), with the *section coefficients*

$$c^f_{a,b} := \frac{|(X_2)_{f,a,b}|}{|\mathrm{Aut}(a)|\,|\mathrm{Aut}(b)|}. \tag{8}$$

In the special case of a Segal space, we can take cardinality of Proposition 1.2.7 to arrive at the following explicit formula for the section coefficients.

PROPOSITION 1.4.2 (See [**48**]). *If X is a locally finite Segal space then*

$$c^f_{a,b} = \frac{|\mathrm{Aut}(f)|\,|\{\varphi \text{ s.t. } \ell_{a,b}(\varphi) \simeq f\}|}{|\mathrm{Aut}(a)|\,|\mathrm{Aut}(b)|}.$$

In the case that X is the fat nerve of a category, $\ell_{a,b}(\varphi) = a\varphi b$ and the term $|\{\varphi \text{ s.t. } \ell_{a,b}(\varphi) \simeq f\}|$ is just the number of isomorphisms $\varphi\colon d_0 a \simeq d_1 b$ such that $a\varphi b \simeq f$.

COROLLARY 1.4.3. *If X is a locally finite Segal space with $X_0 = 1$, then*

$$c^f_{a,b} = \begin{cases} \frac{|\mathrm{Aut}(f))|}{|\mathrm{Aut}(a)||\mathrm{Aut}(b)|} & \text{if } f \simeq ab \\ 0 & \text{if } f \not\simeq ab. \end{cases} \tag{9}$$

Here ab denotes the image of (a, b) under $X_1 \times X_1 \simeq X_2 \xrightarrow{d_1} X_1$.

PROOF. Since X_0 is contractible, the set $\{\varphi\}$ of the proposition is either singleton or empty, depending on whether $ab \simeq f$ or not. □

1.4.4. *'Zeroth section coefficients': the counit.* Let us also say a word about the zeroth section coefficients, i.e. the computation of the counit. If f is not isomorphic to a degenerate simplex then $|\varepsilon|\,(\delta_f) = 0$. In the case f is degenerate, we just remark on two special cases.

- If X is complete (3.1.1), meaning that s_0 is a monomorphism (A.2.4), then $|\varepsilon|\,(\delta_f) = 1$.
- If $X_0 = 1$ then $|\varepsilon|\,(\delta_f) = |\mathrm{Aut}(f)|$.

1.4.5. *Numerical convolution product.* By duality, if X is locally finite, the convolution product descends to the profinite-dimensional vector space $\mathbb{Q}^{\pi_0 X_1}$ obtained by taking cardinality of $\boldsymbol{grpd}^{X_1}$, defining the (numerical) incidence algebra of X, denoted $\mathcal{I}^X$. It follows from the general theory of homotopy linear algebra (see appendix A.4.5 and [**44**]) that the cardinality of the convolution product is the linear dual of the cardinality of the comultiplication. Since it is the same span that defines the comultiplication and the convolution product, it is also the exact same matrix that defines the cardinalities of these two maps. It follows that the structure constants for the convolution product (with respect to the pro-basis $\{\delta^f\}$) are the same as the structure constants for the comultiplication (with respect to the basis $\{\delta_f\}$), i.e. the section coefficients.

1.4.6. *Example.* The strict nerve of a category $\mathscr{C}$ is a decomposition space that is discrete in each degree. The resulting coalgebra at the numerical level (assuming local finiteness) is the coalgebra of Content, Lemay, and Leroux [**23**], and if the category is just a poset, that of Joni and Rota [**59**].

The objective-level incidence algebra of the strict nerve of $\mathscr{C}$ has the convolution product

$$h^a * h^b = \begin{cases} h^{ab} & \text{if } a \text{ and } b \text{ are composable} \\ 0 & \text{else.} \end{cases} \tag{10}$$

For the fat nerve X of $\mathscr{C}$, we find instead

$$h^a * h^b \;=\; \sum_{\varphi\colon d_0 a \simeq d_1 b} h^{a\varphi b} \;=\; \sum_{f\in\pi_0 X_1} \{\varphi \text{ s.t. } a\varphi b \simeq f\}\cdot h^f \tag{11}$$

where the first sum is over all isomorphisms φ from the target of a to the source of b; see [**48**].

To compute the cardinality of this algebra, note first that the cardinality of the representable h^a is generally different from the canonical basis element δ^a: the formula (31) says

$$|h^a| = |\mathrm{Aut}(a)|\ \delta^a,$$

so that (11) becomes

$$\delta^a * \delta^b \;=\; \sum_{f\in\pi_0 X_1} \tfrac{|\{\varphi \text{ s.t. } a\varphi b\simeq f\}|\cdot|\mathrm{Aut}(f)|}{|\mathrm{Aut}(a)|\cdot|\mathrm{Aut}(b)|}\ \delta^f \;=\; \sum_{f\in\pi_0 X_1} c^f_{a,b}\ \delta^f$$

with the section coefficients as given in Proposition 1.4.2.

1.4.7. *Finite support.* The numerical incidence algebra $\mathcal{I}^X$ lives in profinite-dimensional vector spaces, since functions are not required to have finite support; for example, the zeta function does not have finite support for infinite posets or categories. It is also interesting to consider the subalgebra of $\mathcal{I}^X$ consisting of functions with finite support. At the objective level it is the full subcategory $\boldsymbol{grpd}^{X_1}_{\text{fin.sup.}} \subset \boldsymbol{grpd}^{X_1}$, and numerically it is $\mathbb{Q}^{\pi_0 X_1}_{\text{fin.sup.}} \subset \mathbb{Q}^{\pi_0 X_1}$. We have canonical identifications $\boldsymbol{grpd}^{X_1}_{\text{fin.sup.}} \simeq \boldsymbol{grpd}_{/X_1}$, as well as $\mathbb{Q}^{\pi_0 X_1}_{\text{fin.sup.}} \simeq \mathbb{Q}_{\pi_0 X_1}$, but it is important to keep track of which side of duality we are on.

That the decomposition space is locally finite is not the appropriate condition for the convolution and unit to restrict to the functions with finite support. Instead the requirement is that X_1 be locally finite and the maps

$$X_2 \to X_1 \times X_1, \qquad X_0 \to 1$$

be finite. By Lemma 1.2.6 we know that the former map is finite for any Segal space with X_0 *locally* finite, but for the latter X_0 must actually be finite.

1.4.8. *Examples: category algebras.* If X is the *strict* nerve of a category $\mathscr{C}$, then the finite-support convolution algebra is precisely the *category algebra* of $\mathscr{C}$, which is an important notion in representation theory; see [**105**].

Note that since the strict nerve is a Segal space, the formula for the section coefficients are the same as computed above, giving the familiar formula (10). Similarly the formula for the convolution unit is

$$\varepsilon = \sum_x \delta^{\mathrm{id}_x}$$

the sum of all indicator functions of identity arrows: for this sum to be finite we need to require that the category has only finitely many objects.

In the case of the *fat* nerve of a category $\mathscr{C}$, the finiteness condition for having a finite-support convolution is implied by the condition that every object in $\mathscr{C}$ has a finite automorphism group (a condition implied by local finiteness). On the other hand, the convolution unit has finite support precisely when there is only a finite number of isomorphism classes of objects, already a more drastic condition. Compared to the usual category algebra, this 'fat category algebra' has, by (11),

$$h^a * h^b \;=\; \sum_{f\in\pi_0 X_1} \{\varphi \text{ s.t. } a\varphi b \simeq f\} \cdot h^f .$$

Note that an important source of examples of category algebras is given by the path algebra of a quiver Q (see for example [**13**]), that is, the category algebra on the free category on Q. Since there are no automorphisms in a free category, in this case there is no difference between the strict and the fat nerves.

It should be noted that the finite-support incidence algebras are important also outside the setting of category algebras, namely in the case of the Waldhausen $S_\bullet$-construction (see 2.3 below): they are the Hall algebras; see [**88**] and [**45**]. The finiteness conditions are then homological, namely finite Ext^0 and Ext^1.

1.4.9. *Locally discrete decomposition spaces.* In the formula in Proposition 1.4.2 for the section coefficients there are denominators. In very many examples of importance, however, the section coefficients are actually integral. This happens, for instance, when the map $d_1 \colon X_2 \to X_1$ is discrete, that is, has discrete homotopy fibres. Equivalently, the induced group homomorphisms $(d_1)_* \colon \mathrm{Aut}(\sigma) \to \mathrm{Aut}(d_1\sigma)$ are injective for each $\sigma \in X_2$. For the zeroth section coefficients one should also require $s_0 \colon X_0 \to X_1$ to be discrete, but this condition always holds as $(d_1)_*(s_0)_* \colon \mathrm{Aut}(x) \to \mathrm{Aut}(s_0 x) \to \mathrm{Aut}(x)$ is the identity.

We define X to be *locally discrete* when $d_1 \colon X_2 \to X_1$ is a discrete map.

REMARK 1.4.10. In our terminology (1.2.8), 'locally finite' means that $d_1 \colon X_2 \to X_1$ and $s_0 \colon X_0 \to X_1$ are finite maps *and* that X_1 is locally finite. To be consistent with this definition, 'locally discrete' should mean $d_1 \colon X_2 \to X_1$ and $s_0 \colon X_0 \to X_1$ discrete, *and* X_1 a locally discrete groupoid. If we define a groupoid to be locally discrete if all its hom sets are discrete, then every groupoid is locally discrete, and therefore it is not necessary to mention it in the definition.

1.4.11. *Examples.* The fat nerve of a category $\mathscr{C}$ is locally discrete if and only if, in any commutative diagram in $\mathscr{C}$ of the form

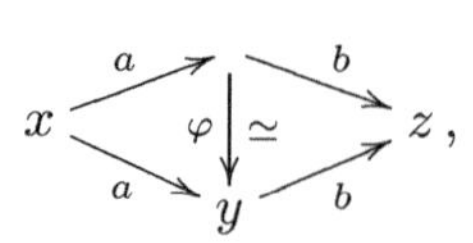

the isomorphism φ is the identity on y. For example, this is the case if $\mathscr{C}$ satisfies any of the following three conditions.

- All the arrows in $\mathscr{C}$ are monos.
- All the arrows in $\mathscr{C}$ are epis.
- All the automorphisms in $\mathscr{C}$ are identities.

Starting from these three cases, many more examples can be derived from the fact that a simplicial groupoid is locally discrete if and only if either of its decalages is; see 1.5.3 and Lemma 1.5.5 below.

As we shall see, examples coming from combinatorics tend to be locally discrete.

1.4.12. *A tiny example: the 'hanger category'.* The following category is perhaps the smallest example of a category whose fat nerve is *not* locally discrete.

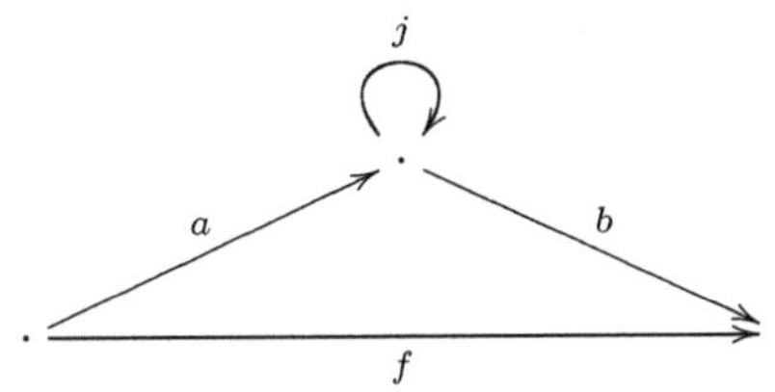

in which

$$ab = f \qquad jj = 1 \qquad aj = a \qquad jb = b.$$

It has

$$\Delta(f) = 1 \otimes f + f \otimes 1 + \frac{a \otimes b}{2}$$

since the factorisation ab admits an involution, given by j.

1.5. CULF functors, coalgebra homomorphisms and bialgebras. An appropriate notion of morphism between decomposition spaces is that of CULF functors [**45**], which we briefly recall. Their importance is that they induce coalgebra homomorphisms between the incidence coalgebras. Two main instances of CULF functors are decalage and monoidal structures. As we shall see, decalage accounts for many reduction procedures in the classical theory of incidence coalgebras. A CULF monoidal structure on a decomposition space is precisely what makes the incidence coalgebra into a bialgebra.

1.5.1. *CULF functors.* A simplicial map $F\colon X \to Y$ is:

- *conservative* if it is cartesian with respect to codegeneracy maps, meaning that we have the pullbacks

$$\begin{array}{ccc} X_n & \xrightarrow{s_i} & X_{n+1} \\ {\scriptstyle F}\downarrow & \lrcorner & \downarrow{\scriptstyle F} \\ Y_n & \xrightarrow[s_i]{} & Y_{n+1} \end{array} \qquad (0 \leq i \leq n);$$

- *ULF* (for Unique Lifting of Factorisations) if it is cartesian with respect to inner coface maps, so we have the pullbacks

$$\begin{array}{ccc} X_{n+1} & \xleftarrow{d_{i+1}} & X_{n+2} \\ {\scriptstyle F}\downarrow & \llcorner & \downarrow{\scriptstyle F} \\ Y_{n+1} & \xleftarrow[d_{i+1}]{} & Y_{n+2} \end{array} \qquad (0 \leq i \leq n);$$

- *CULF* if it is both conservative and ULF, that is, cartesian on all active maps. We shall use the term *CULF functor* even between simplicial groupoids not assumed to be Segal.

If both X and Y are decomposition spaces, then in fact ULF implies CULF [**45**, Proposition 4.2].

In many examples of decomposition spaces, 1-simplices are thought of as arrows: for simplicial maps between Rezk complete Segal spaces (see B.2.3), conservative means not making any arrows invertible, and ULF means inducing a one-to-one correspondence between factorisations of an arrow in X and of its image in Y.

For morphisms of posets, conservative means to preserve $<$, not just $\leq$, while ULF is strictly stronger: it means to induce an isomorphism $[x, x'] \simeq [Fx, Fx']$ on each interval. If the morphism of posets is a full inclusion, then ULF is precisely the same as convex [**49**]: if two elements belong to the subposet then so do all elements between them. Note that an ULF map of posets does not have to be injective: for example, if X is a discrete poset then *any* map $X \to Y$ is ULF.

Given a simplicial map $F\colon X \to Y$ between decomposition spaces, the span $X_1 \xleftarrow{=} X_1 \xrightarrow{F_1} Y_1$ defines a linear functor

$$F_{1!}\colon \boldsymbol{Grpd}_{/X_1} \to \boldsymbol{Grpd}_{/Y_1},$$

which descends to a linear functor $F_{1!}\colon \mathbf{grpd}_{/X_1} \to \mathbf{grpd}_{/Y_1}$ whose cardinality is the linear map $\mathbb{Q}_{\pi_0 X_1} \to \mathbb{Q}_{\pi_0 Y_1}$ given on the basis by $\delta_f \mapsto \delta_{F_1 f}$.

LEMMA 1.5.2 ([**45**]). *If F is CULF, then $F_{1!}$ is a coalgebra homomorphism, meaning that it preserves the comultiplication and counit up to coherent homotopy:*

$$(F_{1!} \otimes F_{1!})\,\Delta_X \;\simeq\; \Delta_Y\, F_{1!}, \qquad\qquad \varepsilon_X \;\simeq\; \varepsilon_Y\, F_{1!}.$$

1.5.3. *Decalage.* An important source of CULF functors is given by decalage. Recall that the *decalage* functor $\mathrm{Dec}_\perp$ on simplicial groupoids forgets the bottom face and degeneracy maps, and shifts the indexing of the groupoids. The unused face map $d_\perp$ provides a natural transformation from the decalage back to the identity functor. We refer to this $d_\perp$ as the *dec map.*

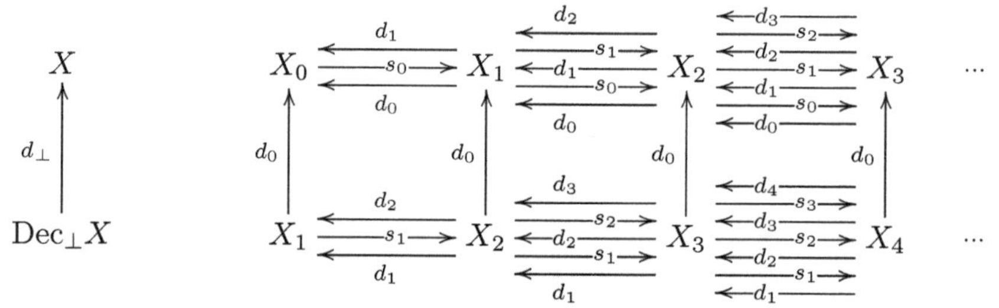

Similarly, the decalage $\mathrm{Dec}_\top$ forgets the top face and degeneracy maps.

Decalage also plays an important role at the theoretical level, as exemplified by the following result.

LEMMA 1.5.4 ([**34**], [**45**], in conjunction with [**37**]). *A simplicial groupoid X is a decomposition space if and only if both* $\mathrm{Dec}_\top X$ *and* $\mathrm{Dec}_\perp X$ *are Segal spaces. Furthermore, in this case the corresponding dec maps $d_\top$ and $d_\perp$ are CULF.*

In particular for any decomposition space X we have a canonical coalgebra homomorphism from the incidence coalgebra of $\mathrm{Dec}_\perp X$ to that of X, and similarly for $\mathrm{Dec}_\top$. This phenomenon appears in many examples.

LEMMA 1.5.5. *The following are equivalent for a decomposition space X.*

(1) *X is locally discrete.*

(2) $\mathrm{Dec}_\perp X$ *is locally discrete.*
(3) $\mathrm{Dec}_\top X$ *is locally discrete.*

PROOF. Note that the dec map is always essentially surjective, since it admits a degeneracy map as a section. Now the result follows from the following lemma. □

LEMMA 1.5.6. *A decomposition space X is locally discrete if it admits an essentially surjective CULF functor $Y \to X$ with Y a locally discrete decomposition space.*

PROOF. The result follows since discreteness is a local property: in a pullback square of groupoids

$$\begin{array}{ccc} E' & \longrightarrow & E \\ {\scriptstyle f'}\downarrow & \lrcorner & \downarrow{\scriptstyle f} \\ B' & \underset{e}{\longrightarrow} & B \end{array}$$

the homotopy fibres $f'^{-1}(b')$ and $f^{-1}(e(b'))$ are equivalent. Thus if e is essentially surjective then f is discrete if and only if f' is discrete. □

1.5.7. *Bialgebras.* Recall that a bialgebra is a coalgebra with a compatible algebra structure, meaning that multiplication and unit are coalgebra homomorphisms. More formally it can be characterised as a monoid object in the category of coalgebras. In Lemma 1.5.2 we saw that a sufficient condition for a simplicial map f between decomposition spaces to induce a coalgebra homomorphism on incidence coalgebras is that f be CULF. Accordingly we define a *monoidal decomposition space* [**45**] to be a decomposition space Z equipped with an associative unital monoid structure given by CULF functors $m\colon Z \times Z \to Z$ and $e\colon 1 \to Z$.

PROPOSITION 1.5.8. *If Z is a monoidal decomposition space then $\mathbf{Grpd}_{/Z_1}$ is naturally a bialgebra, termed its* incidence bialgebra. *Monoidal CULF functors induce bialgebra homomorphisms.*

1.5.9. *Extensivity.* Classically, a category $\mathscr{C}$ with sums is called *extensive* when the natural functor $\mathscr{C}_{/A} \times \mathscr{C}_{/B} \to \mathscr{C}_{/A+B}$ is an equivalence. More generally, a monoidal category $(\mathscr{C}, \otimes, I)$ is called *monoidal extensive* when the natural functor $\mathscr{C}_{/A} \times \mathscr{C}_{/B} \to \mathscr{C}_{/A\otimes B}$ is an equivalence. The fat nerve of a monoidal extensive category is always a monoidal decomposition space. As an example, the category $\mathscr{F}$ of finite sets and all maps is extensive in the classical sense. The category of finite sets and surjections inherits the monoidal structure $+$ from $\mathscr{F}$, but it is no longer the categorical sum (since there are no sum injections). It is still monoidal extensive. We shall come back to this particular example in Subsection 2.4.

LEMMA 1.5.10 ([**45**, Lemma 9.3]). *The Dec of a monoidal decomposition space has again a natural monoidal structure, and the dec map preserves this structure.*

1.5.11. *Example: the Schmitt Hopf algebra of graphs, continued.* The decomposition space of Example 1.1.5 (and 1.2.4) has a canonical monoidal structure given by disjoint union. Recall that X_k is the groupoid of graphs equipped with an ordered partition of the vertex set into k (possibly empty) parts. The disjoint union of two such structures is given by taking the disjoint union of the underlying graphs, with new partition given by joining the two ith parts, for each $1 \leq i \leq k$.

Hence we obtain a simplicial map from $X \times X$ to X. To say that it is CULF is to establish that squares such as

$$\begin{array}{ccc} X_1 \times X_1 & \xleftarrow{d_1} & X_2 \times X_2 \\ {\scriptstyle +}\downarrow & \lrcorner & \downarrow{\scriptstyle +} \\ X_1 & \xleftarrow[d_1]{} & X_2 \end{array}$$

are pullbacks. But a pair of graphs each equipped with a 2-partition can be uniquely reconstructed if we know what the two underlying graphs are (an element in $X_1 \times X_1$) and we know how the disjoint union is partitioned (an element in X_2), provided that we can identify the disjoint union of those two underlying graphs with the underlying graph of the disjoint union (which is to say that the data agree down in X_1). It follows that the resulting incidence coalgebra is also a bialgebra. (Furthermore, this bialgebra has a canonical grading, by the number of vertices, and with respect to this grading it is connected, since the only zero-vertex graph is the empty graph. It is well known that connected graded bialgebras are Hopf [**38**].)

2. Examples

It is characteristic for the classical theory of incidence (co)algebras of posets that most often it is necessary to impose an equivalence relation on the set of intervals in order to arrive at the interesting 'reduced' incidence (co)algebras. This equivalence relation may be simply isomorphism of posets, or equality of length of maximal chains as in binomial posets [**28**], or it may be more subtle order-compatible relations [**29**], [**95**]. Content, Lemay, and Leroux [**23**] remarked that in some important cases the relationship between the original incidence coalgebra and the reduced one amounts to a CULF functor, although they did not make this notion explicit. From our global simplicial viewpoint, we observe that very often these CULF functors arise from decalage, often of a decomposition space that is not a poset and sometimes not even a Segal space.

Recall that for X a locally finite decomposition space, we write $\mathcal{I}_X$ for the incidence coalgebra (with underlying vector space $\mathbb{Q}_{\pi_0 X_1}$), and we write $\mathcal{I}^X$ for the incidence algebra (with underlying profinite-dimensional vector space $\mathbb{Q}^{\pi_0 X_1}$).

2.0.1. *Decomposition spaces for the classical series.* Classically important examples of incidence algebras are power series representations. From the perspective of the objective method, these representations appear as cardinalities of various monoidal structures on species, realised as incidence algebras with groupoid coefficients. We list six examples illustrating some of the various kinds of generating functions listed by Stanley [**96**]; see also Dür [**29**].

(1) *Ordinary generating functions*, the zeta function being $\zeta(z) = \sum_{k \geq 0} z^k$. Such series come from ordered sets and ordinal sum, and the objective incidence algebra is that of ordered species with the ordinary product.
(2) *Exponential generating functions*, the zeta function being $\zeta(z) = \sum_{k \geq 0} \frac{z^k}{k!}$. Objectively, there are two versions: one coming from the standard Cauchy product of species, and one coming from the shuffle product of $\mathbb{L}$-species in the sense of [**8**].
(3) *Ordinary Dirichlet series*, the zeta function being $\zeta(z) = \sum_{k>0} k^{-s}$.

Such series come from ordered sets with the cartesian product.

(4) *Exponential Dirichlet series*, the zeta function being $\zeta(z) = \sum_{k>0} \frac{k^{-s}}{k!}$. Such series come from the Dirichlet product of arithmetic species [**4**], also called the arithmetic product [**78**].

(5) *q-exponential generating series*, with zeta function $\zeta(z) = \sum_{k\geq 0} \frac{z^k}{[k]!}$. Such series come from the Waldhausen $S_\bullet$-construction on the category of finite vector spaces. The incidence algebra is that of q-species with a version of the external product of Joyal–Street [**62**].

(6) *Variant q-exponential series* [**84**], with zeta function $\zeta(z) = \sum_{k\geq 0} \frac{z^k}{|\mathrm{Aut}(\mathbb{F}_q^k)|}$. Such series arise from q-species with the 'Cauchy' product.

Of these examples, only (1) and (3) have trivial section coefficients and come from a Möbius category in the sense of Leroux. We proceed to the details.

2.1. Additive examples. We start with several straightforward examples that serve to reiterate the importance of having incidence algebras of posets, monoids and monoidal groupoids on the same footing, connected by CULF functors, and in particular by decalage.

2.1.1. *Linear orders and the additive monoid.* Let $\mathbf{L}$ denote the nerve of the poset $(\mathbb{N}, \leq)$, and let $\mathbf{N}$ be the nerve of the additive monoid $(\mathbb{N}, +)$. Imposing the equivalence relation 'isomorphism of intervals' on the incidence coalgebra of $\mathbf{L}$ gives that of $\mathbf{N}$, and Content, Lemay, and Leroux [**23**] observed that this reduction is induced by a CULF functor $r \colon \mathbf{L} \to \mathbf{N}$ sending $a \leq b$ to $b - a$. In fact we have the following.

LEMMA 2.1.2. *There is an isomorphism of simplicial sets*

$$\mathrm{Dec}_\perp \mathbf{N} \xrightarrow{\simeq} \mathbf{L}$$

given in degree k by

$$(x_0, \ldots, x_k) \longmapsto [x_0 \leq x_0 + x_1 \leq \cdots \leq x_0 + \cdots + x_k],$$

and the CULF functor r is isomorphic to the dec map

$$d_\perp \colon \mathrm{Dec}_\perp \mathbf{N} \to \mathbf{N}, \qquad (x_0, \ldots, x_k) \mapsto (x_1, \ldots, x_k).$$

The comultiplication on $\boldsymbol{Grpd}_{/\mathbf{N}_1}$ is given by

$$\Delta(\ulcorner n \urcorner) = \sum_{a+b=n} \ulcorner a \urcorner \otimes \ulcorner b \urcorner$$

and, taking cardinality, the incidence coalgebra $\mathcal{I}_{\mathbf{N}}$ is the vector space $\mathbb{Q}_{\mathbb{N}}$ with basis given by the symbols δ_n and comultiplication $\Delta(\delta_n) = \sum_{a+b=n} \delta_a \otimes \delta_b$. The incidence algebra $\mathcal{I}^{\mathbf{N}}$ is the profinite-dimensional vector space $\mathbb{Q}^{\mathbb{N}}$ on the symbols δ^n with convolution product $\delta^a * \delta^b = \delta^{a+b}$, and is isomorphic to the ring of power series in one variable:

$$\begin{aligned} \mathcal{I}^{\mathbf{N}} &\xrightarrow{\simeq} \mathbb{Q}[[z]] \\ \delta^n &\longmapsto z^n \\ (\mathbb{N} \xrightarrow{f} \mathbb{Q}) &\longmapsto \sum f(n)\, z^n. \end{aligned}$$

2.1.3. *Upper dec.* In the previous example, and in most of the following, it is more convenient to work with lower dec. Let us just point out what happens with upper dec. Let $\mathbf{L}^{\text{op}}$ denote the nerve of the opposite poset of $(\mathbb{N}, \leq)$, that is, $(\mathbb{N}, \geq)$. There is a CULF functor $r' \colon \mathbf{L}^{\text{op}} \to \mathbf{N}$ sending $a \geq b$ to $a-b$. We have the following result.

LEMMA 2.1.4. *There is an isomorphism of simplicial sets*

$$\mathrm{Dec}_\top \mathbf{N} \xrightarrow{\simeq} \mathbf{L}^{\text{op}}$$

given in degree k by

$$(x_0, \dots, x_k) \longmapsto [x_0 + \cdots + x_k \geq x_1 + \cdots + x_k \geq \cdots \geq x_{k-1} + x_k \geq x_k],$$

and the CULF functor r' is isomorphic to the dec map

$$d_\top \colon \mathrm{Dec}_\top \mathbf{N} \to \mathbf{N}, \qquad (x_0, \dots, x_k) \mapsto (x_0, \dots, x_{k-1}).$$

In the following examples, this contravariance comes in for all upper decs. It does not play any role until Example 2.5.1.

2.1.5. *Powers.* As a variation of the previous example, fix $k \in \mathbb{N}$ and let $\mathbf{L}^k$ denote the (strict) nerve of the poset $(\mathbb{N}^k, \leq)$ and let $\mathbf{N}^k$ denote the strict nerve of the monoid $(\mathbb{N}^k, +)$. Again there is a CULF functor $\mathbf{L}^k \to \mathbf{N}^k$, and the incidence algebra of $\mathbf{N}^k$ is the power series ring in k variables. The functor is defined by coordinatewise difference, and again it is given by decalage, via a natural identification $\mathbf{L}^k \simeq \mathrm{Dec}_\bot \mathbf{N}^k$. The functor does *not* divide out by isomorphism of intervals, unless $k = 1$, since isomorphic intervals also arise by permutation of coordinates, treated next.

2.1.6. *Symmetric powers.* Let M be a monoid. For fixed $k \in \mathbb{N}$, the power M^k is again a monoid, considered as a decomposition space via its strict nerve X. The symmetric group $\mathfrak{S}_k$ acts on $X_1 = M^k$ by permutation of coordinates, and acts on $X_n = X_1^n = (M^k)^n$ diagonally. There is induced a simplicial groupoid $X/\mathfrak{S}_k$ given by homotopy quotient: in degree n it is the action groupoid $\frac{X_1 \times \cdots \times X_1}{\mathfrak{S}_k}$. Since taking homotopy quotient of a group action is a lowershriek operation, it preserves pullbacks, so it follows that this new simplicial groupoid again satisfies the Segal condition. (It is no longer a monoid, though, since in degree zero we have the one-object groupoid $B\mathfrak{S}_k = 1/\mathfrak{S}_k$, the classifying space of the group $\mathfrak{S}_k$). In general, the quotient map $X \to X/\mathfrak{S}_k$ is a CULF functor which does not arise from decalage.

We now return to the poset $(\mathbb{N}^k, \leq)$ and its nerve $\mathbf{L}^k$ from 2.1.5. The reduced incidence algebra, given by identifying isomorphic intervals, coincides with the incidence coalgebra of $\mathbf{N}^k/\mathfrak{S}_k = (\mathbb{N}^k, +)/\mathfrak{S}_k$. The reduction map is the composite CULF functor

$$\mathbf{L}^k \simeq \mathrm{Dec}_\bot \mathbf{N}^k \longrightarrow \mathbf{N}^k \longrightarrow \mathbf{N}^k/\mathfrak{S}_k.$$

2.1.7. *Injections and the monoidal groupoid of sets under sum.* Let $\mathbf{I}$ be the fat nerve of the category of finite sets and injections, and let $\mathbf{B}$ be the monoidal nerve of the monoidal groupoid $(\mathbb{B}, +, 0)$ of finite sets and bijections (see B.2.4). Dür [**29**] noted that imposing the equivalence relation 'having isomorphic complements' on the incidence coalgebra of $\mathbf{I}$ gives the binomial coalgebra. Again, we can see this reduction map as induced by a CULF functor from a decalage:

LEMMA 2.1.8. *There is an equivalence of simplicial groupoids*

$$\mathrm{Dec}_\perp \mathbf{B} \xrightarrow{\simeq} \mathbf{I}$$

given in degree k *by*

$$(x_0, \dots, x_k) \longmapsto [x_0 \subseteq x_0 + x_1 \subseteq \cdots \subseteq x_0 + \cdots + x_k],$$

and a CULF functor $\mathbf{I} \to \mathbf{B}$ *is given by*

$$d_\perp \colon \mathrm{Dec}_\perp \mathbf{B} \to \mathbf{B}, \qquad (x_0, \dots, x_k) \mapsto (x_1, \dots, x_k).$$

The isomorphism may also be represented diagrammatically using diagrams reminiscent of those in Waldhausen's $S_\bullet$-construction; see Subsection 2.3 below. As an example, both groupoids $\mathbf{I}_3$ and $(\mathrm{Dec}_\perp \mathbf{B})_3 = \mathbf{B}_4$ are equivalent to the groupoid of diagrams

$$\begin{array}{ccccccc}
 & & & & & & x_3 \\
 & & & & & & \downarrow \\
 & & & & x_2 & \longrightarrow & x_2 + x_3 \\
 & & & & \downarrow & & \downarrow \\
 & & x_1 & \longrightarrow & x_1 + x_2 & \longrightarrow & x_1 + x_2 + x_3 \\
 & & \downarrow & & \downarrow & & \downarrow \\
x_0 & \longrightarrow & x_0 + x_1 & \longrightarrow & x_0 + x_1 + x_2 & \longrightarrow & x_0 + x_1 + x_2 + x_3.
\end{array}$$

The face maps $d_i \colon \mathbf{I}_3 \to \mathbf{I}_2$ and $d_{i+1} \colon \mathbf{B}_4 \to \mathbf{B}_3$ both act by deleting the column beginning x_i and the row beginning x_{i+1}. In particular $d_\perp \colon \mathbf{I} \to \mathbf{B}$ deletes the bottom row, sending a sequence of injections to the sequence of successive complements (x_1, x_2, x_3). We will revisit this theme in the treatment of the Waldhausen $S_\bullet$-construction.

From Subsection 1.5 we have the following.

LEMMA 2.1.9. *Both* $\mathbf{I}$ *and* $\mathbf{B}$ *are monoidal decomposition spaces under disjoint union, and* $\mathbf{I} \simeq \mathrm{Dec}_\perp \mathbf{B} \to \mathbf{B}$ *is a monoidal CULF functor inducing a bialgebra homomorphism* $\boldsymbol{Grpd}_{/\mathbf{I}_1} \to \boldsymbol{Grpd}_{/\mathbf{B}_1}$.

Proposition 1.2.7 gives the comultiplication on $\boldsymbol{Grpd}_{/\mathbf{B}_1}$ as

$$\Delta(\ulcorner S \urcorner) = \int^{A,B} \mathrm{Bij}(A+B, S) \cdot \ulcorner A \urcorner \otimes \ulcorner B \urcorner = \sum_{\substack{A,B \subset S \\ A \cup B = S,\ A \cap B = \emptyset}} \ulcorner A \urcorner \otimes \ulcorner B \urcorner.$$

It follows that the convolution product on $\boldsymbol{Grpd}^{\mathbf{B}_1}$ is just the Cauchy product on groupoid-valued species:

$$(F * G)[S] = \sum_{A+B=S} F[A] \times G[B].$$

For the representables, the formula says simply that $h^A * h^B = h^{A+B}$.

The decomposition space $\mathbf{B}$ is locally finite, and taking cardinality gives the classical binomial coalgebra $\mathcal{I}_{\mathbf{B}} = \mathbb{Q}_{\pi_0 \mathbf{B}_1}$, with basis given by the symbols δ_n and

$$\Delta(\delta_n) = \sum_{a+b=n} \frac{n!}{a!\, b!} \, \delta_a \otimes \delta_b.$$

As a bialgebra we have $(\delta_1)^n = \delta_n$, and one recovers the comultiplication from $\Delta(\delta_n) = \big(\delta_0 \otimes \delta_1 + \delta_1 \otimes \delta_0\big)^n$.

Dually, the incidence algebra $\mathcal{I}^{\mathbf{B}}$ is the profinite-dimensional vector space $\mathbb{Q}^{\pi_0 \mathbf{B}_1}$ with basis given by the symbols δ^n and with convolution product

$$\delta^a * \delta^b = \frac{(a+b)!}{a!\,b!}\,\delta^{a+b}.$$

It is isomorphic to the algebra $\mathbb{Q}[[z]]$, where δ^n corresponds to $z^n/n!$ and the cardinality of a species F corresponds to its exponential generating series.

2.1.10. *Finite ordered sets, and the shuffle product of* $\mathbb{L}$*-species.* Let $\mathbf{OI}$ denote (the fat nerve of) the category of finite ordered sets and monotone injections. The resulting incidence coalgebra can be reduced by identifying two monotone injections if they have isomorphic complements, in analogy with Example 2.1.7, yielding in this case the *shuffle coalgebra*. Again, this reduction is an example of decalage. Consider the decomposition space $\mathbf{Z}$ with $\mathbf{Z}_n = \mathbf{OI}_{/\underline{n}}$, the groupoid of arbitrary maps from a finite ordered set S to $\underline{n}$, or equivalently of n-shuffles of S. The incidence coalgebra of this decomposition space is the shuffle coalgebra. This example is subsumed in the theory of restriction species, developed in [**49**]. The section coefficients are the binomial coefficients, but we may now note that on the objective level the convolution algebra is the shuffle product of $\mathbb{L}$-species [**8**].

There is a natural identification

$$\mathbf{OI} \xrightarrow{\simeq} \mathrm{Dec}_\perp \mathbf{Z},$$

which takes a sequence of monotone injections to the list of successive complements. There is also a CULF functor $\mathbf{Z} \to \mathbf{B}$ that takes an n-shuffle to the underlying n-tuple of subsets, and the decalage of this functor is the CULF functor $\mathbf{OI} \to \mathbf{I}$ given by forgetting the order; see [**45**, Example 4.5]. Combining with Lemma 2.1.9, we get the commutative diagram of monoidal decomposition spaces and monoidal CULF functors

$$\begin{array}{ccccc}
\mathbf{OI} & \xrightarrow{\simeq} & \mathrm{Dec}_\perp \mathbf{Z} & \xrightarrow{d_\perp} & \mathbf{Z} \\
\downarrow & & \downarrow & & \downarrow \\
\mathbf{I} & \xrightarrow{\simeq} & \mathrm{Dec}_\perp \mathbf{B} & \xrightarrow[d_\perp]{} & \mathbf{B}.
\end{array}$$

2.1.11. *Words.* Let A be a fixed set, regarded as an alphabet. The slice category $\mathbf{OI}_{/A}$ is the category of finite words in A and subword inclusions; see Lothaire [**76**] and also Dür [**29**]. Again it is naturally identified with the decalage of the A*-coloured shuffle decomposition space* $\mathbf{Z}_A$, which in degree k is the groupoid of A-words (of arbitrary length) equipped with a not-necessarily-order-preserving map to $\underline{k}$. Precisely, the objects are spans of sets

$$\underline{k} \leftarrow \underline{n} \rightarrow A.$$

The dec map $\mathbf{OI}_{/A} \simeq \mathrm{Dec}_\perp \mathbf{Z}_A \to \mathbf{Z}_A$ takes a subword inclusion to its complement word. The incidence algebra $\mathcal{I}^{\mathbf{Z}_A}$ is the Lothaire shuffle algebra of words. Again, it all amounts to observing that A-words admit a forgetful monoidal CULF functor to 1-words, which is just the decomposition space $\mathbf{Z}$ from before, which in turn admits a monoidal CULF functor to $\mathbf{B}$.

Note the difference between $\mathbf{Z}_A$ and the free monoid on A: the latter is like allowing only the trivial shuffles, where the subword inclusions are only concatenation inclusions. In terms of the structure maps $\underline{n} \to \underline{k}$, the free-monoid nerve allows only monotone maps, whereas the shuffle decomposition space allows arbitrary set maps.

2.2. Multiplicative examples.

2.2.1. *Divisibility poset and multiplicative monoid.* In analogy with 2.1.1, let $\mathbf{D}$ denote the (strict) nerve of the divisibility poset $(\mathbb{N}^\times, |)$, and let $\mathbf{M}$ be the strict nerve of the multiplicative monoid $(\mathbb{N}^\times, \cdot)$. Imposing the equivalence relation 'isomorphism of intervals' on the incidence coalgebra of $\mathbf{D}$ gives that of $\mathbf{M}$, and Content, Lemay, and Leroux [**23**] observed that this reduction is induced by the CULF functor $r\colon \mathbf{D} \to \mathbf{M}$ sending $d|n$ to n/d. In fact we have the following.

LEMMA 2.2.2. *There is an isomorphism of simplicial sets*

$$\mathrm{Dec}_\perp \mathbf{M} \xrightarrow{\simeq} \mathbf{D}$$

given in degree k by

$$(x_0, x_1, \ldots, x_k) \longmapsto [x_0|x_0x_1|\ldots|x_0x_1\cdots x_k],$$

and the CULF functor r is isomorphic to the dec map

$$d_\perp\colon \mathrm{Dec}_\perp \mathbf{M} \to \mathbf{M}, \qquad (x_0, \ldots, x_k) \mapsto (x_1, \ldots, x_k).$$

This example can be obtained from 2.1.1 directly, since $\mathbf{M} = \prod_p \mathbf{N}$ and $\mathbf{D} = \prod_p \mathbf{L}$, where the (weak) product is over all primes p. Now $\mathrm{Dec}_\perp$ is a right adjoint, so preserves products, and Lemma 2.2.2 follows from Lemma 2.1.2.

Since $\mathbf{M}_0$ is contractible, we can use Corollary 1.4.3, and since $\mathbf{M}_1$ is discrete, there are no symmetry factors, and we find that the convolution product is

$$\delta^m * \delta^n = \delta^{mn},$$

and the incidence algebra is isomorphic to the Dirichlet algebra:

$$\begin{aligned} \mathcal{I}^{\mathbf{M}} &\longrightarrow \{\sum_{k>0} a_k k^{-s}\} \\ \delta^n &\longmapsto n^{-s} \\ f &\longmapsto \sum_{n>0} f(n) n^{-s}. \end{aligned}$$

2.2.3. *Arithmetic species.* The Dirichlet coalgebra (2.2.1) also has a fatter version: consider now instead the monoidal groupoid $(\mathbb{B}^\times, \times, 1)$ of non-empty finite sets under the cartesian product, and its monoidal nerve $\mathbf{A}$ with $\mathbf{A}_k := (\mathbb{B}^\times)^k$, as in B.2.4, where this time the inner face maps take the cartesian product of two adjacent factors, and the outer face maps project away an outer factor.

The resulting coalgebra structure is

$$\Delta(S) = \sum_{A \times B \simeq S} A \otimes B.$$

Some care is due to interpret this formula correctly: the homotopy fibre of $d_1\colon \mathbf{A}_2 \to \mathbf{A}_1$ over S is the groupoid whose objects are triples (A, B, ϕ) consisting of sets A and B equipped with a bijection $\phi\colon A \times B \xrightarrow{\sim} S$, and whose morphisms are pairs of isomorphisms $\alpha\colon A \xrightarrow{\sim} A'$, $\beta\colon B \xrightarrow{\sim} B'$ such that $\phi' \circ (\alpha \times \beta) = \phi$.

The corresponding incidence algebra $\boldsymbol{grpd}^{\mathbb{B}^\times}$ with the convolution product is the algebra of arithmetic species [**4**] under the Dirichlet product (called the arithmetic product of species by Maia and Méndez [**78**]).

We are in the locally finite situation; since $\mathbf{A}_0$ is contractible, the section coefficients are given directly by Corollary 1.4.3:

$$\delta^m * \delta^n = \frac{(mn)!}{m!n!}\,\delta^{mn}.$$

From this formula we see that the incidence algebra $\mathcal{I}^{\mathbf{A}}$ is isomorphic to the Dirichlet algebra, namely

$$\begin{aligned}\mathcal{I}^{\mathbf{A}} &\longrightarrow \{\sum_{k>0} a_k k^{-s}\}\\ \delta^m &\longmapsto \frac{m^{-s}}{m!}\\ f &\mapsto \sum_{k>0} f(k)\frac{k^{-s}}{k!}.\end{aligned}$$

Such series are the 'exponential' (or modified) Dirichlet series of Baez and Dolan [**4**]. So the incidence algebra zeta function in this setting is

$$\zeta = \sum_{k>0}\delta^k \mapsto \sum_{k>0}\frac{k^{-s}}{k!}$$

(which is not the usual Riemann zeta function).

2.3. Linear examples and the Waldhausen $S_\bullet$-construction. In this subsection, we are concerned with linear versions of the additive examples: instead of starting with finite sets and injections, we look at vector spaces over a finite field $\mathbb{F}_q$, and their linear injections. Instead of binomial coefficients, the linear setting involves the q-binomials $\binom{n}{k}_q = \frac{[n]!}{[k]!\,[n-k]!}$, where $[n]! = \prod_{k=1}^n \frac{q^k-1}{q-1}$. The linear setting is richer: in particular, there is now an essential difference between quotients and complements, which at the level of decomposition spaces is the difference between the Waldhausen $S_\bullet$-construction and the monoidal nerve of direct sums, as we shall see.

2.3.1. *$\mathbb{F}_q$-vector spaces.* Consider a finite field $\mathbb{F}_q$ of q elements. Recall that the automorphism group of an n-dimensional $\mathbb{F}_q$-vector space has cardinality

$$\left|\mathrm{Aut}(\mathbb{F}_q^n)\right| \;=\; \prod_{i=1}^n (q^n - q^{i-1}) \;=\; (q-1)^n\, q^{\binom{n}{2}}\,[n]!\,.$$

Let $\mathbf{W}$ denote the fat nerve of the category $\boldsymbol{vect}^{\mathrm{inj}}$ of finite-dimensional $\mathbb{F}_q$-vector spaces and $\mathbb{F}_q$-linear injections. From this decomposition space we immediately get a coalgebra, which we now describe, but it is not the most interesting.

2.3.2. *Direct sums of $\mathbb{F}_q$-vector spaces and 'Cauchy' product of q-species.* A coalgebra that is the q-analogue of $\mathbf{B}$ can be obtained from the monoidal groupoid $(\boldsymbol{vect}^{\mathrm{iso}}, \oplus, 0)$. Denote by $\mathbf{M}$ the monoidal nerve of $(\boldsymbol{vect}^{\mathrm{iso}}, \oplus, 0)$, in the sense of B.2.4. We compute the section coefficients directly from the definition (8): the fibre of $d_1 \colon \mathbf{M}_2 \to \mathbf{M}_1$ over a vector space V is the groupoid consisting of triples (A, B, φ) where φ is a linear isomorphism $A \oplus B \xrightarrow{\sim} V$. This groupoid projects to

$\boldsymbol{vect}^{\mathrm{iso}} \times \boldsymbol{vect}^{\mathrm{iso}}$: the fibre over (A, B) is discrete, of cardinality $|\mathrm{Aut}(V)|$, giving altogether the section coefficient

$$c^n_{k,n-k} = \frac{\big|\mathrm{Aut}(\mathbb{F}_q^n)\big|}{\big|\mathrm{Aut}(\mathbb{F}_q^k)\big|\,\big|\mathrm{Aut}(\mathbb{F}_q^{n-k})\big|} = q^{k(n-k)}\binom{n}{k}_q .$$

(We could also have invoked Corollary 1.4.3, but using the definition directly has the pedagogical advantage that it also works for the closely related Example 2.3.5 below.) At the objective level, this convolution product corresponds to the 'Cauchy' product of q-species in the sense of Morrison [**84**].

If we let δ_n denote the cardinality of the name of an n-dimensional vector space V, the resulting coalgebra $\mathcal{I}_{\mathbf{M}}$ therefore has comultiplication:

$$\Delta(\delta_n) = \sum_{k \le n} q^{k(n-k)}\binom{n}{k}_q \cdot \delta_k \otimes \delta_{n-k}.$$

In analogy with the discrete case discussed in 2.1.7–2.1.8 there is a canonical simplicial map $\mathrm{Dec}_\perp \mathbf{M} \to \mathbf{W}$, given by sending an $(n+1)$-tuple of vector spaces $(V_0, \ldots, V_n)$ to the sequence of inclusions

$$V_0 \hookrightarrow V_0 \oplus V_1 \hookrightarrow \cdots \hookrightarrow V_0 \oplus \cdots \oplus V_n.$$

But in contrast to Lemma 2.1.8, this simplicial map is not an equivalence: the inverse, which in the discrete case was 'taking complements', does not exist in the linear case (or if it is constructed artificially, for example by reference to euclidean structure, it will mess with the isomorphisms). Let us actually compute $\mathrm{Dec}_\perp \mathbf{M}$.

2.3.3. *Complements as retractions.* Let $\mathbf{W}^{\mathrm{retr}}$ denote the fat nerve of the category whose objects are finite-dimensional $\mathbb{F}_q$-vector spaces and whose morphisms are retracted linear injections

$$V \underset{i}{\overset{r}{\rightleftarrows}} V'$$

Such retracted injections have canonical complements, namely $\ker(r)$. The following analogue of Lemma 2.1.8 is now straightforward to establish.

LEMMA 2.3.4. *There is a canonical equivalence of simplicial groupoids*

$$\mathrm{Dec}_\perp \mathbf{M} \xrightarrow{\simeq} \mathbf{W}^{\mathrm{retr}}$$

given in degree k by

$$(V_0, \ldots, V_k) \longmapsto [V_0 \subseteq V_0 \oplus V_1 \subseteq \cdots \subseteq V_0 \oplus \cdots \oplus V_k]$$

inducing a CULF functor $\mathbf{W}^{\mathrm{retr}} \to \mathbf{M}$.

This discussion shows that altogether $\mathbf{M}$ is not the most interesting viewpoint. We now change perspective from complements to quotients, getting to the more important power series representation with factor $[n]!$ instead of $\big|\mathrm{Aut}(\mathbb{F}_q^n)\big|$, and realise $\mathbf{W}$ as a decalage, in analogy with Lemma 2.1.8.

2.3.5. *q-binomials.* With reference to the incidence coalgebra of $\mathbf{W}$, impose the equivalence relation identifying two injections if their cokernels are isomorphic. This gives the q-binomial coalgebra (see Dür [**29**, 1.54]).

The same coalgebra can be obtained without reduction as follows. Put $\mathbf{V}_0 = 1$, let $\mathbf{V}_1$ be the maximal groupoid of ***vect***, and let $\mathbf{V}_2$ be the groupoid of short exact sequences. The span

$$\begin{array}{ccccc} \mathbf{V}_1 & \longleftarrow & \mathbf{V}_2 & \longrightarrow & \mathbf{V}_1 \times \mathbf{V}_1 \\ E & \longleftarrow\!\shortmid & [E' \to E \to E''] & \longmapsto & (E', E'') \end{array}$$

(together with the span $\mathbf{V}_1 \leftarrow \mathbf{V}_0 \to 1$) defines a coalgebra structure on $\boldsymbol{grpd}_{/\mathbf{V}_1}$ which (after taking cardinality) is the q-binomial coalgebra, without further reduction. The groupoids and maps involved are part of a simplicial groupoid $\mathbf{V}\colon \mathbb{\Delta}^{\mathrm{op}} \to$ ***Grpd***, namely the Waldhausen $S_\bullet$-construction of ***vect***, studied in more detail below, where we will see that this is a decomposition space but not a Segal space. The lower dec of $\mathbf{V}$ is naturally equivalent to the fat nerve $\mathbf{W}$ of the category of injections, and the dec map $d_\perp$ is the reduction map of Dür.

We calculate the section coefficients of $\mathbf{V}$. Since $\mathbf{V}$ is not a Segal space, we cannot invoke Corollary 1.4.3, so we use the definition of section coefficients (8) directly, to find the following standard formula for the *Hall numbers* (see also 2.3.11 below):

$$c^n_{k,n-k} = \frac{|\mathrm{SES}_{k,n,n-k}|}{\left|\mathrm{Aut}(\mathbb{F}_q^k)\right| \left|\mathrm{Aut}(\mathbb{F}_q^{n-k})\right|}.$$

Here $\mathrm{SES}_{k,n,n-k} = (\mathbf{V}_2)_{k,n,n-k}$ is the groupoid of short exact sequences with specified vector spaces of dimensions k, n, and $n-k$. This groupoid is discrete, and it has $\left|\mathrm{Aut}(\mathbb{F}_q^k)\right| \left|\mathrm{Aut}(\mathbb{F}_q^{n-k})\right| \binom{n}{k}_q$ elements. Indeed, there are $\binom{n}{k}_q$ k-dimensional subspaces of the n-dimensional space $\mathbb{F}_q^n$, and hence $\left|\mathrm{Aut}(\mathbb{F}_q^k)\right| \binom{n}{k}_q$ choices for the injection $\mathbb{F}_q^k \hookrightarrow \mathbb{F}_q^n$, and then $\left|\mathrm{Aut}(\mathbb{F}_q^{n-k})\right|$ choices for identifying the cokernel with $\mathbb{F}_q^{n-k}$. Thus

$$c^n_{n,n-k} = \binom{n}{k}_q = \frac{[n]!}{[k]![n-k]!}.$$

This description gives an isomorphism of algebras (see Goldman and Rota [**54**], and also Dür [**29**])

$$\begin{aligned} \mathcal{I}^{\mathbf{V}} &\longrightarrow \mathbb{Q}[[z]] \\ \delta^k &\longmapsto \frac{z^k}{[k]!}. \end{aligned}$$

This algebra is commutative, but an important new aspect is revealed on the objective level: here the convolution product is the external product of q-species of Joyal and Street [**62**]. They show (working with vector-space valued q-species), that this product has a natural non-trivial braiding (which reduces to commutativity upon taking cardinality).

2.3.6. *Waldhausen $S_\bullet$-construction of an abelian category.* The decomposition space with the short exact sequences leading to the Hall algebra is an example of Waldhausen's $S_\bullet$-construction [**104**], a centrepiece of modern K theory, as we now briefly explain.

The Waldhausen $S_\bullet$-construction of an abelian category $\mathscr{A}$ is a simplicial groupoid $S_\bullet\mathscr{A}$, with the following explicit description: $S_0\mathscr{A}$ is a point, $S_1\mathscr{A}$ is the maximal groupoid in $\mathscr{A}$, and $S_2\mathscr{A}$ is the groupoid of short exact sequences in $\mathscr{A}$. More generally, $S_n\mathscr{A}$ is a groupoid of staircase diagrams. For $n = 4$, for example, its objects are diagrams

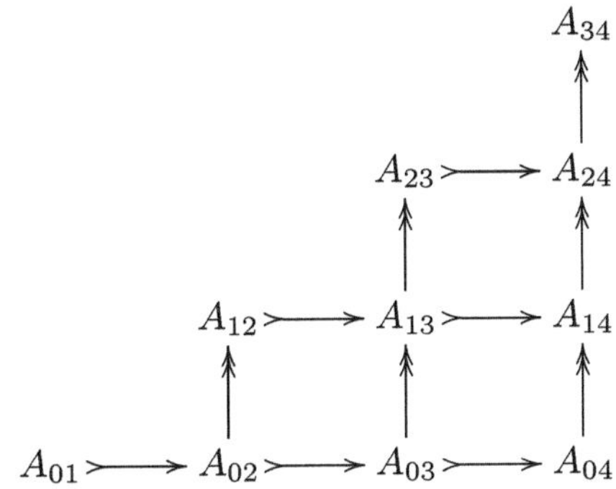

in which each sequence $A_{ij} \to A_{ik} \to A_{jk}$ is exact. The face map d_i deletes all objects containing an i index. The degeneracy map s_i repeats the ith row A_{i*} (if $i < n$) and also the ith column A_{*i} (if $i > 0$), with an additional 0 object at $(i, i+1)$.

A sequence of composable monomorphisms $(A_1 \rightarrowtail A_2 \rightarrowtail \cdots \rightarrowtail A_n)$ determines, up to canonical isomorphism, short exact sequences $A_{ij} \rightarrowtail A_{ik} \twoheadrightarrow A_{jk} = A_{ij}/A_{ik}$ with $A_{0i} = A_i$. Hence the whole diagram can be reconstructed up to isomorphism from the bottom row. (Similarly, since epimorphisms have uniquely determined kernels, the whole diagram can also be reconstructed from the last column.) Precisely we have the following result.

LEMMA 2.3.7. *The projections* $S_{n+1}\mathscr{A} \to \mathrm{Map}([n], \mathscr{A}^{\mathrm{mono}})$ *and* $S_{n+1}\mathscr{A} \to \mathrm{Map}([n], \mathscr{A}^{\mathrm{epi}})$ *are equivalences of groupoids.*

In view of this, one could try to define $S_\bullet\mathscr{A}$ using sequences of monomorphisms instead of staircase diagrams. The bottom face maps would then have to be

$$d_0(A_1 \rightarrowtail A_2 \rightarrowtail \cdots \rightarrowtail A_n) = (A_2/A_1 \rightarrowtail \cdots \rightarrowtail A_n/A_1),$$

but the dependence on choices of quotient objects A_i/A_1 would make it a pseudo-simplicial groupoid rather than a strict simplicial groupoid. In practice it is more convenient to have a strict simplicial groupoid, as accomplished with the staircase diagrams. For the remaining face maps there are no choices involved, and we have the following result.

PROPOSITION 2.3.8. *The equivalences of Lemma* 2.3.7 *assemble into levelwise simplicial equivalences*

$$\mathrm{Dec}_\perp(S_\bullet\mathscr{A}) \simeq \mathbf{N}(\mathscr{A}^{\mathrm{mono}})$$
$$\mathrm{Dec}_\top(S_\bullet\mathscr{A}) \simeq \mathbf{N}(\mathscr{A}^{\mathrm{epi}}).$$

Using Lemma 1.5.4 we now get the following theorem.

THEOREM 2.3.9 ([**34**, 7.3.3], [**45**, 10.10]). *The Waldhausen* $S_\bullet$*-construction of an abelian category* $\mathscr{A}$ *is a decomposition space.*

2.3.10. *Generalised Waldhausen construction.* It was shown by Bergner, Osorno, Ozornova, Rovelli, and Scheimbauer [**9–11**] that in fact every decomposition space arises from an $S_\bullet$-construction, but a more general one, taking as input certain augmented stable double Segal spaces. The classical case is the double Segal space whose horizontal morphisms are the monos and whose vertical morphisms are the epis. See the contributions of Ozornova and Rovelli [**86**] and Rovelli [**91**] for an introduction and further pointers.

2.3.11. *Hall algebras.* The finite-support incidence algebra of a decomposition space X was mentioned in 1.4.7; see [**46**, 7.15] for more details. In order for it to admit a cardinality, the required assumption is that X_1 be locally finite, that X_0 be finite, and that $X_2 \to X_1 \times X_1$ be a finite map. In the case of $X = S_\bullet(\mathscr{A})$ for an abelian category $\mathscr{A}$, this assumption translates into the condition that Ext^0 and Ext^1 be finite (which in practice means 'finite dimension over a finite field'). The finite-support incidence algebra in this case is the *Hall algebra* of $\mathscr{A}$; see Ringel [**89**] and also Schiffmann [**92**], although these sources twist the multiplication by the so-called Euler form.

Hall algebras were one of the main motivations for Dyckerhoff and Kapranov [**34**] to introduce 2-Segal spaces. We refer to their work for development of this important topic, recommending the lecture notes of Dyckerhoff [**30**] as a starting point; see also the contribution of Cooper and Young [**24**] in this volume.

2.4. Faà di Bruno bialgebra and variations. The Faà di Bruno bialgebra, originating with composition of power series, was constructed combinatorially by Doubilet [**27**] by imposing a *type-equivalence* relation on the incidence coalgebra of the partition poset. Joyal [**60**] observed that it can also be realised directly from the category of finite sets and surjections, without the need of a reduction step. Both constructions, and in particular the relationship between them, can be cast elegantly in the framework of decomposition spaces, serving to illustrate many of the characteristic aspects of the theory, such as the use of groupoids and the role of decalage.

2.4.1. *Faà di Bruno from the partition poset.* Fix a finite set of each cardinality, denoted $\underline{0}, \underline{1}, \underline{2}$, etc. Let $\mathcal{P}(\underline{m})$ denote the poset of partitions of the set $\underline{m}$; we write $\rho \leq \pi$ when partition ρ refines partition π. The *partition poset* is by definition the disjoint union of all these, $\mathcal{P} := \sum_{m \in \mathbb{N}} \mathcal{P}(\underline{m})$. The nerve of $\mathcal{P}$ defines a coalgebra (which is furthermore a bialgebra, with multiplication given by disjoint union). More interesting is the reduction of this bialgebra modulo type equivalence. An interval $[\rho, \pi]$ in a poset $\mathcal{P}(\underline{m})$ is said to have *type* $1^{\lambda_1} 2^{\lambda_2} \cdots$ if λ_i is the number of blocks of π that consist of exactly i blocks of ρ. Declare two intervals equivalent if they have the same type. The reduced incidence coalgebra of the poset $\mathcal{P}$ (with reduction given by type equivalence) is the *Faà di Bruno bialgebra.* (We gloss over the multiplicative structure, as it will be clearer in the viewpoint of surjections coming up next.)

2.4.2. *Partitions as surjections.* A partition ρ of $\underline{m}$ can be realised as a surjection $\underline{m} \twoheadrightarrow \underline{n}$, where $\underline{n}$ is the set of blocks. An interval $[\rho, \pi]$, from $\rho\colon \underline{m} \twoheadrightarrow \underline{n}$ to

$\pi \colon \underline{m} \twoheadrightarrow \underline{k}$ say, is then realised as a (unique) comparison surjection λ in a commutative triangle

$$\begin{array}{ccc} & \underline{m} & \\ {}^{\rho}\swarrow & & \searrow^{\pi} \\ \underline{n} & \xrightarrow[\lambda]{} & \underline{k}. \end{array} \tag{12}$$

The equivalence type of the interval $[\rho, \pi]$ is then precisely the isomorphism type of the surjection $\lambda \colon \underline{n} \to \underline{k}$, namely

$$1^{\lambda_1} 2^{\lambda_2} \cdots n^{\lambda_n} \tag{13}$$

if λ has λ_i fibres of cardinality i. Identifying intervals of the same type therefore amounts to forgetting the ambient set $\underline{m}$.

Splitting of intervals, as in the formula for comultiplication in the incidence coalgebra,

$$\Delta([\rho, \pi]) = \sum_{\sigma \in [\rho, \pi]} [\rho, \sigma] \otimes [\sigma, \pi],$$

is now precisely factorisation of such comparison surjections, so as to get

$$\Delta(\underline{n} \twoheadrightarrow \underline{k}) = \sum_{\underline{n} \twoheadrightarrow \underline{s} \twoheadrightarrow \underline{k}} (\underline{n} \twoheadrightarrow \underline{s}) \otimes (\underline{s} \twoheadrightarrow \underline{k}),$$

but some care is needed to count these factorisations correctly. The sum is over isomorphism classes of factorisations $\underline{n} \twoheadrightarrow \underline{s} \twoheadrightarrow \underline{k}$. In detail, consider the *factorisation groupoid* $\mathrm{Fact}(\underline{n} \twoheadrightarrow \underline{k})$, whose objects are factorisations of $\underline{n} \twoheadrightarrow \underline{k}$ into two surjections $\underline{n} \twoheadrightarrow \underline{s} \twoheadrightarrow \underline{k}$, and whose morphisms are bijections $\underline{s} \simeq \underline{s}'$ making the two triangles commute:

$$\begin{array}{ccccc} & \nearrow & \underline{s} & \searrow & \\ \underline{n} & & \downarrow \simeq & & \underline{k}\,. \\ & \searrow & \underline{s}' & \nearrow & \end{array} \tag{14}$$

Then the above sum is over $\pi_0(\mathrm{Fact}(\underline{n} \twoheadrightarrow \underline{k}))$, the set of connected components of the factorisation groupoid. This is Joyal's construction of the Faà di Bruno bialgebra [**60**] (and now the algebra structure is simply given by disjoint union of surjections).

2.4.3. *The decomposition space of surjections.* In decomposition space language, the Faà di Bruno bialgebra is simply the incidence bialgebra of the monoidal decomposition space $\mathbf{S}$ given as the fat nerve of the category of finite sets and surjections. Indeed, it has as $\mathbf{S}_0$ the groupoid of finite sets and bijections, and as $\mathbf{S}_1$ the groupoid whose objects are surjections and whose morphisms are squares

$$\begin{array}{ccc} \underline{n} & \twoheadrightarrow & \underline{k} \\ \sim\downarrow & & \downarrow\sim \\ \underline{n}' & \twoheadrightarrow & \underline{k}'. \end{array}$$

The groupoid $\mathbf{S}_2$ has as objects the composable pairs of surjections, and as morphisms diagrams

$$\begin{array}{ccccc} \underline{n} & \twoheadrightarrow & \underline{s} & \twoheadrightarrow & \underline{k} \\ \wr\downarrow & & \wr\downarrow & & \wr\downarrow \\ \underline{n}' & \twoheadrightarrow & \underline{s}' & \twoheadrightarrow & \underline{k}'. \end{array}$$

The homotopy fibre of $d_1 \colon \mathbf{S}_2 \to \mathbf{S}_1$ over $\lambda \in \mathbf{S}_1$ is equivalent to the subgroupoid whose objects are composable pairs composing to λ and whose morphisms are diagrams (14). It follows readily that the incidence bialgebra is precisely the Faà di Bruno bialgebra à la Joyal.

Note that in the groupoid setting, there is no need to restrict to a skeleton of the category of finite sets and surjections. We may as well work with the whole groupoid, without making choices. The homotopy equivalences take care of 'dividing out', while keeping the correct automorphism data.

2.4.4. *Relationship via decalage.* Having interpreted partitions as surjections, and refinements as factorisations of surjections, one may suspect that the partition poset is the decalage of the surjections nerve. This intuition is almost correct, but not quite: there is a subtle difference related to symmetries (pointed out by Mark Weber), which in turn originates in the fact that the partition poset is based on chosen fixed sets $\underline{m}$. Clearing up this issue will now give us a nice opportunity to see some CULF functors.

We first analyse the relationship between partitions and surjections. Two surjections $\underline{m} \twoheadrightarrow \underline{n}$ and $\underline{m} \twoheadrightarrow \underline{k}$ represent the same partition if there is a bijection $\underline{n} \xrightarrow{\sim} \underline{k}$ making the triangle (12) commute. Consider the category $\mathscr{C}$ whose objects are surjections between the standard sets $\underline{m}$ and whose morphisms are triangles like (12). Since there is at most one arrow λ between any two surjections, this category is equivalent to a poset, and indeed equivalent to the partition poset $\mathcal{P}$. It does contain non-trivial isomorphisms, but all automorphisms are trivial. Since this category $\mathscr{C}$ and the partition poset are equivalent as categories, their fat nerves are (levelwise) equivalent decomposition spaces, and therefore define equivalent coalgebras. Note that for strict posets, the fat nerve is the same thing as the strict nerve.

This category $\mathscr{C}$ sits in a bigger category $\mathscr{D}$ with the same objects (surjections), but where maps between two surjections are allowed to have a non-identity bijection between the domains, instead of just an identity arrow. The categories $\mathscr{C}$ and $\mathscr{D}$ are not equivalent, and their fat nerves are not equivalent, and their bialgebras are not isomorphic—all because of the different amount of symmetry they sport at the surjection domains. However, they have exactly the same type reduction, since the type reduction precisely throws away the surjection domains.

LEMMA 2.4.5. *The inclusion functor $\mathscr{C} \to \mathscr{D}$ is CULF.*

The argument is similar to that of type reduction: CULF-ness is about factorisations of the codomain maps, which is not affected by what happens at the domain.

Now that in $\mathscr{D}$ we have symmetries built in naturally, there is no reason to restrict to the skeleton anymore. We can define an equivalent category $\mathscr{D}'$ whose objects are now surjections between arbitrary finite sets, instead of just those chosen sets $\underline{n}$, and the maps have the same description as in $\mathscr{D}$. Note that this bigger category $\mathscr{D}'$ has a natural interpretation in terms of partitions: suppose we want a notion of partition, but do not wish to restrict attention to those chosen sets $\underline{n}$. We would then have to say when two partitions are considered the same, and more generally what should be the notion of morphism of partitions: the natural notion is to have a bijection at the domain that preserves block membership, i.e. a bijection f such that if t_1 and t_2 belong to the same block, then also ft_1 and ft_2 belong to the same block. Such a bijection induces a unique surjection between the codomains and it follows that this notion of morphism is precisely a refinement. The result is a category rather than a poset: it mixes the poset structure with the invertible maps given by renaming of set elements. Note that a partition in $\mathscr{D}$ or in $\mathscr{D}'$ has more automorphisms than in $\mathscr{C}$: for example a $(2,2)$-partition has an automorphism group of order 8, namely 4 possibilities to permute within the blocks, and 2 possibilities of interchanging the blocks.

With these extra symmetries we have the following.

LEMMA 2.4.6. *The fat nerve of $\mathscr{D}'$ is naturally equivalent to* $\mathrm{Dec}_\perp \mathbf{S}$.

The next result summarises our findings.

PROPOSITION 2.4.7. *Type reduction, and the relationship between the partition poset and the surjections nerve is given by the string of CULF functors*

$$N\mathcal{P} \simeq \mathbf{N}\mathscr{C} \longrightarrow \mathbf{N}\mathscr{D}' \simeq \mathrm{Dec}_\perp \mathbf{S} \xrightarrow{d_\perp} \mathbf{S}.$$

The composite sends a partition to its set of blocks, and sends a refinement to the corresponding surjection as in (12).

The verifications are straightforward. It should be noted that these CULF functors are all monoidal, and hence induce bialgebra homomorphisms by Proposition 1.5.8.

2.4.8. *Faà di Bruno section coefficients.* We work with the decomposition space $\mathbf{S}$. Since $\mathbf{S}$ is monoidal, to describe its section coefficients, it is enough describe the comultiplication of connected surjections $n \twoheadrightarrow 1$. Write A_n for the cardinality of $\ulcorner n \twoheadrightarrow 1 \urcorner$ in $\boldsymbol{grpd}_{/\mathbf{S}_1}$. Since a general surjection $\lambda\colon n \twoheadrightarrow k$ of type $1^{\lambda_1}2^{\lambda_2}\cdots$ is isomorphic to a disjoint union of connected surjections, its cardinality is $A_1^{\lambda_1}A_2^{\lambda_2}\cdots$. We apply the general formula from Proposition 1.4.2. For a given factorisation $\underline{n} \overset{\lambda}{\twoheadrightarrow} \underline{k} \overset{\pi}{\twoheadrightarrow} \underline{1}$, we have to look at the set $\{\varphi\}$ of those isomorphisms $\varphi\colon \underline{k} \xrightarrow{\sim} \underline{k}$ that make the composite surjection $\lambda\varphi\pi$ equal to the original surjection $\underline{n} \twoheadrightarrow \underline{1}$. But since $\underline{1}$ is terminal, all φ will do, so the cardinality of this set is $k!$ (see [**48**] for discussion of this point). With this, the general formula 1.4.2 gives

$$\Delta(A_n) = \sum_{\substack{\underline{n}\twoheadrightarrow\underline{k}\\ \underline{k}\twoheadrightarrow\underline{1}}} \frac{|\mathrm{Aut}(\underline{n} \twoheadrightarrow \underline{1})| \cdot k!}{|\mathrm{Aut}(\underline{n} \twoheadrightarrow \underline{k})| \cdot |\mathrm{Aut}(\underline{k} \twoheadrightarrow \underline{1})|} \; A_1^{\lambda_1}\cdots A_n^{\lambda_n} \otimes A_k.$$

The sum is over all distinct isomorphism classes of pairs of surjections (for fixed n). The formula for the order of the automorphism group of a surjection $\lambda\colon \underline{n} \twoheadrightarrow \underline{1}$ is

$$|\mathrm{Aut}(\underline{n} \twoheadrightarrow \underline{k})| \;=\; \prod_{j=1}^{\infty} \lambda_j!(j!)^{\lambda_j},$$

and in particular $|\mathrm{Aut}(\underline{k} \twoheadrightarrow \underline{1})| = k!$. Altogether we find

$$\Delta(A_n) = \sum_{\underline{n}\twoheadrightarrow\underline{k}} \frac{n!}{\prod_{j=1}^{n} \lambda_j!(j!)^{\lambda_j}} \; A_1^{\lambda_1}\cdots A_n^{\lambda_n} \otimes A_k.$$

For example,

$$\Delta(A_3) = A_3 \otimes A_1 + 3A_1A_2 \otimes A_2 + A_1A_1A_1 \otimes A_3$$
$$\Delta(A_4) = A_4 \otimes A_1 + (4A_1A_3 + 3A_2A_2) \otimes A_2 + 6A_1A_1A_2 \otimes A_3 + A_1A_1A_1A_1 \otimes A_4.$$

The section coefficients, called the Faà di Bruno section coefficients, are the coefficients $\binom{n}{\lambda;k}$ of the Bell polynomials [**38**, (2.5)].

Using a different normalisation we may choose the basis $a_n = A_n/n!$, on which the comultiplication formula becomes

$$\Delta(a_n) = \sum_{\underline{n} \twoheadrightarrow \underline{k}} \frac{k!}{\prod_{j=1}^{n} \lambda_j!} \, a_1^{\lambda_1} \cdots a_n^{\lambda_n} \otimes a_k.$$

For example,

$$\Delta(a_3) = a_3 \otimes a_1 + 2a_1a_2 \otimes a_2 + a_1a_1a_1 \otimes a_3$$
$$\Delta(a_4) = a_4 \otimes a_1 + (2a_1a_3 + a_2a_2) \otimes a_2 + 3a_1a_1a_2 \otimes a_3 + a_1a_1a_1a_1 \otimes a_4.$$

2.4.9. *A decomposition space for the Faà di Bruno* Hopf *algebra.* The Faà di Bruno *Hopf* algebra is obtained by further reduction, classically stated as identifying two intervals in the partition poset if they are isomorphic as posets. This identification is equivalent to forgetting the value of λ_1. There is also a decomposition space that yields this Hopf algebra directly, obtained by quotienting the decomposition space **S** by the same equivalence relation. This means identifying two surjections (or sequences of composable surjections) if one is obtained from the other by taking disjoint union with a bijection. One may think of this as 'levelled forests modulo linear trees'. It is straightforward to check that this reduction respects the simplicial identities so as to define a simplicial groupoid, that it is a monoidal decomposition space, and that the quotient map from **S** is monoidal and CULF.

2.4.10. *Ordered surjections.* Let **OS** denote the fat nerve of the category of finite ordered sets and monotone surjections. It is a monoidal decomposition space under ordinal sum. Hence to describe the resulting comultiplication, it is enough to say what happens to a connected ordered surjection, say $\underline{n} \twoheadrightarrow \underline{1}$, which we denote simply by n: since there are no automorphisms around, we find

$$\Delta(n) = \sum_{k=1}^{n} \sum_{a} a \otimes k,$$

where the second sum is over the $\binom{n-1}{k-1}$ possible surjections $\lambda\colon n \twoheadrightarrow k$. The resulting bialgebra is essentially the (dual) Landweber–Novikov bialgebra in algebraic topology [**83**] (see also [**7**]), the noncommutative Faà di Bruno bialgebra in combinatorics [**14**], and the Dynkin–Faà di Bruno bialgebra in numerical analysis [**85**]; it also comes up in number theory [**55**]. See [**41**] and [**72**] for recent perspectives.

2.5. Trees and graphs. Various bialgebras of trees and graphs can be realised as incidence bialgebras of decomposition spaces that are not Segal. In these cases one can decompose but not compose, as already exemplified in the running example with graphs 1.1.5. In each case the lack of composability is caused by the decomposition destroying information that would have been needed to define a composition. As we shall see (in 2.5.3), it is sometimes possible to 'remedy' this issue to get instead a decomposition space that is Segal, at the price of giving up connectedness of the bialgebra. In the examples based on graphs and trees, this involves keeping 'open-ended' edges, and is intimately related to the theory of operads and related structures (2.5.7).

All the examples of decomposition spaces in this subsection are monoidal under disjoint union, and hence the resulting coalgebras are bialgebras.

2.5.1. *Butcher–Connes–Kreimer Hopf algebra.* A *rooted tree* is a connected and simply connected graph with a specified root vertex; a forest is a disjoint union of rooted trees. The Butcher–Connes–Kreimer Hopf algebra of rooted trees [**22**] is the free algebra on the set of isomorphism classes of rooted trees, with comultiplication defined by summing over certain admissible cuts c:

$$\Delta(T) = \sum_{c \in \text{adm.cuts}(T)} P_c \otimes R_c \,.$$

An admissible cut c is a splitting of the set of nodes into two subsets, such that the second forms a subtree R_c containing the root node (or is the empty forest); the first subset, the complement 'crown', then forms a subforest P_c, regarded as a monomial of trees. (The order of the two factors is dictated by an operadic viewpoint, where leaves are 'in' and the root is 'out', and is further justified in 2.5.7 below.)

Dür [**29**, Ch.IV, §3] gave an incidence-coalgebra construction of the Butcher–Connes–Kreimer coalgebra by starting with the category $\mathscr{C}$ of forests and root-preserving inclusions, generating a coalgebra (in our language the incidence coalgebra of the fat nerve of $\mathscr{C}$), and imposing the equivalence relation that identifies two root-preserving forest inclusions if their complement crowns are isomorphic forests.

Note that to be precise, one must use $\mathscr{C}^{\text{op}}$ instead of $\mathscr{C}$:

$$\mathbf{R} := \mathbf{N}(\mathscr{C}^{\text{op}}).$$

From the viewpoint of the incidence coalgebra this 'op' affects the comultiplication only by reversing the order of the tensor factors. We shall see shortly that the 'op' originates in an upper-dec construction (compare 2.1.3).

We can obtain the Butcher–Connes–Kreimer coalgebra directly from a decomposition space: let $\mathbf{H}_1$ denote the groupoid of forests, and let $\mathbf{H}_2$ denote the groupoid of forests with an admissible cut. More generally, $\mathbf{H}_0$ is defined to be a point, and $\mathbf{H}_k$ is the groupoid of forests with $k-1$ compatible admissible cuts. The groupoids $\mathbf{H}_k$ then assemble into a simplicial groupoid $\mathbf{H}$ in which the inner face maps forget a cut, and the outer face maps project away stuff: $d_\bot$ deletes the crown (everything above the top-most cut) and $d_\top$ deletes the bottom layer (the part of the forest below the bottom-most cut). It is readily seen that $\mathbf{H}$ is not a Segal space: a tree with a cut cannot be reconstructed from its crown and its bottom tree, which is to say that $\mathbf{H}_2$ is not equivalent to $\mathbf{H}_1 \times_{\mathbf{H}_0} \mathbf{H}_1$. It is straightforward to check that it *is* a decomposition space, in fact a symmetric monoidal one under disjoint union, and it also follows from its construction that the resulting bialgebra is the Butcher–Connes–Kreimer Hopf algebra. Note that the decomposition space is graded by the number of nodes (which is precisely the length filtration 3.2.1), and that it is connected since the empty forest is the only forest with zero nodes.

To explain the relationship between the two constructions, note that admissible cuts are essentially the same thing as root-preserving forest inclusions: then the cut is interpreted as the division between the included forest and the forest induced on the nodes in its complement. In this way we see that $\mathbf{H}_k$ is the groupoid of $k-1$ consecutive root-preserving inclusions. Furthermore, there is a natural identification

$$\mathrm{Dec}_{\top}\mathbf{H} \simeq \mathbf{R} = \mathbf{N}(\mathscr{C}^{\mathrm{op}}),$$

where the 'op' occurs since we are dealing with an upper dec, as in 2.1.3. Under this identification, the dec map $\mathrm{Dec}_{\top}\mathbf{H} \to \mathbf{H}$, always a (symmetric monoidal) CULF functor, realises precisely Dür's reduction: on $\mathbf{R}_1 \to \mathbf{H}_1$ it sends a root-preserving forest inclusion to its crown, that is, its complement. More generally, on $\mathbf{R}_k \to \mathbf{H}_k$ it sends a sequence of forest inclusions $F_0 \subset F_1 \subset \cdots \subset F_k$ to

$$F_1 \smallsetminus F_0 \subset \cdots \subset F_k \smallsetminus F_0.$$

2.5.2. *Restriction species and directed restriction species* [**49**]. The Butcher–Connes–Kreimer example shares important characteristics with the graph example of Schmitt, our running example in Section 1 (Examples 1.1.5, 1.2.4, 1.5.11), but where in the graph example there are no constraints on the nature of the cuts, in the tree example, only certain order-respecting cuts are deemed admissible.

Both examples can be subsumed in big families of decomposition spaces, which can be treated uniformly, namely decomposition spaces of restriction species, in the sense of Schmitt [**94**] (see also [**2**]), and decomposition spaces of directed restriction species, introduced and studied in [**49**]. Here we content ourselves with outlining the idea.

A *restriction species* [**94**] is simply a presheaf of the category $\mathbb{I}$ of finite sets and injections. Compared to a classical species [**60**], a restriction species is thus functorial not only on bijections but also on injections, meaning that a given structure on a set induces such structure also on any subset.

Given a restriction species $R\colon \mathbb{I}^{\mathrm{op}} \to \boldsymbol{Set}$, a coalgebra is obtained on the set of isomorphism classes of R-structures with comultiplication

$$\Delta(X) = \sum_{A+B=S} X|A \otimes X|B, \qquad X \in R[S]$$

and counit sending only the empty structures to 1. (This is the construction of Schmitt [**94**].)

It is preferable to work with groupoid-valued species as in [**3**], rather than the traditional set-valued species. Given a (groupoid-valued) restriction species $R\colon \mathbb{I}^{\mathrm{op}} \to \boldsymbol{Grpd}$, we construct a simplicial groupoid $\mathbf{R}$ where $\mathbf{R}_k$ is the groupoid of R-structures with an ordered partition of the underlying set into k (possibly empty) parts. Functoriality on active maps is about joining adjacent parts or inserting an empty part. Functoriality on inert maps is about projecting away outer parts, and is possible precisely because R is a restriction species. This simplicial groupoid can be shown to be a decomposition space, and the resulting incidence coalgebra is the Schmitt coalgebra [**49**]. Furthermore, morphisms of restriction species induce CULF functors and hence coalgebra homomorphisms. A great many species are actually restriction species (such as various classes of graphs, matroids, and posets), providing in this way a large supply of decomposition spaces (which are not generally Segal spaces).

The Butcher–Connes–Kreimer example is subsumed in a large class of examples coming from *directed restriction species*, a notion introduced in [**49**]. Where ordinary restriction species are presheaves on finite sets and injections, directed restriction species are presheaves on the category of finite posets and convex injections. The definition formalises the idea of considering only decompositions compatible with the poset structure in a certain way, as exemplified clearly by the notion of admissible cut.

2.5.3. *Operadic trees and P-trees.* There is an important variation on the Butcher–Connes–Kreimer Hopf algebra, but it is only a bialgebra: instead of considering combinatorial trees one considers operadic trees, i.e. trees with open incoming edges and an open-ended root edge. More generally one can consider P-trees for a finitary polynomial endofunctor P, i.e. trees whose nodes are decorated by the operations of P. For details on this setting, see [**63–65**], [**42**]; it suffices here to note that the notion of P-tree covers many kinds of structured trees, such as planar trees, binary trees, effective trees, linear trees, words, and a large class of inductive data types (W-types).

We wish to copy over the description from combinatorial trees to P-trees to get a simplicial groupoid X, where X_k is the groupoid of P-forests with $k-1$ compatible admissible cuts. There are some notable differences between the two constructions. One is that P-trees include nodeless trees, consisting of a single edge. As a consequence, X_0 is not just a point: it is the groupoid of nodeless forests. (This difference is responsible for getting a bialgebra instead of a Hopf algebra.) A second difference is that while a cut of a combinatorial tree completely removes the edges in question, for P-trees removal of edges is not possible, since it would violate the local structure of the tree (e.g. being binary). Instead the cut operation actually cuts the edge into two open-ended edges. A consequence of this feature is that, unlike $\mathbf{H}$, the simplicial groupoid X is a Segal space by the Key Lemma of [**42**]; see [**72**] for an abstract viewpoint. The 'half-edges' left by the cut provide the data to reconstruct a tree with a cut from its bottom tree and crown by grafting. The overlap is the nodeless forest that constitutes the cut. In this sense the operadic-forest decomposition space X is a 'category' with nodeless forests as objects, and arbitrary forests as morphisms, a forest being seen as a morphism from its leaves to its roots. From this perspective, the decomposition space $\mathbf{H}$ of combinatorial forests is obtained from X by throwing away the object information, i.e. the data governing the composability constraints. These two differences are crucial in the work on Green functions and Faà di Bruno formulae in [**42, 67, 72**].

There is a functor $X \to \mathbf{H}$ from operadic trees or P-trees to combinatorial trees which is *taking core* [**65**]: it amounts to forgetting the P-decoration and shaving off all open-ended edges. This functor is CULF monoidal, and it induces a a bialgebra homomorphism from the bialgebra of operadic trees or P-trees to the Butcher–Connes–Kreimer Hopf algebra of combinatorial trees.

2.5.4. *Note about symmetries.* One *cannot* obtain the same bialgebra of trees (either the combinatorial or the operadic) by taking isomorphism classes in each groupoid X_k: doing so would destroy symmetries that constitute an essential ingredient in the Butcher–Connes–Kreimer bialgebra. Indeed, define a simplicial set Y in which $Y_k = \pi_0(X_k)$, the set of isomorphism classes of forests with k compatible admissible cuts. Consider the tree T

belonging to X_1. The fibre of $d_1\colon X_2 \to X_1$ over T is the (discrete) groupoid of all possible cuts in this tree:

The thing to notice here is that while the second and third cuts are isomorphic as abstract cuts, and therefore get identified in $Y_2 = \pi_0(X_2)$, this isomorphism does not fix the underlying tree T. This means that in the formula for comultiplication of T as an element of X_1 both cuts appear, and there is a total of 5 terms, whereas in the formula for comultiplication of T as an element of Y there are only 4 terms. (Put in another way, the functor $X \to Y$ given by taking components is not CULF.)

It seems that there is no way to circumvent this discrepancy directly at the isomorphism class level: attempts involving ingenious decorations by natural numbers and actions by symmetric groups will almost certainly end up amounting to actually working at the groupoid level, and the conceptual clarity of the groupoid approach seems much preferable.

2.5.5. *Non-commutative versions.* The Butcher–Connes–Kreimer Hopf algebra of combinatorial trees admits a natural non-commutative version, first studied by Foissy [**40**]. It is defined in exactly the same way, but with *ordered* forests of *planar* combinatorial trees. In this case, the decomposition space is monoidal but not symmetric monoidal, giving naturally a non-commutative bialgebra.

The same modification can be applied in the operadic case. Planar operadic trees are precisely M-trees for M the free-monoid monad. More generally, to have planar structure on P-trees is to have a cartesian natural transformation $P \Rightarrow M$ (see [**53**] for details); in this situation there is a non-commutative bialgebra of ordered forests of P-trees.

2.5.6. *Free multicategories* [**53**]. Continuing the previous example, for any polynomial endofunctor P cartesian over M, the groupoid of P-trees is (essentially) discrete, which is to say that it is equivalent to the set of isomorphism classes of P-trees (because the planar structure encoded in the cartesian natural transformation to M fixes the automorphisms). This set is the set of operations of the free monad on P [**53**], [**63**]. Thinking of P as specifying a signature, we can equivalently think of P-trees as operations for the free (coloured) operad on that signature, or as the multi-arrows of the free multicategory on P regarded as a multigraph. To a multicategory there is associated a monoidal category [**58**], whose object set is the free monoid on the set of objects (colours). The decomposition space of P-trees is naturally identified with the (fat) nerve of the (monoidal) category associated to the multicategory of P-trees. (The adjective 'fat' is in parenthesis here because it could be omitted: the categories involved here have no invertible arrows other than the identities, because the multicategory is free.)

2.5.7. *Polynomial monads and operads.* The decomposition space of P-trees for P a polynomial endofunctor (2.5.3) can be regarded as the decomposition space associated to the free monad on P. In fact the construction works for any (cartesian, discrete-finitary) polynomial monad, not just free ones, as we now proceed to explain. This construction has been generalised and subsumed in a more comprehensive setting of relative two-sided bar constructions in [**72**]. Presently we outline, in a more heuristic manner, the construction of a monoidal decomposition space from any coloured operad, and from it a commutative bialgebra. For the numerical version, some finiteness conditions must be assumed.

Coloured operads can be encoded as polynomial monads [**108**]. The combinatorial data of the endofunctor R underlying the monad is a diagram of groupoids

$$I \leftarrow E \rightarrow B \rightarrow I$$

where I is the set (or more generally, groupoid) of colours, B is the groupoid of operations (more precisely the action groupoid of the action of the symmetric groups on the operations), and E is the groupoid of operations with a marked input slot. It follows that $E \rightarrow B$ is a finite map; the fibre over an operation is the set of its input slots. The operad substitution law then amounts to a cartesian monad structure on R, i.e. cartesian natural transformations $R \circ R \Rightarrow R \Leftarrow \mathrm{Id}$ subject to axioms.

Following the graphical interpretation given in [**70**], one can regard I as the groupoid of decorated unit trees (i.e. trees without nodes), and B as the groupoid of corollas (i.e. trees with exactly one node) decorated with B on the node and I on the edges, compatibly. The arity of a corolla labeled by $b \in B$ is then the cardinality of the fibre E_b.

We can now form a simplicial groupoid X in which X_0 is the groupoid of disjoint unions of decorated unit trees, X_1 is the groupoid of disjoint unions of decorated corollas, and where more generally X_n is the groupoid of R-forests of height n. For example, X_2 is the groupoid of R-forests of height 2, which equivalently can be described as configurations consisting of a disjoint unions of bottom corollas whose leaves are decorated with other corollas, in such a way that the roots of the decorating corollas match the leaves of the bottom corollas. This groupoid can more formally be described as the free symmetric monoidal category on $R(B)$ (the endofunctor R applied to B). Similarly, X_n is the free symmetric monoidal category on $R^{n-1}(B)$. The outer face maps project away the top or bottom layer in a forest of height n. For example $d_0 \colon X_1 \rightarrow X_0$ sends a disjoint union of corollas to the disjoint union of their root edges, while $d_1 \colon X_1 \rightarrow X_0$ sends a disjoint union of corollas to the forest consisting of all their leaves. The active face maps (i.e. inner face maps) join two adjacent layers by means of the monad multiplication on R. The degeneracy maps insert unary corollas by the unit of the monad. Associativity of the monad law ensures that this simplicial groupoid is actually a category object and a Segal space [**72**]. The operation of disjoint union makes this decomposition space symmetric monoidal, and altogether an incidence bialgebra results from the construction.

The example (2.5.3) of P-trees (for P a polynomial endofunctor) and admissible cuts is an example of this construction, namely corresponding to the free monad on P: indeed, the operations of the free monad on P form the groupoid of P-trees, which now plays the role of B. Trees of height n in which each node is decorated by objects in B is the same thing as P-trees equipped with $n-1$ compatible admissible cuts, and grafting of P-trees (as prescribed by the active face maps in 2.5.3) is precisely the monad multiplication in the free monad on P.

It should be stressed that while the decomposition space of a free operad is always automatically locally finite, the case of a general operad is not automatically so. This condition must be imposed separately if numerical examples are to be extracted.

Another subexample is the case where the monad is the terminal reduced monad Comm, which is the free-commutative-semimonoid monad. In this case, the resulting category object in groupoids is equivalent to the fat nerve of the category of surjections (as in 2.4), so the associated bialgebra is the classical Faà di Bruno bialgebra. The main achievement of [**72**] is to show that the Faà di Bruno formula for the comultiplication in the classical Faà di Bruno bialgebra generalises to incidence bialgebras of arbitrary operads and polynomial monads (the free case having been established previously in [**42**]).

2.5.8. *Progressive graphs and free PROPs.* The constructions in 2.5.3 readily generalise from trees to progressive graphs (although the attractive polynomial interpretation does not). By a progressive graph we understand a finite directed graph with a certain number of open input edges, a certain number of open output edges, and not allowed to contain an oriented cycle; see [**66**] for a categorical formalism. In particular, the set of vertices of a progressive graph has a natural poset structure. The progressive graphs form a groupoid $\mathbf{G}_1$. We allow graphs without vertices, these form a groupoid $\mathbf{G}_0$. Let $\mathbf{G}_2$ denote the groupoid of progressive graphs with an *admissible cut*, by which we mean a partition of the set of vertices into two disjoint parts, a down-set V_1 and an up-set V_2. This partition determines a set of edges, called the *cut*, consisting of the edges that connect a vertex in V_1 with a vertex in V_2, the out-edges of V_0, the in-edges of V_2, and the edges of G that are both in-edges and out-edges. The two vertex sets V_1 and V_2 induce new progressive graphs $G|V_1$ and $G|V_2$, by including all edges incident to the given vertex set, and including in both cases also the whole cut set, which becomes the new set of output edges for $G|V_1$ and the new set of input edges for $G|V_2$. Similarly, let $\mathbf{G}_k$ denote the groupoid of progressive graphs with $k-1$ compatible admissible cuts, just like we did for forests. One can check that these groupoids assemble into a simplicial groupoid $\mathbf{G}$, readily verified to be a decomposition space and in fact a Segal space. The progressive graphs form the set of operations of the free PROP with one generator in each input/output degree (m, n). Decorating data for progressive graphs are called tensor schemes in [**61**], and the progressive graphs decorated by a tensor scheme form the set of operations of the free (coloured) PROP on the tensor scheme. The same construction is important for the operational semantics of Petri nets [**69**]. In fact, the construction works for any PROP, not just free ones, in analogy with the passage from trees and free operads to arbitrary operads (2.5.7). Note that disjoint union (or the monoidal structure underlying any PROP) makes the resulting incidence coalgebras into bialgebras.

Bialgebras of progressive graphs have been studied in the context of Quantum Field Theory by Manchon [**79**]. Certain decorated progressive graphs, and the resulting bialgebra, have been studied by Manin [**81, 82**] in the theory of computation: his graphs are decorated by operations on partial recursive functions and switches.

2.5.9. *Hereditary species and directed hereditary species.* In incidence coalgebras of categories or (directed) restriction species the left and the right tensor factor of the comultiplication are on equal footing. As we have just seen, incidence bialgebras of operads of various kinds have a monomial in the left-hand tensor factor and linear terms in the right-hand tensor factor, reflecting the many-to-one nature of operads. Another large class of combinatorial bialgebras with this feature comes from Schmitt's hereditary species [**94**]. Just like ordinary species are presheaves on

$\mathbb{B}$ and restriction species are presheaves on $\mathbb{I}$, hereditary species are functors on the category of finite sets and partial surjections, that is, spans where the backward leg is injective and the forward leg is surjective. We are thus talking about structures that can be restricted along injections and pushed forth along surjections.

A prototypical example is the hereditary species of graphs: the comultiplication of a graph is given by summing over partitions of the vertex sets; then on each block there is an induced graph structure (as in restriction species), and these graphs are formally multiplied together to form a monomial placed in the left-hand tensor factor. The right-hand tensor factor receives the graph structure induced on the quotient, that is, the set of blocks. Carlier [**16**] showed how a hereditary species H induces a monoidal decomposition space H whose incidence bialgebra is the one constructed by Schmitt [**94**]. The construction is similar to the two-sided bar construction of an operad: the decomposition space H has in degree 1 the groupoid of (families of) H-structures, and in degree 2 the groupoid of (families of) H-structures with a partition on the underlying set. The middle face map d_1 forgets the partition; the bottom face map d_0 returns the H-structure induced on the quotient set, and the top face map d_2 returns the family of H-structures given by the blocks. Just as for operads, it is this nature of the top face maps that necessitates working with families of structures rather than single structures.

Cebrian and Forero [**21**] gave a directed version of this construction, starting with a new notion of directed hereditary species, which is a functor on the category of posets and partial monotone contractions. Just as the case of graphs is the paradigmatic example of hereditary species, posets and trees form canonical examples of directed hereditary species. In particular, there are monoidal decomposition spaces of finite posets (giving a certain bialgebra of finite topologies first studied by Fauvet, Foissy, and Manchon [**36**]), trees (giving the Calaque–Ebrahimi-Fard–Manchon bialgebra of trees [**15**]), or linear trees (giving a version of the Faà di Bruno bialgebra as in Subsection 2.4, with the alternative normalisation mentioned in 2.4.8, in turn a symmetrised version of the example in 2.4.10).

Both for hereditary species and directed hereditary species, an important feature is that there is an ordinary restriction species present at the same time, and in this way there are *two* different comultiplications. It is shown in [**16**] and [**21**] that these structures form what is called a *comodule bialgebra*: it is about two bialgebras with the same underlying algebra but with two different comultiplications, and such that one comultiplication co-distributes over the other from one side. The interest in these structures comes mostly from their appearance in various branches of analysis; see Manchon [**80**] and Foissy [**39**].

Both hereditary species and directed hereditary species have a strong operad flavour, since there is one output (the quotient) and many inputs (the blocks). However, they are rarely operads; instead they are so-called *operadic categories* in the sense of Batanin and Markl [**6**]. An operadic analogue of the tree examples of Cebrian and Forero was given by Kock [**68**], leading to other decomposition-space constructions of comodule bialgebras. The difference is essentially the difference between combinatorial trees (as in 2.5.1) and operadic trees (as in 2.5.3).

2.6. Symmetric functions. The ring Sym of symmetric functions is the subring of the ring of power series in countably many variables, $\mathrm{Sym} \subset \mathbb{Q}[[x_1, x_2, \ldots]]$, consisting of the power series that are of bounded degree and are invariant under permutation of the variables; see Stanley [**98**].

Given an integer partition $\lambda \vdash n$, let $m_\lambda(\underline{x})$ be the symmetric function given by the following prescription: list all *distinct* permutations of the vector λ, then apply these as exponents to all $x_i\ x_j\ x_k \dots$ with $i < j < k < \dots$, and take the sum of the resulting monomials. For example, the integer partition 331 has three permutations, namely 331, 313, 133, and if we restrict to the alphabet with three variables, we obtain

$$m_{331} = x_1^3 x_2^3 x_3^1 + x_1^3 x_2^1 x_3^3 + x_1^1 x_2^3 x_3^3 \,. \tag{15}$$

The symmetric functions $m_\lambda(\underline{x})$ form an additive basis for Sym, termed the monomial basis.

2.6.1. *Towards an objective theory: the groupoid of surjections.* We outline some developments of an objective theory of symmetric functions. Further details are forthcoming in [**51**, **52**]. As in 2.4.3 we take the basic object to be the groupoid of surjections $\mathbf{S}_1$ rather than the set of integer partitions.

The theory of symmetric functions is huge, and even setting up the standard bases and base changes is a big task. In the following we content ourselves to outline the monomial basis and the elementary basis, and the base change between the two. This already illustrates the flavour of the objective theory, and reveals some subtleties, as we now explain.

2.6.2. *Symmetric functions in the M-basis.* The decomposition space Λ^M encoding the comultiplication of symmetric functions in the monomial basis is given by

$$\Lambda^M_r := \left\{ \begin{array}{l} \underline{n} \\ \big\downarrow \lambda \\ \underline{k} \xrightarrow{\ \ell\ } r \end{array} \right\}$$

So $\Lambda^M_0 = 1$ as only the empty surjection splits into 0 parts, and Λ^M_1 is the groupoid of surjections λ, while Λ^M_2 is the groupoid of surjections λ with a splitting or *layering* ℓ of the codomain into two parts, which we can picture more explicitly as

$$\begin{array}{ccccc} \underline{n}' & \dashrightarrow & \underline{n} & \dashleftarrow & \underline{n}'' \\ {\scriptstyle \lambda'=d_0(\lambda,\ell)} \big\downarrow & & {\scriptstyle d_1(\lambda,\ell)} \big\downarrow {\scriptstyle \lambda} & & \big\downarrow {\scriptstyle \lambda''=d_2(\lambda,\ell)} \\ \underline{k}' & \longrightarrow & \underline{k} & \longleftarrow & \underline{k}'' \end{array} \qquad \begin{array}{l} \text{sum splitting} \\ \\ \text{sum splitting } \ell. \end{array} \tag{16}$$

The bottom row is a sum-splitting diagram, meaning a pair of injective maps whose images are disjoint and that is jointly surjective. Note that then the top row is a sum splitting too, and that the data of the top row and the surjections λ', λ'' are implied (taking preimages) by λ and the bottom row.

Let M_λ denote the cardinality of $\ulcorner\lambda\urcorner\colon 1 \to \Lambda^M_1$. The simplicial groupoid Λ^M is a locally discrete decomposition space in the sense of 1.4.9 and we have

$$\Delta(M_\lambda) = \sum_{\underline{k}'+\underline{k}''=\underline{k}} M_{\lambda'} \otimes M_{\lambda''}$$

where the summation is over all splittings (16).

2.6.3. *Example comultiplication.* In the comultiplication of M_{331} there are 8 terms, corresponding to the $2^3 = 8$ splittings of the codomain. We get

$$\Delta(M_{331}) = M_{331} \otimes 1 + M_{33} \otimes M_1 + 2\, M_{31} \otimes M_3 + 2\, M_3 \otimes M_{31} + M_1 \otimes M_{33} + 1 \otimes M_{331}.$$

2.6.4. *Comparison with classical normalisation.* Experts will notice here that this comultiplication is different from the comultiplication of the usual monomial symmetric functions m_λ, where

$$\Delta(m_{331}) = m_{331} \otimes 1 + m_{33} \otimes m_1 + m_{31} \otimes m_3 + m_3 \otimes m_{31} + m_1 \otimes m_{33} + 1 \otimes m_{331}.$$

In fact we are dealing with a different normalisation, which was first studied by Doubilet [**26**]. Precisely, the relation with the classical monomial symmetric functions is given by

$$M_\lambda = (\lambda_1!\lambda_2!\lambda_3!\cdots)\, m_\lambda$$

if the type of the surjection is $1^{\lambda_1}2^{\lambda_2}\ldots$ as in (13).

2.6.5. *Polynomial semantics.* Let A be a set, here playing the role of an alphabet. A *polynomial functor* in A-many variables is a diagram

$$A \overset{s}{\leftarrow} T \to B \to 1$$

which defines a functor

$$\begin{array}{rcl} \mathbf{grpd}_{/A} & \longrightarrow & \mathbf{grpd} \\ X{\to}A & \longmapsto & \displaystyle\sum_{b\in B}\prod_{e\in T_b} X_{se} \;=\; \sum_{b\in B}\prod_{a\in A}(X_a)^{T_{a,b}}. \end{array}$$

Note that with B infinite, it is more like a power series.

To the monomial symmetric function M_λ we assign the polynomial functor defined by the diagram of sets

$$A \leftarrow T_\lambda \to B_\lambda \to 1$$

where A is our alphabet which could be finite or could be $\mathbb{N}$, and T_λ and B_λ are the sets of possible diagrams

$$T_\lambda = \left\{ \begin{array}{ccc} 1 & \longrightarrow & \underline{n} \\ & & \downarrow\!\!\downarrow \lambda \\ & & \underline{k} \hookrightarrow A \end{array} \right\}, \qquad B_\lambda = \left\{ \begin{array}{c} \underline{n} \\ \downarrow\!\!\downarrow \lambda \\ \underline{k} \hookrightarrow A \end{array} \right\}.$$

The map from T_λ to A returns the element singled out by the composite, and the map to B_λ forgets which element of $\underline{n}$ was singled out.

In other words, the polynomial associated to λ consists of one monomial for each monomorphism $\underline{k} \hookrightarrow A$, formed using the fibres n_i (for $i \in k$) as the exponents:

$$M_\lambda \longmapsto \sum_{j:\, k \hookrightarrow A} x_{j_1}^{n_1} \cdots x_{j_k}^{n_k}.$$

2.6.6. *Example.* Consider M_{331} and take the alphabet $A = \{x_1, x_2, x_3\}$. In contrast to the recipe for classical symmetric polynomials in (15) we keep the exponents vector fixed, and then let the variables come in all orders, so as to get

$$M_{331} = x_1^3x_2^3x_3^1 + x_1^3x_3^3x_2^1 + x_2^3x_1^3x_3^1 + x_2^3x_2^3x_1^1 + x_3^3x_1^3x_2^1 + x_3^3x_2^3x_1^1.$$

This is consistent with the normalisation $M_{331} = 2\, m_{331}$, where the factor $2 = 1!\,0!\,2!$ appears since the surjection has type $1^1 2^0 3^2$.

2.6.7. *Elementary symmetric functions.* The classical elementary basis for Sym is given by defining

$$e_n := m_{11\ldots 1}$$

and then defining

$$e_\lambda := e_{n_1} \cdots e_{n_k}.$$

In the objective theory we define E_λ directly in terms of surjections, with comultiplication given by splitting the domain. There is a locally discrete decomposition space Λ^E with

$$\Lambda^E_r := \left\{ \begin{array}{ccc} \underline{n} & \longrightarrow & r \\ \big\downarrow & & \\ \underline{k} & & \end{array} \right\}.$$

Thus $\Lambda^E_0 = 1$ and Λ^E_1 is the groupoid of surjections as before, while now Λ^E_2 is the groupoid of surjections with a splitting of the domain into two parts, which we can picture more explicitly as

$$\begin{array}{ccccc} \underline{n}' & \longrightarrow & \underline{n} & \longleftarrow & \underline{n}'' \\ \big\downarrow & & \big\downarrow & & \big\downarrow \\ \underline{k}' & \longrightarrow & \underline{k} & \longleftarrow & \underline{k}'' \end{array} \qquad \begin{array}{l} \text{sum splitting} \\ \\ \text{cover} \end{array}$$

where the top row is a sum-splitting diagram. The bottom row then consists of two injections that jointly cover k.

2.6.8. *Example.* There are 8 terms in $\Delta(E_{21})$, since there are 8 ways to split the domain. If the split is inside a block, that block splits into two:

$$\Delta(E_{21}) = 1 \otimes E_{21} + 2E_{11} \otimes E_1 + E_1 \otimes E_2 + 2E_1 \otimes E_{11} + E_2 \otimes E_1 + E_{21} \otimes 1.$$

Experts in symmetric functions will notice that this comultiplication differs from that of classical elementary symmetric functions, where

$$\Delta(e_{21}) = 1 \otimes e_{21} + e_{11} \otimes e_1 + e_1 \otimes e_2 + e_1 \otimes e_{11} + e_2 \otimes e_1 + e_{21} \otimes 1.$$

2.6.9. *Another normalisation.* For the elementary symmetric functions, as for the monomial basis, there is a scalar factor relating the classical and our objective version. The relationship here says $E_2 = 2\,e_2$ and $E_{21} = 2\,e_{21}$, and in general

$$E_\lambda \;=\; n_1! \cdots n_k!\, e_\lambda.$$

Again, the reason is the polynomial semantics: to E_λ represented by a surjection $\lambda\colon \underline{n} \twoheadrightarrow \underline{k}$ we assign the polynomial functor $A \leftarrow T \rightarrow B \rightarrow 1$, where B is the set of diagrams

$$\begin{array}{cc} \underline{n} & \xrightarrow{\text{loc.inj.}} A \\ \big\downarrow & \\ \underline{k} & \end{array}.$$

The map $\underline{n} \to A$ is required to be locally injective, which means that it is injective on each fibre.

For example, if the alphabet A has 3 elements, then $e_{21} = m_{11} m_1$ is a sum of 3×3 monomials, while $E_{2,1}$ has 18 terms: 6 choices of an injective map $\underline{2} \hookrightarrow A$ and 3 of an injective map $\underline{1} \hookrightarrow A$.

This normalisation was perhaps first studied by Doubilet [**26**].

2.6.10. *Change of basis: elementary symmetric functions in terms of monomial symmetric functions.* One can write every elementary symmetric function as a sum of monomial symmetric functions

$$e_\lambda = \sum_{\lambda\wedge\pi=\mathrm{id}} m_\pi. \tag{17}$$

Traditionally the summation is over 0/1 matrices whose row and column totals give the partitions λ and π. There are several ways to approach this formula in a more objective manner. For example, consider the sum over all injections $\underline{n} \hookrightarrow \underline{k}\times\underline{b}$ whose components are surjections $\lambda\colon \underline{n} \twoheadrightarrow \underline{k}$ and $\pi\colon \underline{n} \twoheadrightarrow \underline{b}$. Note that two surjections λ and π with common domain are jointly injective if and only if they are *transversal partitions*, that is, their meet $\lambda \wedge \pi$ is the identity surjection. Alternatively, an injection $\underline{n} \hookrightarrow \underline{k} \times \underline{b}$ can be regarded as a relation between $\underline{k}$ and $\underline{b}$, and is termed an *effective relation* when its components are surjective.

The change of basis (17) is best expressed as a coalgebra homomorphism, which means at the level of decomposition spaces that we describe a IKEO-CULF span of decomposition spaces

$$\Lambda^E \xleftarrow{\text{ikeo}} W \xrightarrow{\text{culf}} \Lambda^M.$$

Recall that IKEO maps are those that induce coalgebra homomorphisms contravariantly, while CULF maps induce coalgebra homomorphisms covariantly.

In this case, the middle decomposition space W has W_1 the groupoid of effective relations, or transversal partitions

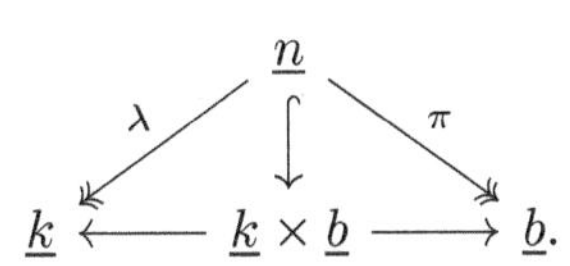

Given two partitions λ and π, fitting them into an effective relation is a yes/no question, namely whether the resulting map $(\lambda,\pi)\colon \underline{n} \to \underline{k} \times \underline{b}$ is injective.

The span

$$(\Lambda^E)_1 \leftarrow W_1 \rightarrow (\Lambda^M)_1$$

implements the map we aim for:

$$E_\lambda \mapsto \sum_{\lambda\wedge\pi=\mathrm{id}} M_\pi. \tag{18}$$

There is now the natural question whether this assignment is comultiplicative. It will be so if we can fit W_1 into a simplicial object (necessarily a decomposition space) that is IKEO over Λ^E and CULF over Λ^M. Guessing what such a decomposition space should be is a question of knowing what it is in degree 1 (which we do, since we already know what it should do), and then use the CULF condition and the IKEO condition to arrange the rest. The result in this case is

$$W_r = \{\underline{k} \twoheadleftarrow \underline{n} \twoheadrightarrow \underline{b} \rightarrow r\}$$

where the part $\underline{k} \twoheadleftarrow \underline{n} \twoheadrightarrow \underline{b}$ is an effective relation.

There are evident simplicial projection maps to Λ^E and Λ^M. It is not difficult to show that the simplicial map $W \to \Lambda^M$ is CULF, while it requires some work to show that the simplicial map $\Lambda^E \leftarrow W$ is IKEO. The span thus defines a coalgebra homomorphism, whose cardinality is (18).

2.6.11. *Discussion of non-invertibility of the base change.* The span is not invertible! However, after taking cardinality, it becomes invertible, and the inverse is given by a formula involving Möbius inversion in the lattice of partitions; see Doubilet [**26**]. The upshot is that at the objective level, we can see more detail: there is not a single decomposition space of symmetric functions, but rather one for each basis (all of them having the groupoid of surjections in simplicial degree 1). These decomposition spaces are not equivalent, because while there is maybe an IKEO-CULF span in one direction to express a change of coordinates, the inverse change of coordinates would involve minus signs (coming from some instance of Möbius inversion), and it cannot exist at the objective level. (It exists in the same way as Möbius inversion exists: only in virtue of formal differences; see Section 3 below.)

2.6.12. *Bialgebra structure.* As well as the comultiplication of symmetric functions, there is also a multiplication. At the objective level, the two together should be encoded by certain bisimplicial spaces $B\colon (\mathbb{\Delta}\times\mathbb{\Delta})^{\mathrm{op}} \to \boldsymbol{Grpd}$ that are decomposition spaces in each direction, and with a certain compatibility condition. The idea is then that the bialgebra is $\boldsymbol{grpd}_{/B_{11}}$ with the horizontal direction used to define the comultiplication and the vertical direction used to define the multiplication. The compatibility condition expresses the multiplicativity of the comultiplication (or equivalently the comultiplicativity of the multiplication.) In practice, such bisimplicial spaces often arise as CULF-monoidal decomposition spaces, meaning that one of the two directions is actually a Segal monoid (rather than a more general decomposition space). We have already seen several examples.

2.6.13. *Bialgebra structure (CULF monoidal structure) on Λ^E.* The two decomposition spaces of symmetric functions we have described, Λ^M and Λ^E, fit into bisimplicial groupoids in this way. In fact, one is the transpose of the other. From the viewpoint of Λ^E, for example, we have a CULF-monoidal decomposition space with monoidal structure given by disjoint union of surjections, so it is straightforward to multiply in the E-basis:

$$E_{\lambda'} \cdot E_{\lambda''} = E_{\lambda' \sqcup \lambda''}$$

where $\lambda' \sqcup \lambda''$ is the disjoint union $\underline{n}' + \underline{n}'' \twoheadrightarrow \underline{k}' + \underline{k}''$.

If we write out the monoidal nerve of this monoidal operation in the way we did with $\mathbb{B}$ in 2.1.7, it is precisely a question of splitting the codomain. As a bisimplicial space $\Lambda^E_{\bullet\bullet}$, we have Λ^E_{ij} the groupoid of surjections with the domain split into i parts and the codomain split into j parts. There are no compatibility requirements on the splittings of domain and codomain (and this independence is essentially the statement that the monoidal structure is CULF). So for Λ^E, the multiplication is simply given by a monoidal structure.

2.6.14. *Bisimplicial groupoid for Λ^M.* For Λ^M, in contrast, we have the same bisimplicial groupoid, but transposed, so that we use the monoidal direction to define the comultiplication, and use the 'decomposition-space' direction for the multiplication. This means that the formula for multiplication in the M-basis is less straightforward. Given $\lambda'\colon \underline{n}' \twoheadrightarrow \underline{k}'$ and $\lambda''\colon \underline{n}'' \twoheadrightarrow \underline{k}''$, we need to find all ways to fit it into a diagram

$$\begin{array}{ccccc} \underline{n}' & \dashrightarrow & \underline{n} & \dashleftarrow & \underline{n}'' \\ \downarrow\!\!\!\downarrow & & \downarrow\!\!\!\downarrow & & \downarrow\!\!\!\downarrow \\ \underline{k}' & \dashrightarrow & \underline{k} & \dashleftarrow & \underline{k}'' \end{array}$$

such that the top row is a sum-inclusion diagram, which forces $n = n' + n''$. We can list all such diagrams by first writing the sum $\lambda' \sqcup \lambda'' \colon \underline{n}' + \underline{n}'' \twoheadrightarrow \underline{k}' + \underline{k}''$, and then postcomposing with all quotient maps $q \colon \underline{k}' + \underline{k}'' \twoheadrightarrow \underline{k}$ whose components are both injective.

2.6.15. *Example.* For any positive numbers a, b, c, d, e we have

$$\begin{aligned} M_{a,b,c} * M_{d,e} \quad = \quad & M_{a,b,c,d,e} \\ + M_{a+d,b,c,e} + M_{a+e,b,c,d} &+ M_{b+d,a,b,e} + M_{b+e,a,c,d} + M_{c+d,a,b,e} + M_{c+e,a,b,d} \\ + M_{a,b+d,c+e} + M_{a,b+e,c+d} &+ M_{b,a+d,c+e} + M_{b,a+e,c+d} + M_{c,b+d,a+e} + M_{c,b+e,a+d}. \end{aligned}$$

To arrive at this expression we have to enumerate all possible quotient maps q in

$$\begin{array}{ccccc} \underline{3} & \hookrightarrow & \underline{3} + \underline{2} & \hookleftarrow & \underline{2} \\ & \searrow & \downarrow\!\!\!\downarrow q & \swarrow & \\ & & \underline{k}\,. & & \end{array}$$

We can choose to join nothing, in just 1 way, or join one point in $\underline{3}$ with one point in $\underline{2}$, in 6 possible ways, or join two points in $\underline{3}$ with two points in $\underline{2}$, also in 6 ways. Altogether we get the 13 terms displayed.

2.6.16. *Multiplicativity of the change of basis.* The change of basis expressed by the IKEO-CULF span must be not only comultiplicative but also multiplicative, meaning that the intermediate decomposition space W should also be bisimplicial and equipped with bisimplicial maps

$$\Lambda^E_{\bullet\bullet} \leftarrow W \rightarrow \Lambda^M_{\bullet\bullet}.$$

This span must be not only row-wise IKEO-CULF, as we already saw, but also column-wise CULF-IKEO, meaning that the corresponding homomorphism on incidence coalgebras is furthermore a homomorphism of bialgebras.

2.6.17. *Outlook.* We have briefly outlined two (double) decomposition spaces, one corresponding to the monomial basis and one corresponding to the elementary basis, and explained how to express elementary symmetric functions in terms of monomial ones by way of an intermediate double decomposition space and IKEO-CULF spans. While these constructions give a hint at what the decomposition-space approach to symmetric functions looks like, we have only scratched the surface here. The other combinatorial bases (complete homogeneous, power-sum, and the forgotten basis) admit similar decomposition-space interpretations (with similar non-standard normalisations). Unfortunately, the Schur basis, which is perhaps the most interesting basis, for its many connections to representation theory and geometry, is not easy to describe as a decomposition space.

Next, related Hopf algebras, such as that of quasi-symmetric functions, non-commutative symmetric functions, free quasi-symmetric functions (the Malvenuto–Reutenauer Hopf algebra), word quasi-symmetric functions, also admit decomposition space interpretations [**51**]. The decomposition space for the Hopf algebra of quasi-symmetric functions as well as the word quasi-symmetric functions have received some interest recently [**57**], as they are examples of free decomposition spaces (meaning that they are left Kan extended from $\triangle_{\text{inert}}$, as briefly touched upon in 3.3.7 below; see also Hackney's contribution to this volume [**56**]). The bialgebra of quasi-symmetric functions is the terminal object in the category of graded connected bialgebras with a zeta function, by a theorem of Aguiar, Bergeron, and Sottile [**1**]. An objective version of this result has been established recently by Hackney, Kock, and Steinebrunner.

3. Möbius inversion

3.1. Completeness, and Möbius inversion at the objective level. We are interested in the invertibility of the zeta functor (see 1.3.3) under the convolution product (see 1.3.2). Unfortunately, at the objective level it can practically *never* be convolution invertible, because the inverse μ should always be given by an alternating sum

$$\mu = \Phi_{\text{even}} - \Phi_{\text{odd}}.$$

We do not have minus signs available, but the sign-free equation

$$\zeta * \Phi_{\text{even}} = \varepsilon + \zeta * \Phi_{\text{odd}}$$

holds, as we proceed to recall. In the category case [**23, 73**], Φ_{even} is given by the even-length chains of non-identity arrows, that is, by the nondegenerate simplices of even dimension, and similarly for Φ_{odd}. To make sense of nondegeneracy for more general decomposition spaces we need to recall the notion of completeness from [**46**].

A simplex in any simplicial groupoid is degenerate when it is in the image of a degeneracy map. 'Nondegenerate' should mean to be in the complement of the image, but this description is only well behaved for monomorphisms of groupoids, i.e. maps that are fully faithful as functors; see A.2.4.

3.1.1. *Completeness and nondegeneracy.* A decomposition space is *complete* if $s_0 \colon X_0 \to X_1$ is mono [**46**]. It follows that all other degeneracy maps in X are also mono (see [**45**]).

For a complete decomposition space X we define $\vec{X}_n \subset X_n$ to be the full subgroupoid of nondegenerate n-simplices, i.e. those not in the image of any of the degeneracy maps. More importantly, in a decomposition space one can measure whether a simplex is nondegenerate on its principal edges: it is nondegenerate if and only if all its principal edges are [**46**, Corollary 2.14]. Hence it really just boils down to defining nondegenerate 1-simplices: define $\vec{X}_1 \subset X_1$ to be the complement of the monomorphism $s_0 \colon X_0 \to X_1$.

3.1.2. *Examples and non-example.* Every discrete decomposition space (such as a strict nerve) is complete, since any map between sets that admits a retraction is a monomorphism. Also every Rezk complete Segal space is complete in the sense of 3.1.1. In particular, fat nerves of categories are complete.

To see an example of a non-complete decomposition space, let G be a nontrivial group, and write BG for the same group considered as a groupoid with one object. Now consider the simplicial groupoid X with $X_n = (BG)^n$. Here $s_0 \colon 1 \to BG$ is not a monomorphism, as the trivial subgroupoid of BG is not a full subgroupoid.

3.1.3. *'Phi' functors.* We define Φ_n to be the linear functor given by the span

$$X_1 \longleftarrow \vec{X}_n \longrightarrow 1.$$

If $n = 0$ then $\vec{X}_0 = X_0$ by convention, and Φ_0 is given by the span

$$X_1 \longleftarrow X_0 \longrightarrow 1.$$

That is, Φ_0 is the linear functor ε. Note that $\Phi_1 = \zeta - \varepsilon$, and is denoted η in the classical literature [**23, 90**]. The minus sign makes sense here, since X_0 and $\vec{X}_1$, representing ε and Φ_1, define complementary full subgroupoids of X_1, representing ζ.

Computing convolution with the functors Φ_n is really about knowing how the groupoids $\vec{X}_n$ behave under various pullbacks. These calculations are carried out in detail in [**46**], leading to the following results.

LEMMA 3.1.4 ([**46**, Lemma 3.6]). *For a complete decomposition space, we have*

$$\Phi_n \quad = \quad (\Phi_1)^n \quad = \quad (\zeta - \varepsilon)^n,$$

the nth convolution product of Φ_1 with itself.

PROPOSITION 3.1.5. *For a complete decomposition space X, the square*

$$\begin{array}{ccc} \vec{X}_1 + \vec{X}_2 & \longrightarrow & X_2 \\ \downarrow & & \downarrow \\ X_1 \times \vec{X}_1 & \longrightarrow & X_1 \times X_1 \end{array}$$

is a pullback.

These results are special cases of [**46**, Lemma 3.5]. The proposition can be read as saying that if a 2-simplex σ has its second principal edge nondegenerate, then there are two possibilities for the first principal edge: either it is degenerate and the whole simplex σ is determined by the second principal edge (an element of $\vec{X}_1$), or it is nondegenerate and the whole simplex σ is nondegenerate (an element of $\vec{X}_2$).

From this lemma and its higher-dimensional analogues, it is not difficult to prove the following key result.

PROPOSITION 3.1.6 ([**46**, Proposition 3.7]). *The linear functors Φ_n satisfy the explicit equivalences of linear functors*

$$\zeta * \Phi_n \quad = \quad \Phi_n + \Phi_{n+1} \quad = \quad \Phi_n * \zeta.$$

Now let

$$\Phi_{\text{even}} := \sum_{n \text{ even}} \Phi_n, \qquad \Phi_{\text{odd}} := \sum_{n \text{ odd}} \Phi_n.$$

THEOREM 3.1.7 ([**46**, Theorem 3.8]). *For a complete decomposition space, the following Möbius inversion principle holds (explicit equivalences of linear functors):*

$$\begin{aligned} \zeta * \Phi_{\text{even}} &= \varepsilon + \zeta * \Phi_{\text{odd}}, \\ \Phi_{\text{even}} * \zeta &= \varepsilon + \Phi_{\text{odd}} * \zeta. \end{aligned}$$

PROOF. This result follows immediately from the previous proposition: all four linear functors are in fact equivalent to $\sum_{r \geq 0} \Phi_r$. □

For these results there is no need for finiteness conditions: there is no problem in taking infinite sums of groupoids. In the following subsection, however, we must impose finiteness conditions before we can take cardinality and recover Möbius inversion at the level of vector spaces and (co)algebras over $\mathbb{Q}$.

3.2. Length and Möbius decomposition spaces. If X is a complete and locally finite decomposition space, then by Proposition A.4.3 the linear functors

$$\Phi_r \colon \boldsymbol{Grpd}_{/X_1} \to \boldsymbol{Grpd}$$

are finite for each $r \geq 0$ and descend to linear functors

$$\Phi_r \colon \mathbf{grpd}_{/X_1} \to \mathbf{grpd}.$$

This property is not enough to guarantee finiteness of the sum of all those Φ_r and hence allow the Möbius inversion formula to descend to the vector-space level. We also need to assume that, for each $f \in X_1$, there is an upper bound on the dimension of a nondegenerate n-simplex with long edge f. This condition is important in its own right, as it is the condition for the existence of a length filtration 3.2.1, useful in many applications. When X is the nerve of a category, the condition says that for each arrow f, there is an upper bound on the number of non-identity arrows in a sequence of arrows composing to f. We are led to the following definition.

3.2.1. *Length.* A complete decomposition space X is *of locally finite length* if, for each $a \in X_1$, the fibres $F_a^{(n)}$ of $d_1^{n-1} \colon \vec{X}_n \to X_1$ over a are empty for n sufficiently large.

The *length* of a is the greatest n for which $F_a^{(n)} \neq \varnothing$. It defines a filtration on the incidence coalgebra. If X is a Segal space, the length of a is the longest factorisation of a into nondegenerate $a_i \in \vec{X}_1$.

3.2.2. *Example.* The incidence coalgebra of $(\mathbb{N}^2, +)/\mathfrak{S}_2$ (see 2.1.6) is the simplest example we know of in which the length filtration does not agree with the coradical filtration; see Sweedler [**100**] for this notion. The elements $(1,1)$ and $(2,0) \simeq (0,2)$ are clearly of length 2. On the other hand, the element

$$P := (1,1) - (2,0) - (0,2)$$

is primitive, meaning

$$\Delta(P) = (0,0) \otimes P + P \otimes (0,0)$$

and is therefore of coradical filtration degree 1. (Note that in $(\mathbb{N}^2, +)$ it is not true that P is primitive: it is the symmetrisation that makes the $(0,1)$ terms cancel out in the computation, to make P primitive.)

3.2.3. *Möbius condition.* A complete decomposition space X is *Möbius* if it is locally finite and of locally finite length, that is, for each a, the fibre $F_a^{(n)}$ is finite and eventually empty.

Note that for posets, 'locally finite' already implies 'locally finite length', so the Möbius condition is not needed separately in the poset case. If X is the strict nerve of a category, then it is Möbius in our sense if and only if it is Möbius in the sense of Leroux [**75**].

Classically, it is known that a Möbius category in the sense of Leroux does not have non-identity invertible arrows [**73**, Lemma 2.4]. Similarly, by [**46**, Corollary 8.7], if a Möbius decomposition space X is a Segal space then it is Rezk complete, meaning that all invertible arrows are degenerate; see B.2.3.

LEMMA 3.2.4. *A complete decomposition space X is Möbius if and only if X_1 is locally finite and the restricted composition map*

$$\sum_r {d_1}^{r-1} \colon \sum_r \vec{X}_r \to X_1$$

is finite.

Thus, if X is Möbius, the linear functors Φ_{even} and Φ_{odd} also descend to

$$\Phi_{\text{even}}, \Phi_{\text{odd}} \colon \boldsymbol{grpd}_{/X_1} \to \boldsymbol{grpd}$$

and their cardinalities are elements $|\Phi_{\text{even}}|, |\Phi_{\text{odd}}| : \mathbb{Q}_{\pi_0 X_1} \to \mathbb{Q}$ of the incidence algebra. We can therefore take the cardinality of the abstract Möbius inversion formula of Theorem 3.1.7 to arrive at the following result.

THEOREM 3.2.5. *If X is a Möbius decomposition space, then the cardinality of the zeta functor, $|\zeta| : \mathbb{Q}_{\pi_0 X_1} \to \mathbb{Q}$, is convolution invertible with inverse $|\mu| := |\Phi_{\text{even}}| - |\Phi_{\text{odd}}|$:*

$$|\zeta| * |\mu| = |\varepsilon| = |\mu| * |\zeta| .$$

3.3. Möbius functions and cancellation. We compute the Möbius functions in some of our examples. While the formula $\mu = \Phi_{\text{even}} - \Phi_{\text{odd}}$ seems to be the most general and uniform expression of the Möbius function, it is often not the most economical. At the numerical level, it is typically the case that much more practical expressions for the Möbius functions can be computed with different techniques. The formula $\Phi_{\text{even}} - \Phi_{\text{odd}}$ should not be dismissed on these grounds, though: it must be remembered that it constitutes a natural 'bijective' account, valid at the objective level, in contrast to many of the elegant cancellation-free expressions in the classical theory which are often the result of formal algebraic manipulations, often power-series representations.

Comparison with the economical formulae raises the question whether these cancellations can also be realised at the objective level. This question can sometimes be answered positively, by exhibiting an explicit cancellation between Φ_{even} and Φ_{odd}, which in turn may or may not be given by a *natural* bijection.

Once a more economical expression has been found for some Möbius decomposition space X, it can be transported back along any CULF functor $f \colon Y \to X$ to yield also more economical formulae for Y.

3.3.1. *Natural numbers.* For the decomposition space $\mathbf{N}$ (see 2.1.1), the incidence algebra is $\boldsymbol{grpd}^{\mathbb{N}}$, with basis given by the representables h^n, and with convolution product

$$h^a * h^b = h^{a+b}.$$

To compute the Möbius functor, we have

$$\Phi_{\text{even}} = \sum_{r \text{ even}} (\mathbb{N} \smallsetminus \{0\})^r,$$

hence $\Phi_{\text{even}}(\underline{n})$ is the set of ordered compositions of the ordered set $\underline{n}$ into an even number of parts, or equivalently

$$\Phi_{\text{even}}(n) = \{n \twoheadrightarrow \underline{r} \mid r \text{ even}\},$$

the set of monotone surjections. In conclusion, with an abusive sign notation, the Möbius functor is

$$\mu(\underline{n}) = \sum_{r \geq 0} (-1)^r \{\underline{n} \twoheadrightarrow \underline{r}\}.$$

At the numerical level, this formula simplifies to

$$\mu(n) = \sum_{r \geq 0} (-1)^r \binom{n-1}{r-1} = \begin{cases} 1 & \text{for } n = 0 \\ -1 & \text{for } n = 1 \\ 0 & \text{else} \end{cases}$$

(remembering that $\binom{-1}{-1} = 1$, and $\binom{k}{-1} = 0$ for $k \geq 0$).

On the other hand, since the incidence algebra is isomorphic to the power series ring under the identification $|h^n| = \delta^n \leftrightarrow z^n \in \mathbb{Q}[[z]]$, and since the zeta function corresponds to the geometric series $\sum_n x^n = \frac{1}{1-x}$, we find that the Möbius function is $1 - x$, which corresponds to the functor $\delta^0 - \delta^1$.

At the objective level, there is indeed a cancellation of groupoids taking place. It amounts to an equivalence of the Phi-groupoids restricted to $n \geq 2$:

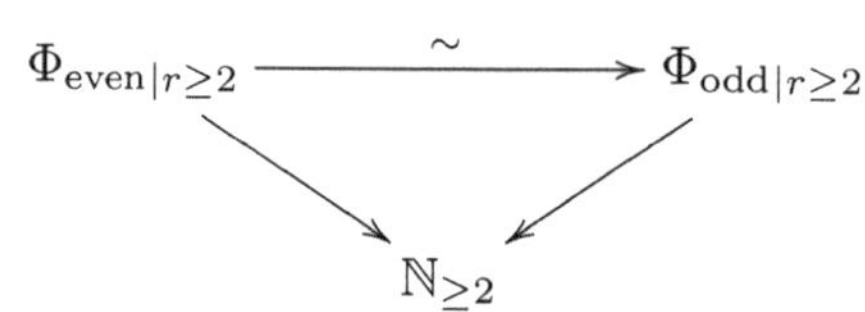

which cancels out most of the terms, leaving us with the much more economical Möbius function

$$\delta^0 - \delta^1$$

supported on $\mathbb{N}_{\leq 1}$. Since $\mathbb{N}$ is discrete, this equivalence (just a bijection) can be established fibrewise, as follows.

For each $n \geq 2$ there is a natural fibrewise bijection

$$\Phi_{\text{even}}(n) \simeq \Phi_{\text{odd}}(n).$$

To see this, encode the elements $(x_1, x_2, \ldots, x_k)$ in $\Phi_{\text{even}}(n)$ (and $\Phi_{\text{odd}}(n)$) as binary strings of length n and starting with 1 as follows: each coordinate x_i is represented as a string of length x_i whose first bit is 1 and whose other bits are 0, and all these strings are concatenated. In other words, thinking of the element $(x_1, x_2, \ldots, x_k)$ as a ordered partition of the ordered set n, in the binary representation the 1-entries mark the beginning of each part. (The binary strings must start with 1 since the first part must begin at the beginning.) For example, with $n = 8$, the element $(3, 2, 1, 1, 1) \in \Phi_{\text{odd}}(8)$, is encoded as the binary string 10010111. Now the bijection between $\Phi_{\text{even}}(n)$ and $\Phi_{\text{odd}}(n)$ can be taken simply to flip the second bit in the binary representation. In the example, 10010111 is sent to 11010111, meaning that $(3, 2, 1, 1, 1) \in \Phi_{\text{odd}}(8)$ is sent to $(1, 2, 2, 1, 1, 1) \in \Phi_{\text{even}}(8)$. Because of this cancellation (which is possible when $n \geq 2$, as we need the second bit in order to flip), the difference $\Phi_{\text{even}} - \Phi_{\text{odd}}$ is the same as $\delta_0 - \delta_1$, which is the cancellation-free formula.

The minimal solution $\delta^0 - \delta^1$ can also be checked immediately at the objective level to satisfy the defining equation for the Möbius function:

$$\zeta * \delta^0 = \zeta * \delta^1 + \delta^0.$$

This equation says

$$\begin{array}{ccc} \mathbb{N}\times\{0\} & = & (\mathbb{N}\times\{1\})+\{0\} \\ {\scriptstyle\text{add}}\big\downarrow & & \big\downarrow{\scriptstyle\text{add+incl}} \\ \mathbb{N} & & \mathbb{N}. \end{array}$$

In conclusion, the classical formula lifts to the objective level.

3.3.2. *Finite sets and bijections.* Already for the next example (2.1.7), that of the monoidal groupoid $(\mathbb{B}, +, 0)$, whose incidence algebra is the algebra of species under the Cauchy convolution product [**2**], the situation is more subtle.

Similarly to the previous example, we have $\Phi_r(S) = \operatorname{Surj}(S, \underline{r})$, but this time we are dealing with arbitrary surjections, as S is just an abstract set. Hence the Möbius functor is given by

$$\mu(S) = \sum_{r\geq 0} (-1)^r \operatorname{Surj}(S, \underline{r}).$$

Numerically, the incidence algebra is just the power series ring $\mathbb{Q}[[z]]$; see 2.1.7 Since this time the zeta function is the exponential $\exp(z)$, the Möbius function is the series $\exp(-z)$, corresponding to

$$\mu(n) = (-1)^n.$$

The economical Möbius function suggests the existence of the following equivalence at the groupoid level:

$$\mu(S) = \int^r (-1)^r h^r(S) \;\simeq\; \mathbb{B}_{\text{even}}(S) - \mathbb{B}_{\text{odd}}(S),$$

where

$$\mathbb{B}_{\text{even}} = \sum_{r \text{ even}} \mathbb{B}_{[r]} \quad \text{and} \quad \mathbb{B}_{\text{odd}} = \sum_{r \text{ odd}} \mathbb{B}_{[r]}$$

are the full subgroupoids of $\mathbb{B}$ consisting of the even and odd sets, respectively. However, it seems that such an equivalence is not possible, at least not over $\mathbb{B}$: while we are able to exhibit a bijective proof, this bijection is *not* natural, and hence does not assemble into a groupoid equivalence.

PROPOSITION 3.3.3. *For a fixed set S, there are monomorphisms $\mathbb{B}_{\text{even}}(S) \hookrightarrow \Phi_{\text{even}}(S)$ and $\mathbb{B}_{\text{odd}}(S) \hookrightarrow \Phi_{\text{odd}}(S)$, and a residual bijection*

$$\Phi_{\text{even}}(S) - \mathbb{B}_{\text{even}}(S) = \Phi_{\text{odd}}(S) - \mathbb{B}_{\text{odd}}(S).$$

This bijection is not *natural in S, though, and hence does not constitute an isomorphism of species, only an equipotence of species* [**8**].

COROLLARY 3.3.4. *For a fixed S there is a bijection*

$$\mu(S) \simeq \mathbb{B}_{\text{even}}(S) - \mathbb{B}_{\text{odd}}(S)$$

but it is not *natural in S.*

PROOF OF THE PROPOSITION. The map $\mathbb{B}_{\text{even}} \to \mathbb{B}$ is a monomorphism of groupoids (A.2.4), so for each set S of even cardinality there is a single element to subtract from $\Phi_{\text{even}}(S)$. The groupoid Φ_{even} has as objects finite sets S equipped with a surjection $S \twoheadrightarrow \underline{k}$ for some even k. If S is itself of even cardinality n, then among such partitions there are $n!$ possible partitions into n parts. If there were given a total order on S, among these $n!$ n-block partitions, there is one for which the order of S agrees with the order of the n parts. We would like to subtract that one and then establish the required bijection. This can be done fibrewise: over a given n-element set S, we can establish the bijection by choosing first a bijection $S \simeq \underline{n} = \{1, 2, \ldots, n\}$, the totally ordered set with n elements.

For each n, there is an explicit bijection

$$\{\textit{surjections } p\colon \underline{n} \twoheadrightarrow \underline{k} \mid k \textit{ even}, p \textit{ not the identity map}\}$$

$$\leftrightarrow$$

$$\{\textit{surjections } p\colon \underline{n} \twoheadrightarrow \underline{k} \mid k \textit{ odd}, p \textit{ not the identity map.}\}$$

Indeed, define first the bijection on the subsets for which $p^{-1}(1) \neq \{1\}$, i.e. the element 1 is not alone in the first block. In this case the bijection goes as follows. If the element 1 is alone in a block, join this block with the previous block. (There exists a previous block as we have excluded the case where 1 is alone in block 1.) If 1 is not alone in a block, separate out 1 to a block on its own, coming just after the original block. For example, the following surjections match up under the bijection.

$$(34, 1, 26, 5) \leftrightarrow (134, 26, 5).$$

For the remaining case, where 1 is alone in the first block, we just leave it alone, and treat the remaining elements inductively, considering now the case where the element 2 is not alone in the second block. In the end, the only case not treated is the case where for each j, we have $p^{-1}(j) = \{j\}$, that is, each element is alone in the block with the same number. This case is precisely the identity map that was excluded explicitly in the bijection. (Note that for each n, this case only appears on one of the sides of the bijection, as either n is even or n is odd.) □

In fact, already subtracting the groupoid $\mathbb{B}_{\text{even}}$ from Φ_{even} is not possible naturally. We would first have to find a monomorphism $\mathbb{B}_{\text{even}} \hookrightarrow \Phi_{\text{even}}$ over $\mathbb{B}$. But the automorphism group of an object $\underline{n} \in \mathbb{B}$ is $\mathfrak{S}_n$, whereas the automorphism group of any overlying object in Φ_{even} is a proper subgroup of $\mathfrak{S}_n$. In fact it is the subgroup of those permutations that are compatible with the surjection $\underline{n} \twoheadrightarrow \underline{k}$. So locally the fibration $\Phi_{\text{even}} \to \mathbb{B}$ is a group monomorphism, and hence it cannot have a section. So in conclusion, we cannot even realise $\mathbb{B}_{\text{even}}$ as a full subgroupoid in Φ_{even}, and hence it does not make sense to subtract it.

One may note that it is not logically necessary to be able to subtract the redundancies from Φ_{even} and Φ_{odd} in order to find the economical formula. It is enough to establish directly (by a separate proof) that the economical formula holds, by actually convolving it with the zeta functor. At the object level the simplified Möbius function would be the groupoid

$$\mathbb{B}_{\text{even}} - \mathbb{B}_{\text{odd}}.$$

We might try to establish directly that

$$\zeta * \mathbb{B}_{\text{even}} = \zeta * \mathbb{B}_{\text{odd}} + \varepsilon,$$

which should be a groupoid equivalence over $\mathbb{B}$, but again we can only establish this equivalence fibrewise. On the left-hand side the fibre over an n-element set S consists of an arbitrary set and an even set whose disjoint union is S. In other words it suffices to give an even subset of S. Analogously on the right-hand side it amounts to giving an odd subset of S, and in the special case of $S = \emptyset$ the empty set itself also belongs to the fibre on the right, thanks to the summand ε.

For a fixed nonempty set S, there is an explicit bijection between even subsets of S and odd subsets of S.

Indeed, fix an element $s \in S$. The bijection consists of adding s to the subset U if it does not belong to U, and removing it if it already belongs to U. Clearly this changes the parity of the set. Since the bijection involves the choice of a basepoint, it seems impossible to lift it to a natural bijection.

3.3.5. *Restricting Möbius formulae along CULF functors.* Once a more economical Möbius function has been found for a decomposition space X, it can be exploited to yield more economical formulae for any decomposition space Y with a CULF functor to X. This is the content of the following straightforward lemma:

LEMMA 3.3.6. *Suppose that for the complete decomposition space X we have found a Möbius inversion formula $\mu_X = \Psi_1 - \Psi_0$, that is*

$$\zeta_X * \Psi_0 = \zeta_X * \Psi_1 + \varepsilon.$$

Then for every decomposition space CULF over X, say $f\colon Y \to X$, we have the formula $\mu_Y = f^*\Psi_1 - f^*\Psi_0$, that is

$$\zeta_Y * f^*\Psi_0 = \zeta_Y * f^*\Psi_1 + \varepsilon.$$

3.3.7. *Free decomposition spaces.* In most of the examples treated, the length filtration 3.2.1 is actually a grading. Recall from [**46**, 6.20] that a grading amounts to having a simplicial map $X \to B\mathbb{N}$ to the nerve of $(\mathbb{N}, +)$. In the rather special situation when this map is CULF, the economical Möbius function formula

$$\mu = \delta^0 - \delta^1$$

for $B\mathbb{N}$ induces the same formula for the Möbius functor of X. This condition is rather restrictive; in fact, for nerves of categories, it is satisfied only for free categories on directed graphs (see Street [**99**]). More generally, decomposition spaces admitting a CULF functor to $B\mathbb{N}$ are precisely the *free decomposition spaces* [**57**], meaning that they are simplicial groupoids obtained by left Kan extension along the inclusion functor $j\colon \mathbb{\Delta}_{\text{inert}} \to \mathbb{\Delta}$; see also [**56**].

Given $A\colon \mathbb{\Delta}_{\text{inert}}^{\text{op}} \to \boldsymbol{Grpd}$, the formula for $X = j_!(A)$ begins with $X_0 = A_0$ and $X_1 = \sum_{k\in\mathbb{N}} A_k$. The CULF functor $X \to B\mathbb{N}$ is given as the simplicial map $j_!(A) \to j_!(1) = B\mathbb{N}$, which is always CULF [**57**]. The decomposition space $X = j_!(A)$ now inherits the economical Möbius functor from $B\mathbb{N}$, by Lemma 3.3.6. In detail, for each $n \in \mathbb{N}$ there is a linear span $X_1 \leftarrow A_n \to 1$, denoted δ^n, consisting of all the arrows of length n, and the economical Möbius functor is $\delta^0 - \delta^1$, which is essentially $A_0 - A_1$, with reference to the original $A\colon \mathbb{\Delta}_{\text{inert}}^{\text{op}} \to \boldsymbol{Grpd}$.

Many combinatorial coalgebras of deconcatenation type are incidence coalgebras of free decomposition spaces. The simplest example is the free monoid on a set S, i.e. the monoid of words in the alphabet S. The economical Möbius function is then $\delta^0 - \delta^1$, where $\delta^1 = \sum_{s\in S} \delta^s$. In the power series ring, with a variable z_s for each letter $s \in S$, it is the series $1 - \sum_{s\in S} z_s$. A slightly more elaborate example is the decomposition space of quasi-symmetric functions (briefly mentioned in 2.6.17): it is free on $B\vec{\mathbb{N}}\colon \mathbb{\Delta}_{\text{inert}}^{\text{op}} \to \boldsymbol{Grpd}$, and it follows that we have

$$\mu(p) = \begin{cases} 1 & \text{for } p\colon \emptyset \twoheadrightarrow \emptyset, \\ -1 & \text{for } p\colon n \twoheadrightarrow 1, \\ 0 & \text{for } p\colon n \twoheadrightarrow k \text{ with } k \geq 2. \end{cases}$$

(Note that the length grading (by codomain of a surjection) is not the usual grading of quasi-symmetric functions, which is instead by the domain of a surjection.)

3.3.8. *Decomposition spaces over* **B** (2.1.7). Similarly, if a decomposition space X admits a CULF functor $\ell\colon X \to \mathbf{B}$ (which may be thought of as a 'length function with symmetries') then at the numerical level and at the objective level, locally for each object $S \in X_1$, we can pull back the economical Möbius 'functor' $\mu(n) = (-1)^n$ from **B** to X, yielding the numerical Möbius function on X

$$\mu(f) = (-1)^{\ell(f)}.$$

An example is the coalgebra of graphs 1.2.4 of Schmitt [**94**]: the functor from the decomposition space of graphs to **B** which to a graph associates its vertex set is CULF. Hence the Möbius function for this decomposition space is

$$\mu(G) = (-1)^{|V(G)|}.$$

This argument works for any restriction species [**49**].

3.3.9. *Finite vector spaces.* We calculate the Möbius function in the incidence algebra of the Waldhausen decomposition space of $\mathbb{F}_q$-vector spaces; see 2.3.5. In this case, Φ_r is the groupoid of strings of $r-1$ nontrivial injections. The fibre over V is the discrete groupoid of strings of $r-1$ nontrivial injections whose last space is V. This fibre is precisely the set of nontrivial r-flags in V, i.e. flags for which the r consecutive codimensions are nonzero. In conclusion,

$$\mu(V) = \sum_{r=0}^{n} (-1)^r \{ \text{ nontrivial } r\text{-flags in } V\}.$$

(In principle, $\mu(V)$ is a groupoid, but since we have fixed V, it is just a discrete groupoid: a flag inside a fixed vector space has no automorphisms.)

The number of flags with codimension sequence p is the q-multinomial coefficient

$$\binom{n}{p_1, p_2, \ldots, p_r}_q.$$

In conclusion, at the numerical level we find

$$\mu(V) = \mu(n) = \sum_{r=0}^{n} (-1)^r \sum_{\substack{p_1+\cdots+p_r=n \\ p_i>0}} \binom{n}{p_1, p_2, \ldots, p_r}_q.$$

On the other hand, it is classical that from the power-series representation (2.3.5) one gets the numerical Möbius function

$$\mu(n) = (-1)^n q^{\binom{n}{2}}.$$

While the equality of these two expressions can readily be established at the numerical level (for example via a zeta-polynomial argument as explained below), we do not know of an objective interpretation of the expression $\mu(n) = (-1)^n q^{\binom{n}{2}}$. Realising the cancellation on the objective level would require first of all to being able to impose extra structure on V in such a way that among all nontrivial r-flags, there would be $q^{\binom{r}{2}}$ special ones!

3.3.10. *Faà di Bruno.* Recall (from 2.4) that the incidence bialgebra of the fat nerve of the monoidal category of finite sets and surjections is the Faà di Bruno bialgebra. Since clearly ζ and ε are multiplicative, also μ is multiplicative, i.e. determined by its values on the connected surjections. The general formula gives

$$\mu(\underline{n} \twoheadrightarrow \underline{1}) = \sum_{r=0}^{n} (-1)^n \boldsymbol{Tr}(n,r),$$

where $\boldsymbol{Tr}(n,r)$ is the (discrete) groupoid of trees with n leaves, height r, and no trivial layer (in fact, more precisely, strings of r nontrivial surjections composing to $n \twoheadrightarrow 1$), and where the minus sign is abusive notation for splitting into even and odd.

On the other hand, classical theory (see Doubilet, Rota, and Stanley [**28**]) gives the following 'connected Möbius function':

$$\mu(n) = (-1)^{n-1}(n-1)!.$$

In conjunction, the two expressions yield the following combinatorial identity:

$$(-1)^{n-1}(n-1)! \;=\; \sum_{r=0}^{n} (-1)^r \left|\boldsymbol{Tr}(n,r)\right|.$$

We do not know how to realise the cancellation at the objective level. It would require first developing a bit further the theory of monoidal decomposition spaces and incidence bialgebras.

3.3.11. *Zeta polynomials.* For a complete decomposition space X, we can classify the r-simplices according to their degeneracy type, writing

$$X_r = \sum_{k=0}^{r} \binom{r}{k} \vec{X}_k,$$

where the binomial coefficient is an abusive shorthand for that many copies of $\vec{X}_k$, embedded disjointly into X_r by specific degeneracy maps; see [**46**, 2.6] for details. Now we fibre over a fixed arrow $f \in X_1$, to obtain

$$(X_r)_f = \sum_{k=0}^{\infty} \binom{r}{k} (\vec{X}_k)_f, \tag{19}$$

where we have now allowed ourselves to sum to infinity, but for fixed f of finite length it is still a finite sum.

The *'zeta polynomial'* of a decomposition space X is the function

$$\begin{aligned} \zeta^r(f)\colon X_1 \times \mathbb{N} &\longrightarrow \boldsymbol{Grpd} \\ (f,r) &\longmapsto (X_r)_f \end{aligned}$$

assigning to each arrow f and $r \in \mathbb{N}$ the ∞-groupoid of r-simplices with long edge f. For fixed $f \in X_1$ of finite length ℓ, this groupoid is a polynomial expression in r, as witnessed by (19). In this case, at the numerical level, we can substitute $r = -1$ into it to find:

$$\zeta^{-1}(f) = \sum_{k=0}^{\infty} (-1)^k \Phi_k(f).$$

Hence $\zeta^{-1}(f) = \mu(f)$, as the notation suggests.

In some cases there is a polynomial formula for $\zeta^r(f)$. For example, in the case $X = (\mathbb{N},+)$ of 2.1.1 we find $\zeta^r(n) = \binom{n+r-1}{n}$, and therefore $\mu(n) = \binom{n-2}{n}$, in agreement with the other calculations (of this trivial example). In the case $X = (\mathbb{B}, +)$ of 2.1.7, we find $\zeta^r(n) = r^n$, and therefore $\mu(n) = (-1)^n$ again.

Sometimes, even when a formula for $\zeta^r(n)$ cannot readily be found, the (-1)-value can be found by a power-series representation argument. For example in the case of the Waldhausen $S_\bullet$-construction of ***vect*** (2.3.5), we have that $\zeta^r(n)$ is the set of r-flags of $\mathbb{F}_q^n$ (allowing trivial steps). We have

$$\zeta^r(n) = \sum_{\substack{p_1+\cdots+p_r=n \\ p_i \geq 0}} \frac{[n]!}{[p_1]! \cdots [p_r]!},$$

and therefore

$$\sum_{n=0}^{\infty} \zeta^r(n) \frac{z^n}{[n]!} = \left(\sum_{n=0}^{\infty} \frac{z^n}{[n]!} \right)^r,$$

Now $\zeta^{-1}(n)$ can be read off as the nth coefficient in the inverted series $\left(\sum_{n=0}^{\infty} \frac{z^n}{[n]!}\right)^{-1}$. In the case at hand, these coefficients are $(-1)^n q^{\binom{n}{2}}$, as we already saw.

3.4. Tools for calculation of Möbius functors. We mention three high-level tools for calculating Möbius functions, and wrap up with a few open ends.

3.4.1. *Carlier's Rota formula for bicomodules.* A classical formula of Rota [**90**] compares the Möbius functions of two posets related by a Galois connection. Carlier [**17**] has generalised Rota's formula to the setting of decomposition spaces, where it concerns a notion of adjunction, or more generally certain bicomodule configurations: An X–Y–bicomodule configuration is a bisimplicial groupoid B equipped with an augmentation column X and an augmentation row Y, and such that B is Segal in every row and every column; furthermore B is stable, which is a pullback condition on top vertical faces against top horizontal faces, and on bottom vertical faces against bottom horizontal faces, and also the augmentation maps must be CULF. Such a bicomodule configuration has an incidence bicomodule over the incidence coalgebras of X and Y.

One application of Carlier's Rota formula concerns the relationship between restriction species and directed restriction species, briefly treated in 2.5.2. The decomposition space $\mathbb{I}$ corresponding to the terminal restriction species has Möbius function $(-1)^n$ for a set with n elements. Since every restriction species is CULF over $\mathbb{I}$, a similar formula exists for general restriction species. Carlier [**18**] sets up a Möbius bicomodule interpolating between $\mathbb{I}$ and the decomposition space $\mathbb{C}$ of finite posets (corresponding to the terminal directed restriction species), and applies the generalised Rota formula [**17**] to calculate the Möbius function of any directed restriction species to be $\mu(Q) = (-1)^n$ if the underlying poset of Q is discrete with n elements, and zero otherwise. By a CULF argument, this approach also leads to a formula for the Möbius function of the decomposition space of the free monad on a polynomial endofunctor P as in 2.5.6 (the bialgebra of P-trees), namely $\mu(T) = (-1)^n$ if the forest T consists of n corollas (and possibly some isolated edges), and zero otherwise.

3.4.2. *Antipodes.* The (signed) span representing the Möbius function

$$X_1 \leftarrow \sum_k (-1)^k \vec{X}_k \rightarrow 1 \tag{20}$$

admits an elegant variation in the case where X is a CULF monoidal decomposition space. Rather than having 1 at the end of the span, we can have X_1 so as to define an endomorphism instead of a functional. The more refined span is

$$X_1 \leftarrow \sum_k (-1)^k \vec{X}_k \rightarrow X_1, \tag{21}$$

where the right-hand map is

$$\sum_k (-1)^k \vec{X}_k \rightarrow \sum_k (-1)^k \prod_{i=1}^{k} \vec{X}_1 \overset{\otimes}{\rightarrow} X_1,$$

returning the monoidal product of all the principal edges of a k-simplex. This span can be seen as a decomposition-space version of the antipode formula of Takeuchi [**101**] and Schmitt [**93**]; see [**19**]. It is not a true antipode unless X_0 is contractible (so that the incidence bialgebra becomes connected), but even in the case where X_0 is not contractible, this mock antipode S has some merit. For example, it can still be used to calculate the Möbius function as $\zeta \circ S$ (since the span (21) composed with the zeta span gives back the span (20)).

3.4.3. *Crapo's complementation formula.* Another classical tool for calculating Möbius functions is Crapo's complementation formula, originally established for lattices [**25**] but generalised to general finite posets by Björner and Walker [**12**]. It concerns the situation where a poset X (respectively, a decomposition space) has a convex subposet (respectively, sub-decomposition space) C. It reads

$$\mu^X = \mu^{X \smallsetminus C} + \mu^X * \zeta^C * \mu^X$$

(with self-explanatory notation). This formula has been established recently [**50**] at the objective level for any Möbius decomposition space X, and provides in particular a bijective proof for the Björner–Walker theorem. It should be mentioned that it is more difficult to find interesting applications of this formula for decomposition spaces: at the moment, the only known applications are the original poset applications of [**12**].

In the context of the present discussions, the formula is more interesting for the questions it raises. For $C = X$, the formula reads $\mu = \mu * \zeta * \mu$. Numerically this fact is immediate, as μ is convolution inverse to ζ. But objectively there are *two* different cancellations establishing the formula: one from $(\mu * \zeta) * \mu$ and one from $\mu * (\zeta * \mu)$. What is more interesting is that the proof of the Crapo formula gives yet another two different cancellations. Altogether it is pressing to get to learn more about the structure of cancellations in general.

3.4.4. *Product formula?* Classically, a very useful tool for calculating Möbius functions is the product formula: it states, for two locally finite posets P and Q,

$$\mu_{P \times Q} = \mu_P \times \mu_Q.$$

For example, it calculates the classical Möbius function from number theory, which is the Möbius function of the divisibility poset $(\mathbb{N}^\times, |)$. Since this is the infinite (but finitely supported) product of copies of the poset $(\mathbb{N}, \leq)$, one for each prime, the classical formula for the Möbius function for $(\mathbb{N}^\times, |)$

$$\mu(n) = \begin{cases} (-1)^r & \text{if } n \text{ is the product of } r \text{ distinct primes} \\ 0 & \text{else} \end{cases}$$

follows from the product formula and the Möbius function for $(\mathbb{N}, \leq)$

$$\mu(k) = \begin{cases} 1 & \text{for } k = 0 \\ -1 & \text{for } k = 1 \\ 0 & \text{else} \end{cases}$$

from 3.3.1. Unfortunately, it is not easy to derive any such formula at the objective level, since there is no easy description of the nondegenerate simplices of a product of decomposition spaces in terms of the nondegenerate simplices of the factors. Elaborate cancellations seem to be required, and at the moment it is not known how to handle this.

We finish with a kind of non-example, which raises further interesting questions.

EXAMPLE 3.4.5. Consider the strict nerve of the category

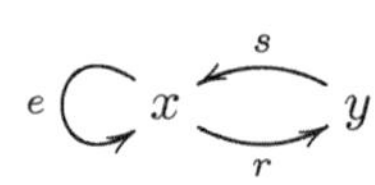

in which $r \circ s = \mathrm{id}_y$, $s \circ r = e$ and $e \circ e = e$. Since this decomposition space X is locally finite (in fact finite), it defines a vector-space coalgebra, in fact a finite-dimensional one. One can check by linear algebra (see Leinster [**74**, Ex.6.2]), that this coalgebra has Möbius inversion. On the other hand, X is not of locally finite length, because the identity arrow id_y can be written as an arbitrary long string $\mathrm{id}_y = r \circ s \circ \cdots \circ r \circ s$. In particular X is not a Möbius decomposition space. So we are in the following embarrassing situation: on the objective level, X has Möbius inversion (as it is complete), but the formula does not have a cardinality. At the same time, at the numerical level Möbius inversion exists nevertheless. Since inverses are unique if they exist, it is therefore likely that the infinite Möbius inversion formula of the objective level admits some drastic cancellation at this level, yielding a finite formula, whose cardinality is the numerical formula. Unfortunately, so far we have not been able to pinpoint such a cancellation.

Appendix A. Groupoids

A.1. Homotopy theory of groupoids. We briefly recall the necessary basic notions of groupoids and their homotopy cardinalities.

A.1.1. *Groupoids.* A *groupoid* is a small category in which all the arrows are invertible. A map of groupoids is just a functor. Let ***Grpd*** denote the category of groupoids and maps.

Intuitively we consider groupoids as sets with built-in symmetries. While a group models symmetry automorphisms of one object, groupoids model automorphisms and isomorphisms between several objects.

A.1.2. *Homotopy equivalences.* A homotopy between two maps of groupoids is just a natural transformation of functors. A map of groupoids $f\colon X \to Y$ is called a *homotopy equivalence* when there exists a pseudo-inverse $g\colon Y \to X$, meaning that the two composites are homotopic to the identities: $g \circ f \simeq \mathrm{id}_X$ and $f \circ g \simeq \mathrm{id}_Y$. Just as for categories, homotopy equivalences can also be characterised as functors that are essentially surjective and fully faithful.

Homotopy equivalence is the appropriate notion of sameness for groupoids, and it is important that all the notions involved be invariant under homotopy equivalence.

We adopt the convention that all notions in the paper are the homotopy invariant ones: outside this appendix we usually say *equivalence, finite, discrete, trivial, cartesian, pullback, fibre, sum, colimit* and *monomorphism* instead of 'homotopy equivalence', 'homotopy finite', 'homotopy discrete', etc, for the notions defined below. It is essential that the word 'homotopy' is understood throughout.

A.1.3. *Connectedness, discreteness.* A groupoid X is *connected* if $\mathrm{obj}(X)$ is non-empty and the set $\mathrm{Hom}_X(x,y)$ is non-empty for all $x, y \in X$. A maximal connected subgroupoid of X is termed a *component* of X and denoted $[x]$ or $X_{[x]}$, where x is some object in the component. The set of components is denoted $\pi_0(X)$. We denote by $\pi_1(X,x)$ the automorphism group $\mathrm{Aut}_X(x) = \mathrm{Hom}_X(x,x)$. A groupoid X is *homotopy discrete* if $\pi_1(X,x)$ is trivial for all x, and *contractible* if it is homotopy discrete and also connected. This means homotopy equivalent to a point, i.e. the terminal groupoid 1.

A.1.4. *Finiteness.* A groupoid X is *locally finite* if $\pi_1(X,x)$ is finite for every x, and is *(homotopy) finite* if in addition $\pi_0(X)$ is finite. We denote by **grpd** the category of finite groupoids.

A.1.5. *Pullbacks.* The *homotopy fibre product* of maps $f\colon G \to B$ and $g\colon E \to B$ is the groupoid $H = G \times_B E$ whose objects are triples (x, y, φ) consisting of $x \in G$, $y \in E$, and $\varphi\colon fx \to gy$ in B, and whose arrows $(x', y', \varphi') \to (x, y, \varphi)$ are pairs $(\beta, \varepsilon) \in \mathrm{Hom}_G(x', x) \times \mathrm{Hom}_E(y', y)$ such that $\varphi \circ f(\beta) = g(\varepsilon) \circ \varphi'$. There are canonical projections p, q:

$$\begin{array}{ccc} H & \xrightarrow{q} & E \\ {\scriptstyle p}\downarrow & \lrcorner & \downarrow{\scriptstyle g} \\ G & \xrightarrow[f]{} & B. \end{array} \tag{22}$$

The diagram does not commute on the nose, but the third components of objects $a = (x, y, \varphi)$ provide a natural isomorphism $\{\varphi\colon fp(a) \cong gq(a)\}$. We say a square (22) is *homotopy cartesian* or a *homotopy pullback* if H is homotopy equivalent to the homotopy fibre product $G \times_B E$ given explicitly above. The map p is sometimes called the *pullback of g along f* and denoted $f^*(g)$.

A.1.6. *Fibres.* The *homotopy fibre* E_b of a map $p\colon E \to B$ over an object b of B is the homotopy pullback of p along the map $\ulcorner b \urcorner\colon 1 \to B$ that picks out the element b:

$$\begin{array}{ccc} E_b & \longrightarrow & E \\ \downarrow & \lrcorner & \downarrow \\ 1 & \xrightarrow[\ulcorner b \urcorner]{} & B. \end{array}$$

A.1.7. *Loops.* The *loop groupoid* $\Omega_b B$ of a groupoid B at an object b is given by the homotopy pullback $1 \times_B 1$ of the inclusion $\ulcorner b \urcorner\colon 1 \to B$ along itself. This groupoid is discrete: it has $\mathrm{Aut}_B(b)$ as its set of objects, and only the identity isomorphisms.

A.2. Slices and the fundamental equivalence.

A.2.1. *Slices.* We shall need homotopy slices, sometimes called weak slices. First recall the usual notion of slice category. If $\mathscr{C}$ is a category, and $I \in \mathscr{C}$, then the usual slice category $\mathscr{C}/I$ is the category whose objects are morphisms $X \to I$

in $\mathscr{C}$ and whose arrows are commutative triangles

$$\begin{array}{ccc} X' & \longrightarrow & X \\ & \searrow \quad \swarrow & \\ & I. & \end{array}$$

We are concerned instead with groupoid-enriched categories $\mathscr{C}$, i.e. categories such that the arrows between each pair of objects X, Y define a groupoid $\mathrm{Map}(X,Y)$ instead of just a set, and the composition law is given by groupoid maps instead of just functions. Thus, between two parallel arrows $X \rightrightarrows Y$ there may be invertible 2-cells. The *homotopy slice category* $\mathscr{C}_{/I}$ then has as objects the morphisms $X \to I$; its arrows are triangles with a 2-cell

$$\begin{array}{ccc} X & \longrightarrow & X' \\ & \searrow \ \Rightarrow \ \swarrow & \\ & I. & \end{array} \tag{23}$$

The basic example is $\mathscr{C} = \boldsymbol{Grpd}$ with 2-cells given by homotopies between maps (that is, the natural isomorphisms).

A.2.2. *Homotopy sums and Grothendieck construction.* For a map $p\colon E \to B$, each isomorphism $\beta\colon b' \to b$ in B induces an equivalence of homotopy fibres $\beta_*\colon E_{b'} \to E_b$, sending an object $(1, e, \varphi : pe \cong b')$ to $(1, e, \beta\varphi : pe \cong b)$. Thus the homotopy fibres of $p\colon E \to B$ form a B-indexed family of groupoids, that is, a functor $E_{(-)}$ from B to the category $\boldsymbol{Grpd}$ of groupoids.

The *homotopy sum* of any B-indexed family of groupoids $E\colon B \to \boldsymbol{Grpd}$ is the groupoid given by the *homotopy colimit* of this functor, which may be defined by the *Grothendieck construction* and denoted $\int^{b\in B} E_b$. Its objects are pairs (b, e) with $b \in B$ and of $e \in E_b$, and isomorphisms $(b', e') \to (b, e)$ are pairs (β, ε) of isomorphisms $\beta\colon b' \to b$ in B and $\varepsilon\colon \beta_* e' \to e$ in E_b.

The Grothendieck construction of any family $E\colon B \to \boldsymbol{Grpd}$ comes equipped with a canonical projection to B whose homotopy fibres give back the original family E up to homotopy equivalence. Conversely, for any map $E \to B$, the homotopy sum of its homotopy fibres E_b is homotopy equivalent, over B, to E. Thus we have the following.

THEOREM A.2.3 (Fundamental Equivalence). *There is a canonical equivalence between the categories of groupoids over a fixed groupoid B and that of groupoid-valued functors from B*

$$\boldsymbol{Grpd}_{/B} \simeq \boldsymbol{Grpd}^B,$$

given by taking homotopy fibres and the Grothendieck construction.

A.2.4. *Monomorphisms.* A map $E \to B$ is a *homotopy monomorphism* if each homotopy fibre E_b is empty or contractible. Up to homotopy equivalence, such a map is the inclusion of some collection of connected components of B, that is, the Grothendieck construction of an indicator function $B \to \{\varnothing, 1\} \subset \boldsymbol{Grpd}$. Note that in general neither $\ulcorner b \urcorner\colon 1 \to B$ nor the diagonal map $B \to B \times B$ are homotopy mono.

A.2.5. *Finite maps.* A map is *homotopy finite* if each homotopy fibre is homotopy finite. A pullback of any homotopy monic or finite map is again homotopy mono or finite.

A.2.6. *Families.* The homotopy sum of an I-indexed family in $\boldsymbol{Grpd}^B$ is defined as the homotopy sum of the corresponding object of $\boldsymbol{Grpd}^{I\times B}$, composed with the projection,

$$E \longrightarrow I \times B \longrightarrow B.$$

Homotopy sums of I-indexed families in $\boldsymbol{Grpd}_{/B}$ are defined similarly. We regard the maps $\ulcorner b \urcorner \colon 1 \to B$, for $[b] \in \pi_0 B$ as a *basis* of $\boldsymbol{Grpd}_{/B}$, in analogy with vector spaces. *Scalar multiples* $A\ulcorner b \urcorner$ of basis elements in $\boldsymbol{Grpd}_{/B}$ are given by $A \to 1 \xrightarrow{\ulcorner b \urcorner} B$.

LEMMA A.2.7. *Any $f \colon E \to B$ in $\boldsymbol{Grpd}_{/B}$ may be expressed as a linear combination of basis elements as follows:*

$$f \;\simeq\; \int^{e\in E} \ulcorner f(e) \urcorner \;\simeq\; \int^{b\in B} E_b \ulcorner b \urcorner.$$

A.3. Linear functors.

A.3.1. *Basic slice adjunction.* Taking homotopy pullback along a morphism of groupoids $f \colon B' \to B$ defines a functor between the slice categories

$$f^* \colon \boldsymbol{Grpd}_{/B} \to \boldsymbol{Grpd}_{/B'}.$$

The functor f^* has a homotopy left adjoint, given by postcomposition:

$$f_! \colon \boldsymbol{Grpd}_{/B'} \to \boldsymbol{Grpd}_{/B}.$$

The homotopy adjointness is expressed by natural equivalences of mapping groupoids

$$\mathrm{Map}_{/B}(f_! E', E) \simeq \mathrm{Map}_{/B'}(E', f^* E). \tag{24}$$

LEMMA A.3.2 (Beck–Chevalley). *For any homotopy pullback square* (22), *the functors*

$$q_! \, p^*, \; g^* f_! \colon \boldsymbol{Grpd}_{/G} \to \boldsymbol{Grpd}_{/E}$$

are naturally homotopy equivalent.

A.3.3. *Spans and linear functors.* A pair of groupoid maps $A \xleftarrow{r} G \xrightarrow{f} B$ is termed a *span* between A and B, and induces a functor between the slice categories by pullback and postcomposition

$$f_! \, r^* \colon \boldsymbol{Grpd}_{/A} \longrightarrow \boldsymbol{Grpd}_{/B}.$$

A functor $\boldsymbol{Grpd}_{/A} \longrightarrow \boldsymbol{Grpd}_{/B}$ is *linear* if it is homotopy equivalent to one arising from a span in this way. Composites of linear functors

$$(sq)_!(rp)^* \colon \; \boldsymbol{Grpd}_{/A} \xrightarrow{\;f_! r^*\;} \boldsymbol{Grpd}_{/B} \xrightarrow{\;s_! g^*\;} \boldsymbol{Grpd}_{/C}$$

are again linear as follows by the Beck–Chevalley Lemma A.3.2 applied to the composite span

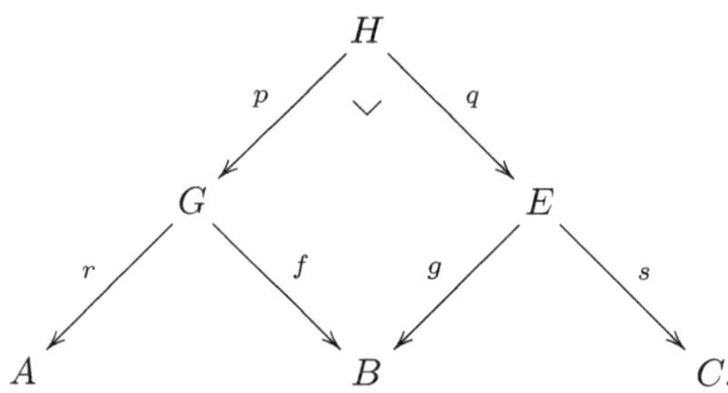

We write $\boldsymbol{LIN}$ for the monoidal 2-category of all slice categories $\boldsymbol{Grpd}_{/B}$ and linear functors between them, with the tensor product induced from the cartesian product

$$\boldsymbol{Grpd}_{/A} \otimes \boldsymbol{Grpd}_{/B} \ := \ \boldsymbol{Grpd}_{/A\times B}.$$

The neutral object is $\boldsymbol{Grpd} \simeq \boldsymbol{Grpd}_{/1}$, playing the role of the ground field.

A.3.4. *Duality.* The functor category $\boldsymbol{Grpd}^S$ is the *linear dual* of the slice category $\boldsymbol{Grpd}_{/S}$, since there is an equivalence (see [**44**, §2.11])

$$\boldsymbol{Grpd}^S \ \simeq \ \boldsymbol{LIN}(\boldsymbol{Grpd}_{/S}, \boldsymbol{Grpd}).$$

A span $A \leftarrow G \rightarrow B$ defines both a linear functor $\boldsymbol{Grpd}_{/A} \rightarrow \boldsymbol{Grpd}_{/B}$ and the dual linear functor $\boldsymbol{Grpd}^B \rightarrow \boldsymbol{Grpd}^A$. In particular the span $1 \leftarrow G \rightarrow S$ may be thought of as an element of $\boldsymbol{Grpd}_{/S}$, and its transpose $S \leftarrow G \rightarrow 1$ as an element of $\boldsymbol{Grpd}^S$.

There is a canonical pairing

$$\boldsymbol{Grpd}_{/S} \times \boldsymbol{Grpd}^S \rightarrow \boldsymbol{Grpd}$$

$$\langle \ulcorner t \urcorner , h^s \rangle = \mathrm{Hom}(s,t) = \begin{cases} \Omega_s(S) & (s \cong t) \\ \varnothing & (s \ncong t) \end{cases} \tag{25}$$

The maps $\ulcorner t \urcorner \colon 1 \rightarrow S$ (or the spans $1 \leftarrow 1 \rightarrow S$) form the canonical basis of the slice category, and the representable functors $h^s = \mathrm{Hom}(s,-)\colon S \rightarrow \boldsymbol{Grpd}$ (or the spans $S \leftarrow 1 \rightarrow 1$) form the canonical basis for the dual.

A.4. Cardinality.

A.4.1. *Cardinality of groupoids.* The cardinality of a finite groupoid X is given by

$$|X| \ := \ \sum_{[x]\in\pi_0(X)} \frac{1}{|\pi_1(X,x)|} \ \in \ \mathbb{Q}.$$

where $\pi_1(X,x)$ is the automorphism group $\mathrm{Aut}(x)$ of the object x in the groupoid X, and the norm signs on the right refer to the order of the group. Homotopy equivalent groupoids have the same cardinality. For any component of a locally finite groupoid B we have

$$\left|B_{[b]}\right| \ = \ |\pi_1(B,b)|^{-1} \ = \ |\Omega_b(B)|^{-1}. \tag{26}$$

For any function $q\colon \pi_0 B \to \mathbb{Q}$, we use the notation

$$\int^{b\in B} q(b) \;:=\; \sum_{[b]\in\pi_0 B} |B_{[b]}|\, q(x) \;=\; \sum_{[b]\in\pi_0 B} \frac{q(b)}{|\pi_1(B,b)|},$$

chosen to resemble the notation for the Grothendieck construction, since for any map $E \to B$ from a finite groupoid we have, by [**44**, Lemma 3.5],

$$|E| = \int^{b\in B} |E_b|.$$

The case of the map $\ulcorner b \urcorner\colon 1 \to B$ is just equation (26).

A.4.2. *Global cardinality.* A span $A \xleftarrow{r} G \xrightarrow{f} B$, and the corresponding linear functor $\boldsymbol{Grpd}_{/A} \to \boldsymbol{Grpd}_{/B}$, are termed *finite* if the map r is finite (that is, the homotopy fibres of r are finite). Being now concerned with finite groupoids, the slices we want to consider are $\boldsymbol{grpd}_{/A}$, where objects are maps $X \to A$ with X finite and A *locally* finite. If A is not itself finite, they are not properly speaking slice categories but rather comma categories; however we will continue to call them slices.

PROPOSITION A.4.3 ([**44**, 4.3]). *Let A, B, G be locally finite groupoids and $A \leftarrow G \to B$ a finite span. Then the induced finite linear functor $\mathbf{Grpd}_{/A} \to \mathbf{Grpd}_{/B}$ restricts to*

$$\mathbf{grpd}_{/A} \to \mathbf{grpd}_{/B}. \tag{27}$$

To a slice category $\boldsymbol{grpd}_{/A}$, with A locally finite, we associate the vector space $\mathbb{Q}_{\pi_0 A}$ with canonical basis $\{\delta_a\}_{[a]\in\pi_0 A}$. To the finite linear functor (27), we associate the linear map

$$\begin{aligned} \mathbb{Q}_{\pi_0 A} &\longrightarrow \mathbb{Q}_{\pi_0 B} \\ \delta_a &\mapsto \sum_{[b]\in\pi_0 B} \big|B_{[b]}\big|\, |G_{a,b}|\, \delta_b \;=\; \int^{b\in B} |G_{a,b}|\, \delta_b \end{aligned} \tag{28}$$

where $G_{a,b}$ are the fibres of the map $G \to A\times B$ defined by the span. This process is functorial [**44**, Proposition 8.2], and defines what we call *meta* or *global cardinality*

$$\|\ \|\colon \underrightarrow{\boldsymbol{lin}} \to \boldsymbol{Vect}$$

from the category $\underrightarrow{\boldsymbol{lin}}$ of slice categories $\boldsymbol{grpd}_{/A}$ (with A locally finite) and finite linear functors.

A.4.4. *Local cardinality.* To each object $p\colon G \to B$ in $\boldsymbol{grpd}_{/B}$ (with B locally finite) we can associate a vector $|p\colon G \to B|$ in $\mathbb{Q}_{\pi_0 B}$, called the *relative* or *local cardinality*:

$$|p| \;:=\; \sum_{[b]\in\pi_0 B} \big|B_{[b]}\big|\, |G_b|\, \delta_b \;=\; \int^{b\in B} |G_b|\, \delta_b.$$

Note that p determines a finite linear functor via $1 \leftarrow G \to B$, and the local cardinality $|p|$ is just the image of 1 under the global cardinality $\mathbb{Q} \to \mathbb{Q}_{\pi_0 B}$. It follows that local cardinality respects the action of finite linear functors L:

$$|L(p)| \;=\; \|L\|(|p|).$$

The local cardinality of the basis object $\ulcorner b \urcorner\colon 1 \to B$ in $\boldsymbol{grpd}_{/B}$ is just the basis vector δ_b in $\mathbb{Q}_{\pi_0 B}$, by (26).

To simplify notation we write $|L|$ for $\|L\|$ when the meaning is clear from the context, and say just 'cardinality' rather than meta, global, relative or local cardinality.

A.4.5. *Cardinality of the dual.* Dually we define cardinality of finite-groupoid valued functors (see A.3.4) as a map

$$|\ |: \boldsymbol{grpd}^S \to \|\boldsymbol{grpd}^S\| = \mathbb{Q}^{\pi_0 S},$$

where $\mathbb{Q}^{\pi_0 S}$ is the function space, the profinite-dimensional vector space with pro-basis given by the characteristic functions δ^s.

Finite spans $A \leftarrow G \to B$ define linear maps $\boldsymbol{grpd}^B \to \boldsymbol{grpd}^A$, whose cardinality is defined using the same matrix as in (28) above:

$$\begin{aligned} \mathbb{Q}^{\pi_0 B} &\longrightarrow \mathbb{Q}^{\pi_0 A} \\ \delta^b &\mapsto \sum_{[a]\in\pi_0 A} \left|B_{[b]}\right| \left|G_{a,b}\right| \delta^a. \end{aligned} \tag{29}$$

An element $g \in \boldsymbol{grpd}^S$ is represented by a finite span $S \leftarrow G \to 1$ (using the fundamental equivalence) and has cardinality

$$|g| \;=\; \|(S \leftarrow G \to 1)\| \left(\delta^1\right) \;=\; \sum_{[s]\in\pi_0 S} |g(s)|\,\delta^s. \tag{30}$$

The cardinality of the representable functor h^s in $\boldsymbol{grpd}^S$ is thus

$$|h^s| \;=\; \|(S \leftarrow 1 \to 1)\| \left(\delta^1\right) \;=\; |\Omega_s(S)|\,\delta^s \tag{31}$$

and the 'objective pairing' (25) is consistent with the classical pairing

$$\langle |\ulcorner t \urcorner|, |h^s| \rangle \;=\; \langle \delta_t, |\Omega_s(S)|\,\delta^s \rangle \;=\; \langle \delta_t, \delta^s \rangle\, |\Omega_s(S)| \;=\; |\langle \ulcorner t \urcorner, h^s \rangle|\,.$$

Appendix B. Simplicial groupoids and fat nerves

In this appendix, we provide some background material on simplicial groupoids and fat nerves. The general notion of simplicial set (originally termed a *complete semi-simplicial complex*) has been widely used in homotopy theory since the work of Eilenberg, Kan and others in the 1950s, owing its utility on one hand to the fact that simplicial sets are a model for topological spaces up to homotopy by way of the singular functor, and on the other hand because it receives a fully faithful functor from the category of small categories, namely the nerve; see B.1.7 below. The theory of ∞-categories, the common generalisation of spaces up to homotopy and categories, exploits the simplicial setting in a crucial way.

Any poset can naturally be regarded as a category, hence we may talk about posets in terms of their nerves. In combinatorics, however, it is common to view posets as simplicial *complexes* instead of simplicial sets, associating to a poset its order complex. The simplicial complexes that arise in this way have a canonical order on each simplex, and due to this they can be regarded as special kinds of simplicial sets, characterised by the property that n-simplices are completely determined by their vertex sets. Although such simplicial sets are of a simple kind, the subcategory they form is not as nice as the category of simplicial sets (which is a presheaf topos). For the purposes of the present undertakings, it is crucial to work with simplicial sets.

In this short appendix we recall the basic definitions, contrasting with simplicial complexes.

B.1. Simplicial sets and nerves.

B.1.1. *The simplex category (the topologist's Delta).* Let $\mathbb{\Delta}$ be the *simplex category*, whose objects are the finite nonempty standard ordinals

$$[n] = \{0 < 1 < \cdots < n\},$$

and whose arrows are the order-preserving maps between them. These maps are generated by the injections $\partial^i\colon [n-1] \to [n]$ that skip the value i, termed *coface* maps, and the surjections $\sigma^i\colon [n+1] \to [n]$ that repeat i, termed *codegeneracy* maps. The relations between these generators are called *cosimplicial identities*, and are dual to the simplicial identities listed below.

B.1.2. *Simplicial sets.* A *simplicial set* is a functor $X\colon \mathbb{\Delta}^{\text{op}} \to \textbf{\textit{Set}}$. One writes X_n for the image of $[n]$, and d_i, s_i for the images of ∂^i, σ^i. The elements of X_n are called *n-simplices.*

Explicitly, a simplicial set X is thus a sequence of sets X_n $(n \geq 0)$ together with *face maps* $d_i\colon X_n \to X_{n-1}$ and *degeneracy maps* $s_i\colon X_n \to X_{n+1}$, $(0 \leq i \leq n)$,

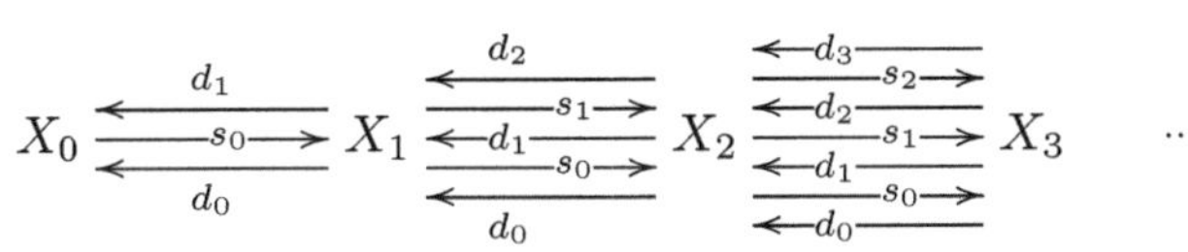

subject to the *simplicial identities*: $d_i s_i = d_{i+1} s_i = 1$ and

$$d_i d_j = d_{j-1} d_i, \quad d_{j+1} s_i = s_i d_j, \quad d_i s_j = s_{j-1} d_i, \quad s_j s_i = s_i s_{j-1} \qquad (i < j).$$

B.1.3. *Simplicial maps.* A *simplicial map* $F\colon X \to Y$ between simplicial sets is given by a sequence of maps $(F_n\colon X_n \to Y_n)_{n\geq 0}$ commuting with face and degeneracy maps, that is, a natural transformation between the functors X and Y. A simplicial map is *cartesian* with respect to an order-preserving map $[m] \to [n]$ in $\mathbb{\Delta}$ if the associated naturality square is a pullback; see 1.5.1 for examples.

B.1.4. *Simplicial complexes.* A *simplicial complex* K consists of a set V of vertices together with a collection $S(K)$ of nonempty subsets of V, termed the *simplices* of K, satisfying the following.

- All the one-element subsets of V are simplices of K.
- Any non-empty subset of a simplex of K is a simplex of K.

A map between simplicial complexes is a function between their vertex sets such that the image of each simplex is a simplex.

B.1.5. *Locally ordered simplicial complexes.* Certain simplicial complexes can be regarded as simplicial sets, but some ordering is necessary so as to have well-defined face maps. We call a simplicial complex *locally ordered* if there is specified a linear order on each simplex, in such a way that all inclusions preserve these orders. (The terminology *hierarchical simplicial complex* is used by Ehrenborg [**35**].) A map of locally ordered simplicial complexes is a map of simplicial complexes whose restriction to each simplex is order preserving. This defines a category ***LOSC***.

To each locally ordered simplicial complex K there is associated canonically a simplicial set X whose n-simplices are sequences $(v_0, v_1, \ldots, v_n)$ of elements in V, permitting repetitions, which as a set is required to form a simplex $F \in S(K)$, and which as a sequence is required to be non-decreasing for the linear order in the simplex F. This construction can be described more formally as follows. Each linear order $[n] \in \mathbb{\Delta}$ can be regarded as a locally ordered simplicial complex, defining a functor $\mathbb{\Delta} \to \boldsymbol{LOSC}$. Now the simplicial set X assigned to K has

$$X_n := \mathrm{Hom}_{\boldsymbol{LOSC}}([n], K).$$

This definition automatically accounts for the face and degeneracy maps, simply induced by precomposition with the coface maps and codegeneracy maps $[m] \to [n]$ in $\mathbb{\Delta}$.

This assignment defines a fully faithful functor from locally ordered simplicial complexes to simplicial sets. (Note that allowing repetition in the sequences is necessary for the assignment to be functorial in maps of locally ordered simplicial complexes, because these are allowed to send a simplex to a simplex of lower dimension.)

B.1.6. *The order complex and the nerve of a poset.* The *order complex* of a poset P is the simplicial complex whose vertices are the elements of P and whose n-simplices are those subsets that form n-chains $v_0 < \cdots < v_n$ in the poset. The order complex is naturally locally ordered since each simplex is a total order, and its associated simplicial set is usually termed the *nerve* of the poset. The definition of the nerve extends to more general categories as follows.

B.1.7. *Strict nerve.* The *nerve* of a category $\mathscr{C}$ is the simplicial set

$$N\mathscr{C} : \mathbb{\Delta}^{\mathrm{op}} \to \boldsymbol{Set}$$

whose set of n-simplices is the set of sequences of n composable arrows in $\mathscr{C}$ (allowing identity arrows). The face maps are given by composing arrows (for the inner face maps) and by discarding arrows at the beginning or the end of the sequence (outer face maps). The degeneracy maps are given by inserting an identity map in the sequence. By regarding the total order $[n]$ as a category, we see that a sequence of n composable arrows in $\mathscr{C}$ is the same thing as a functor $[n] \to \mathscr{C}$, and more formally the n-simplices can be described as

$$(N\mathscr{C})_n = \mathrm{Fun}([n], \mathscr{C}),$$

and in particular we see that the face and degeneracy maps of $N\mathscr{C}$ are given simply by precomposition with the coface and codegeneracy maps in $\mathbb{\Delta}$.

B.2. Simplicial groupoids, fat nerves, and Segal spaces.

B.2.1. *Simplicial groupoids.* For any category $\mathscr{E}$, one can talk about simplicial objects $X : \mathbb{\Delta}^{\mathrm{op}} \to \mathscr{E}$. Thus, in the case of the category of groupoids, a *simplicial groupoid* is a sequence of groupoids X_n, $n \geq 0$, and face and degeneracy maps $d_i : X_n \to X_{n-1}$, $s_i : X_n \to X_{n+1}$, $(0 \leq i \leq n)$, subject to the simplicial identities above.

B.2.2. *Fat nerve of a small category.* Important examples of simplicial groupoids are given by the *fat nerve* of a small category $\mathscr{C}$. Here X_n is the groupoid of all composable sequences $a_0 \xrightarrow{\alpha_1} a_1 \xrightarrow{\alpha_2} \ldots \xrightarrow{\alpha_n} a_n$ of n arrows in $\mathscr{C}$, that is,

$$X_n = \{\text{functors } \alpha : [n] \to \mathscr{C}\}.$$

In the case of the classical *strict nerve* this is just a set, or a discrete groupoid; in the *fat nerve*, X_n includes all natural isomorphisms $\alpha \to \alpha'$, which are diagrams of the form

$$\begin{array}{ccccccccc}
\cdot & \longrightarrow & \cdot & \longrightarrow & \cdot & \longrightarrow & \cdots & \longrightarrow & \cdot \\
\downarrow \wr & & \downarrow \wr & & \downarrow \wr & & & & \downarrow \wr \\
\cdot & \longrightarrow & \cdot & \longrightarrow & \cdot & \longrightarrow & \cdots & \longrightarrow & \cdot .
\end{array}$$

This description can be given categorically in terms of the functor category, but allowing only invertible natural transformations:

$$(\mathbf{N}\mathscr{C})_n = \mathrm{Fun}([n], \mathscr{C})^{\mathrm{iso}}.$$

As in the previous cases (B.1.5), B.1.7, this definition automatically accounts for face and degeneracy maps by precomposition. In particular, $d_0 \colon X_1 \to X_0$ assigns to an arrow its codomain, and $d_1 \colon X_1 \to X_0$ assigns to an arrow its domain.

Since X_2 is by definition the groupoid of composable pairs of arrows, we have $X_2 \simeq X_1 \times_{X_0} X_1$. Here the fibre product is

$$\begin{array}{ccc} X_2 & \xrightarrow{d_0} & X_1 \\ {\scriptstyle d_2}\downarrow & \lrcorner & \downarrow{\scriptstyle d_1} \\ X_1 & \xrightarrow[d_0]{} & X_0 \end{array} \tag{32}$$

expressing the composability condition: only those pairs of arrows such that the target of the first matches the source of the second.

In particular, $d_1 \colon X_2 \to X_1$ is the composition map. Also, $d_0 \colon X_2 \to X_1$ assigns to a composable pair the second arrow, and $d_2 \colon X_2 \to X_1$ assigns to a composable pair the first arrow. (Here we are referring to the order of composition, as in a-followed-by-b, and not the order used when writing this as $b \circ a$.)

Note that if $\mathscr{C}$ is just a poset, then it has no invertible arrows except the identities. Therefore, the notions of strict and fat nerve coincide.

B.2.3. *Rezk complete Segal spaces.* A simplicial groupoid is a *Segal space*, or a *Segal groupoid*, if $X_2 \simeq X_1 \times_{X_0} X_1$, as in (32), and in general the canonical Segal map

$$X_n \longrightarrow X_1 \times_{X_0} X_1 \times_{X_0} \cdots \times_{X_0} X_1$$

is an equivalence for each $n \geq 1$.

Consider the contractible groupoid generated by one isomorphism $0 \xrightarrow{\sim} 1$, and its strict nerve J. A Segal space X is *Rezk complete* if the map $J \to *$ induces an equivalence of groupoids $\mathrm{Map}(*, X) \to \mathrm{Map}(J, X)$, which in turn means that $s_0 \colon X_0 \to X_1$ is fully faithful and has as its essential image the arrows that admit left and right quasi-inverses. More intuitively, the condition expresses the idea that up to homotopy there are no other weakly invertible arrows than those coming from X_0 via the degeneracy map s_0.

The Rezk complete Segal spaces are precisely those simplicial groupoids that are levelwise-equivalent to the fat nerve of a category.

B.2.4. *Monoidal groupoids.* A monoidal groupoid is a monoidal category $(\mathscr{C}, \otimes, I)$ that happens to be a groupoid. For these, one can define the *monoidal nerve*, which is essentially a simplicial groupoid $X \colon \mathbb{\Delta}^{\mathrm{op}} \to \boldsymbol{Grpd}$. One takes X_0 to be a singleton and takes X_1 to be the groupoid itself, and more generally

$$X_n = \underbrace{\mathscr{C} \times \cdots \times \mathscr{C}}_{n \text{ factors}}.$$

The outer face maps just project away an outer factor. The inner face maps use the monoidal structure $\otimes \colon \mathscr{C} \times \mathscr{C} \to \mathscr{C}$ on two adjacent factors. The degeneracy maps insert a unit object. All these constructions are completely canonical, given the monoidal structure. The only problem is that the simplicial identities do not hold on the nose, due to the fact that the monoidal structure is not assumed to be strict. The diagram is therefore not literally speaking a simplicial groupoid, but only a pseudo-functor $\mathbb{\Delta}^{\mathrm{op}} \to \boldsymbol{Grpd}$.

While this issue may be a slight annoyance sometimes, it is not actually important for the purpose of this work: for the sake of defining a homotopy-coherently coassociative coalgebra structure on $\boldsymbol{Grpd}_{/X_1}$, a pseudo-functor is just as good as a strict functor. Another thing is that one can alternatively invoke strictification theorems (see Mac Lane [**77**, §XI.3, Theorem 1]): any monoidal category is equivalent to a strict monoidal category. The monoidal nerve of the strictification of a monoidal groupoid is then a simplicial groupoid on the nose, equivalent to the original monoidal nerve.

References

[1] M. Aguiar, N. Bergeron, and F. Sottile, *Combinatorial Hopf algebras and generalized Dehn-Sommerville relations*, Compos. Math. **142** (2006), no. 1, 1–30, DOI 10.1112/S0010437X0500165X. MR2196760

[2] M. Aguiar and S. Mahajan, *Monoidal functors, species and Hopf algebras*, CRM Monograph Series, vol. 29, American Mathematical Society, Providence, RI, 2010. With forewords by Kenneth Brown and Stephen Chase and André Joyal, DOI 10.1090/crmm/029. MR2724388

[3] J. C. Baez and J. Dolan, *From finite sets to Feynman diagrams*, Mathematics unlimited—2001 and beyond, Springer, Berlin, 2001, pp. 29–50. MR1852152

[4] J. C. Baez and J. Dolan, *Zeta functions*. `http://ncatlab.org/johnbaez/Zeta+functions` (2010).

[5] J. C. Baez, A. E. Hoffnung, and C. D. Walker, *Higher dimensional algebra VII: Groupoidification*, Theory Appl. Categ. **24** (2010), No. 18, 489–553. MR2770073

[6] M. Batanin and M. Markl, *Operadic categories and duoidal Deligne's conjecture*, Adv. Math. **285** (2015), 1630–1687, DOI 10.1016/j.aim.2015.07.008. MR3406537

[7] H.-J. Baues, *The cobar construction as a Hopf algebra*, Invent. Math. **132** (1998), no. 3, 467–489, DOI 10.1007/s002220050231. MR1625728

[8] F. Bergeron, G. Labelle, and P. Leroux, *Combinatorial species and tree-like structures*, Encyclopedia of Mathematics and its Applications, vol. 67, Cambridge University Press, Cambridge, 1998. Translated from the 1994 French original by Margaret Readdy; With a foreword by Gian-Carlo Rota. MR1629341

[9] J. E. Bergner, A. M. Osorno, V. Ozornova, M. Rovelli, and C. I. Scheimbauer, *2-Segal sets and the Waldhausen construction*, Topology Appl. **235** (2018), 445–484, DOI 10.1016/j.topol.2017.12.009. MR3760213

[10] J. E. Bergner, A. M. Osorno, V. Ozornova, M. Rovelli, and C. I. Scheimbauer, *2-Segal objects and the Waldhausen construction*, Algebr. Geom. Topol. **21** (2021), no. 3, 1267–1326, DOI 10.2140/agt.2021.21.1267. MR4299667

[11] J. E. Bergner, A. M. Osorno, V. Ozornova, M. Rovelli, and C. I. Scheimbauer, *Comparison of Waldhausen constructions*, Ann. K-Theory **6** (2021), no. 1, 97–136, DOI 10.2140/akt.2021.6.97. MR4283095

[12] A. Björner and J. W. Walker, *A homotopy complementation formula for partially ordered sets*, European J. Combin. **4** (1983), no. 1, 11–19, DOI 10.1016/S0195-6698(83)80003-1. MR694463

[13] M. Brion, *Representations of quivers* (English, with English and French summaries), Geometric methods in representation theory. I, Sémin. Congr., vol. 24, Soc. Math. France, Paris, 2012, pp. 103–144. MR3202702

[14] C. Brouder, A. Frabetti, and C. Krattenthaler, *Non-commutative Hopf algebra of formal diffeomorphisms*, Adv. Math. **200** (2006), no. 2, 479–524, DOI 10.1016/j.aim.2005.01.005. MR2200854

[15] D. Calaque, K. Ebrahimi-Fard, and D. Manchon, *Two interacting Hopf algebras of trees: a Hopf-algebraic approach to composition and substitution of B-series*, Adv. in Appl. Math. **47** (2011), no. 2, 282–308, DOI 10.1016/j.aam.2009.08.003. MR2803804

[16] L. Carlier, *Hereditary species as monoidal decomposition spaces, comodule bialgebras, and operadic categories*, Int. Math. Res. Not. IMRN **8** (2022), 5745–5780, DOI 10.1093/imrn/rnaa260. MR4410767

[17] L. Carlier, *Incidence bicomodules, Möbius inversion and a Rota formula for infinity adjunctions*, Algebr. Geom. Topol. **20** (2020), no. 1, 169–213, DOI 10.2140/agt.2020.20.169. MR4071369

[18] L. Carlier, *Möbius functions of directed restriction species and free operads, via the generalised Rota formula*, Mediterr. J. Math. **18** (2021), no. 4, Paper No. 170, 20, DOI 10.1007/s00009-021-01778-6. MR4280546

[19] L. Carlier and J. Kock, *Antipodes of monoidal decomposition spaces*, Commun. Contemp. Math. **22** (2020), no. 2, 1850081, 15, DOI 10.1142/S0219199718500815. MR4077086

[20] P. Cartier and D. Foata, *Problèmes combinatoires de commutation et réarrangements* (French), Lecture Notes in Mathematics, No. 85, Springer-Verlag, Berlin-New York, 1969. MR239978

[21] A. Cebrian and W. Forero. *Directed hereditary species and decomposition spaces.* Preprint, `arXiv:2211.07721`.

[22] A. Connes and D. Kreimer, *Hopf algebras, renormalization and noncommutative geometry*, Comm. Math. Phys. **199** (1998), no. 1, 203–242, DOI 10.1007/s002200050499. MR1660199

[23] M. Content, F. Lemay, and P. Leroux, *Catégories de Möbius et fonctorialités: un cadre général pour l'inversion de Möbius* (French, with English summary), J. Combin. Theory Ser. A **28** (1980), no. 2, 169–190, DOI 10.1016/0097-3165(80)90083-7. MR563554

[24] B. Cooper and M. Young. Hall algebras via 2-Segal spaces. This volume.

[25] H. H. Crapo, *The Möbius function of a lattice*, J. Combinatorial Theory **1** (1966), 126–131. MR193018

[26] P. Doubilet, *On the foundations of combinatorial theory. VII. Symmetric functions through the theory of distribution and occupancy*, Studies in Appl. Math. **51** (1972), 377–396, DOI 10.1002/sapm1972514377. MR429577

[27] P. Doubilet, *A Hopf algebra arising from the lattice of partitions of a set*, J. Algebra **28** (1974), 127–132, DOI 10.1016/0021-8693(74)90027-1. MR335568

[28] P. Doubilet, G.-C. Rota, and R. Stanley, *On the foundations of combinatorial theory. VI. The idea of generating function*, Proceedings of the Sixth Berkeley Symposium on Mathematical Statistics and Probability (Univ. California, Berkeley, Calif., 1970/1971), Univ. California Press, Berkeley, CA, 1972, pp. 267–318. MR403987

[29] A. Dür, *Möbius functions, incidence algebras and power series representations*, Lecture Notes in Mathematics, vol. 1202, Springer-Verlag, Berlin, 1986, DOI 10.1007/BFb0077472. MR857100

[30] T. Dyckerhoff, *Higher categorical aspects of Hall algebras*, Building bridges between algebra and topology, Adv. Courses Math. CRM Barcelona, Birkhäuser/Springer, Cham, 2018, pp. 1–61. MR3793857

[31] T. Dyckerhoff. Cyclic polytopes, orientals, and correspondences: some aspects of higher Segal spaces. This volume. `arXiv:2505.02051`.

[32] T. Dyckerhoff and M. Kapranov, *Crossed simplicial groups and structured surfaces*, Stacks and categories in geometry, topology, and algebra, Contemp. Math., vol. 643, Amer. Math. Soc., Providence, RI, 2015, pp. 37–110, DOI 10.1090/conm/643/12896. MR3381470

[33] T. Dyckerhoff and M. Kapranov, *Triangulated surfaces in triangulated categories*, J. Eur. Math. Soc. (JEMS) **20** (2018), no. 6, 1473–1524, DOI 10.4171/JEMS/791. MR3801819

[34] T. Dyckerhoff and M. Kapranov, *Higher Segal spaces*, Lecture Notes in Mathematics, vol. 2244, Springer, Cham, 2019, DOI 10.1007/978-3-030-27124-4. MR3970975

[35] R. Ehrenborg, *On posets and Hopf algebras*, Adv. Math. **119** (1996), no. 1, 1–25, DOI 10.1006/aima.1996.0026. MR1383883

[36] F. Fauvet, L. Foissy, and D. Manchon, *The Hopf algebra of finite topologies and mould composition* (English, with English and French summaries), Ann. Inst. Fourier (Grenoble) **67** (2017), no. 3, 911–945, DOI 10.5802/aif.3100. MR3668763

[37] M. Feller, R. Garner, J. Kock, M. U. Proulx, and M. Weber, *Every 2-Segal space is unital*, Commun. Contemp. Math. **23** (2021), no. 2, Paper No. 2050055, 6, DOI 10.1142/S0219199720500558. MR4201031

[38] H. Figueroa and J. M. Gracia-Bondía, *Combinatorial Hopf algebras in quantum field theory. I*, Rev. Math. Phys. **17** (2005), no. 8, 881–976, DOI 10.1142/S0129055X05002467. MR2167639

[39] L. Foissy. *Algebraic structures associated to operads.* Preprint, `arXiv:1702.05344`.

[40] L. Foissy, *Les algèbres de Hopf des arbres enracinés décorés. I* (French, with English and French summaries), Bull. Sci. Math. **126** (2002), no. 3, 193–239, DOI 10.1016/S0007-4497(02)01108-9. MR1905177

[41] I. Gálvez-Carrillo, R. M. Kaufmann, and A. Tonks, *Three Hopf algebras from number theory, physics & topology, and their common background I: operadic & simplicial aspects*, Commun. Number Theory Phys. **14** (2020), no. 1, 1–90, DOI 10.4310/CNTP.2020.v14.n1.a1. MR4060970

[42] I. Gálvez-Carrillo, J. Kock, and A. Tonks, *Groupoids and Faà di Bruno formulae for Green functions in bialgebras of trees*, Adv. Math. **254** (2014), 79–117, DOI 10.1016/j.aim.2013.12.015. MR3161093

[43] I. Gálvez-Carrillo, J. Kock, and A. Tonks, *Decomposition spaces, incidence algebras and Möbius inversion III: The decomposition space of Möbius intervals*, Adv. Math. **334** (2018), 544–584, DOI 10.1016/j.aim.2018.03.018. MR3828744

[44] I. Gálvez-Carrillo, J. Kock, and A. Tonks, *Homotopy linear algebra*, Proc. Roy. Soc. Edinburgh Sect. A **148** (2018), no. 2, 293–325, DOI 10.1017/S0308210517000208. MR3777576

[45] I. Gálvez-Carrillo, J. Kock, and A. Tonks, *Decomposition spaces, incidence algebras and Möbius inversion I: Basic theory*, Adv. Math. **331** (2018), 952–1015, DOI 10.1016/j.aim.2018.03.016. MR3804694

[46] I. Gálvez-Carrillo, J. Kock, and A. Tonks, *Decomposition spaces, incidence algebras and Möbius inversion II: Completeness, length filtration, and finiteness*, Adv. Math. **333** (2018), 1242–1292, DOI 10.1016/j.aim.2018.03.017. MR3818099

[47] I. Gálvez-Carrillo, J. Kock, and A. Tonks, *Decomposition spaces, incidence algebras and Möbius inversion III: The decomposition space of Möbius intervals*, Adv. Math. **334** (2018), 544–584, DOI 10.1016/j.aim.2018.03.018. MR3828744

[48] I. Gálvez-Carrillo, J. Kock, and A. Tonks, *Decomposition spaces, incidence algebras and Möbius inversion II: Completeness, length filtration, and finiteness*, Adv. Math. **333** (2018), 1242–1292, DOI 10.1016/j.aim.2018.03.017. MR3818099

[49] I. Gálvez-Carrillo, J. Kock, and A. Tonks, *Decomposition spaces and restriction species*, Int. Math. Res. Not. IMRN **21** (2020), 7558–7616, DOI 10.1093/imrn/rny089. MR4176832

[50] I. Gálvez-Carrillo, J. Kock, and A. Tonks, *Decomposition spaces, incidence algebras and Möbius inversion III: The decomposition space of Möbius intervals*, Adv. Math. **334** (2018), 544–584, DOI 10.1016/j.aim.2018.03.018. MR3828744

[51] I. Gálvez-Carrillo, J. Kock, and A.Tonks. Decomposition spaces of quasi-symmetric functions. Unpublished/in preparation.

[52] I. Gálvez-Carrillo, J. Kock, and A. Tonks. Decomposition spaces of symmetric functions. Unpublished/in preparation.

[53] N. Gambino and J. Kock, *Polynomial functors and polynomial monads*, Math. Proc. Cambridge Philos. Soc. **154** (2013), no. 1, 153–192, DOI 10.1017/S0305004112000394. MR3002590

[54] J. Goldman and G.-C. Rota, *On the foundations of combinatorial theory. IV. Finite vector spaces and Eulerian generating functions*, Studies in Appl. Math. **49** (1970), 239–258, DOI 10.1002/sapm1970493239. MR265181

[55] A. B. Goncharov, *Galois symmetries of fundamental groupoids and noncommutative geometry*, Duke Math. J. **128** (2005), no. 2, 209–284, DOI 10.1215/S0012-7094-04-12822-2. MR2140264

[56] P. Hackney. The decomposition space perspective. This volume. `arXiv:2409.19061`.

[57] P. Hackney and J. Kock, *Free decomposition spaces*, Collect. Math. **76** (2025), no. 3, 575–599, DOI 10.1007/s13348-024-00446-8. MR4950314

[58] C. Hermida, *Representable multicategories*, Adv. Math. **151** (2000), no. 2, 164–225, DOI 10.1006/aima.1999.1877. MR1758246

[59] S. A. Joni and G.-C. Rota, *Coalgebras and bialgebras in combinatorics*, Stud. Appl. Math. **61** (1979), no. 2, 93–139, DOI 10.1002/sapm197961293. MR544721

[60] A. Joyal, *Une théorie combinatoire des séries formelles* (French, with English summary), Adv. in Math. **42** (1981), no. 1, 1–82, DOI 10.1016/0001-8708(81)90052-9. MR633783

[61] A. Joyal and R. Street, *The geometry of tensor calculus. I*, Adv. Math. **88** (1991), no. 1, 55–112, DOI 10.1016/0001-8708(91)90003-P. MR1113284

[62] A. Joyal and R. Street, *The category of representations of the general linear groups over a finite field*, J. Algebra **176** (1995), no. 3, 908–946, DOI 10.1006/jabr.1995.1278. MR1351369

[63] J. Kock, *Polynomial functors and trees*, Int. Math. Res. Not. IMRN **3** (2011), 609–673, DOI 10.1093/imrn/rnq068. MR2764874

[64] J. Kock. *Data types with symmetries and polynomial functors over groupoids.* In *Proceedings of the 28th Conference on the Mathematical Foundations of Programming Semantics (Bath, 2012)*, vol. 286 of Electr. Notes in Theoret. Comp. Sci., pp. 351–365, 2012. arXiv:1210.0828.

[65] J. Kock, *Categorification of Hopf algebras of rooted trees*, Cent. Eur. J. Math. **11** (2013), no. 3, 401–422, DOI 10.2478/s11533-012-0152-1. MR3016312

[66] J. Kock, *Graphs, hypergraphs, and properads*, Collect. Math. **67** (2016), no. 2, 155–190, DOI 10.1007/s13348-015-0160-0. MR3484016

[67] J. Kock, *Polynomial functors and combinatorial Dyson-Schwinger equations*, J. Math. Phys. **58** (2017), no. 4, 041703, 38, DOI 10.1063/1.4977012. MR3632541

[68] J. Kock, *The incidence comodule bialgebra of the Baez-Dolan construction*, Adv. Math. **383** (2021), Paper No. 107693, 79, DOI 10.1016/j.aim.2021.107693. MR4235203

[69] J. Kock, *Whole-grain Petri nets and processes*, J. ACM **70** (2023), no. 1, Art. 1, 58. MR4529841

[70] J. Kock, A. Joyal, M. Batanin, and J.-F. Mascari, *Polynomial functors and opetopes*, Adv. Math. **224** (2010), no. 6, 2690–2737, DOI 10.1016/j.aim.2010.02.012. MR2652220

[71] J. Kock and D. I. Spivak, *Decomposition-space slices are toposes*, Proc. Amer. Math. Soc. **148** (2020), no. 6, 2317–2329, DOI 10.1090/proc/14834. MR4080877

[72] J. Kock and M. Weber, *Faà di Bruno for operads and internal algebras*, J. Lond. Math. Soc. (2) **99** (2019), no. 3, 919–944, DOI 10.1112/jlms.12201. MR3977896

[73] F. W. Lawvere and M. Menni, *The Hopf algebra of Möbius intervals*, Theory Appl. Categ. **24** (2010), No. 10, 221–265. MR2720184

[74] T. Leinster, *Notions of Möbius inversion*, Bull. Belg. Math. Soc. Simon Stevin **19** (2012), no. 5, 911–935. MR3009022

[75] P. Leroux. *Les catégories de Möbius.* Cahiers Topol. Géom. Diff. **16** (1976), 280–282.

[76] M. Lothaire, *Combinatorics on words*, Cambridge Mathematical Library, Cambridge University Press, Cambridge, 1997. With a foreword by Roger Lyndon and a preface by Dominique Perrin; Corrected reprint of the 1983 original, with a new preface by Perrin, DOI 10.1017/CBO9780511566097. MR1475463

[77] S. Mac Lane, *Categories for the working mathematician*, 2nd ed., Graduate Texts in Mathematics, vol. 5, Springer-Verlag, New York, 1998. MR1712872

[78] M. Maia and M. Méndez, *On the arithmetic product of combinatorial species*, Discrete Math. **308** (2008), no. 23, 5407–5427, DOI 10.1016/j.disc.2007.09.062. MR2459362

[79] D. Manchon, *On bialgebras and Hopf algebras or oriented graphs*, Confluentes Math. **4** (2012), no. 1, 1240003, 10, DOI 10.1142/S1793744212400038. MR2921530

[80] D. Manchon, *A review on comodule-bialgebras*, Computation and combinatorics in dynamics, stochastics and control, Abel Symp., vol. 13, Springer, Cham, 2018, pp. 579–597. MR3967397

[81] Yu. I. Manin, *A course in mathematical logic for mathematicians*, 2nd ed., Graduate Texts in Mathematics, vol. 53, Springer, New York, 2010. Chapters I–VIII translated from the Russian by Neal Koblitz; With new chapters by Boris Zilber and the author, DOI 10.1007/978-1-4419-0615-1. MR2562767

[82] Y. I. Manin, *Renormalization and computation I: motivation and background* (English, with English and French summaries), OPERADS 2009, Sémin. Congr., vol. 26, Soc. Math. France, Paris, 2013, pp. 181–222. MR3203373

[83] J. Morava, *Some examples of Hopf algebras and Tannakian categories*, Algebraic topology (Oaxtepec, 1991), Contemp. Math., vol. 146, Amer. Math. Soc., Providence, RI, 1993, pp. 349–359, DOI 10.1090/conm/146/01232. MR1224924

[84] K. E. Morrison, *An introduction to q-species*, Electron. J. Combin. **12** (2005), Research Paper 62, 15, DOI 10.37236/1959. MR2180799

[85] H. Munthe-Kaas, *Lie-Butcher theory for Runge-Kutta methods*, BIT **35** (1995), no. 4, 572–587, DOI 10.1007/BF01739828. MR1431350

[86] V. Ozornova and M. Rovelli. Waldhausen's $S_\bullet$-construction. This volume.

[87] T. Poguntke. *Equivariant motivic Hall algebras.* Preprint, arXiv:1808.04165.
[88] C. M. Ringel, *Hall algebras*, Topics in algebra, Part 1 (Warsaw, 1988), Banach Center Publ., vol. 26, PWN, Warsaw, 1990, pp. 433–447. MR1171248
[89] C. M. Ringel, *Hall algebras and quantum groups*, Invent. Math. **101** (1990), no. 3, 583–591, DOI 10.1007/BF01231516. MR1062796
[90] G.-C. Rota, *On the foundations of combinatorial theory. I. Theory of Möbius functions*, Z. Wahrscheinlichkeitstheorie und Verw. Gebiete **2** (1964), 340–368 (1964), DOI 10.1007/BF00531932. MR174487
[91] M. Rovelli. The Waldhausen construction as an equivalence between stable augmented double Segal spaces and 2-Segal spaces. This volume.
[92] O. Schiffmann, *Lectures on Hall algebras* (English, with English and French summaries), Geometric methods in representation theory. II, Sémin. Congr., vol. 24, Soc. Math. France, Paris, 2012, pp. 1–141. MR3202707
[93] W. R. Schmitt, *Antipodes and incidence coalgebras*, J. Combin. Theory Ser. A **46** (1987), no. 2, 264–290, DOI 10.1016/0097-3165(87)90006-9. MR914660
[94] W. R. Schmitt, *Hopf algebras of combinatorial structures*, Canad. J. Math. **45** (1993), no. 2, 412–428, DOI 10.4153/CJM-1993-021-5. MR1208124
[95] W. R. Schmitt, *Incidence Hopf algebras*, J. Pure Appl. Algebra **96** (1994), no. 3, 299–330, DOI 10.1016/0022-4049(94)90105-8. MR1303288
[96] R. P. Stanley, *Generating functions*, Studies in combinatorics, MAA Stud. Math., vol. 17, Math. Assoc. America, Washington, DC, 1978, pp. 100–141. MR513004
[97] R. P. Stanley, *Enumerative combinatorics. Vol. I*, The Wadsworth & Brooks/Cole Mathematics Series, Wadsworth & Brooks/Cole Advanced Books & Software, Monterey, CA, 1986. With a foreword by Gian-Carlo Rota, DOI 10.1007/978-1-4615-9763-6. MR847717
[98] R. P. Stanley, *Enumerative combinatorics. Vol. 2*, Cambridge Studies in Advanced Mathematics, vol. 62, Cambridge University Press, Cambridge, 1999. With a foreword by Gian-Carlo Rota and appendix 1 by Sergey Fomin, DOI 10.1017/CBO9780511609589. MR1676282
[99] R. Street, *Categorical structures*, Handbook of algebra, Vol. 1, Handb. Algebr., vol. 1, Elsevier/North-Holland, Amsterdam, 1996, pp. 529–577, DOI 10.1016/S1570-7954(96)80019-2. MR1421811
[100] M. E. Sweedler, *Hopf algebras*, Mathematics Lecture Note Series, W. A. Benjamin, Inc., New York, 1969. MR252485
[101] M. Takeuchi, *Free Hopf algebras generated by coalgebras*, J. Math. Soc. Japan **23** (1971), 561–582, DOI 10.2969/jmsj/02340561. MR292876
[102] B. Toën, *Derived Hall algebras*, Duke Math. J. **135** (2006), no. 3, 587–615, DOI 10.1215/S0012-7094-06-13536-6. MR2272977
[103] T. Walde. *Hall monoidal categories and categorical modules.* Preprint, arXiv:1611.08241.
[104] F. Waldhausen, *Algebraic K-theory of spaces*, Algebraic and geometric topology (New Brunswick, N.J., 1983), Lecture Notes in Math., vol. 1126, Springer, Berlin, 1985, pp. 318–419, DOI 10.1007/BFb0074449. MR802796
[105] P. Webb, *An introduction to the representations and cohomology of categories*, Group representation theory, EPFL Press, Lausanne, 2007, pp. 149–173. MR2336640
[106] M. Weber, *Generic morphisms, parametric representations and weakly Cartesian monads*, Theory Appl. Categ. **13** (2004), No. 14, 191–234. MR2116333
[107] M. Weber, *Familial 2-functors and parametric right adjoints*, Theory Appl. Categ. **18** (2007), No. 22, 665–732. MR2369114
[108] M. Weber, *Operads as polynomial 2-monads*, Theory Appl. Categ. **30** (2015), Paper No. 49, 1659–1712. MR3431299
[109] M. B. Young, *Relative 2-Segal spaces*, Algebr. Geom. Topol. **18** (2018), no. 2, 975–1039, DOI 10.2140/agt.2018.18.975. MR3773745

Universidad de Málaga and IMTECH-UPC
Email address: imma.galvez@uma.es

University of Copenhagen, Universitat Autònoma de Barcelona, and Centre de Recerca Matemàtica
Email address: joachim.kock@uab.cat

Universidad de Málaga
Email address: at@uma.es

838 **Julia E. Bergner, Joachim Kock, and Maru Sarazola, Editors,** Higher Segal Spaces and Applications, 2026

835 **Mee Seong Im and Tony Shaska, Editors,** Artificial Intelligence and Mathematics Research, 2026

834 **Mee Seong Im, Hanna E. Makaruk, Bach Nguyen, and Robert Owczarek, Editors,** Inverse Problems, Geometric and Topological Aspects of Representation Theory, and Low-Dimensional Topology, 2026

832 **Nils Bruin, David Kohel, and Chloe Martindale, Editors,** Arithmetic, Geometry, Cryptography and Coding Theory, 2026

831 **Irena Swanson and Kei-ichi Watanabe, Editors,** Lectures on Tight Closure and Its Applications, 2025

830 **O. Bezushchak, Yu. Drozd, P. Etingof, V. Futorny, R. Grigorchuk, and E. Zelmanov, Editors,** Modern Algebra Volume 2, 2025

829 **O. Bezushchak, Yu. Drozd, P. Etingof, V. Futorny, R. Grigorchuk, and E. Zelmanov, Editors,** Modern Algebra Volume 1, 2025

828 **Hafedh Herichi, Maria Rosaria Lancia, Therese Basa Landry, Anna Rozanova-Pierrat, and Steffen Winter, Editors,** Fractal Geometry in Pure and Applied Mathematics, 2025

827 **Carmen Caprau, J. Scott Carter, Neslihan Gügümcü, and Sam Nelson, Editors,** Algebraic Structures in Knot Theory, 2025

826 **Ahmed Laghribi and André Leroy, Editors,** Algebraic Structures and Applications, 2025

825 **Saurabh Verma, María A. Navascués, and Amit Priyadarshi, Editors,** Advances in Dimension Theory, Fractal Functions and Measures, 2025

824 **Hanna E. Makaruk and Robert M. Owczarek, Editors,** Inverse Problems: In Memory of Professor Zbigniew Oziewicz, 2025

823 **Mikhail Bershtein, Anton Dzhamay, and Andrei Okounkov, Editors,** The Versatility of Integrability, 2025

822 **Ahmad Barhoumi, Roozbeh Gharakhloo, and Andrei Martinez-Finkelshtein, Editors,** Recent Developments in Orthogonal Polynomials, 2025

821 **Bogdan D. Suceavă, Adara M. Blaga, Marian Ioan Munteanu, Cezar Oniciuc, Shoo Seto, Gabriel-Eduard Vîlcu, and Lihan Wang, Editors,** Recent Advances in Differential Geometry and Related Areas, 2025

820 **Sudeb Mitra, Rakesh Kumar Parmar, Saminathan Ponnusamy, and Toshiyuki Sugawa, Editors,** Geometric Function Theory and Related Topics, 2025

819 **Howard S. Cohl, Roberto S. Costas-Santos, and Robert S. Maier, Editors,** Applications and q-Extensions of Hypergeometric Functions, 2025

818 **Howard S. Cohl, Roberto S. Costas-Santos, and Robert S. Maier, Editors,** Classical Hypergeometric Functions and Generalizations, 2025

817 **Mikhail Khovanov, Joshua Sussan, and Anton Zeitlin, Editors,** From Representation Theory to Mathematical Physics and Back, 2025

816 **Luca F. Di Cerbo and Laurenţiu G. Maxim, Editors,** Geometry and Topology of Aspherical Manifolds, 2025

815 **Daniel Orr and Joshua Jeishing Wen, Editors,** Macdonald Theory and Beyond, 2025

814 **Susan Montgomery and Siu-Hung Ng, Editors,** Quantum Groups, Hopf Algebras, and Applications, 2025

813 **Thomas Creutzig, Julia Plavnik, and David Ridout, Editors,** Quantum Symmetries, 2025

For a complete list of titles in this series, visit the
AMS Bookstore at **www.ams.org/bookstore/conmseries/**.

Kolibri 360 GmbH
Ferdinand-Jühlke-Straße 7
99095 Erfurt, Deutschland
produktsicherheit@kolibri360.de